NANO SCIENCE AND CARBON NANOTUBES

Contents

Foreword

A good book is essential for an understanding of the fundamentals of a subject and development of ability to apply the knowledge in real life situations. This book covers about carbon nanotubes, their synthesis and wide applications, including properties in separate chapters in details which are very useful. The treatment is at a level commensurate with the understanding of research scholars, educationists and even by students who are known to learn about nano science and technology.

The text is very lucid and the illustrations add immeasurably to the value of the book. This book is well written using simple language, consistent with prevision and brevity.

I am sure that this book will prove quite useful for increasing knowledge on nano.

DR. S.M. ISHTIAQUE
Professor,
Department of Textile Technology
Indian Institute of Technology, Delhi
Hauz Khas, New Delhi - 110 016

Preface

Nanotechnology is one of the important areas for technology development.

It has created considerable interest in many countries and is currently facing new, yet long-range challenges.

Antibiotics, aircraft, satellites, nuclear weapons, television, mass production, computers, all are the familiar revolutions of twenty-first century technology, with their growing consequences for human life and the Earth itself, have emerged within living memory.

The prerequisite for any substantial progress in nanotechnology is dependent on the existence or the supply with new nanomaterials. The specific properties of nanomaterials are based primarily on surface and quantum effects. An important task in this area is to understand how materials behave when sample sizes are close to 00 nm for example nanofibrils that are 10 to 100 times smaller in diameter than conventional textile fibers that are used for medical. In comparison to a human hair which is 80,000 nm in diameter, the nanofibers are 1,000 times smaller in diameter. When the characteristic length scale of the microstructure is in the 1-50 nm range, it becomes comparable with the critical length scales of physical phenomena resulting in the so-called "size and shape effects."

Phenomena occurring on this length scale are of interest to physicists, chemists, biologists, electrical and mechanical engineers, and computer scientists, making research in nanotechnology a frontier activity in materials science.

Nanotechnologies can be regarded as genuinely interdisciplinary, and have prompted the collaboration between researchers in previously disparate areas to share knowledge, tools and techniques. An understanding of the physics and chemistry of matter and processes at the nanoscale is relevant to all scientific disciplines, from chemistry and physics to biology, engineering and medicine. Indeed, it could be argued, that evolutionary developments in each of these fields towards investigating matter at increasingly small size scales has now come to be known as 'nanotechnology'.

Nanoscience and nanotechnologies are widely seen as having huge potential to bring benefits in areas as diverse as drug development, water decontamination, information and communication technologies, and the production of stronger, lighter materials.

Nanosized components such as particles, fibers, tubes, and composites based on carbon, are covered in this book. Specially, it comprises chapters on carbon nanotubes and nanofibers carbon nanotubes.

As carbon nanotubes have various physical properties and are an important material which can be broadly used for various advanced industries such as electronic and information technology 21st century, they are globally in the limelight.

In recent years, significant progress in understanding the physics of carbon nanotube electronic devices and in identifying potential applications has occurred. In a nanotube, low bias transport can be nearly ballistic across distances of several hundred nanometers. Deposition of high-κ gate insulators does not degrade the carrier mobility. The conduction and *valence* bands are symmetric, which is advantageous for complementary applications. The bandstructure is direct, which enables optical emission. Because of these attractive features, carbon nanotubes are receiving much attention. In this book simulation approaches are developed and applied to understand carbon nanotube and to explore device engineering issues for better transistor performance.

Although I have given briefly the History of Nanotechnology; I have included a chapter on Nano Science and Technology whereas I have dealt in detail on Carbon Nanotubes.

The wide variety of topics covered in this book (on Carbon Nanotubes) and other volumes will be of great interest for professionals working in fundamental and applied research. It is a valuable reference source for scientists, graduate students, post-doctoral fellows, engineers, industrial researchers, and other professionals working in the fields of nanomaterials, nanotechnology, materials science, chemistry, physics, polymer science, engineering, and biosciences. The detailed chapters on Carbon Nanotubes presented herein are expected to have commercial applications in many areas such as in electronics fields. On the other hand, significant investment in research will be needed not only to develop this new area of materials science to make progress in the utilization of the new materials, but also to assess and avoid the risks associated with introducing unknown matter into the biosphere and ecosphere.

New Delhi

A.S. BHATIA

PART I

NANO SCIENCE AND TECHNOLOGY

HISTORY OF NANOTECHNOLOGY

The topic of nanotechnology was touched upon by "There's Plenty of Room at the Bottom," a talk given by physicist Richard Feynman at an American Physical Society meeting at Caltech on December 29, 1959. Feynman described a process by which the ability to manipulate individual atoms and molecules might be developed, using one set of precise tools to build and operate another proportionally smaller set, so on down to the needed scale. He also noted, scaling issues would arise from the changing magnitude of various physical phenomena: gravity would become less important, surface tension and Van der Waals attraction would become more important, etc. This basic idea appears feasible, and exponential assembly enhances it with parallelism to produce a useful quantity of end products.

Feynman announced two challenges, and he offered a prize a $1000 for the first individuals to solve the each one. The first challenge involved the construction of a nanomotor, which, (to Feynman's surprise), was achieved by November of 1960 by William McLellan. The second challenge involved the possibility of scaling down letters small enough so as to be able to fit the entire Encyclopedia Britannica on the head of a pin; this prize was claimed in 1985 by Tom Newman.

In 1965 Gordon Moore observed that silicon transistors were undergoing a continual process of scaling downward, an observation which was later codified as Moore's law. Since his observation, minimum feature sizes of transistors have decreased from 10 micrometers to the 45-65 nm range in 2007. The term "nanotechnology" was first defined by Professor Norio Taniguchi in 1974.

"'Nano-technology' mainly consists of the processing of, separation, consolidation, and deformation of materials by one atom or one molecule." Since that time the definition of nanotechnology has generally been extended upward in size to include features as large as 100 nm. Additionally, the idea that nanotechnology embraces structures exhibiting quantum mechanical aspects, such as quantum dots, has been known into the definition.

Nanotechnology and nanoscience got started in the early 1980s with two major developments; the birth of cluster science and the invention of the scanning tunneling microscope (STM). This development led to the discovery of fullerenes in 1986 and carbon nanotubes a few years later.

In another development, the synthesis and properties of semiconductor nanocrystals was studied. This led to a fast increasing number of metal oxide nanoparticles of quantum dots. The atomic force microscope was invented five years after the STM was invented. The AFM uses atomic force to see the atoms.

At present in 2007 the practice of nanotechnology embraces both stochastic approaches (in which, for example, supramolecular chemistry creates waterproof pants) and deterministic approaches where in single molecules (created by stochastic chemistry) are manipulated on substrate surfaces (created by stochastic deposition methods) by deterministic methods comprising nudging them with STM or AFM probes and causing simple binding or cleavage reactions to occur. The dream of complex, deterministic molecular nanotechnology remains elusive.

NANOTECHNOLOGY— A REVOLUTION

Approximately 20-25 years ago scientists and engineers began discussing a technological revolution that would be as dramatic and far-reaching to society as the industrial revolution - the nanotechnology revolution. At first the primary promoters of the nanotechnology revolution were considered eccentric at best, and a little crazy at worst. However, their ideas and visions are becoming accepted by the mainstream intellectual, scientific and engineering communities.

The systematic study, manipulation, and modification of atoms and molecules and having nanometer-sized dimensions began several hundred years ago. Society has benefited greatly because chemists can use chemical reactions to combine several types of atoms to create new types of molecules. With the advent of quantum physics, physicists, chemists and biologists routinety studied the spectra of atoms and molecules. Biochemists discovered the usefulness of all types of molecules from proteins to enzymes to DNA several decades ago.

Clearly, nanotechnology has captured the attention of the scientific community, the media and now the public. For some, it conjures up images of fabulous new materials, lighter and stronger than steel. Others envision microscopic robots that clean plaque from our arteries and tartar from our teeth.

Scientists studying molecular nanotechnology are interested in ways in which we might manipulate individual atoms and molecules to build things with almost unimaginable precision. Building things with atomic precision is an amazing contrast to the manufacturing techniques of today, which are very crude when examined at the molecular level.

The fundamental knowledge gained through nanoscience and developments in nanotechnology will certainly accelerate over the next several decades. With the control of materials at the nanometer dimension, engineers are already able to create new types of products and services. As an example the smallest transistors made today in a factory are about 20 nm gate length. With future nanotechnology advancements engineers will be able to make chips that have transistors that are two or three nanometers wide and density in billions of transistors per chip.

Today, cosmetic manufacturers use liposomes with diameters of a few tens of nanometer to reduce the dehydration of skin.

That the nanotechnology revolution will result in the creation of new types of products and services will greatly benefit our lives. Nanoscale science and technology in general covers synthesis characterization, studying properties, processing, fabrication and applications of the nanomaterials in the field of electronic medicine, healthcare, energy, biotechnology and environmental studies.

Today, making a smooth edge requires manufacturers to grind away and discard trillions of atoms. Manufacturers today frequently create objects by first creating larger objects and then removing excess material by grinding, chiseling or sanding. Components that seen to fit together precisely are really billion of atoms out of alignment. Poorly fitting components wear faster, require costly lubrication and eventually lead to mechanical breakdown.

From a molecular perspective, even the most precise of today's mechanical components fit poorly. These factors plague modern manufacturing techniques causing enormous waste, pollution and imprecision.

Enter molecular nanotechnology. The ability to precisely pick and place individual atoms and molecules will allow us 'to create objects with almost perfect efficiency. Right now, scientists are picking up and moving individual atoms of silicon and other materials using scanning tunneling microscopy. STM uses an amazingly sharp tip, usually made of tungsten, to push atoms around on a surface.

It is by no means the only field concerned with atoms and molecules. In different ways, the disciplines of physics, chemistry and biology have long dealt with atoms and molecules, their behaviour and their manipulation and quantum mechanics is already firmly established as the science of the absolutely small, however nanotechnology is generally considered harder to pin down, it is a general capability that impacts on many scientific disciplines.

This is only the beginning. As the technology improves, we will be able to move hundreds, thousands and then millions of atoms at will to precisely construct objects of virtually any size, shape and material.

Objects created using molecular nanotechnology will be precise to within the size of a molecule. Machined components will no longer abrade one another because they will fit together with molecular precision. Moreover, they will have been made with virtually no waste or pollution. They will have been made one atom at a time with readily available components like carbon oxygen and nitrogen as their building blocks.

Nanotech factories will transform today's crude, inefficient, polluting factories into factories able to build many different things with little or no modification economically and cleanly.

In addition to making cleaner products, we'll be able to make smaller ones. Much smaller. When, we can place individual atoms, we will be able to construct devices at an almost unimaginable scale.

Equipped with nanotechnology design tools that resemble today's computer-aided-design software, engineers of the future will design devices smaller than, can be seen with the naked eye.

Certainly, many scientists have designed and fabricated structures for decades via chemical synthesis. During the last decade, however, developments the areas of surface microscopy, silicon fabrication, biochemistry, physical chemistry, and computational engineering have converged to provide remarkable capabilities for understanding, fabricating and manipulating structures at the atomic level. The semiconductor industry, for example, has relied on developing smaller and smaller features in large silicon wafers to fabricate computer chips. In contrast, using a nanoscience approach, one can self-assemble chains of molecules to replace wires on conventional computer chips, and it allows the semiconductor industry to produce revolutionary computer chips that are not only smaller, but also faster and more powerful than anything that exists today. The building blocks in the biological world are nanometer-sized molecules such as proteins and sugars that, when assembled into intermediate length-scaled objects, determine and control biological function.

The primary driving force for miniaturization has come from the electronics industry, which aimed to develop tools to create smaller (and therefore faster and more complex) electronic devices on silicon chips.

In addition to developing nanoelectronics, many other features of the biosynthetic process lend themselves to devising nanotechnological materials. To do its work, nature uses highly sophisticated processes, for example, selection; self-organization; and self-assembly to provide an enormous range of "bio"-materials that ultimately form cells, tissues and organs. These materials exhibit remarkable,' powers of memory, replication, self-healing and self-repair.

Medical science will be able to create devices small enough to enter the body's bloodstream to repair damage and treat diseases at the cellular level. For example, instead of treating cancer using chemotherapy that weakens the entire body, nanotechnology will one day produce medical devices that can identify cancer cells and repair or destroy them with no damage to healthy tissue.

This manufacturing revolution will also give rise to starling advances in material science. Instead of steel or aluminum being mined from the ground, new materials will be made from carbon atoms in the form of nanotubes and related structures. A single-walled, carbon nanotube is a strawlike structure with a one-atom-thick wall of carbon atoms. Lighter, stronger and more flexible than steel, carbon nanotubes are believe by many to be the most promising of all nanomaterials.

Researcher also have developed miniaturized biosensors that can detect bioagents and markers for disease by mimicking cellular membranes and depositing these membranes onto optical chips. The surface of these this membrane-based sensors look like the natural target of a biological agent, receptor molecules that decorate the surface of a cell membrane. By copying nature's functions using nanoscaled materials, the useful properties of sensors can be optimized permitting entirely. New approaches to, for example, the early detection of disease.

Living cells are perfect examples of nanoscale devices that assemble structures one atom or molecule at a time. DNA is copied, proteins are formed and complex hormones are manufactured by cellular devices far more complex than the most advanced manufacturing processes we have today.

An acorn, for example, uses the energy within it to read the "computer program"–the DNA–to sprout roots and leaves to gather more energy from the soil and the sun. The computer program tells the acorn to rearrange the atoms in soil, air and water to produce an oak tree–a material far more complex than today's material science can produce.

If scientists can create a nanoscale assembly device with only a fraction of the capabilities of the simplest biological cell, commercial nanotechnology will be a reality.

Imagine a nanoscale robotic arm, much like its larger cousin on today's automobile assembly line. Only this robotic arm would have a "sticky" tip suitable for picking up and placing individual atoms. This robotic arm's first task, under computer control, would be to make a copy of itself, thus resulting in two robotic arms. Those two would make two more. The four would become eight. Then 16 and so on.

In a short while, the assembly device would contain millions of microscopic robotic arms, all capable of picking and placing atoms. Once that point is reached, the computer software will begin instructing the robotic arms to build the required product in parallel with a million times more bandwidth than the single arm. This process is called parallel assembly.

Far from being the stuff of science fiction, nanotechnology promises to be the next technological revolution, with cleaner, cheaper and more flexible manufacturing capabilities than exist anywhere in the world today. Enhanced materials will be lighter, stronger and more flexible than steel; nanosized devices will improve our quality and quantity of life.

When the ideas and concepts that are discussed as part of the nanotechnology revolution are fully implemented, what is possible? At this point many of the possible advancements that are discussed seem like science fiction.

1. All of recorded history will fit in a package that will fit in our pockets. This includes all written documents, all music and all movies.
2. Our world will be safer because the computers and sensing systems that fit in a package the size of a pill will be able to warn us of dangers.
3. Life will be extended because we can create systems and modules that replicate the functions and systems in our bodies.
4. New types of "quantum computers" will make calculations billions of times faster than today's digital computers.
5. We can create new types of molecules with the mechanical assembly of chemical systems instead of today's assembly by thermodynamic chemical reactions.

The nanotechnology revolution is being driven by a number of developments, ideas, and technical advancements. The primary driving forces behind the nanotechnology revolution are:

Instruments that Measure & Manipulate Atoms and Molecules (Microscopy, will AFM, SPM etc)

The invention of the Scanning Tunneling Microscope permitted us for the first time to see single atoms on a surface. Before this, using techniques based on electromagnetic radiation, it was possible to view and create images of lattices of many molecules. For example, with x-ray techniques it is possible to recreate the positions of atoms in a complex matrix or lattice. With tunneling electronic microscopes (TEM) it is possible to directly image atoms in a lattice. However, these techniques rely on the scattering of electromagnetic radiation from a collection of atoms, and thus cannot see single atoms.

Another important innovation is the laser "tweezer". By using the momentum of photons it is possible to isolate in a single location collections of several hundred molecules or atoms. Before this invention the possibility of isolating a few molecules or even a few hundred molecules was not considered possible.

The Drive to make smaller Computer Chips and higher Density Information Storage

The storage of information is considered an essential advancement of modern civilization. At first, recording information and ideas on written paper was a great achievement; books and newspapers allowed the flow of knowledge and information throughout the world. Today information is stored digitally and transmitted electronically. Digital bits with dimensions of less than a micron are stored on magnetic disks and compact discs.

What is the Atomic Force Microscopes Contribution to Nanotechnology?

Measurement

An atomic force microscope (AFM) creates a highly magnified three dimensional I image of a surface: The magnified image is generated by monitoring the motion of an atomically sharp probe as it is scanned across a surface. With the AFM it is possible to directly view features on a surface having a few nanometer-sized dimensions including single atoms and molecules on a surface. This gives scientists and engineers an ability to directly visualize nanometer-sized objects and to measure the dimensions of the surface features.

With an atomic force microscope it is possible to measure more than the physical dimensions of a surface. This is because there is a "physical" interaction of the probe with a surface. An example is that by lightly pushing against a surface with the probe, it is possible to measure how hard the surface is. Also the ease by which the probe glides across a surface is a measure of the surface "friction".

Modification

Just as a pen is used for writing on a paper's surface, it is possible to write on a surface with an atomic force microscope. This new type of "lithography" results in a completely new method for making surface modifications at the nanometer scale. It is already possible to modify surfaces by physically scratching the surface, by directly depositing molecules on a surface, and by using electric fields to modify surfaces. Presently the use of the AFM is in a very exploratory phase, but showing tremendous promise. One of the important technological issues that must be solved is the writing speed of the AFM lithography systems.

Manipulation

With an AFM probe it is possible to directly move objects across a surface. The objects may be pushed, rolled around, or even picked up by the probe. With such methods it is possible to create nanometer sized objects. One of the important aspects of using an AFM for direct manipulation is the user interface that is used for generating the motions of the probe. There are interfaces that measure the locations of a particle, such as microspheres on a surface, and then automatically move the spheres into a pre-established location. In another type of interface, called the nanomanipulator, the motion of the probe follows the motion of your hand. When you move your hand up and down, the probe moves up and down. Such an interface also allows users to "feel" and "touch" a surface.

Nanotechnology is sometimes considered as just one of four converging technologies, the other three being biotechnology and biomedicine, information technology, and cognitive science.

Alternatively all may be considered as part of nanotechnology.

The question of why nanotechnology is often stated to be revolutionary. Consider three distinct aspects (Figure below): indirect, direct and conceptual. By indirect is meant the progressive miniaturization of existing technologies, which opens up new areas of application for these technologies. Direct refers to the application of novel, nanoengineered artifacts, either to enhance the performance of existing processes and materials, or for wholly novel purposes. Finally, there is the conceptual aspect of nanotechnology, in which all materials and processes are considered from a molecular or even atomic viewpoint, as in living systems, in which complicated molecules (like proteins) are broken down into their constituent amino acids, which are then used for the templated synthesis of new proteins. The artificial counterpart of this process is largely untouched territory. Entirely novel integrated manufacturing life cycles await development, in which extreme energy economy and the absence of unpleasant waste products will be prominent. Furthermore, the conceptual nano-viewpoint offers the possibility of a new understanding of the world, its structures and its processes.

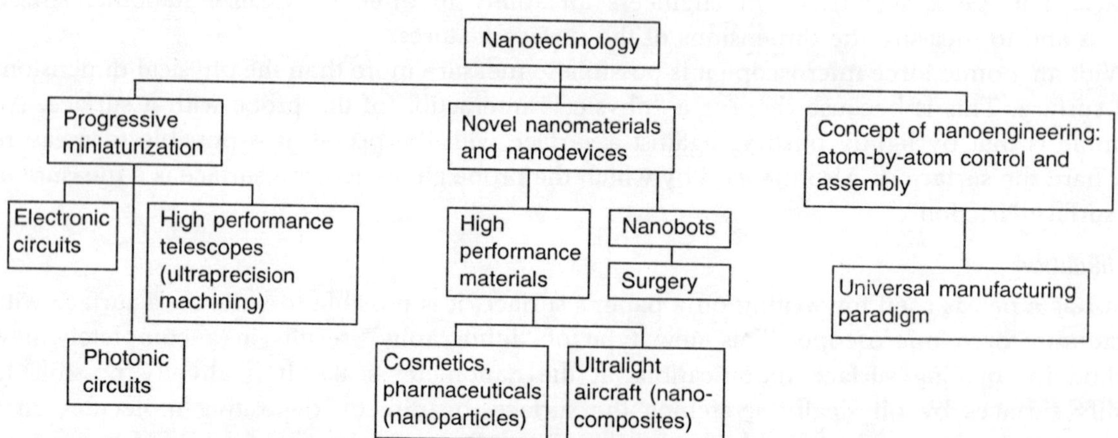

From left to right, the indirect, direct and conceptual branches of nanotechnology, with examples.

Indirect nanotechnology is enabling technology. Miniaturization can quantitatively enhance performance, and when a quantitative change is big enough, it becomes qualitative.

NANO IN THE NEAR FUTURE

These applications of nanotechnology in creating novel fibers and textiles are either already materializing or likely will be in the future.

- Supersensitive biofilters made of fibers capable of filtering out viruses, bacteria, and hazardous particles;
- Nanolayers applied to natural fibers and then made into protective clothing for firefighters, emergency responders, and military personnel that selectively block hazardous gases and minuscule contaminants but allows air and moisture to flow through;
- Lightweight smart textiles for hikers, athletes, and environmentally sensitive individuals;
- Fibers that control the movement of medicine to administer time released antibacterial and antiallergenic compounds; for example gloves that deliver arthritis medicine or antibacterial sheets in hospitals;
- Magnetic nanoparticles embedded inside a garment or paper document to create a unique signature that can be scanned to detect counterfeit currency or fake passports;
- Sensor that could swab a food or surgical preparation surface to immediately detect the presence of hazardous bacteria;
- Biodegradable fibers saturated with time-released pesticides that could be planted with seeds as an alternative to spraying pesticides;
- Doilies, seat cushions, or wall hangings used in airplanes that would continually absorb particles or gases or other airborne biohazards.

CURRENT USES OF NANOTECHNOLOGY INCLUDE:

- Sunscreens based around microfine particles are already in use in Europe, including the UK. Microfine and nanoparticles of zinc oxide absorb UV light while allowing visible light to pass through. These sunscreens appear translucent, but act as effective absorbers of UV light.
- Nanomaterials are currently used for very thin coatings–for example, self cleaning windows. The film made out of titanium dioxide nanoparticles speeds up the breakdown of dirt and bacteria in the presence of water and sunlight which is then washed off the glass more easily.
- Nanocomposites are used to make car bumpers that are claimed to be 60% lighter and twice as resistant to scratches and bumps than usual.

NANO SCIENCE AND TECHNOLOGY

SECTION-I

WHAT IS NANO?

The prefix 'nano' is derived from the greek word for dwarf. It is a unit of measurement. In metric system of measurement a nanometer is one billionth of a meter; 10^{-9} or 0.000000001 m. Example: A single human hair is approximately 80000 nm wide (Fig. 1.1), therefore, a nanometer is 80000 times smaller than the width of a human hair. A red blood cell is approximately 7000 nm wide and a water molecule is 0.3 nm across.

WHAT IS NANO TECHNOLOGY?

It is the science of extremely tiny; and the study and use of materials and devices that operate at unimaginable nanometer scale.

People are interested in the nanoscale (which is defined to be from 100nm down to the size of atoms (approximately 0.2 nm)) because it is at this scale that the properties of materials can be very different from those at a larger scale.

The properties of materials can be different at the nanoscale for two main reasons. First, nanomaterials have a relatively larger surface area when compared to the same mass of material produced in a larger form. This an make materials more chemically reactive (in some cases materials that are inert in their larger form are reactive when produced in their nanoscale form), and affect their strength or eiectrical properties. Second, quantum effects can begin to dominate the behaviour of matter at the nanoscale - particularly at the lower end - affecting the optical,' electrical and magnetic behaviour of materials. Materials can be produced that are nanoscale in one dimension (for example, very thin surface coatings), in two dimensions (for example, nanowires and nanotubes) or in all three dimensions (for example, nanoparticles).

The wide-ranging definitions cut across many traditional scientific disciplines. The only feature common to the diverse activities characterised as 'nanotechnology' is the tiny dimensions on which they operate. Therefore it is more appropriate to refer it as 'nanotechnologies'.

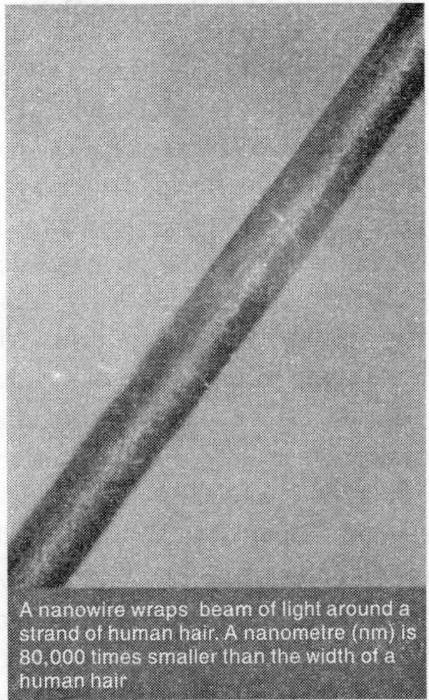

A nanowire wraps beam of light around a strand of human hair. A nanometre (nm) is 80,000 times smaller than the width of a human hair

Fig. 3.1

Nanotechnology is the technology of the 21st century. The work at nanometer scale has brought together many research and technological disciplines: physics, chemistry, medicine, biotechnology, computer science, electronics and material science.

Nanotechnology is the creation of USEFUL/FUNCTIONAL materials, devices and systems through control of matter, on the molecular level, on the nanometer length scale (1-100 nm) and exploitation of novel phenomenon and properties (physical, chemical, biological) at that length scale and fabrication of devices within that size range. At these dimensions the prperties of a material are transformed, sometimes quite radically. Properties of materials can be very different at this level; for instance they can be more chemically reactive, have greater strength or conduct electricity more effectively. Take the difference between graphite, which is used in the lead of pencils, and diamond-although both are made up of the same element (carbon), the way in which the atoms are arranged in each is slightly different, giving the dramatic difference between the two. While in the case of graphite and diamond this happens naturally, nanotechnology offers the opportunity to make such differences by design.

Techniques developed in the microelectronics industry have also enabled the miniaturisation of small mechanical moving devices (MEMS), which in turn have lead to research into NEMS. MEMS technology seeks to exploit and extend the capabilities that have been provided by silicon integrated circuit manufacturing from one of making chips for electronic signal processing to the provision of on-chip sensing and/or actuatio through the use of moving mechanical parts. Some MEMS technologies are starting to attain maturity in manufacture (for example, MEMS accelerometers are used widely in air-bag sensors). However, there are currently difficulties in the reproducible large-scale manufacture of more complex MEMS systems. Although not strictly a 'nanotechnology' as defined in this report, MEMS, NEMS and the technologies used to make them are used extensively in techniques that can access and exploit the nanoscale (such as SPMs or lab-on-a-chip and biosensing). The reducing dimensional tolerances (less than 100 nm) being provided by modern lithographic patterning techniques are now enabling the production of structures of such small dimensions that they are becoming a legitimate part of nanotechnologies in their own right.

WHAT IS NANOSCIENCE?

It is the study of arrangement of atoms and molecules (the basic units) which affect the properties of materials; with the aim of producing new or enhanced materials or products. In other words *Nanoscience is the study of phenomena and manipulation (measure, sec, predict and make) of materials at atomic, molecular and macromolecular scales and to exploit the novel properties which differ significantly from those at a larger scale.*

Nanoscience is concerned with understanding the above effects and their influence on the properties of material. Nanotechnologies aim to exploit these effects to create structures, devices and systems with novel properties and functions due to their size.

In some senses, nanoscience and nanotechnologies are not new. Many chemicals and chemical processes have nanoscale features–for example, chemists have been making polymers, large molecules made up of tiny nanoscalar subunits, for many decades.

Nanotechnologies have been used to create the tiny features on computer chips for the past several years. The natural world also contains many examples of nanoscale structures, from milk (a nanoscale colloid) to sophisticated nanosized and nanostructured proteins that control a range of biological activities, such as flexing muscles, releasing energy and repairing cells.

In recent years sophisticated tools have been developed to investigate and manipulate matter at the nanoscale, which have greatly affected our understanding of the nanoscale world. A major step in this direction was the invention of the scanning tunnelling microscope (STM) in 1982, and the atomic force microscope (AFM) in 1986. These tools use nanoscale probes to image a surface with atomic resolution, and are also capable of picking up, sliding or dragging stoms or molecules around on surfaces to build rudimentary nanostructures. The use of these tools is not restricted to engineering, but has been adopted across a range of disciplines. AFM, for example, is routinely used to study biological molecules such as proteins.

Nanoscience and nanotechnologies encompass a broad and varied range of materials, tools and approaches to manipulate and produce nanomaterials, commonly categorised as either 'top-down' or 'bottom-up'. 'Top-down' techniques involve starting with a block of material, and etching or milling it down to the desired shape, whereas 'bottom-up' involves the assembly of smaller sub-units (atoms or molecules) to make a larger structure. The main challenge for top-down manufacture is the creation of increasingly small structures with sufficient accuracy, whereas for bottom-up manufacture, it is to make structures large enough, and of sufficient quality, to be of use as materials.

WHAT IS NANOBIOTECHNOLOGY?

It refer to materials and processes at the nanometre scale that are based on biological, biomimetic or biologically-inspired molecules, and nanotechnological devices used to monitor or control biological processes, e.g. in medicine. An example of the former is the optically switched optical switch incorporating the biomolecule bacteriorhodopsin and an example of the latter is the biochip–an array of known DNA fragments used to capture unknown DNA from a sample.

WHY DO MATERIALS BEHAVE DIFFERENTLY AT THE NANOSCALE?

The size range that holds so much interest is typically from 100 nm down to the atomic level (approximately 0.2 nm), because it is in this range (particularly at the lower end) that materials can have different or enhanced properties compared with the same materials at a larger size. The two main reasons for this change in behaviour are an increased relative surface area, and the dominance of quantum effects. An increase in surface area (per unit mass) will result in a corresponding increase in chemical reactivity, making some nanomaterials useful as catalysts to improve the efficiency of fuel cells and bateries. As the size of matter is reduced to tens of nanometres or less, quantum effects can begin to play a role, and these can significantly change a material's optical, magnetic or electrical properties. In some cases, size-dependent properties have been exploited for centuries. For example, gold and silver nanoparticles (particles of diameter less than 100 nm have been used as coloured pigments in stained glass and ceramics. Depending on their size, gold particles can appear red, blue or gold in colour. The challenge for the ancient (al) chemists was to make all nanoparticles the same size (and hence the same colour), and the production of single-size nanoparticles. Also other effects such as surface tension or 'stickness' are important, which also effect physical and chemical properties. For liquid or gaseous environments Brownian motion, which describes the random movement of larger particles or molecules owing to their bombardment by smaller molecules and atoms, is also important. This effect makes control of individual atoms or molecules in these environments extremely difficult.

Carbon (the graphite in pencil lead, for example) is relatively soft; but carbon in the form of nanotubes is a hundred times stronger than steel. Hold a match to an aluminium can and it will not burn; but aluminium in the form of nanometre-sezed particles is an exlosive material used as a catalyst in rocket fuels.

THE THREE NANOTECHNOLOGIES

"Dry" nanotechnology, which derives from surface science and physical chemistry, focuses on fabrication of structures in carbon (for example, fullerenes and nanotubes), silicon, and other inorganic materials. "Dry" techniques admit use of metals and semiconductors. The active conduction electrons of these materials make them too reactive to operate in a "wet" environment, but these same electrons provide the physical properties that make "dry" nanostructures promising as electronic, magnetic, and optical devices. Another objective is to develop "dry" structures that possess some of the same attributes of the self-assembly that the wet ones exhibit.

"Wet" nanotechnology, which is the study of biological systems that exist primarily in a water environment. The functional nanometer-scale structures of interest here are genetic material, membranes, enzymes are other cellular components. The success of this nanotechnology is amply demonstrated by the existence living organisms whose form, function, and evolution are governed by the interactions of nanometer-scale structures.

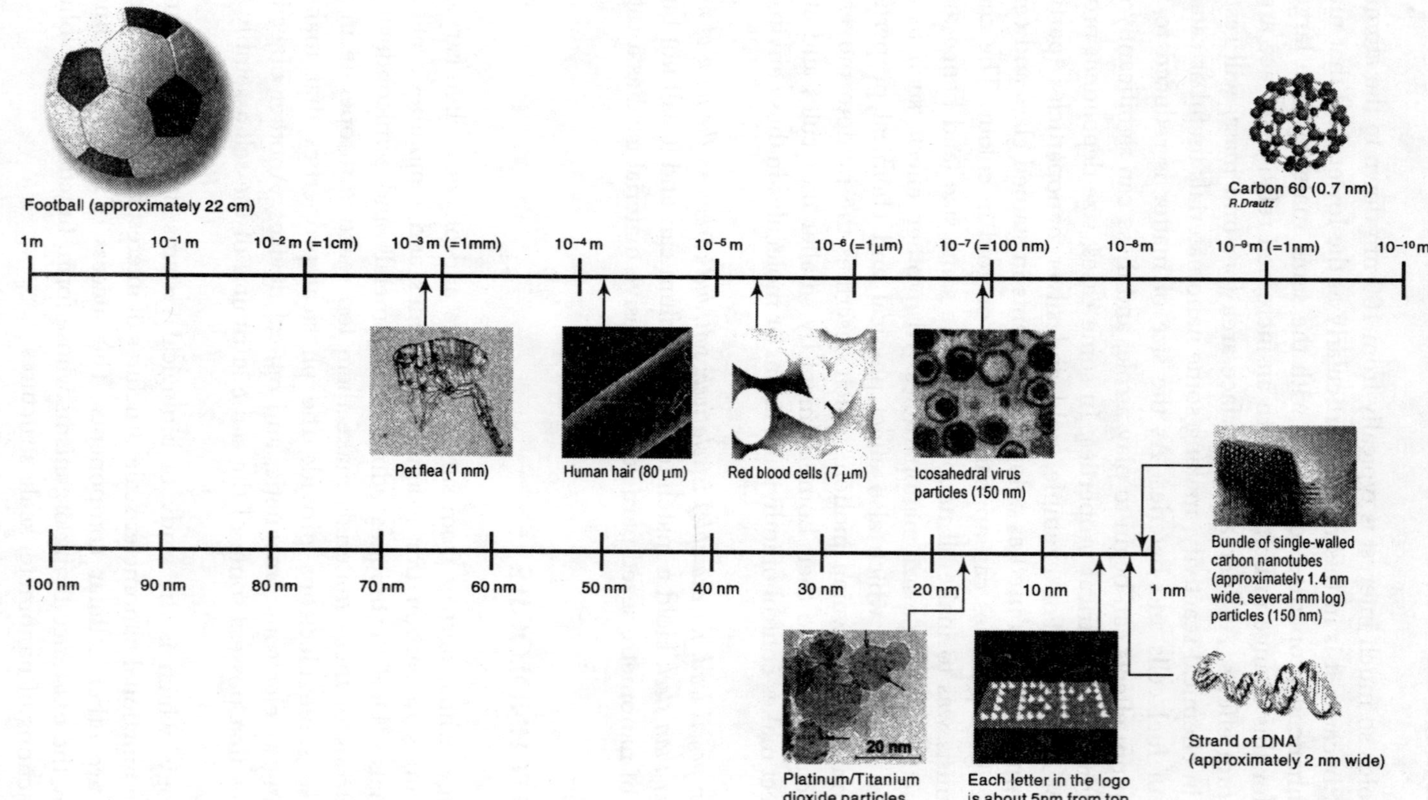

Fig. 3.2 *Length scale showing the nanometre in context. The length scale at the top ranges from 1 m to 10^{-10} m, and illustrate the size of a football compared to a carbon 60 (C_{60}) molecule, also known as a buckyball. For comparison the world is approximately one hundred million times larger than a football, which is in turn one hundred million times larger than a buckyball. The section from 10^{-7} m (100 nm) to 10^{-9} m (1 nm) is expanded below. The lengthscale of interest for nanoscience and nanotechnologies is from 100 nm down to the atomic scale–approximately 0.2 nm. Atoms are below a nanometre in size, whereas many molecules, including some proteins, range from a nanometre upwards*

Computational nanotechnology, which permits the modeling and simulation of complex nanometer-scale structures. The productive and analytical power of computation is critical to success in nanotechnology: nature required several hundred million years to evolve a functional "wet" nanotechnology; the insight provided by computation should allow us to reduce the development time of a working "dry" nanotechnology to a few decades, and it will have a major impact on the "wet" side as well.

So, Dry, Wet and Computational nanotechnology are three nanotechnologies that are highly interdependent.

SECTION-II

NANOSCIENCE OPPORTUNITIES AND UNCERTAINTIES

In this section some developments in nanoscience and nanotechnologies are given, and possible future applications are highlighted.

As nanoscience and nanotechnologies cover such a wide range of fields (from chemistry, physics and biology, to medicine, engineering and electronics) here they are considered in four broad categories: nanomaterials; nanometrology; electronics, optoelectronics and information and communication technology; and bio-nanotechnology and nanomedicine.

NANOMATERIALS

A key driver in the development of new and improved materials, has been the ability to control their structure at smaller and smaller scales in any one of the three dimensions. The overall properties of materials as diverse as paints and silicon chips are determined by their structure at the micro and nanoscales.

By definition, nanomaterials as those which have structured components with at least one dimension less than 100 nm. Materials that have one dimension in the nanoscale (and are extended in the other two dimensions are layers, such as a thin films or surface coatings, eg: computer chips.

Materials that are nanoscale in two dimensions (and extended in one dimension) include nanowires and nanotubes whereas materials that are nanoscale in three dimensions are nano-particles, example precipitates, colloids and quantum dots (tiny nano-particles of semiconductor materials). Nanocrystalline materials, made up of nanometre-sized grains, also fall into this category.

The properties of nanomaterials, which differ significantly from other materials are increased relative surface area, and quantum effects. These can change or enhance properties such as reactivity, strength and electrical characteristics. As a particle decreases in size, a greater proportion of atoms are found at the surface compared to those inside. For example, a particle of size 30 nm has 5% of its atoms on its surface, at 10 nm 20% of its atoms, and at 3 nm 50% of its atoms. Thus nanoparticles have a much greater surface area per unit mass compared with larger particles. As growth and catalytic chemical reactive than the same mass of material made up of larger particles. Besides surface-area effects, quantum effects can begin to dominate the properties of matter as size is reduced to the nanoscale. These can affect the optical, electrical and magnetic behaviour of materials, particularly as the structure or particle size approaches the smaller end of the nanoscale. Materials that exploit these effects include quantum dots, and quantum well lasers for optoelectronics.

For other materials such as crystalline solids, as the size of their structural components decreases, there is much greater interface area within the material, this can greatly affect both mechanical and electrical properties. For example, most metals are made up of small crystalline grains; the boundaries between the grain slow down or arrest the propagation of defects when the material is stressed, thus giving it strength. If these grains can be made very small, or even nanoscale in size, the interface area within the material greatly increases, which enhances its strength eg: Nanocrystalline nickle is as strong as hardened steel. Thus understanding surfaces and interfaces is a challenge for those working on nanomaterials.

Below are some examples of nanomaterials and the range of nanoscience that is aimed at understanding their properties. As will be seen, the behaviour of some nanomaterials is well understood, whereas others present greater challenges.

(A) NANOSCALE IN ONE DIMENSION

Thin Films, Layers and Surfaces

One-dimensional nanomaterials, such as thin films and engineered surfaces, have been developed and used for decades in fields such as electronic device manufacture, chemistry and engineering. In the silicon integrated-circuit industry, for example, many devices rely on thin films for their operation, and control of film thicknesses approaching the atomic level is routine. Monolayers (layers that are one atom or molecule deep) are also routinely made and used in chemistry. The formation and properties of these layers are resonably well understood from the atomic level upwards, even in quite comples layers (such as lubricants). Advances are being made in the control of the composition and smoothness of surfaces, and the growth of films.

(B) NANOSCALE IN TWO DIMENSIONS

Two dimensional nanomaterials such as tubes and wires have generated considerable interest among the scientific community, particularly, their novel electrical and mechanical properties.

Carbon Nanotubes

Carbon nanotubes (CNTs) were first observed by Sumio lijima in 1991. CNTs are extended tubes of rolled graphene sheets. There are two types of CNT: single-walled (one tube) or multi-walled (several concentric tubes) (Figure 3.3). Both of these are typically a few nanometres in diameter and several micrometres (10^{-6} m) to centimetres long. CNTs have assumed an important role in the context of nanomaterials, because of their novel chemical and physical properties.

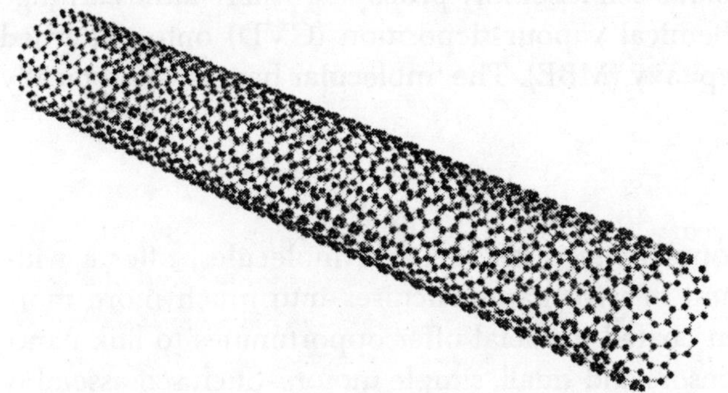

(a) *Schematic of a single-walled carbon nanotube (SWNT)*

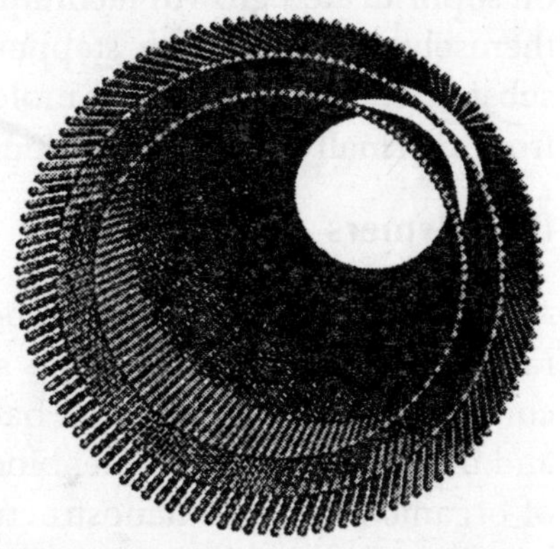

(b) *Schematic of a multi-walled carbon nanotube (MWNT)*

Fig. 3.3

They are mechanically very strong (their Young's modulus is over 1 terapascal, making CNTs as stiff as diamond), flexible (about their axis), and can conduct electricity extremely well (the helicity of the graphene sheet determines whether the CNT is a semiconductor of metallic). All of these remarkable properties give CNTs a range of potential applications: for example, in reinforced composites, sensors, nanoelectronics and display devices.

CNTs can be grown by several techniques, which are discussed in a seperate chapter in this book. However, the selective and uniform production of CNTs with specific dimensions and physical properties is yet to be achieved. The potential similarity in size and shape between CNTs and asbestos fibres has led to concerns about their safety.

Inorganic Nanotubes

Inorganic nanotubes and inorganic fullerene-like materials based on layerd compounds such as molybdenum disulphide were discovered shortly after CNTs. They have excellent tribological (lubricating) properties, resistance to shockwave impact, catalytic reactivity, and high capacity for hydrogen and lithium storage, which suggest a range of promising applications. Oxide-based nanotubes (such as titanium dioxide) are being explored for their applications in catalysis, photo-catalysis and energy storage.

Nanowires

Nanowires are ultrafine wires or linear arrays of dots, formed by self-assembly. They can be made from a wide range of materials. Semiconductor nanowires made of silicon, gallium nitride and indium phosphide have denonstrated remarkable optical, electronic and magnetic characteristics (for example, silica nanowires can bend light around very tight corners). Nanowires have potential applications in high-density data storage, either as magnetic read heads or as patterned storage media, and electronic and opto-electronic nanodevices, for metallic interconnects of quantum devices and nanodevices. The preparation of these nanowires relies on sophisticated growth techniques, which include self-assembly processes, where atoms arrange themselves naturally on stepped surfaces, chemical vapour deposition (CVD) onto patterned substrates, electroplating or molecular beam epitaxy (MBE). The 'molecular beams' are typically from thermally evaporated elemental sources.

Biopolymers

The variability and site recognition of biopolymers, such as DNA molecules, offer a wide range of opportunities for the self-organization of wire nanostructures into much more more complex patterns. The DNA backbones when coated in metal offer opportunities to link nano- and biotechnology example, biocompatible sensors and small, simple motors. Such self-assembly of organic backbone nanostructures is often controlled by weak interactions, such as hydrogen bonds, hydrophobic, or van der Waals interactions (generally in aqueous environments) and hence requires quite different synthesis strategies to CNTs.

(C) NANOSCALE IN THREE DIMENSIONS

Nanoparticles

Nanoparticles are often defined as particles of less than 100 nm in diameter that exhibit new or enhanced size-dependent properties (such as chemical reactivity and optical behaviour) that they exhibit compared with larger particles of the same materials. For example, titanium dioxide and zinc oxide become transparent at the nanoscale, however are able to absorb and reflect UV light, and have found application in sunscreens. Nanoparticles have a range of potential applications: in the shot-term in new cosmetics, textiles and paints; in the longer term, in methods of targeted drug delivery where they could be to used deliver drugs to a specific site in the body. Nanoparticles can also be arranged into layerss on surfaces, providing a large surface area and hence enhanced activity, relevant to a range of potential applications such as catalysts.

Nano particles exist in the nature as the products of photochemical and volcanic activity, and created by plants and algae. They have also been created as products of combustion and food cooking, and more recently from vehicle exhausts.

Fullerenes (carbon 60)

A new class of carbon material was discovered called carbon material was discovered called carbon 60 (C_{60}). These are spherical molecules about 1 nm in diameter, comprising 60 carbon atoms arranged as 20 hexagons and 12 pentagons: the configuration of a football. The C_{60} species was named 'Buckminsterfullerene' in recognition of the architect Buckminster Fuller, who was well-known for building geodesic domes, and the term fullerenes was then given to any closed carbon cage. In 1990, a technique to produce larger quantities of C_{60} was developed by resistively heating graphite rods in a helium atmospere. Several applications are envisaged for fullerenes, such as miniature 'ball bearings' to lubricate surfaces, drug delivery vehicles and in electronic circuits.

Dendrimers

Dendrimers are spherical polymeric molecules, formed through a nanoscale hierarchical self-assembly process. There are many types of dendrimer; the smallest is several nanometres in size. Dendrimers are used in conventional applications such as coatings and inks, but they also have a range of interesting properties which could lead to useful applications. For example, dendrimers can act as nanoscale carrier molecules and as such could be used in drug delivery. Environmental clean-up could be assisted by dendrimers as they can trap metal ions, which could then be filtered out of water with ultra-filtration techniques.

Quantum Dots

Nanoparticles of semiconductors (quantum dots) were initially created in the early 1980s. If semiconductor particles are made small enough, quantum effects come into play, which limit the energies at which electrons and holes (the absence of an electron) can exist in the particles. As energy is related to wavelength (or colour), this means that the optical properties of the particle can be finely tuned depending on its size. Thus, particles can be made to emit or absorb specific wavelengths (colours) of light, merely by controlling their size. Quantum dots have found applications in composites, solar cells (Gratzel cells) and fluorescent biological labels (for example to trace a biological molecule) which use both the small particle size and tuneable energy levels. Recent advances in chemistry have resulted in the preparation of monolayer-protected, high-quality, monodispersed, crystalline quantum dots as small as 2 nm in diameter, which can be conveniently treated and processed as a typical chemical reagent.

APPLICATION OF NANOMATERIALS

Below are some main applications of nanomaterials.

Sunscreens and Cosmetics

Nanosized titanium dioxide and zinc oxide are used in sunscreens, as they absorb and reflect ultraviolet (UV) rays and yet are transparent to visible light and so are more appealing to the consumer Nanosized iron oxide is present in some lipsticks as a pigment.

Composites

An important use of nanoparticles and nanotubes is in composites materials that combine one or more separate components and which are designed to exhibit overall the best properties of each component. This multi-functionality applies not only to mechanical properties, but extends to optical, electrical and magnetic ones. Currently, carbon fibres and bundles of multi-walled CNTs are used in polymers to control or enhance conductivity, with applications such as antistatic packaging.

A particular type of nanocomposite is where nanoparticles act as fillers in a matrix; for example, carbon black used as a filler to reinforce car tyres, particles of carbon black can range from tens to hundreds of nanometres in size.

Clays

Clays containing naturally occurring nanoparticles have long been important as construction materials. Clay particle based composites–containing plastics and nano-sized flakes of clay–are used in car bumpers.

Coatings and Surfaces

Coatings with thickness controlled at the nano or atomic scale have been in routine production example: in MBE or metal oxide CVD for optoelectronic devices, or in catalytically active and chemically functionalized surfaces, the self-cleaning window (which is coated in highly activated titanium dioxide, highly hydrophobic (water repellent) and antibacterial, and coatings based on nanoparticulate oxides that catalytically destroy chemical agents.

Wear and scratch-resistant hard coatings are significantly improved by nanoscale intermediate layers (or multilayers) between the hard outer layer and the substrate material. The intermediate layers give good bonding and graded matching of elastic and thermal properties, thus improving of elastic and thermal properties, thus improving adhesion. A range of enhanced textiles, such as breathable, waterproof and stain-resistant fabrics, have been enabled by the improved control of porosity at the nanoscale and surface roughness in a variety of polymers and inorganics.

Tougher and Harder Cutting Tools

Cutting tools made of nanocrystalline materials, such as tungsten carbide, tantalum carbide and titanium carbide, are more wear and erosion-resistant, and last longer than their conventional (large-grained) counterparts. They are finding applications in the drills used to bore holes in circuit boards.

Paints

Nanoparticles in paints improve their performance, by making them lighter and giving them different properties. Thinner paint coatings ('lightweighting'), when used on aircraft, would reduce its weight.

New types of fouling-resistant coatings, eg: in marine, paint could be developed and are urgently needed as alternatives to tributyl tin (TBT), now that the ecological impacts of TBT have been recognised. Anti-fouling surface treatment is also valuable in process applications such as heat exchange, thus leading to energy savings. If they can be produced at sufficiently low cost, fouling-resistant coatings could be used in piping for domestic and industrial water systems.

Anti-fouling coatings could reduce the use of biocides, including chlorine.

Applications for nanoparticles lie in paints that change colour in response to change in temperature of chemical environment, or paints that have reduced infra-red absorptivity and so reduce heat loss.

Nanoparticles react with pollutants in soil and groundwater and transform them into harmless compounds the large surface area and high surface ractivity of iron nanoparticles were exploited to transform chlorinated hydrocarbons (some of which are believed to be carcinogens) into less harmful end products in groundwater.

Fuel Cells

Engineered surfaces are essential in fuel cells, where the external surface properties and the pore structure affect performance. The hydrogen used as the immediate fuel in fuel cells may be generated from hydrocarbons by catalytic reforming, usually in a reactor module associated directly with the fuel cell. The potential use of nano-engineered membranes to intensity catalytic processes could enable higher-efficiency, small-scale fuel cells. These could act as distributed sources of electrical power.

Batteries

With the growth in portable electronic equipment (mobile phones, navigation devices, laptop computers, remote sensors), there is great demand for lightweight, high-energy density batteries. Nanocrystalline materials synthesized by sol-gel techniques are candidates for separator plates in batteries because of their foam-like (aerogel) structure, which can hold considerably more energy than conventional ones. Nickel-metal hydride batteries made of nanocrystalline nickel and metal hydrides are envisioned to require less frequent recharging and to last longer because of their large grian boundary (surface) area.

Displays

High brightness flat-panel displays, as used in television screens and computer monitors, is driving the development of some nanomaterials. Nanocrystalline zinc selenide, zinc sulphide, cadmium sulphide and lead telluride synthesized by sol-gel techniques (a process for making ceramic and glass materials, involving the transition from a liquid 'sol' phase to a solid 'gel' phase) are candidates for the next generation of light-emitting phosphors. CNTs are being investigated for low voltage field-emission displays; their strength, sharpness, conductivity and inertness make them potentially very efficient and long-lasting emitters.

Catalysts

Nanoparticles have a high surface area, and hence provide higher catalytic activity. It is possible to synthesise metal nanoparticles in solution in the presence of a surfactant to form highly ordered monodisperse films of the catalyst nanoparticles on a surface. This allows more uniformity in the size and chemical structure of the catalyst, which in turn leads to greater catalytic activity and the production of fewer by-products.

Carbon Nanotube Composites

CNTs have exceptional mechanical properties, particularly high tensile strength and light weight. An area of application would be in nanotube-reinforced composites, with performance beyond carbon-fibre composites. One current limit to the introduction of CNTs in composites is the problem of structuring the tangle of nanotubes in a well-ordered manner so that use can be made of their strength. Generating strong bonding between CNTs and the matrix, to give good overall composite performance and retention during wear or erosion of composites. The surfaces of CNTs are smooth and relatively unreactive, and so tend to slip through the matrix when it is stressed. One approach that is being explored to prevent this slippage is the attachment of chemical side-groups to CNTs, effectively to form 'anchors'. Another limiting factor is the cost of production of CNTs. However, the potential benefits of such light, high strength material in numerous applications for transportation are such that significant further research is likely.

Lubricants

Nanospheres of inorganic materials could be used as lubricants by acting as nanosized 'ball bearings'. The controlled shape makes them more durable than conventional solid lubricants and wear additives. It is claimed that nanoparticles reduce friction between metal surfaces, particularly at high normal loads. So their first applications is in high-performance engines and drivers; this could include the energy sector as well as transport and this type of lubricant is also effective even if the metal surfaces are not highly smooth.

Magnetic Materials

It has been shown that magnets made of nanocrystalline yttrium-samarium-cobalt grains possess unusual magnetic properties due to their extremely large grain interface area (high coercivity can be obtained because magnetization/flips cannot easily propagate past the grain boundaries). This could lead to applications in motors, analytical instruments like magnetic resonance imaging (MRI), used widely in hospitals, and microsensors. Overall magnetisation, however, is currently limited by the ability to align the grains' direction of magnetisation.

Nanoscale-fabricated magnetic materials also have application in data storage. Devices such as computer hard disks depend on the ability to magnetize small areas of a spinning disk to record information. If the area required to record one piece of information can be shrunk in the nanoscale (and can be written and read reliably), the storage capacity of the disk can be improved dramatically.

Medical Implants

Medical implants, such as orthopaedic implants and heart valves, are made of titanium and stainless steel alloys, primarily because they are bio-compatible. In some cases these metal alloys may wear out within the lifetime of the patient. Nanocrystalline zirconium oxide (zirconia) is hard, wear-resistant, bio-corrosion resistant and bio-compatible. It therefore presents an attractive alternative material for implants. It and other nanoceramics can also be made as strong, light aerogels by sol-gel techniques. Nanocrystalline silicon carbide is a candidate material for artificial heart valves primarily because of its low weight, high strength and inertness.

Machinable Ceramics

Ceramics are hard, brittle and difficult to machine. However, with a reduction in grain size to the nanoscale, ceramic ductility can be increased. Zirconia, normally a hard, brittle ceramic, has even been rendered superplastic (for example, able to be deformed up to 300% of its original length). Nanocrystalline ceramics, such as silicon nitride and silicon carbide, have been used in such automotive applications as high-strength springs, ball bearings and valve lifters, because they can be easily formed and machined, as well as exhibiting excellent chemical and high-temperature properties. They are also used as components in high-temperature furnaces. Nanocrystalline ceramics can be pressed into complex net shapes and sintered at significantly lower temperatures than conventional ceramics.

Water Purification

Nano-engineered membranes could potentially lead to more energy-efficient water purification processes, by desalination by reverse osmosis.

Military Battle Suits

Enhanced nanomaterials form the basis of a state-of-the-art 'battle suit' that is being developed by the Institu1e of Soldier Nanotechnologies at Massachusetts Institute of Technology, USA. A short-term development is likely to be energy-absorbing materials that will withstand blast waves; longer-term are those that incorporate sensors to detect or respond to chemical and biological weapons (for example, responsive nanopores that 'clpse' upon detection of a biological agent). There is speculation that developments could include materials which monitor physiology-while a soldier is still on the battlefield, and uniforms with potential medical applications, such as splints for broken bones.

Bio-nanotechnology and Nanomedicine

Applications of nanotechnologies in medicine are especially promising, and areas such as disease diagnosis, drug delivery targeted at specific sites in the body and molecular imaging are being intensively investigated and some products are undergoing clinical trials. Nanocrystalline silver, which is known to have antimicrobial properties, is being used in wound dressings. Applications of nanoscience and nanotechnologies are also leading to the production of materials and devices such as scaffolds for cell and tissue engineering, and sensors that can be used for monitoring aspects of human health. Many of the applications may not be realised for many years. In a longer term, the development of nanoelectronic systems that can detect and process information could lead to the development of an artificial retina or cochlea. Progress in the area of bio-nanotechnology will build an understanding of natural biological structures on the molecular scale, such as proteins.

Health and Environmental Impacts

Concerns have been expressed that the properties of nanoscale particles in certain applications (such as high surface reactivity and the ability to cross cell membranes), have negative health and environmental impacts. Many nanotechnologies pose no new risks to health and almost all the concerns relate to the potential impacts of deliberately manufactured nanoparticles and nanotubes that are free rather than fixed to or within a material. Only a few chemicals are being manufactured in nanoparticulate form on an industrial scale and exposure to free manufactured nanoparticles and nanotubes is currently limited to some workplaces and a small number of cosmetic uses. The likelihood of nanoparticles or nanotubes being released from products in which they have been fixed or embedded (such as composites) is expected to be low but manufacturers are recommended to assess this potential exposure risk for the lifecycle of the product.

The effects of inhaling free manufactured nanoparticles and exposure to other small particles such as the pollutant nanoparticles known to be present in large numbers in urban air, and the mineral dusts in some workplaces. The evidence suggests that at least some manufactured nanoparticles will be more toxic per unit of mass than larger particles of the same chemical. This toxicity is related to the surface area of nanoparticles (which is greater for a given mass than that of larger particles) and the chemical reactivity of the surface (which could be increased or decreased by the use of surface coatings). It seems that nanoparticles will penetrate cells more readily than larger particles.

It is likely that new manufactured nanoparticles, be introduced into humans in doses, sufficient to cause the health effects that have been associated with the nanoparticles in polluted air. However, some may be inhaled in certain workplaces in significant amounts and steps should be taken to minimise exposure. Toxicological studies have investigated nanoparticles of low solubility and low surface activity. Newer nanoparticles with characteristics that differ substantially from these should be treated with particular caution. The physical characteristics of carbon and other nanotubes mean that they may have toxic properties similar to those of asbestos fibres. Until further toxicological studies have been undertaken, human exposure to airborne nanotubes in laboratories and workplaces should be restricted.

If nanoparticles penetrate the skin they might facilitate the production of reactive molecules that could lead to cell damage. There is some evidence to show that nanoparticles of titanium dioxide (used in some sun protection products) do not penetrate the skin but it is not clear whether the same conclusion holds for individuals whose skin has been damaged by sun or by common diseases such as eczema. There is insufficient information about whether other nanoparticles used in cosmetics (such as zinc oxide) penetrate the skin. Much of the information relating to the safety of these ingredients has been carried out by industry and is not published in the open scientific literature.

Important information about the fate and behaviour of nanoparticles that penetrate the body's defences can be gained from researchers developing nanoparticles for targeted drug delivery. So, there should be collaboration between these researchers and those investigating the toxicity of other nanoparticles and nanotubes. The safety testing of these novel drug delivery methods must consider the toxic properties specific to such particles, including their ability to affect cells and organs distant from the intended target of the drug.

There is virtually no information available about the effect of nanoparticles on species other than humans or about how they behave in the air, water or soil, or about their ability to accumulate in food chains. Until more is known about their environmental impact we are keen that the release of nanoparticles and nanotubes to the environment is avoided as far as possible. Specifically, as a precautionary measure the factories and research laboratories should treat manufactured nanoparticles and nanotubes as if they were hazardous and reduce them from waste streams and that the use of free nanoparticles in environmental applications such as remediation of groundwater be prohibited.

There is some evidence to suggest that combustible nanoparticles might cause an increased risk of explosion because of their increased surface area and potential for enhanced reaction. Until this hazard has been properly evaluated this risk should be managed by taking steps to avoid large quantities of these nanoparticles becoming airborne.

Research into the hazards and exposure pathways of nanoparticles and nanotubes is required to reduce the many uncertainties related to their potential impacts on health, safety and the environment. This research must keep pace with the future development of nanomaterials. An interdisciplinary centre (perhaps from existing research institutions) should undertake research into the toxicity, epidemiology, persistence and bioaccumulation of manufactured nanoparticles and nanotubes, to work on exposure pathways and to develop measurement methods. The centre should liaise closely with regulators and with other researchers in the world.

Social and Ethical Impacts

If it is difficult to predict the future direction of nanoscience and nanotechnologies and the timescale over which particular developments will occur, it is even harder to predict what will trigger social and ethical concerns. In the short to medium term concerns are expected to focus on two basic questions: 'Who controls uses of nanotechnologies?' and 'Who benefits from uses of nanotechnologies?' These questions are not unique to nanotechnologies but past experience with other technologies demonstrates that they will need to be addressed.

The perceived opportunities and threats of nanotechnologies often stem from the same characteristics. For example, the convergence of nanotechnologies with information technology, linking complex networks of remote sensing devices with significant computational power, could be used to achieve greater personal safety, security and individualised healthcare and to allow businesses to track and monitor their products. It could equally be used for covert surveillance, or for the collection and distribution of information without adequate consent. As new forms of surveillance and sensing are developed, further research and expert legal analysis might be necessary to establish whether current regulatory frameworks and institutions provide appropriate safeguards to individuals and groups in society. In the military context, too, nanotechnologies hold potential for both defence and offence and will therefore raise a number of social and ethical issues.

There is speculation that a possible future convergence of nanotechnologies with biotechnology, information and cognitive sciences could be used for radical human enhancement.

NANOTECHNOLOGY AND CHEMISTRY

Chemistry is a powerful contender for the nanotechnology label. Nevertheless, there is an essential difference in approach. While chemistry also deals with the manipulation of molecules, and hence is no stranger to nanometre dimensions, the chemist does not control systems in the way that the engineer does. Molecules mostly reside in their free energy minima and it requires ingenuity, intuition and luck to steer their precursors

Fig. 3.4 *Xenon atoms spelling out the IBM logo. Each letter is about 5 nm from top to bottom*

along the desired end product. Chemistry thus lacks the dictatorial element of control over matter as practised by the engineer.

That is not to say that chemistry is not tremendously important to nanotechnology, but until now that importance has been rather negative, in the sense that chemical reality often thwarts the nanotechnologist's desire. A good illustration of this is Eigler and Schweizer's now famous experiment in which they manipulated molecules of zenon on a surface to form the letters 'IBM'. (see figure 3.3) They attempted to generate structures in dictatorial engineering mode, but were ultimately limited by chemistry: they could not put the atoms wherever they wanted, but only at certain lattice sites of the underlying surface, even at the ultralow temperature (4 K, i.e. just four degrees centigrade above absolute zero) at which they had to work. At present it looks as though chemistry may impose fundamental limitations on the freedom of the nanotechnologist to manipulate matter at the atomic and molecular scales.

NANOTECHNOLOGY AND BIOLOGY

Biology is considered to provide living proof of principle of nanotechnology. Biological structures at macromolecular and supramolecular scales are apparently assembled using the principles of self-assembly so eagerly sought by the nanotechnologist, and these structures, mostly protein-based, often combine extraordinary lightness with extraordinary strength, and may be miniature mechanisms of marvellous complexity. Two examples of amazing mechanical devices are the so-called F_1ATPase enzyme that uses a proton gradient across the membrane in which it is embedded to synthesize that universal biological energy provider, adenosine triphosphate (ATP); and the so-called type III secretion system (TTSS)–a spherical assembly of needles found on the surface of certain pathogenic bacteria, and used to inject poison into their targets.

NANOMETROLOGY

The science of measurement at the nanoscale is called nanometrology. Nanotechnologies, however defined, cannot progress independently of progress in nanometrology.The ability to measure and characterise materials (determine their size, shape and physical properties) at the nanoscale is vital if nanomaterials and devices are to be produced to a high degree of accuracy and reliability and the applications of nanotechnologies are to be realised. Nanometrology includes length or size measurements (where dimensions are typically given in nanometres and the measurement uncertainty is often less than 1 nm) as well as measurement of force, mass, electrical and other properties. As techniques for making these measurements advance, so too does the understanding of nanoscale behaviour and therefore the possibility of improving materials, industrial processes and reliability of manufacture. The instruments for making such measurements are many and varied eg: electron beam, scanning probe techniques and optical twizeers. The solutions developed for nanometrology problems can often be exploited elsewhere. For example, the concept of the AFM, a key nanometrology tool, has bad a direct influence on lithographic processes and techniques for molecular manipulation.

Instruments are available that can make sufficiently precise measurements. There are a number of sensor technologies and instruments with nanometre, or better, sensitivity for measuring length that repeat well if used carefully, including the scanning probe and electron microscopes and some optical devices, sophisticated users of atomic force microscopes can produce large variations in their measurements of the same artefacts.

Length and Force Measurement

Although there is not currently an international standard that can be applied to them, calibration artefacts are becoming available for AFMs for length measurement calibration in each of the three dimensions. Such artefacts can also be used for tip characterization, as the exact shape of the fine tip which scans across the surface can strongly influence length measurements, particularly when the tip becomes blunt. Excellent realisations of the metre, mainly through optical and xray interferometry, that can calibrate transfer standards with uncertainties of the order of nanometres. when working with sufficiently large wave-fronts has been achieved. However, comparative methods (akin to the interferometric measurement of optical lenses and mirrors) become ineffective when working with very small objects comparable to the wavelength of the photons being used. Hence, it is not yet possible to accurately determine dimensions or shape in all axes. Many surface characterization methods (especially for the electronics industries) directly exploit comparison via x-ray (occasionally optical reflection or diffraction.

Along with length measurement, force measurement (measured in Newtons (N)) is likely to become an important area of nanometrology, The control of probe stiffness and geometry will improve if quantitative measurements of surface mechanical properties are made, particularly while measuring biological and other soft materials. The need to accurately measure the elasticity of protein and nucleotide molecules, to determine bond strength and all properties of the molecules. There is a large, and growing, need for force characterisation in the pico- to micro newton (10^{-12} to 10^{-6} N) range. Currently, no fully satisfactory techniques are available either for secondary standards or transfer artefacts. Unlike length measurement, there is also a lack of readily available and applicable force or mass instrumentation with sensitivity adequatte for engineering on the nanometre scale. AFM cantilevers have nano-newton force sensitivity, but their calibration tends to be through indirect calculation from their dimensions. Some nano-indentation instruments for hardness measurement use micro-electromechanical systems (MEMS).

Measurement of Single Molecules

Measurements of single organic molecules and of structures such as single-wall nanotubes have been made, provided the molecules can be anchored to a substrate. Electron microscope and AFM/STM determinations of shape are done in many research laboratories. There is interest in molecule stiffness, in effect producing a tensile test curve in which jumps indicate the breaking and by inference the location of various types of bond in folded proteins and nucleotides.

APPLICATIONS

Metrology forms the basis of the semiconductor industry. Shrinking feature sizes, tighter control of device electrical parameters and new interconnect materials provides the main challenges for physical metrology methods. To achieve desired device scaling, metrology tools must be capable of measurement of properties at atomic distances.

The developing capabilities of semiconductor processing, particularly the ever-reducing dimensions that can be defined using lithographic tools, are being combined with the techniques developed for MEMS device fabrication to enable the manufacture of electro-mechanical components with sub 100 nm dimensions. The exploitation of these structures in nano-electromechanical systems (NEMS) devices has produced some interesting and exciting developments in the field of nanometrology, example, the measurement of the quantum of thermal conductance was achievd by Schwab et al (2000); and a device for making ultra-thin silicon cantilevers with atto-newton (10^{-18} N). sensitivity was made by another group. These devices have potential applications in the characterisation of single molecule properties and are examples of how the field of NEMS is increasing the capabilities of nanometrology.

SECTION-III

ELECTRONICS, OPTOELECTRONICS AND INFORMATION AND COMMUNICATION TECHNOLOGY (ICT)

Information technology (IT) has impact the lives of many people around the world.

It is the desire, to share information, wherein the printed word, images or sounds. This requires a technology that can absorb and process information on one side of the planet and deliver it almost instantaneously to the other in a form that is immediately accessible. Such a technology places enormous pressure on advances in processing and storing information, and on transmitting it and converting it from and to a human readable form. It also increasingly requires secure encryption of information so that access to information can ultimattely be restricted to particular individuals.

The market size of the IT industry is increasing due to the trend for miniaturisation the physical size of the computer chip has remained virtually unchanged as it is the transistor and all the circuitry associated with it that has shrunk dramatically.

Nanoscience research in ICT *(information and communication technology)* shares many of the same goals as for other applications of nanotechnologies: an improved understanding of nanoscale properties of materials and devices, advances in fabrication and process technology to satisfy increasingly stringent dimensional tolerances, and exploration of alternative technologies

There is no doubt that the ICT sector has effectively driven a large proportion of nanoscience, indeed, the first use of the word nanotechnology was in relation to ultra thin layers of relevance to the then up-.and-coming semiconductor industry. Since then, the research into all aspects of semiconductor device fabrication, from fundamental physics to process technology, has dominated the nanoscience landscape and will continue so to do. Decreasing device scales has added further impetus to the truly nanoscalar aspects this global research activity.

The dominant role of miniaturisation in the evolution of the computer chip is reflected in terms of a length. The current 130 nm technology node that produces the Intel Xeon processor defines the size of the DRAM *(dynamic random access memory)* half-pitch (half the distance between two adjacent metal wires in a memory cell). This is turn places a requirement on the lithography, process technology and metrology required to manufacture a working device to this tolerance. As a comparison, the 1971 Intel 4004 chip used 10,000 nm technology; the chips of 2007 and 2013 will require 65 nm rd 32 nm technology, respectively.

Furthermore, it is not simply the DRAM half-pitch that is on the nanometre scale. All the technology that goes into the metrology and production of chips has been at the sub-nanometre atomic level. The variety of tools that support the IT industry includes computer modelling of advanced devices and materials atom by atom, microscopies that can image single atoms, metrologies that can define the absolute position of a single atomic defect over a 30 cm diameter wafer (the substrate used for computer chips), thin-film growth processes that can produce layers of material with atomic precision, and lighographies that can 'write features, such as the DRAM cell, with an accuracy of sub- 10 nm.

A technology that has necessarily developed in tandem with IT is that of memory of data storage. This can be divided into two quite different types: solid-state memory such as DRAM that a processor chip would use or flash memory for storing images in a digital camera; and disk-based memory such as the magnetic hard drives as found in all computers. Solid-state memory essentially uses the same processes and technology as the computer chip, with very similar design rules and a similar emphasis on packing more memory into a given. area to increase total memory per device. The development of the hard disk drive, however, has taken a quite different route in evolution as it is based on reading and writing information magnetically to a spinning disk It is therefore primarily mechanical, or more strictly electro-mechanical, and presents quite different technical challenges. Once again, however, the importance of length scales is paramount as the ideal disk drive is one that has the minimal physical size with a massive ability to store data. Although the individual bits of magnetic information that are written onto the disk drive to give it the high-density storage are currently smaller than 100 nm, the constraints related to this nanotechnology on other aspects of the drive require fabrication of components with even greater precision. The importance of this nanotechnology in the related compact disk (CD) and digital versatile disk (DVD) drives that are now commonplace is equally ubiquitous.

An other crucial element of the IT revolution, opto-electronics, relates to devices that rely on converting electrical signals to and from light for data transmission, for displays for optica-based sensing and, in the future, for optical-based computing.

Although some opto-electronic devices do not depend so critically on miniaturisation as computer chips do, there is nevertheless a similar trend towards miniaturisation, with some existing components, such as quantum-well lasers and liquid crystal displays, requiring nanometre precision in their fabrication.

NANOMANUFACTURING AND THE INDUSTRIAL APPLICATION OF NANOTECHNOLOGIES

Industrial Applications

Current applications are mainly in the areas of determining the properties of materials, the production of chemicals, precision manufacturing and computing. In mobile phones for instance, materials involving nanotechnologies have and are being developed for use in advanced batteries, electronic packaging and in displays. The total weight of these materials constitutes a very small fraction of the whole product but is responsible for most of the functions that the devices offer. In the longer term, many more areas may be influenced by nanotechnologies although there will be significant challenges in scaling up production from the research laboratory to mass manufacturing.

It is also hoped that nanotechnologies will enable more efficient approaches to manufacturing which will produce a host of multi-functional materials in a cost-effective manner, with reduced resource use and waste. However, it is important that claims of likely environmental benefits are assessed for the entire lifecycle of a material or product, from its manufacture through its use to its eventual disposal.

Hopes have been expressed for the development and use of mechanical nano-machines which would be capable of producing materials (and themselves) atom-by-atom. Alongside such hopes for self-replicating machines, fears have been raised about the potential for these machines to go out of control, produce unlimited copies of themselves, and consume all available material on the planet in the process (the so called 'grey goo' scenario).

Characterisation

The characterisation of materials - the determination of their shape, size, distribution, mechanical and chemical properties - is an important part of the industrial process. It serves two broad purposes; as quality control, and as part of the research and development of new processes, materials and products. Industrialists believed that a nanotechnology 'breakthrough' had occurred in the tools used to observe and measure properties and processes at the nanoscale level. Sophisticated tools, such as the STM, AFM and TEM, enable surface and interfacial characterisation of materials at the nanoscale, allowing individual atoms to be observed and analysed.

Fabrication Techniques

There are a wide variety of techniques that are capable of creating nanostructures with various degrees of quality, speed and cost. These manufacturing approaches fall under two categories *'bottom-up'*, and *'top-down'*.

A diagram Fig. 3.5 illustrating some of the types of materials and products that these two approaches are used for is shown on next page.

Bottom-up Manufacturing

Bottom-up manufacturing involves the building of structures, atom-by-atom or molecule-by-molecule. The wide variety of approaches towards achieving this goal can be split into three categories; chemical synthesis, self-assembly, and positional assembly. Positional assembly is the only technique in which single atoms or molecules can be placed deliberately one-by-one. More typically, large numbers of atoms, molecules or particles are used or created by chemical synthesis, and then arranged through naturally occurring processes into a desired structure.

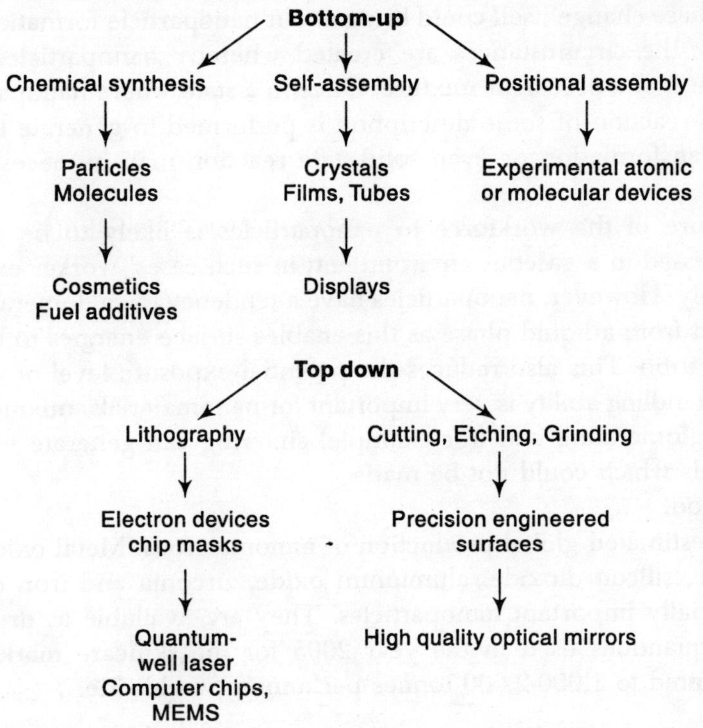

Fig. 3.5 *The use of bottom-up and top-down techniques in manufacturing*

(a) Chemical Synthesis

Chemical synthesis is a method of producing raw materials, such as molecules or particles, which can then be used either directly in products in their bulk disordered form, or as the building blocks of more advanced ordered materials, produced using the techniques outlined in sections (b) and (c) below.

A generic process by which nanoparticles may be produced by chemical synthesis is shown in Figure 3.6.

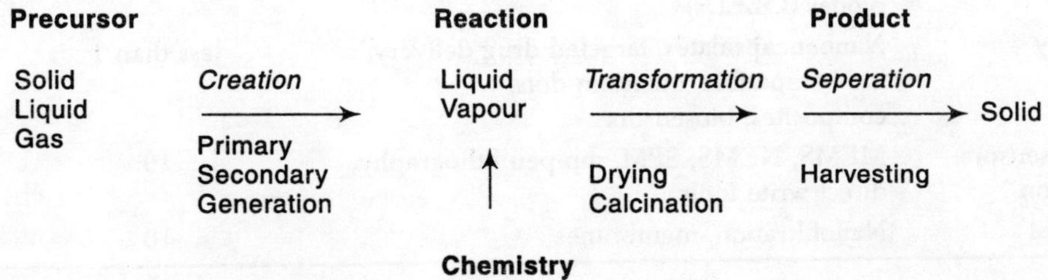

Fig. 3.6 *The generic processes that are involved in the production of nanoparticles*

The precursor phase is the starting point, and the material can be in any physical state (or multi phase) or spatial arrangement to other components. The first step is the creation of a new phase or state where the nanoparticles either form or can be formed by a chemical step. In other words, the phase change itself could bring about nanoparticle formation (rare but possible) although generally the circumstances are created whereby nanoparticles can be made, for example vaporisation of a precursor mixture. Once in a state where nanoparticles can be made, usually a chemical reaction of some description is performed to generate the desired material. A further phase transformation or even solid-state reaction may be necessary to produce the final product.

Potential exposure of the workforce to nanoparticles is likely to be greatest when these materials are processed in a gaseous environment; in such cases' worker exposure will need to be monitored closely. However, nanoparticles have a tendency to agglomerate, and are therefore often manufactured from a liquid phase as this enables surface energies to be better controlled, reducing agglomeration. This also reduces the potential exposure level of workers.

Processing and handling ability is very important for nanomaterials: mixing nanoscale particles together before agglomerating and (for example) sintering can generate wholly new complex nanophase materials which could not be made
by any other method.

Table 3.1 gives estimated global production of nanomaterials. Metal oxides such as titanium dioxide, zinc oxide, silicon dioxide: aluminium oxide, zirconia and iron oxide, are currently the most commercially important nanoparticles. They are available as dry powders or liquid suspensions. The quantities used in the year 2005 for the skincare market sectors (titanium dioxide etc.) amounted to 1,000-2,000 tonnes per annum worldwide.

TABLE 3.1 Estimated global production

Application	*Material/device*	*Estimated production*	
		Present	*2005–2010*
Structural applications	Ceramics, catalysts, composites, coatings, thin films, powders, metals	10	10^3
Skincare products	Metal oxides (titanium dioxide, zinc oxide, iron oxide)	10^3	10^3
ICT	Single wall nanotubes, nano electronics, opto-electro materials (titanium dioxide, zinc oxide, iron oxide), organic light-emitting diodes (OLEDs)	10	10^2
Biotechnology	Nanoencapsulates, targeted drug delivery, bio-compatible, quantum dots, composites, biosensors	less than 1	1
Instruments, sensors, characterisation	MEMS, NEMS, SPM, dip-pen lithography, direct write tools	10	10^2
Environmental	Nanofiltration, membranes	10	

Although the world market for nanoparticles is expected to increase during the next few years, to provide perspective, it is worth noting that the global production rate of all chemicals is around 400 M tonnes per annum and so chemicals in nanoparticulate form account for only a tiny fraction of the total (around 0.01 %) currently produced. Nanoscalar inorganic, metallic or semiconductor material often will have multifunctionality, which enables it to be used across many industry sectors. Zinc oxide, for example, will have more commercial use as an optoelectronic material (for displays or advanced solar and photovoltaic cells) where it will be fixed in the final product, than as an ingredient for skincare products, where particles will be free.

(b) Self Assembly

Self assembly is a bottom-up production technique in which atoms or molecules arrange themselves into ordered nanoscale structures by physical or chemical interactions between the units. The formation of salt crystals and snowflakes, with their intricate structure, are examples of self-assembly processes. Although self assembly has occurred in nature for thousands of years, the use of self assembly in industry is relatively new. There is an economic and environmental interest in processes through which materials or product components essentially form themselves, creating less waste and using less energy. However, current understanding extends only to the creation of fairly rudimentary systems. Improved understanding of thermodynamic and kinetic processes at the nanoscale, enabled through advances in the characterisation techniques and improved computer modelling are expected to aid the development of more complex systems. One potential processing technique involves the use of an external force or field (for example, electric or magnetic) to accelerate the often slow self-assembly process, which is attractive in an industrial context. This is known as directed self assembly.

CNTs are generating interest within industry because of their remarkable properties. Potential applications include composites, conductive plastics, sensors, batteries and fuel cells. CNTs can be grown by several techniques, such as the laser ablation of metal-doped graphite targets, carbon are discharge, and the pyrolysis of hydrocarbons over metal catalysts. However, because of a lack of understanding of the growth mechanism, the selective and uniform production of CNTs with specific dimensions and physical properties has yet to be achieved.

(c) Positional Assembly

It is a bottom-up technique whereby atoms, molecules or clusters are deliberately manipulated and positioned one-by-one. Techniques such as SPM for work on surfaces, or optical tweezers in free space, are used for this. Positional assembly is extremely laborious and is currently not suitable as an atomic-scale industrial process the utility and strength of SPM in industry lie in their ability to characterise and measure surfaces with atomic-level precision, rather than as fabrication tools.

The fact that structures can be fabricated atom-by-atom has lead to speculation that tiny nanoscale machines could be made which could be used in parallel to manufacture materials atom-by-atom. The idea is to fabricate one or a few machines (or assemblers) that would first make copies of themselves, and then go on to make materials in parallel, in principle solving the problem of slow production speed.

Top-down Manufacturing

Top-down manufacturing involves starting with a larger piece of material and etching, milling or machining a nanostructure from it by removing material (as for example, in circuits on microchips). This .can be done by using techniques such as precision engineering and lithography, and has been developed and refined by the semiconductor industry. Top-down methods offer reliability and device, complexity, although they are generally higher in energy usage and produce more waste than bottom-up methods.

(a) Precision Engineering

In general, ultra-precision engineering and manufacture underpin much of the micro-electronics industry in everything from the production of the flat low-damage semiconductor wafers used as substrates for computer chips, to the mechanical stages used to position the wafers, to the manufacture of the precision optics used to print the patterns on the wafers. In addition, the techniques of ultra-precision engineering. are used in a variety of consumer products such as computer hard disks, CD and DVD Players.

Ultra-precision machine tools can now achieve very high performance in terms of both the accuracy with which form can be defined (up to 1 part in 10^7, or better than 100 nm over distances of tens of centimetres) and the surface finishes that can be achieved (0.5-1 nm root mean square surface roughness), although these are currently on simple shape surfaces and with low output levels capability, which is bringing benefits in several areas has been achieved through a combination of advances. These include: the use of advanced materials for cutting tools, based on diamond or cubic boron nitride; very stiff, precise machine tool structures; new linear and rotary bearing designs employing fluid films; and sensors for size control combined with numerical control and advanced servo-drive technologies. Very precise process and temperature control is needed to achieve this performance (the latter being of the order of ± 0.01 K).

There are strong drivers to reduce tolerances in engineering, including miniaturisation, improved wear and reliability characteristics, automated assembly and greater interchangeability, reduced waste and requirement for re-work. As the trend towards miniaturisation continues, research and the industrial application of energy beam processing methods will increase, driven in particular by the electronics and computer industries. Techniques such as electron beam lithography (EBL), focused ion beam (FIB), reactive ion etching (RIE) and femtosecond pulsed laser ablation are becoming more accurate and cheaper to apply in a production context. Some examples of future applications of high-precision engineering are given below.

- ICT: the machines used to fabricate chips depend fundamentally upon the use of ultra-high precision techniques for their manufacture and nanometrology techniques for their operation. The manufacture of larger-diameter semiconductor wafers with improved flatness and reduced sub-surface damage should lead to improved device yields and reduced costs.
- Optics: innovative ductile-mode grinding processes together with electrolytic in-process dressing (ELID) should result in the elimination of polishing in the production of high-quality optical devices. This is likely to be of particular importance in the production of

the optics for extra-large astronomical telescopes such as the proposed 50 m and 100 m systems, which will consist of many individually figured segments.

- Transport: precision-machined parts should be more reliable, because of reduced wear, requiring fewer replacement parts and less energy consumption. For example, the ability to produce surfaces with controlled textures through finishing to 10 nm average roughness followed by laser surface treatment is expected to lead to improved power transmission trains with losses through slip reduced by up to 50%. Precision manufacturing is predicted to lead to weight reductions in airframe wings and to improve the performance of internal combustion chambers.
- Medical: it is hoped that the use of ultra-precision machining techniques to produce improved surface finishes on prosthetic implants should lead to lower wear and better reliability.

It is hoped that advances in precision engineering will enable the reduction of environmental impacts by, for example, reducing the use of lubricants. However, for any particular product, the whole life cycle needs to be taken into account before it can be established

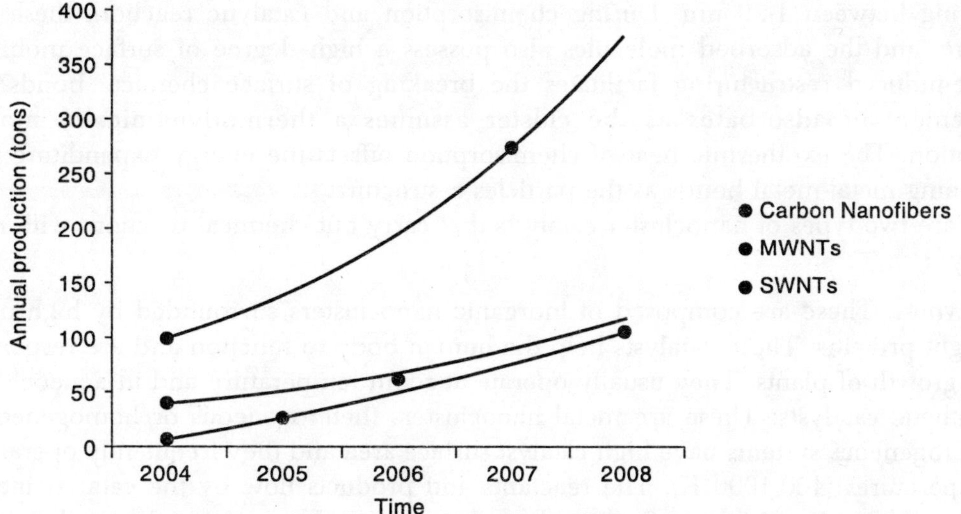

Fig. 3.7 *Estimated future global production of nanotubes (cientifica 2004)*

(b) Lithography

Manufacturing in the ICT sector predominantly involves lithographic processes that pattern a semiconductor wafer in a sequence of fabrication steps. Lithography involves the patterning of a surface through exposure to light, ions or electrons, and then subsequent etching and/or deposition of material on to that surface to produce the desired device. The ability to pattern features in the nanometre range is fundamental to the success of the IT industry. The main lithographic tools can be conveniently separated into methods that use a focused beam of electrons or ions to write patterns, and those that rely on the projection of light through a mask to define a pattern over a complete semiconductor wafer. Electron- and ion-based methods are both capable of making sub-10 nm structures (with electron beam lithography having the greatest routine resolution), but they are too slow to be used directly in production. Optical lithography is used for production of semiconductor devices. Although it does not have the resolution of the beam-based techniques, it provides rapid throughput and cost-effective manufacture. Electron beam lithography is primarily used to fabricate the masks used for optical lithography, and ion beam techniques are mostly used to repair masks and for specialist device application.

The requirement for ever-shrinking device structures has placed enormous technical demands on optical lithographic process, as the nanostructures have length scales similar to or less than the wavelength of the illuminating light (ultraviolet).

SECTION-IV

CATALYSIS IN NANOSCIENCE

Catalysts are usually nanoparticles composed of clusters of atoms, often metals, with particle size varying between 1-20 nm. During chemisorption and catalytic reaction, these particles restructure, and the adsorbed molecules also possess a high degree of surface mobility. The adsorbate-induced restructuring facilitates the breaking of surface chemical bonds and the reammgement of adsorbates as the cluster assumes a thermodynamicalIy most-stable configuration. The exothermic heat of chemisorption offsets the energy expenditure involved in weakening metal-metal bonds as the particles restructure.

There are two types of nanocluster catalysts that carry out chemical reactions with rates and selectivity.

1. Enzymes: These are composed of inorganic nanoclusters surrounded by high molecular weight proteins. These catalysts help the human body to function and are responsible for the growth of plants. They usually operate at room temperature and in aqueous solution.
2. Synthetic catalysts: These are metal nanoclusters (heterogeneous or homogeneous). The heterogeneous systems have high catalyst surface area and they frequiently operate at high temperatures (400-1000 K). The reactants and products flow by the catalyst interface in the gas phase, if possible, to facilitate high flow rates and removal of the product molecules from the catalyst bed

Figure 3.8 below shows schematically the nanoscale size of catalyst systems. Figure 3.9 shows several enzyme catalyst active sites, which are inorganic nanoclusters containing iron, that carry out important oxidation or reduction chemistry.

Catalytic Activity and Selectivity

There are only two multi-path catalytic reactions. These use surface science approach to explore selectivity. These are CO hydrogenation and the conversion of n-hexane or n-heptane to branched isomers, aromatics and other products (reforming). There are six important

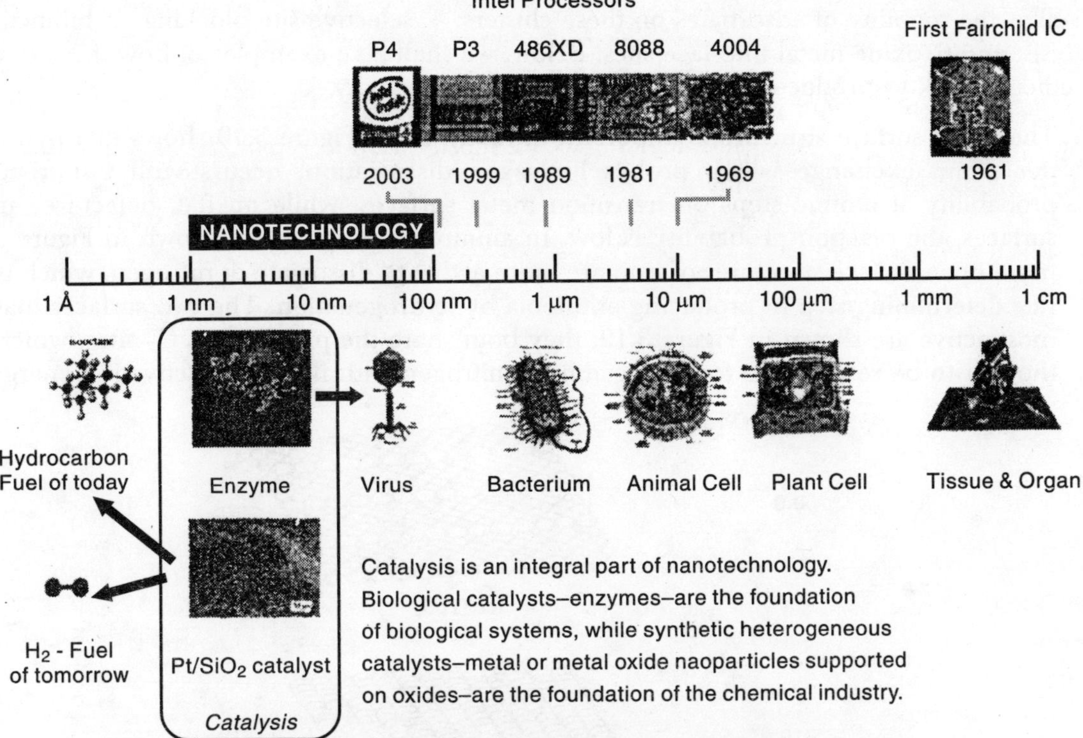

Fig. 3.8 *Catalysts and the nanometer regime*

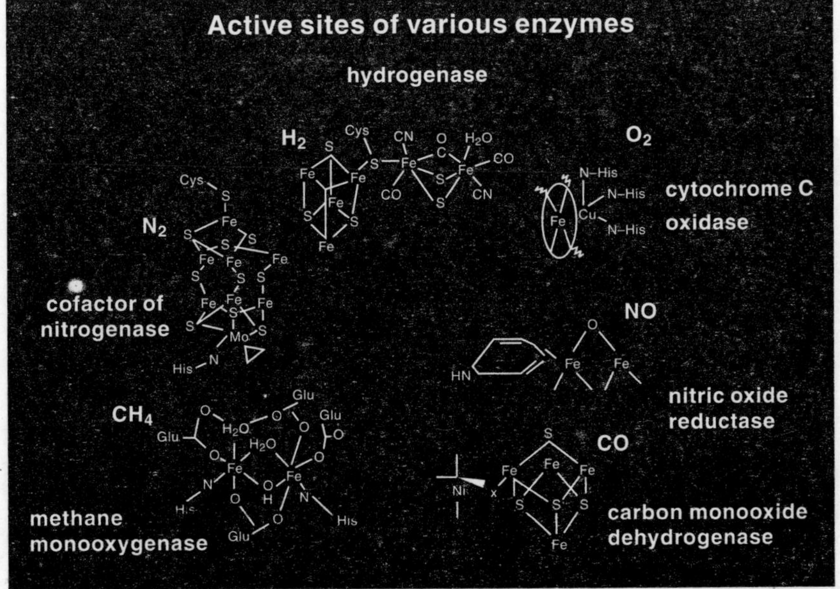

Fig. 3.9 *Enzyme catalyst active sites*

identifiable features that influence both catalyst activity and selectivity. These are 1. metal surface structure; 2. bonding modifier additives; 3.mobility of the metal clusters to restructure as well as the mobility of adsorbates on these clusters; 4. selective site blocking; 5. bifunctional catalysis; and 6. oxide-metal interface sites. Below, we shall give examples of how these catalyst ingredients work to produce improved reactivity and selectivity.

1. The metal surface structure is important in many ways. Figure 3.10 shows that hydrogen deuterium exchange which probes hydrogen dissociation, occurs with unit reaction probability at atomic steps on transition metal surfaces, while on flat, defect-free metal surfaces, the reaction probability is low. In ammonia synthesis,' as shown in Figure 3.11, more open surface structures of iron are more active to dissociate di-nitrogen, which is the rate-determining step in producing ammonia by hydrogenation. The two surfaces that are most active are shown in Figure 3.12; they both share the presence of C_7 sites, which are thought to be responsible for dissociating di-nitrogen with near zero activation energy.

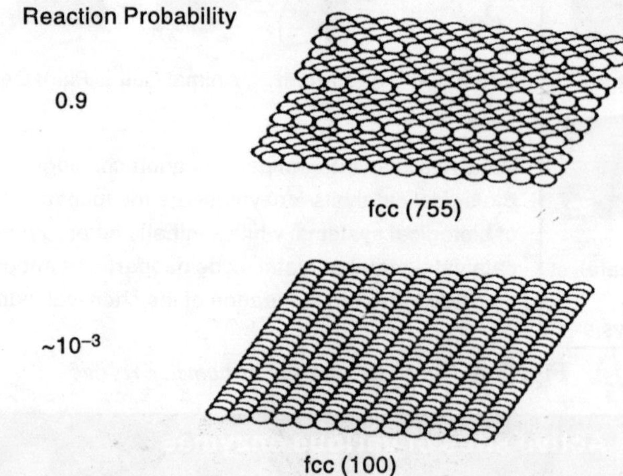

Fig. 3.10 *Metal structure sensitivity of hydrogen/deuterium exchange*

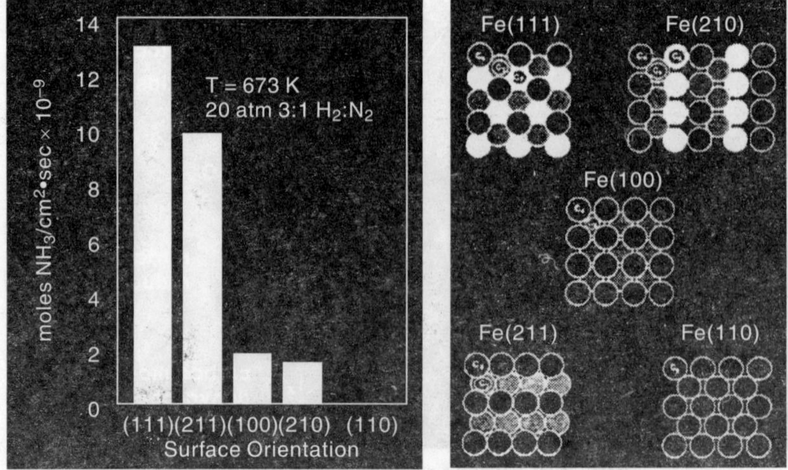

Fig. 3.11 *Structure sensitivity of ammonia synthesis on iron*

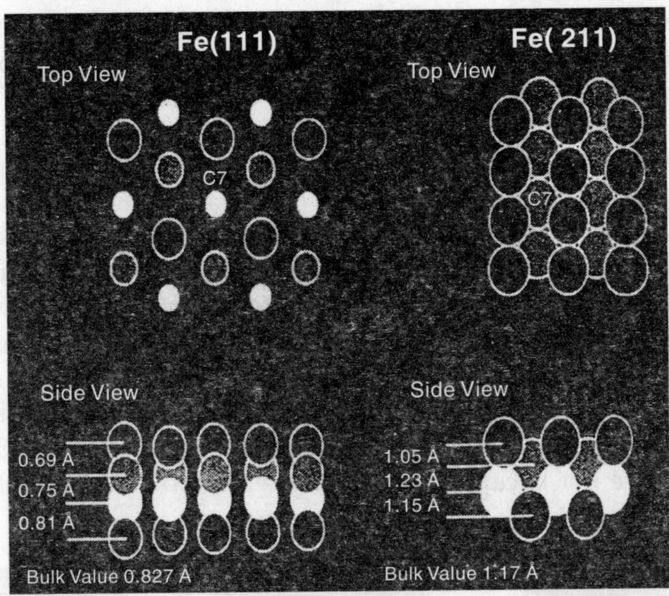

Fig. 3.12 *Surface structure of active (clean) iron surfaces*

2. By adding a small amount of potassium to iron, the turnover rate for producing ammonia from nitrogen and hydrogen could be increased by 30-fold (Figure 3.13). The reason for

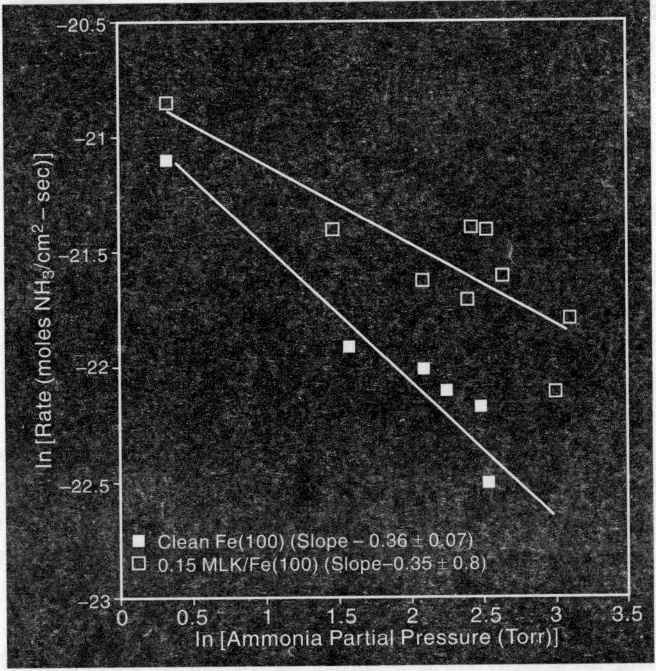

Fig. 3.13 *The effect of potassium as a promoter on the ammonia pressure dependence for ammonia synthesis on iron. Potassium is the bonding modifier of adsorbed ammonium*

this is that potassium is an electron donor, as is ammonia, which is the product of this catalytic reaction. The presence of potassium weakens the ammonia bonding to the surface; its heat of adsorption to iron is reduced by 10.5 $kJmol^{-1}$. Thus, due to this repulsive donor-donor interaction, -ammonia's residence time is shorter on the surface, and therefore it does not inhibit the turnover.

3. Figure 3.14 shows that during ethylene hydrogenation, the adsorbate-covered surface is disordered as shown by high pressure scanning turmeling microscopy. Upon poisoning this reaction with carbon monoxide, ordered structures form and the reaction stops. This clearly indicates that poisoning a catalyst surface stops the mobility of adsorbates; this mobility is a key component of catalytic activity. The active platinum surface is covered with strongly adsorbed ethylidyne molecules that restructure the metal surface around the bonding sites. While these adsorbates do not turn over, their role is to prepare the surface for catalytic hydrogenation. Weakly adsorbed, π-bonded ethylene that occupy the remaining metal sites that are not occupied by ethylidyne, hydrogenate rapidly and is the reaction intermediate responsible for turnover.

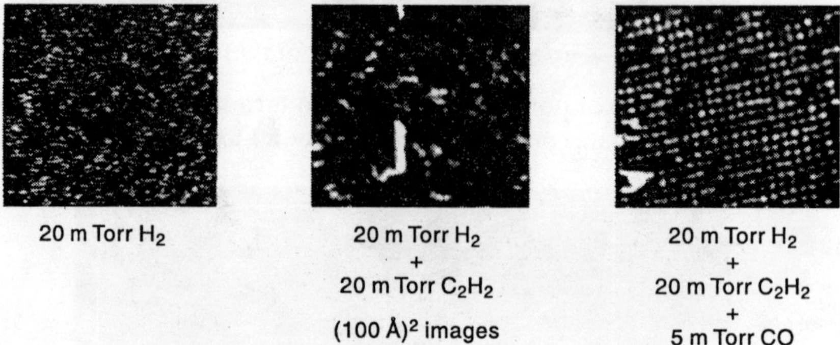

Fig. 3.14 *STM image of the disordered addsorbate-covered surface of Rh(111) during ethylene hydrogenation and the subsequent ordering of the surcace after CO poisoning*

4. Figure 3.15 shows the pore structures of zeolites used to inhibit access of larger molecules to the active sites inside the micropores by restricting the pore size of high surface area porous catalysts. By decreasing the pore size, smaller molecules can enter and undergo chemical change, while the bigger molecules cannot. This way, excellent product selectivity can be obtained, such as using zeolites as acid-base catalysts.
5. Bifunctional catalysis means that sequential or parallel catalytic reactions can occur on more than one type of sites on complex surfaces. Often, a transition-metal site is used to dissociate hydrogen and oxygen, but organic rearrangement reactions may occur on oxide surfaces that provide acid-base catalytic sites for isomerization, alkylation, and dehydrocylization. This is shown schematically in Figure 3.16.
6. Oxide-metal interface sites are often thought to be important in catalysis. This can be shown by placing nickel on various oxide supports, as shown in Figure. 3.17. The oxide supports cannot-carry .out CO hydrogenation reactions; however, when nickel is placed on both oxide supports, the activity and selectivity change dramatically, depending on the type of oxide used.

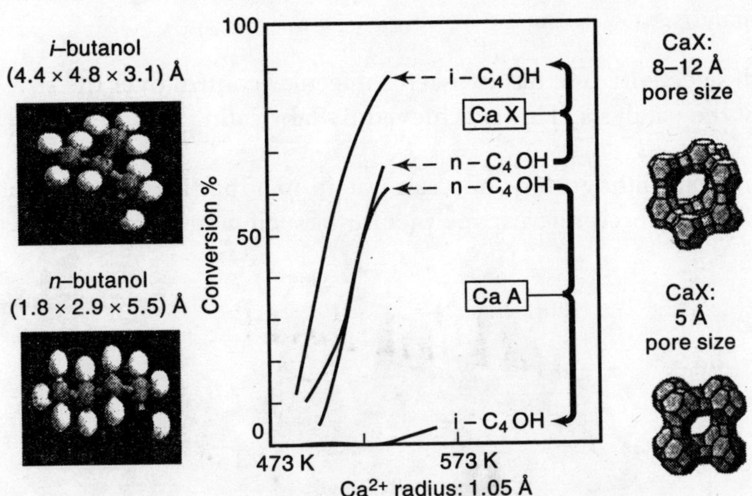

Fig. 3.15 *Dehydration of n-butanol, without reaction of i-butanol, due to pore size constraints on 5A Linde zeolite*

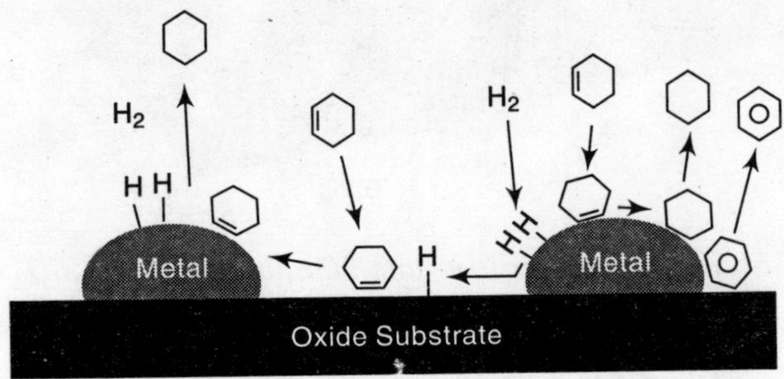

Fig. 3.16 *The sequential and parallel catalytic reactions that can occur on the multiple sites of a bifunctional catalyst*

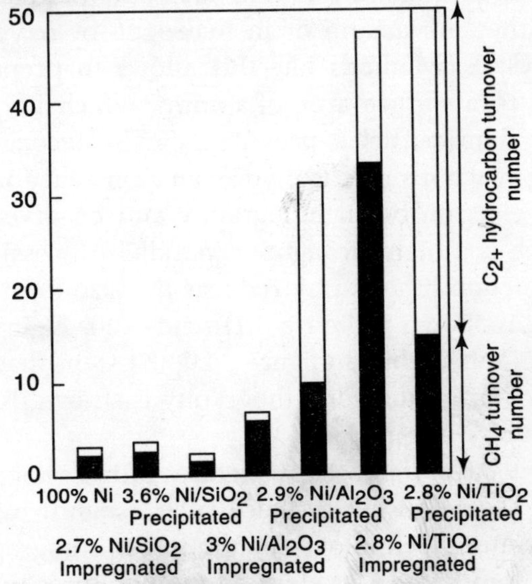

Fig. 3.17 *The effect on activity and selectivity of the oxide support on CO hydrogenation over Ni catalysts at 525 K*

Fabrication of Catalysts

To obtain high selectivity one has to assert molecular control over the size, location, structure, and promoters of the catalysts. This is achieved by fabricating 2-dimensional and 3-dimensional catalysts.

Using electron beam lithography, one can obtain nanoparticles of platinum in the size range of 5-50 nm. An atomic force microscope picture of such a structure is shown in Figure. 3.18.

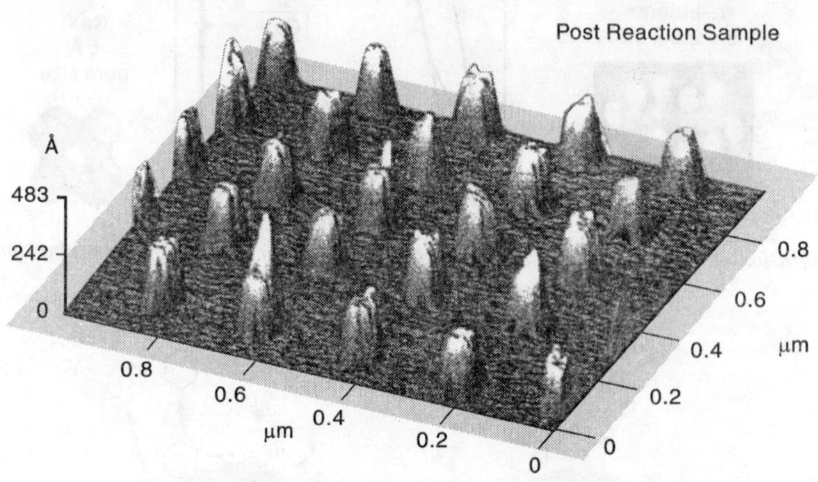

Fig. 3.18 *AFM image of 50 nm Pt nanoparticles with 200 nm spacing on SiO_2, fabricated by electron beam lithography*

Using electron microscopy, both in the back reflection and in the transmission modes, one can image the nanoparticle array and the structure of each nanoparticle. It was found that upon heating these nanoparticles either in vacuum or in hydrogen or oxygen, they restructured to form highly perfect single crystals. Methods like this allows to prepare 10^9 nanoparticles in about one day. This amounts to a surface area of 1 mm^2, which is quite small for studies of reaction selectivity. About 10^{11} nanoparticles, provide a cm^2 surface area, which is adequate to study, for example, reforming reactions or CO hydrogenation reactions. However, that would take 100 days to produce by electron beam lithography and be a very expensive project. So photolithography is used which is a high throughput (parallel process), low cost technique.

Another method called reduction lithography reduces the size from 600 nm to 7 nm.

The scheme of this process is shown in figure 3.19 and 3.20. Figure 3.19 shows the overall scheme of size reduction lithography, while in figure 3.20, the steps that lead to a 50% reduction in size through the first cycle of size reduction lithography is shown. Repetition of such a cycle several times permits us to reach 7 nm in size.

Figure 3.21 and 3.22 show electron microscopy pictures of the nanowires produced this way, before and after the first size reduction cycle. Once such a nanowire or nanodot mould has been produced, polymer imprinting technology (nanoimprint lithography) is used to press the silicon nanowires or nanodots into PMMA at elevated temperatures under pressure to produce

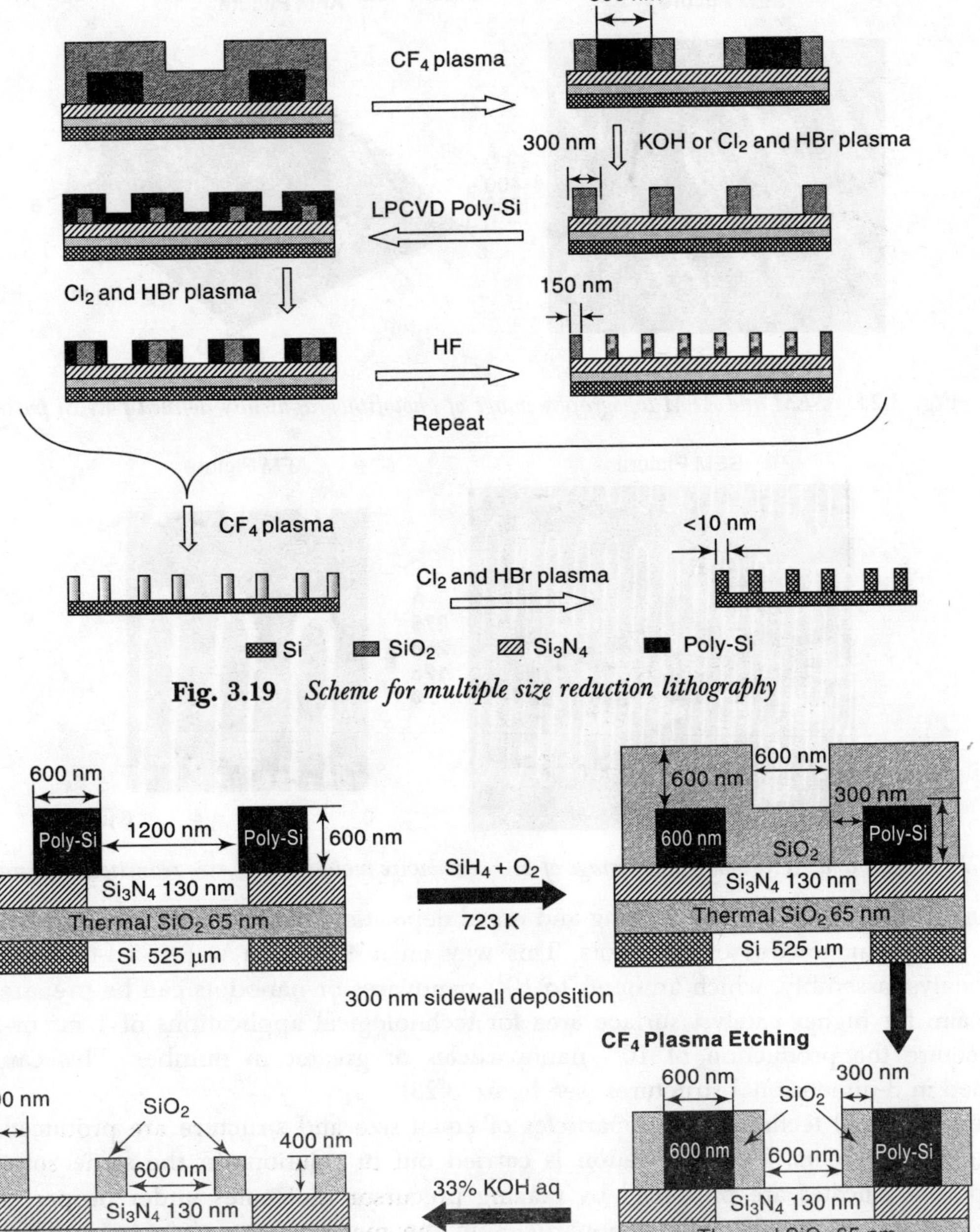

Fig. 3.19 *Scheme for multiple size reduction lithography*

Fig. 3.20 *Focus on the steps of size reduction lithography that results in a 50% reduction in feature size*

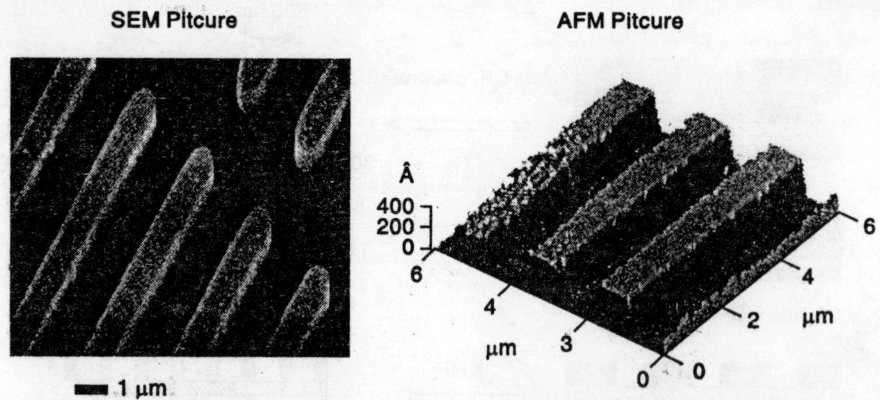

Fig. 3.21 *SEM and AFM topography image of photolithographically-defined poly-Si pattern*

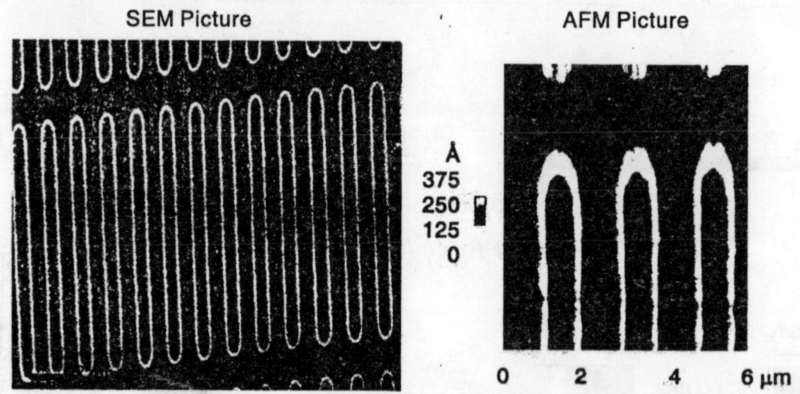

Fig. 3.22 *SEM and AFM topography image of SiO_2 nanowire mold after 1st size reduction lithography process*

an image, which upon plasma etching and metal deposition, leads to the formation of transition metal catalyst nanowires or nanodots. This way on a 4″ silicon wafer, a l-cm^2 surface area nanocatalyst assembly, which amounts to 10^{11} nanowires or nanodots can be prepared.

To aim for higher catalyst surface area for technological applications of 1 m^2 or higher, it will require the production of 10^{15} nanoparticles or greater in number. This can only be obtained in 3-dimensional structures (see figure 3.23)

Using a sol-gel technique, nanoparticles of equal size and structure are produced, that are capped by polymers. This fabrication is carried out in solution. In the same solution, then mesoporous silicates are produced by placing precursor molecules under proper conditions. These silicates encapsulate the nanoparticles in the mesopores as they form. This way, are placed large concentrations of transition metal nanoparticles inside mesoporous solids to achieve high loading as well as precise size and surface structure control. These approaches are aimed toward producing nanoparticles for very high selectivity catalytic reactions.

It should be noted that enzymes exhibit 100% selectivity in important catalytic reactions, and these enzymes operate at room temperature and in aqueous solutions.

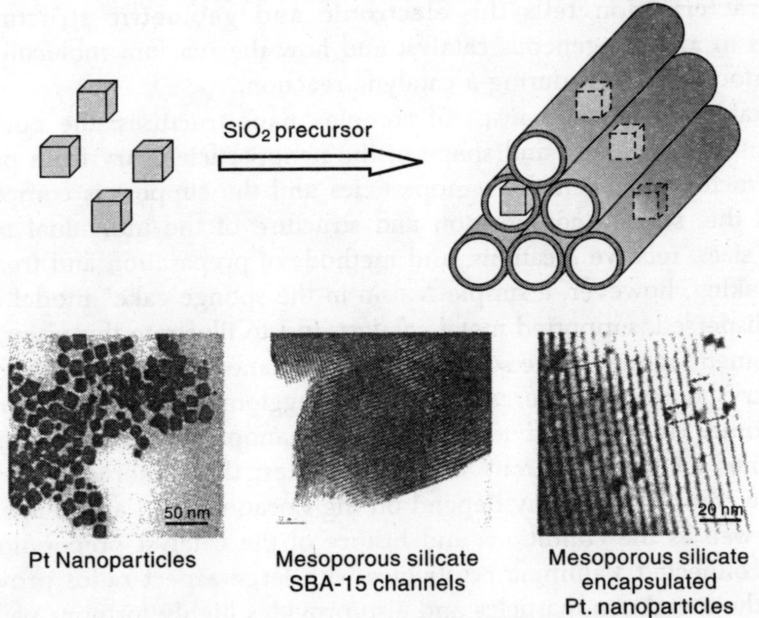

Fig. 3.23 *TEM image of Pt nanoparticles encapsulated in mesoporous silicate*

Heterogeneous Catalysts

Catalysis is the science of chemical reactions on an atomic or molecular level. It is also defined as a process of carilitated self-assembly of atoms or molecules; and acts as the facilitator. Heterogeneous catalysis refers to chemical reactions that occur on a surface; it involves solid catalysts and gaseous or liquid reactants. Heterogeneous catalysts used in commercial processes are often chemically and physically complex systems.

Nanoclusters and nanoparticles are key components of supported catalysts, which is one of the most important classes of heterogeneous catalysts. The- use of nanoparticles is not just for significantly increasing the total surface area of a catalyst; nanoparticles also possess unique electronic, magnetic, optical, and catalytic properties. Kinks, steps, corners, and different surface planes on a nanoparticle provide various specific sites that can bind and/or dissociate reactant molecules. The reactivity of a nanoparticle depends on its size, shape, surface composition, and surface atomic arrangement. Each nanocluster with a specific number of atoms has its own intrinsic chemical and physical properties and the size-dependent quantum effects can become dominant.

The behavior of nanostructured heterogeneous catalysts can be described from their physical characteristics, size and shape of nanoparticles. The performance of a heterogeneous catalyst depends on the size distribution, of the active nanoparticles, the surface area distribution, or the volume distribution of the individual nanoparticles to describe the physical nature of distribution of the individual nanoparticles to describe the physical nature of heterogeneous catalysts. Size- and shape-controled synthesis of nanoparticles is critical to developing nanostructured catalysts.

Catalyst characterization tells the electronic and geometric structure of the active nanocomponents in a heterogeneous catalyst and how the reactant molecules interact with or modify these nanocomponents during a catalytic reaction.

Supported catalysts generally consist of complex nanostructures: the pore structure of the support is non-umform; the size and shape of the nanoparticles vary from particle to particle; the interfacial structure between the nanoparticles and the support is complex and is poorly understood; and the. surface composition and structure of the individual nanoparticles may depend on their sizes, relative locations, and methods of preparation and treatment.

Generally speaking, however, a simple "raisin in the sponge cake" model may conceptually describe a well-dispersed, supported metal catalyst. Just to illustrate this point, Fig. 3.24a shows an idealized schematic model of the structure of metal nanoparticles supported on high-surface-area γ-alumina crystallites. The pores within large agglomerates of γ-alumina crystallites are formed by randomly stacking individual γ-alumina nanoplatelets or nanoneedles. The metal nanoparticles generally have different shapes and sizes; their interaction with the γ-alumina crystallites can be complex and may depend on the Location, size, and shape of the individual nanoparticles as well as the conditions and history of the catalyst preparation and treatment. The use of interconnected γ-alumina crystallites with large aspect ratios provides high-surface area to disperse the metal nanoparticles and also provides highly tortuous pathways to prevent migration and agglomeration of metal nanoparticles during the catalyst preparation processes and during the catalytic chemical reactions. Figure 3.24 b is a high-angle annular dark-field (HAADF) image of an industrial Pd/γ-Al_2O_3, catalyst; it clearly shows numerous interconnected pores of various sizes and the high tortuosity as well as the non-uniform size distribution of the palladium nanoparticles (bright dots in the image). The accessibility of the palladium nanoparticles by the reactant molecules is determined by the relative locations of the nanoparticles with respect to the interconnected pores of the alumina support. Therefore, not only the size distribution but also the spatial distribution of metal nanoparticles is critical to understanding

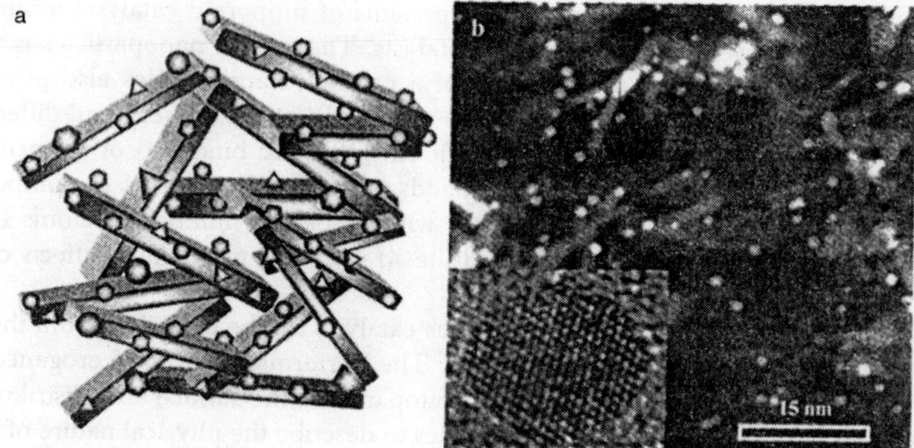

Fig. 3.24 *A schematic model of a γ-Al_2O_3 supported Pd catalyst (a) and an experimental high-angle annular dark-field image of an industrial Pd/γ-Al_2O_3 catalyst showing Pd nanoparticles and the γ-Al_2O_3 nanoplatelets (b). The inset in (b) shows an atomic resolution image of a Pd nanoparticle.*

the performance of supported metal catalysts. In practice, each commercial catalyst has its own unique physicochemical properties; and often these properties are difficult to be controlled or characterized.

The next level of characterization of the catalyst shown in 3.24b is to examine the shape, arrangement of surface atoms, and the electronic structure of the individual Pd nanoparticles. The inset in Fig. 3.24b is a high-resolution electron microscopy (HREM image, showing the atomic structure and the shape of the Pd nanoparticle. The binding energy of the Pd surface atoms can vary at different sites of the Pd nanoparticle.

The interfaces between the Pd nanoparticles and the γ-Al_2O_3 nanoplatelets should also be characterized, understood, and preferably controlled. A complete characterization of industrial catalysts requires extensive resources and is often extremely expensive and slow because of the complex nature of the industrial catalysts; this is why we still do not understand the behavior of most of the heterogeneous catalysts used commercially.

Among many characterization techniques, advanced electron microscopy techniques are the only techniques that can provide information on the individual nanocomponents of industrial heterogeneous catalysts. All the other techniques (e.g., X-ray diffraction, X-ray absorption spectroscopy, infrared spectroscopy, nuclear magnetic resonance spectroscopy, X-ray photoelectron spectroscopy, etc.) provide information either averaged over millions to trillions of nanocomponents or they (e.g., scanning tunneling microscopy and atomic force microscopy techniques) require stringent conditions on the samples to be examined. Therefore, advanced electron microscopy techniques are indispensable for understanding the properties of heterogeneous catalysts (see below) and for guiding the development of nanostructured industrial catalysts.

Electron Microscopy In Developing Nanostructured Catalysts

Electron microscopy techniques are clearly widely used for characterizing nanoparticle systems. Advanced electron microscopy techniques, especially atomic resolution electron microscopy and spectroscopy techniques, are indispensable for studying the fundamental properties of nanoparticles and for guiding the synthesis of nanoparticles and nanostructured catalysts with desired structural properties.

The one most single important feature of a modem electron microscope is its versatility: atomic resolution images, diffraction patterns from nanometer regions, and nanometer-scale spectroscopy data can be obtained either simultaneously or sequentially from the same region of the specimen. The availability of the various imaging, diffraction, and spectroscopy techniques within a single instrument makes the modem electron microscope a powerful tool for characterizing the physicochemical nature of nanoscale sysrems, especially nanostructured heterogeneous catalysts.

When an electron beam interacts with a thin specimen inside an electron microscope, a variety of electron, photon, phonon, and other signals can be generated. Figure shows a schematic diagram illustrating the common signals that are produced by strong electron-solid interactions. There are three types of transmitted electrons: 1) non-scattered electrons, 2) elastically scattered electrons, and 3) inelastically scattered electrons. There are also three types of electrons that can be emitted from the electron-entrance surface of the specimen: 1) secondary electrons with energies < 50 eV, 2) Auger electrons produced by the decay of the excited atoms, and 3) backscattered electrons that have energies close to those of the incident electrons.

All these signals can be used to form images or diffraction patterns of the specimen or can be analyzed to provide spectroscopic information. For example, electron energy-loss spectroscopy (EELS), which is based on the energy analysis of the inelastically scattered electrons, can provide information on the electronic structure, oxidation states, and chemical composition of catalysts on an atomic or sub-nanometer scale.

The de-excitation of atoms that are excited by the primary electrons also produces continuous and characteristic X-rays as well as visible light. These signals can be utilized to provide qualitative and quantitative information on the elements present in the regions of interest. X-ray energy dispersive spectroscopy (XEDS), for example, is a powerful tool for identifying elements that are present in a heterogeneous catalyst. XEDS spectra data is transferred into quantitative data describing different phases and phase distributions in heterogeneous catalysts.

Transmission Electron Microscopy in Heterogeneous Catalysts

Atomic resolution imaging can now be routinely achieved in the modern transmission electron microscope (TEM) and has been widely applied to charactenzing the atomic structure of a plethora of materials. Although scanning tunneling microscopy and atomic force microscopy techniques can also provide atomic resolution information and have proved invaluable for studying the fundamental processes of heterogeneou catalysis. Most heterogeneous catalysts can be directly examined in an electron microscope.

Atomic structure of heterogeneous catalysts such as facets, surface steps, and surface reconstructions can be visualized by using a technique called surface profile imaging;[1.17] in the profile-imaging mode, a thin crystal is aligned such that the direction of the electron beam is parallel to a low-index zone-axis of the crystal. As an example, Fig. 3.25 shows an atomic resolution profile image of a small crystal of an antimony-doped tin oxide (ATO) nanocatalyst. The ATO nanocatalyst is of interest in the oxidation of propylene to acrolein and the selective oxidation of olefins. The image was recorded at the optimum defocus value so that the atomic columns appear as black dots in the image.[1] Surface steps, rearrangement of surface atoms (indicated by the double arrows), facets, and an internal defect, which is probably caused by the incorporation of Sb dopants, are clearly revealed. The inset shows the corresponding digital diffractogram clearly revealing that the Sb-doped SnO_2 nanocrystallite has a rutile-type structure. The effect of the antimony doping and the calcination temperature on the growth of SnO_2 nanocrystallites as well as the oxidation states of the Sb dopant has recently been studied by the combined use of HREM, HAADF, and EELS techniques.

Atomic structure and structural evolution of many heterogeneous catalysts can be examined by the profile imaging method. Surface reconstructions of small gold particles and adsorbate-induce restructuring of Pt metal śurfaces were investigated by the HREM technique. HREM examination of fresh and used catalysts can often yield information on the surface modification of metallic nanoparticles as well as their sintering behavior during catalytic reactions.

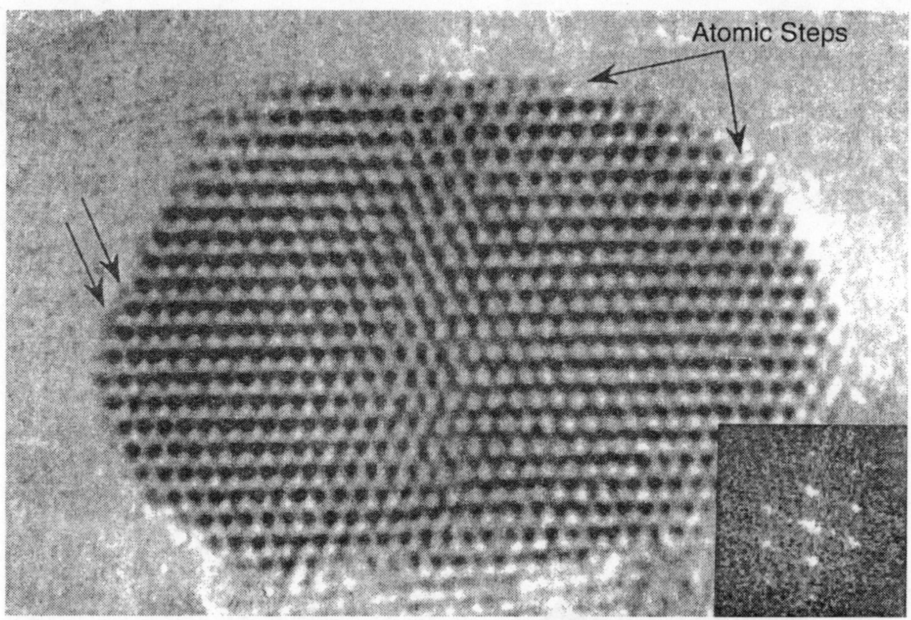

Fig. 3.25 *HREM image of an antimony-doped SnO_2 nanopanicle shows surface steps, facets, restructuring of surface atoms, and an internal defect. The inset shows the corresponding digital diffractogram clearly revealing that the nanocrystal has a rutile-type structure.*

In supported metal catalysts, the performance of a catalyst can be directly related to the available number of active sites of the metal nanoparticles; thus, it is important to determine their sizes and size distributions. For nanoparticles larger than about 3 nm in diameter, XRD techniques, especially when synchrotron sources are used, can provide the required information; however, the sizes provided by the XRD technique reflects the sizes of the single crystallites which may not be the sizes of the nanoparticles. For nanoparticles less than 3 nm in diameter, it is difficult, if not impossible, to obtain accurate information about the size distributions of the metal nanoparticles by the XRD technique. On the other hand, small particles, clusters, or even single atoms can be directly observed in the modem electron microscope. In the following, we will discuss the particle visibility, structure and shape of metal nanoparticles, the measurement of lattice parameters of small nanoparticles, and the limitations of applying the HREM techniques to the study of nanoparticles and supported metal catalysts.

Scanning Transmission Electron Microscopy in Heterogeneous Catalysts

Scanning transmission electron microscopes (STEM) are tailor-made to provide structural, chemical, and morphological information on heterogeneous catalysts on a nanometer or atomic level. A remarkable capability of a STEM instrument is the formation of a high-brightness electron probe with diameters < 0.2 nm at 100 keV and as small as 0.13 nm at 300 keV. To form such an electron nanoprobe, the use of a field-emission gun (FEG) is necessary to provide sufficient signal strength for viewing and recording STEM images, spectra, and diffraction patterns. Another attractive feature of STEM is its great flexibility in the detection system. Different signals (see and Fig. 3.26) generated from the same sample area can be collected either independently or simultaneously and can be analyzed in parallel to yield complementary information.

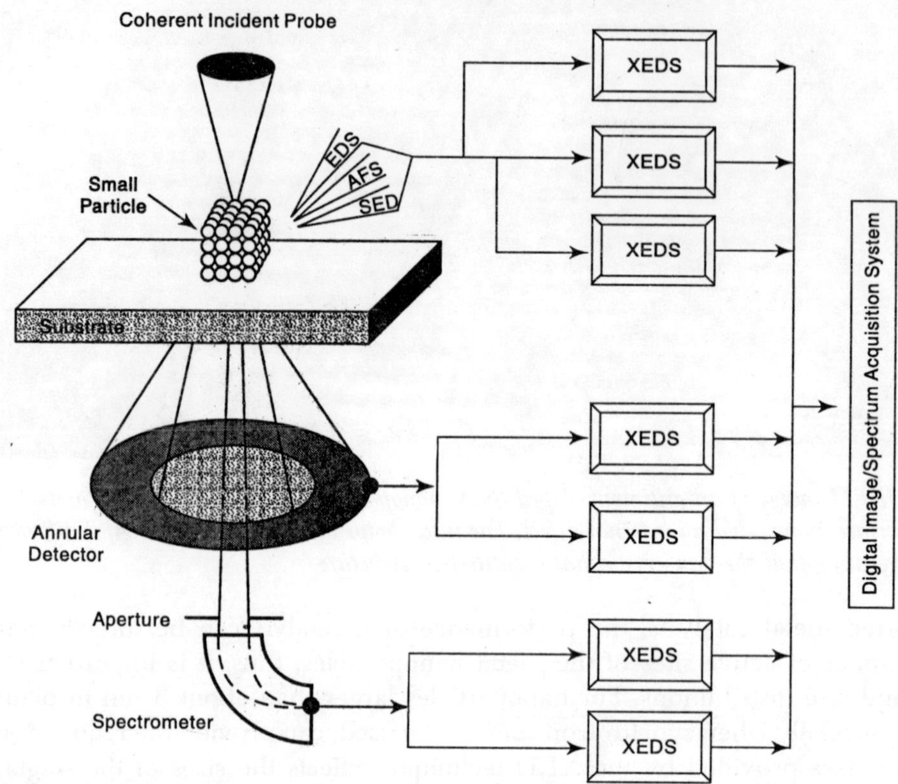

Fig. 3.26 *Schematic diagram illustrates the available signals and detectors in a dedicated scanning transmission electron microscope. XEDS–X-ray energy dispersive spectroscopy; AES–Auger electron spectroscopy; SAM–scanning Auger microscopy; SES–secondary electron spectroscopy; SEM–secondary electron microscopy; HAADF–high-angle annular dark-field; CEND–coherent electron nanodiffraction; PEELS–parallel electron energy-loss spectroscopy; BF– bright field; DF–dark field*

Unlike in TEM, where a stationary, parallel electron beam is used to form images, the STEM is a mapping device. In a STEM instrument, a fine electron probe, formed by using a strong objective lens to de-magnify a small FEG electron source, is scanned over a specimen in a two-dimensional raster. Signals generated from the specimen are detected, amplified, and used to modulate the brightness of a second electron beam that is scanned synchronously with the STEM electron probe across a cathode-ray-tube (CRT) display. Therefore, a specimen image is mapped onto the CRT display for observation or . is digitally stored in a computer. The resolution of STEM images is primarily determined by the incident probe size, the stability of the microscope, and the inherent properties of the signal generation processes.

Multiple imaging and analytical detectors have been developed. to simultaneously collect several signals that can be displayed individually or combined in perfect register with each other. This unique capability makes the STEM a powerful nanoanalytical tool since multiple views of a sample, in imaging, diffraction, and spectroscopy modes, can be collected, analyzed, and compared in a single pass of the incident electron beam. Figure 3.26 illustrates the available detectors commonly used for collecting imaging and analytical signals in high-resolution STEM instruments.

By collecting high-angle scattered electrons with an annular detector, HAADF images (or Z-contrast images) can be formed to provide information about structural variations across the sample at an atomic level. EELS and XEDS can give quantitative data describing changes of elemental composition, electronic structure, or state of oxidation associated with inhomogeneous structures of the sample. The combination of XEDS and EELS with HAADF imaging technique can provide detailed information on the composition, chemistry, and electronic and crystal structure of heterogeneous catalysts with atomic resolution and high sensitivity.

By collecting and analyzing secondary electron (SE) and Auger electron (AE) signals we can extract information on surface topography and surface chemistry of heterogeneous catalysts with nanometer spatial resolution. By positioning an electron nanoprobe at the area of interest, diffraction patterns from nanometer or subnanometer regions can be acquired to provide local crystallographic structure of nanoparticles. The structural relationship of nanoparticles to the surrounding materials and, in some cases, the shape of nanoparticles, can also be extracted from nanodiffraction patterns. The powerful combination of atomic resolution HAADF imaging with nano-spectroscopy and nano-diffiaction techniques has proved invaluable in characterizing heterogeneous catalysts, especially supported metal and alloy catalysts.

PART II

CARBON NANOTUBES

FULLERENES OR BUCKYBALLS

In 1985, scientists made an exciting breakthrough in chemistry, in an experiment on carbon molecules in space. It revolutionized the world of science. Called buckminsterfullerene, or the buckyball, (for short), they are a family of carbon allotropes. They are molecules composed entirely of carbon, in the form of a hollow sphere, ellipsoid, or tube. It is the roundest and most symmetrical large molecule known to man, named after American architect R. Buckminster Fuller, who designed a geodesic dome with the same fundamental symmetry.

Spherical fullerenes are sometimes called **buckyballs** because the C_{60} variant is often compared in appearance to a typical white and black soccer football such as the Telstar of 1970. Cylindrical fullerenes are called **Carbon nanotubes or buckytubes.**

Before this discovery, there were only two known forms of "pure" carbon to be found on earth. The first is graphite, the slippery, soft, metallic-grey material used in-pencils and second is a form of pure carbon: diamond, a rigid crystal. Also made with pure carbon is the Fullerene or buckyball which have similar structure to graphite and is composed of a sheet of linked hexagonal rings, but they contain 12 regular pentagonal and 30 regular hexagonal rings that prevent the sheet from being planar. They consist of 60 carbon atoms, all linked together in a coordinated fashion to form a hollow, geodesic dome with bonding strains equidistributed among 60 carbon atoms (Figure 4.1).

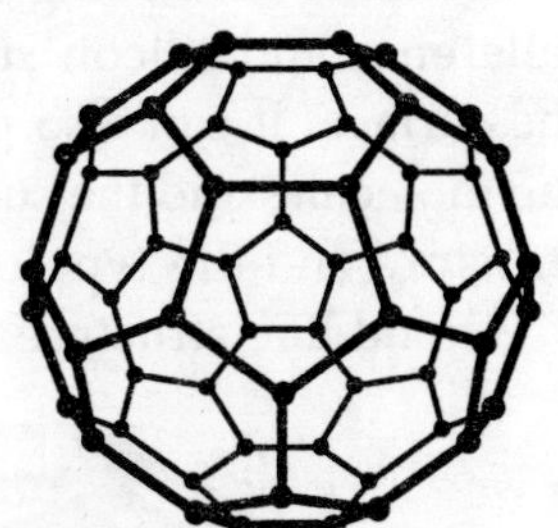

Buckminsterfullerene C_{60}

Fig. 4.1

Buckminsierfullerene (IUPAC name (C_{60}-I_h) [5,6] fullerene) Fig. (4.1) is the smallest fullerene in which no two pentagons share an edge (which can be destabilizing). It is also the most common in terms of natural occurrence, as it can often be found in soot.

A common method used to produce fullerenes is to send a large current between two nearby graphite electrodes in an inert atmosphere. The resulting carbon plasma arc between the electrodes cools into sooty residue from which many fullerenes can be isolated.

Structure of Fullerene

The structure of C_{60} is a truncated $T = 3$ icosahedron, which resembles a round soccer ball of the type made of hexagons and pentagons, with a carbon atom at the corners of each hexagon and a bond along each edge.

The C_{60} molecule does not bond readily to other atoms or molecules, as all bonds are to another carbon atom. Fullerenes are molecules which consist of a network of sp^2 hybridised carbons in pentagons and hexagons which form a spherical molecule.

The C_{60} molecule has two bond lengths. The 6:6 ring bonds (between two hexagons) can be considered "double bonds" and are shorter than the 6:5 bonds (between a hexagon and a pentagon). Some of the electrons are delocalized over the entire molecule–

Benzene is flat and many of its derivatives also tend to stack in flatsheets. Spherical buckyballs literally add a new dimension to the chemistry of such aromatic compounds. It is ten times larger than benzene.

The structure of a fullerene is a trivalent convex polyhedron with pentagonal and hexagonal faces (see Figure 4.2. In graph theory, the term **fullerene** refers to any 3-regular, planar graph with all faces of size 5 or 6 (including the external face). It follows from Euler's polyhedron formula, $|V| - |E| + |F| = 2$, (where $|V|$, $|E|$, $|F|$ indicate the number of vertices, edges, and faces), that there are exactly 12 carbon pentagon rings in a fullerene to close the graphite hexagonal network.

The smallest fullerene is the dodecahedron–the unique C_{20}, There are no fullerenes with 22 vertices. The number of fullerenes C_{2n} grows with increasing n = 12, 13, 14 ... For instance, there are 1812 non-isomorphic fullerenes C_{60}. Note that only one of the C_{60}'s, the buckminsterfullerene alias truncated icosahedron, has no pair of adjacent pentagons (the smallest such fullerene). To further illustrate the growth, there are 214, 127, 713 non-isomorphic fullerenes C_{200}, 15, 655, 672 of which have no adjacent pentagons.

Scanning tunnelling microscopy (STM) provides structural information on work involving fullerenes and silicon surfaces. Such as adsorption sites and the ordering of the molecules on the surface. It can also give us an insight into the interactions involved between molecules and the molecules and the surface. Synchrotron radiation studies provide information on the electronic structure and the type of bonding that is involved.

Abond-line structure of buckyballs are shown below:

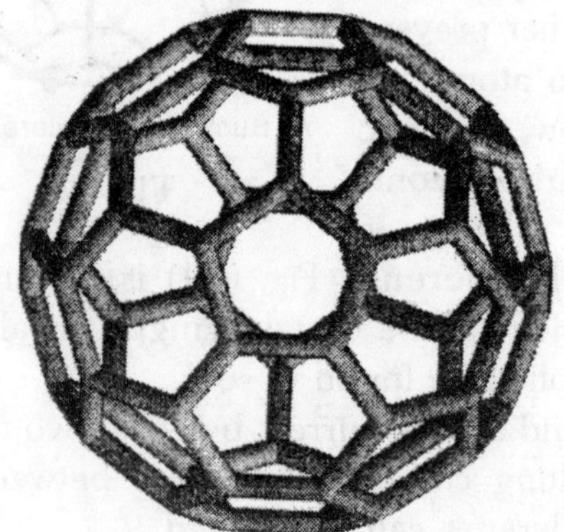

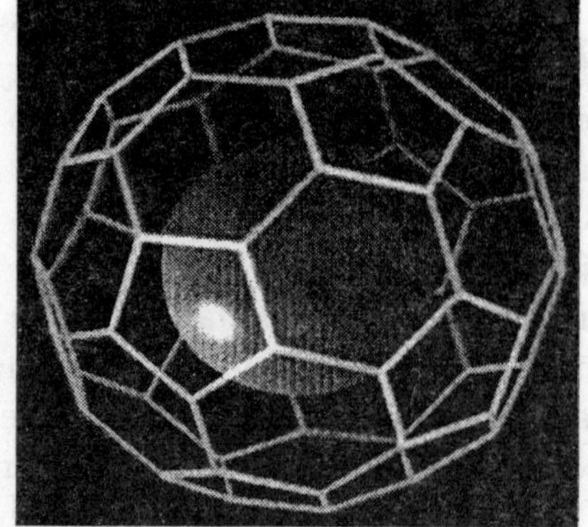

More bucky balls

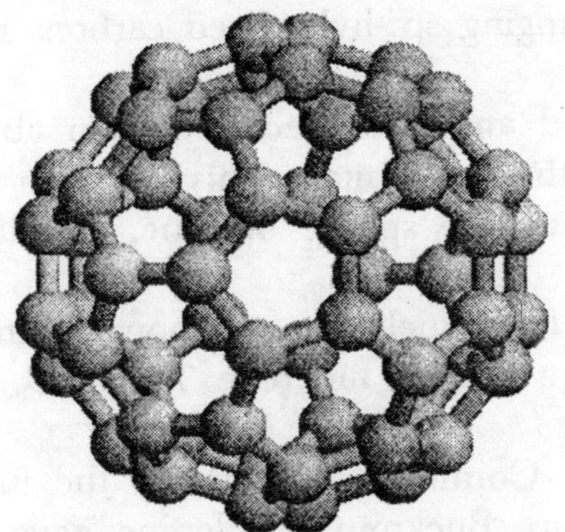

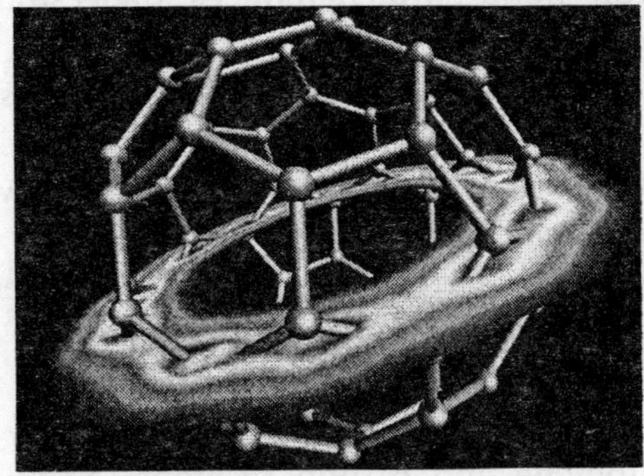

Fig. 4.2 *Bucky ball structures (C_{60}) has 20 hexagons and 12 pentagons.*

PROPERTIES OF FULLERENES

For the past decade, the chemical and physical properties of fullerenes have been a hot topic in the field of research and development, and are likely to continue to be for a long time.

The buckyball is the only molecule of a single atom to form a hollow spheroid, and it spins at over one hundred million times per second. "there are 174 ways that The buckyball, can vibrate." It is the largest possible symmetric molecule, because sixty is the largest number of proper rotations in the icosahedral group. That, in turn, is the largest point group-the largest group where symmetry operations: rotations, reflections etc. leave one point unmoved. Buckyball are also formed via shining a laser onto graphite but under different conditions to nanotubes. The buckyball, being the roundest of round molecules, is also quite resistant to high speed collisions. In fact, the buckyball can withstand slamming into a stainless steel plate at 15,000 mph, merely bouncing back, unharmed. When compressed to 70 percents original size, the buckyball becomes more than twice as hard as diamond.

In addition to openingup new fields on chemistry C_{60} also shows interesting physical properties. It is resistant to shock and it has been suggested that as a lubricant, there is even evidence of superconductivity and it may provide the added ingredient that makes diamond films more practical. "Depending on the compound, the buckyball can serve as a conductor, insulator, semiconductor, or superconductor."

It is exceedingly rugged and very stable, capable of surviving the temperature extremes of outer space. At first, however, the molecule was a mystery wrapped in an enigma. But when a convenient way of making this molecule, also known as C_{60}, was discovered, it set off an explosion of research among chemists, physicists, and materials scientists to uncover the molecule's secrets.

A whole family of related molecules, including C_{70}, C_{84} and other "fullerenes"–clusters as small as C_{28} and as large as a postulated C_{240}.(These unusual molecules turn out to have extraordinary chemical and physical properties.

Fullerenes are stable, but not totally unreactive. The sp^2-hybridized carbon atoms, which are at their minimum energy in planar graphite, must be bent to form the closed sphere or tube, which produces angle strain. The characteristic reaction of fullerenes is electrophilic addition at

6, 6-double bonds, which reduces angle strain by changing sp^2-hybridized carbons into sp^3-hybridized ones.

The change in hybridized orbitals causes the bond angles to decrease from about 120 degrees in the sp^2 orbitals to about 109.5 degrees in the sp^3 orbitals. This decrease in bond angles allows for the bonds to bend less when closing the sphere or tube, and thus, the molecule becomes more stable.

Other atoms can be trapped inside fullerenes to form inclusion compounds known as endohedral fullerenes. An unusual example is the egg shaped fullerene $Tb_3N@C_{84}$, which violates the isolated pentagon rule.

Fullerenes are sparingly soluble *in* many solvents. Common solvents for the fullerenes include toluene and carbon disulfide. Solutions of pure Buckminsterfullerene have a deep purple color. Solutions of C_{70} are a reddish brown. The higher fullerenes C_{76} to C_{84} have a variety of colors. C_{76} has two optical forms, while other higher fullerenes have several structural isomers. Fullerences are the only known allotrope of carbon that can be dissolved in common solvents at room temperature.

Some fullerene structures are not soluble because they have a small bandgap between the ground and excited states. These include the small fullerenes C_{36} and C_{50}. The C_{72} structure is also in this class, but the endohedral version with a trapped lanthanide-group atom is soluble due to the interaction of the metal atom and the electronic states of the fullerene. Small bandgap fullerences are highly reactive and bind to other fullerenes or to soot particles.

Solvents that are able to dissolve a fullerene extract mixture (C_{60}/C_{70}) are listed below in order from highest solubility. The value in parentheses is the approximate saturated concentration in mg. ml^{-1}.

1. 1,2,4-trichlorobenzene (20)
2. carbon disulfide (12)
3. toluene (3.2)
4. benzene (1.8)
5. chloroform (0.5)
6. carbon tetrachloride (0.4)
7. cyclohexane (0.054)
8. *n*-hexane (0.046)
9. tetrahydrofuran (0.037)
10. acetonitrile (0.02)
11. methanol (0.0009)

The reactivity increases by attaching active groups to the surface of fullerenes.

Buckminsterfullerene does not exhibit "superaromaticity": that is, the electrons in the hexagonal rings do not delocalize over the whole molecule.

A spherical fullerene of *n* carbon atoms has *n* pi-bonding electrons. These should try to delocalize over the whole molecule. The quantum mechanics of such an arrangement should be like one shell only of the well-known quantum mechanical structure of a single atom, with a stable filled shell for $n = 2$. 8, 18, 32, 50, 98, 128, etc, i.e. twice a perfect square; but this series does not include 60. As a result, C_{60} in water tends to pick up two more electrons and become an anion.

The C_{60} molecule is injurious to organisms. By introducing fullerenes into water at concentrations of 0.5 parts per million it is found that largemouth bass suffered a 17-fold increase in cellular damage in the brain tissue after 48 hours. The damage was of the type lipid peroxidation, which is known to impair the functioning of cell membranes. There were also inflammatory changes in the liver and activation of genes related to the making of repair enzymes.

It is also found that C_{60} when dissolved in water binds to the spirals in DNA molecules in an aqueous environment, causing the DNA to deform, potentially interfering with its biological functions and possibly causing long-term negative side effects in people and other living organisms.

Pristine C_{60} can be suspended in water at low concentrations as large clusters often termed nC_{60}. These clusters are spherical clumps of C_{60} between 250-350 nm in diameter. Thus, nC_{60} represents a different chemical entity than solutions of C_{60} in which the fullerenes exist as individual molecules.

It is also found that nC_{60} is moderately toxic to water fleas and juvenile largemouth bass at concentrations in water of around 800 ppb. In the field of nanotechnology, heat resistance and superconductivity are some of the more heavily studied properties.

Malleability is another property of buckyballs–"the buckyball is resilient beyond any particle that's been known," having sustained no damage after being hurtled at a stainless steel plate at 15,000 mph. Resilience is an asset in creating rocket fuels, as they must undergo extreme pressures. In addition, the buckyball's compressibility up to seventy percent of its original volume might tend to create a new type of shock absorber, able to withstand 15,000 mph collision unscathed. Either that, or the buckyballs could be compressed and held, which could take the place of industrial diamonds, as a compressed buckyball is at least twice as rigid as diamond. This characteristic could prove to be suitable in either metals or plastics, creating a super-strong, relatively cheap armor.

Chemical Properties of Fullerenes

As fullerene or C_{60} is soccer-ball-shaped or I_h with 12 pentagons and 20 hexagons. So, according to Euler's theorem these 12 pentagons are required for closure of the carbon network consisting of *n* hexagons and C_{60} is the first stable fullerene because it is the smallest possible to obey this rule. In this structure none of the pentagons make contact with each other. Both C_{60} and its relative C_{70} obey this so-called **isolated pentagon rule (IPR).** The next homologue C_{84} has 24 IPR isomers of which several are isolated and another 51, 568 non-IPR isomers. Non-IPR fullerenes have thus far only been isolated as endohedral fullerenes such as Tb3N@C_{84} with two fused pentagons at the apex of an egg-shaped cage.

Because of the molecule's spherical shape the carbon atoms are highly pyramidalized, which has far-reaching consequences for reactivity. It is estimated that strain energy constitutes 80% of the heat of formation. The conjugated carbon atoms respond to deviation from planarity by orbital rehybridization of the sp^2 orbitals and pi orbitals to a $sp^{2.27}$ orbital with a gain in p-

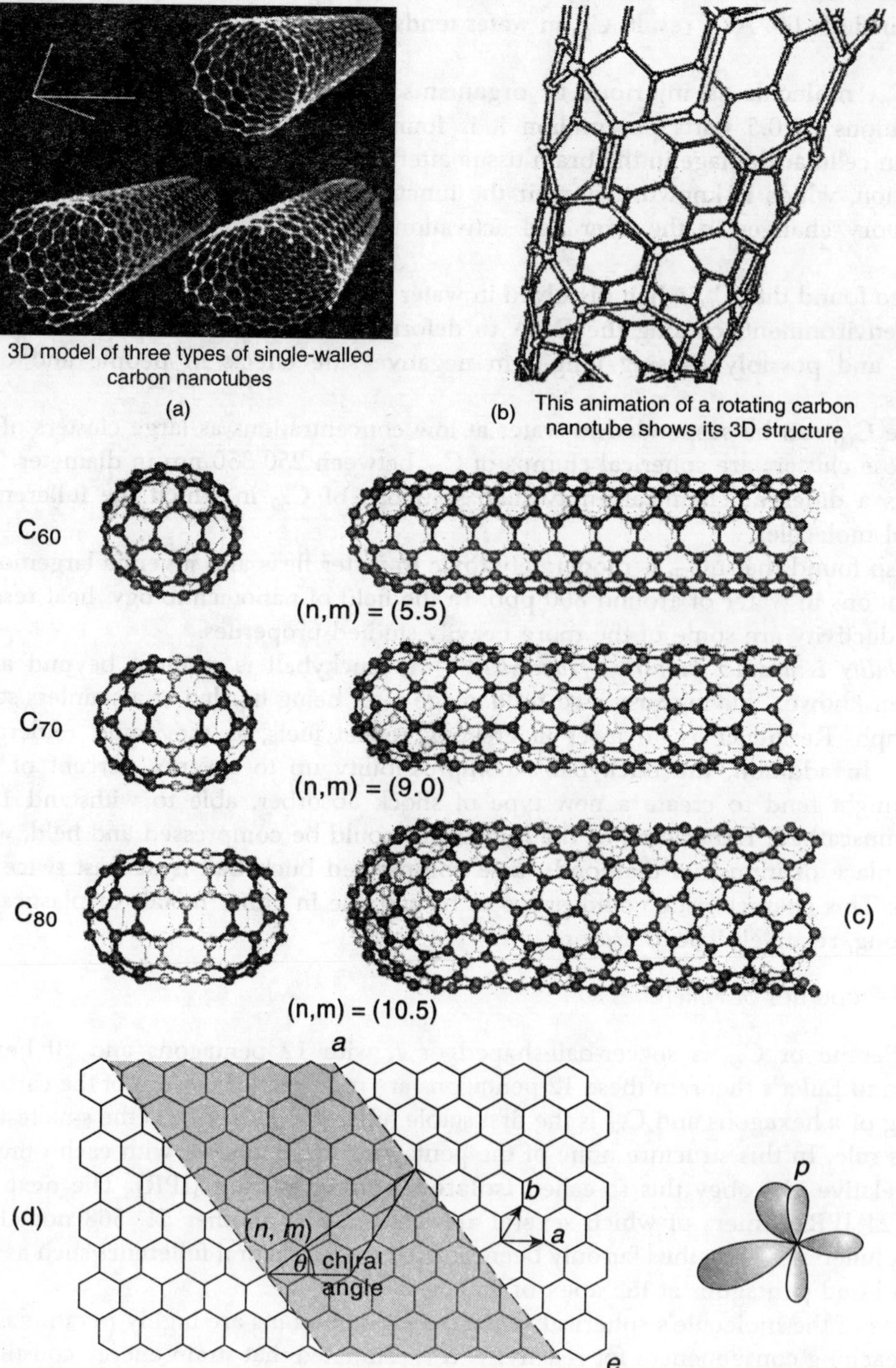

Fig. 4.3 *Fullerence/Nanotubes*

character. The p lobes extend further outside the surface than they do into the interior of the sphere and this is one of the reasons a fullerene is electronegative. The other reason is that the empty low-lying pi* orbitals also have high s character.

The double bonds in fullerene are not all the same. Two groups can be identified: 30 so-called [6,6] double bonds connect two hexagons and 60 [5,6] bonds connect a hexagon and a pentagon Fig. 4.4. Of the two the [6,6] bonds are shorter with more double-bond character and therefore a hexagon is often represented as a cyclohexatriene and a pentagon as a pentalene or [5] radialene. In other words, although the carbon atoms in fullerene are all conjugated the superstructure is not a super aromatic compound. The X-ray diftTaction bond length values are 135.5 pm for the [6, 6] bond and 146.7 pm for the [5, 6] bond.

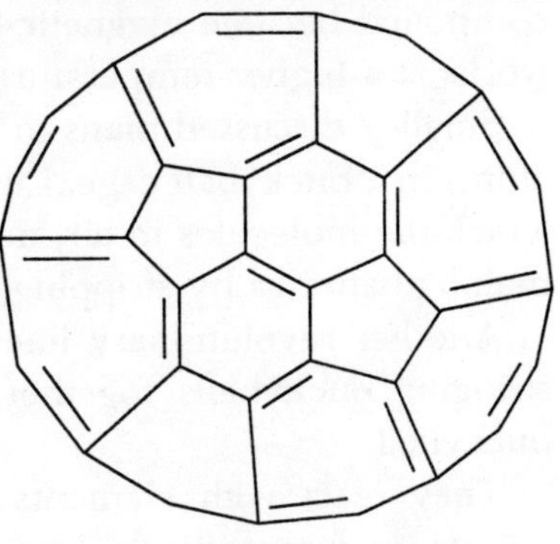

Fig. 4.4

C_{60} fullerene has 60 pi electrons but a closed shell configuration requires 72 electrons. The fullerene is able to acquire the missing electrons by reaction with potassium to form first the $K_6C_{60}{}^{6-}$ salt and then the $K_{12}C_{60}{}^{12-}$. In this compound the bond length alternation observed in the parent molecule has vanished.

Applications of Fullerenes

Perhaps the most exciting characteristic of the buckyball is that it is hollow on the inside, that all elements in the periodic table will fit inside. This could create any number of practical uses, the most notable being in the field of medicine. Drugs could be administered molecularly, or more importantly, individual radioactive molecules could be contained within the buckyball for specific treatment of cancer, compared to chemotherapy, which bombards the patient with low level (yet relatively large quantities of) radiation.

A fuzzyball is created by adding single atoms to the outside of the buckyball. One example predicted would be the fully hydrogenated buckyball, $C_{60}H_{60}$. This would cause the buckyball to take on a fuzzy appearance, hence the name. There are many applications to fuzzyballs, one of which could create a slicker substance than teflon. A fully fluorinated buckyball would create the slickest molecular lubricant known to man, $C_{60}F_{60}$. The uses for a molecular lubricant are boundless, limited only by our imagination.

A pure buckyball is an insulator while the buckyball becomes a superconductor with the addition of a few atoms, including potassium. However, if too many potassium atoms are added to a single buckyball, it will become insulating instead of superconducting. Three potassium atoms create the best conditions for superconductivity. K_3C_{60} is a stable metallic crystal with a face-centered cubic form of buckyballs, with potassium ions filling the gaps between buckyballs. This potassium buckide is the first molecular metal that is totally three-dimensional. Therefore, metallic buckides could be produced as a totally free superconductor, not linear, but in all three dimensions.

The buckyball also exhibits ferromagnetic properties Pierre-Marc. Allemand af the University of California at Santa Barbara, combined buckyballs with a reactant which readily donated its electrons to the carbon molecules. The compound was then cooled; then warmed again. The compound became magnetic when it was warmed to 16 kelvins, and only one organic magnet works at a higher temperature.

Smalley discussed plans to create a buckyball battery, created wrapping lithium and fluorine atoms in a buckyball cage. Lithium and fluorine create energy when combined, but oxygen will attack the molecules in air, unless shielded by a buckyball shell. Some scientists also suggested making batteries by stripping away electrons from the buckyball.

Another revolutionary line of thought is to create a whole new line of plastics, created by stringing buckyballs together, and hanging individual atoms or chemical groupd off each, buckyball.

They react with elements from across the periodic table and with the chemical species known as free radicals–key to the polymerization processes widely used in industry–thus opening up the fullerenes to the manipulative magic of organic chemists. When a fullerene is "doped" by inserting just the right amount of potassium or cesium into empty spaces with in the crystal, it becomes a superconductor–the best organic superconductor known. More important, because C_{60} is a relatively simple system, it may help physicists master the still mysterious theory of high-temperature superconductivity. Speculation and some hard work on potential applications began almost immediately after the discovery of buckyballs. Possible applications of interest to industry include optical devices; chemical sensors and chemical separation devices; production of diamonds and carbides as cutting tools or hardening agents; batteries and other electrochemical applications, including hydrogen storage media; drug delivery systems and other medical applications; polymers, such as new plastics; and catalysts. Catalysts, in fact, appear to be a natural application for fullerenes, given their combination of rugged structure and high reactivity. Experiments suggest that fullerenes which incorporate alkali metals possess catalytic properties resembling those of platinum. The C_{60} molecule can also absorb large numbers of hydrogen atoms–almost one hydrogen for each carbon–without disrupting the buckyball structure. This property suggests that fullerenes may be a better storage medium for hydrogen than metal hydrides, the best current material, and hence possibly a key factor in the development of new batteries and even of non-polluting automobiles based on fuel cells. A thin layer of the C_{70} fullerene, when deposited on a silicon chip, seems to provide a vastly improved template for growing thin films of diamond.

USES OF FULLERENE

1. Fullerenes–are used in armors as they are as hard or harder than diamond.
2. The use of buckyballs practically utilizes ancient alchemy to turn buckyballs into diamonds. French and Argentine chemists have found a way to subject buckyballs to steep pressure gradients, which in turn cause the buckyballs to change into diamonds. Graphite also can be turned into diamond crystals, but that requires a higher temperature and higher pressure gradients. The method with buckyballs can be done at room temperature.

   Already present in a number of products are carbon fibers. Graphite fibers are used in such products as tennis rackets, golf clubs, bicycles, and airplane parts-all products which stress strength and durability. Carbon, or graphite, fibers are already the strongest fibers known, which is why the discovery of buckytubes is so exciting. The buckytube is an elongated fullerene, hollow in the center as the buckyball. It is basically a buckyball stretched out to form a tube. This tube creates a fiber even stronger than graphite fiber, since it is a single molecule. Again, industry could be revolutionized.

3. Fullerenes are also used in medicine: It binds to specific antibiotics to the structure to target resistant bacteria and even target certain cancer cells such as melanoma.
4. It can also be used as a light-activated antimicrobial agents.

FULLERENE REACTIONS

Fullerenes tend to react as electrophiles. An additional dividing force is relief of strain when double bonds become saturated. Key in this type of reaction is the level of functionalization i.e. monoaddition or multiple additions and in case of multiple additions their topological relationships (new substituents huddled together or evenly spaced). Some fullerene chemistry involves the opening of fullerenes by breaking several of the double bonds. The aim is to be able to insert small molecules through the hole, for instance hydrogen in endohedral hydrogen fullerene.

- Fullerenes react as electrophiles with a host of nucleophiles in nucleophilic additions. The intermediary formed carbanion is captured an electrophile. Examples of nucleophiles are Grignard reagents and organolithium reagents. For example the reaction of C_{60} with methylmagnesium chloride stops quantitatively at the penta-adduct with the methyl groups centered around a cyclopentadienyl anion which is subsequently protonated. Another nucleophilic reaction is the Bingel reaction.
  - Fullerene reacts with chlorobenzene and aluminum chloride in a Friedel-Gwfts alkylation type reaction. In this hydroarylation the reaction product is the 1, 2-addition adduct (Ar-CC-H).
  - The [6,6] bonds offullerenes react as dienes or dienophiles in cycloadditions for instance Diels-Alder reactions. 4-membered rings can be obtained by [2+2] cycloadditions for instance with benzyne. An example of a 1, 3-dipolar cycloaddition to a 5-membered ring is the Prato reaction.
  - Fullerenes are resistant to hydrogenation, the completely hydrogenated $C_{60}H_{60}$ is hypothetical because of large strain but hexahydrides are known
  - Although more difficult than reduction, oxidation of fullerene is possible for instance with oxygen and osmium tetraoxide
  - Fullerenes react in electrophilic additions as well. The reaction with bromine can add up to 24 bromine atoms to the sphere.
  - Fullerenes react with carbenes to methanofullerenes

ADSORPTION OF FULLERENES ON CLEAN Si SURFACES

Clean silicon surfaces, e.g. Si(111)–7 × 7 and Si(100)–2 × 1, are highly reactive due to their exposed dangling bonds. When fullerene molecules are first deposited onto these substrates they do not diffuse and form chemical bonds with the surface. On higher coverages, close to a monolayer, the molecules exhibit only short range ordering, as the molecule-surface interaction dominates over the intermolecular interaction. On greater than monolayer coverages the molecules can diffuse across the first layer to interact with each other and form islands.

Terminated silicon surfaces, such as Ag/Si(111) and Si(100)–2 × 1 :H, are less reactive than the clean surfaces as the reconstructions involved saturation of the dangling bonds Fig. (4.5). The adsorption of fullerenes on these surfaces Fig. (4.3)is more interesting as the molecules are able to diffuse and interact with each other to form islands.

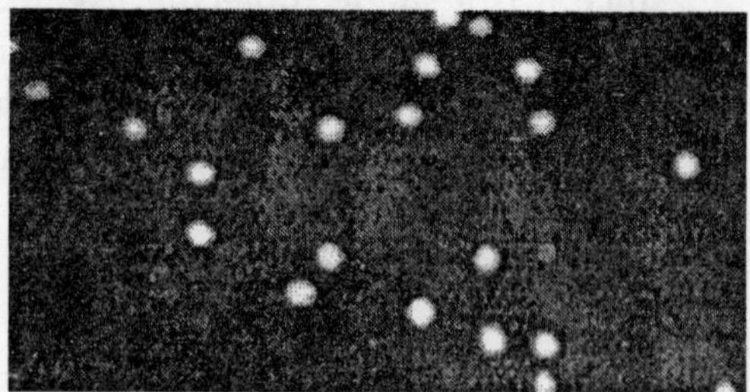

Fig. 4.5 *Adsorption of fullerenes on terminated Si surface*

Theoretical Calculations of Fullerenes

All atoms, molecules and crystals vibrate. Sometimes there can be a strong coupling between these vibrations and the motion of the electrons in these materials. The electrons in the C_{60} molecule are known to be sensitive to these vibrations. It is believed that this vibronic coupling plays an important role in determining various aspects of the behaviour of C_{60} molecules and related fullerene compounds. Modes of vibration between fullerene molecules (intramolecular modes) probably play an important role in the relatively high-temperature superconductivity of certain fulleride compounds.

Ab initio and molecular orbital methods are used to study the binding geometries of a variety organic molecules on the Si(100)-(2 × 1) surface. The organic molecules studied include ethene, benznene, 1, 4-cyclohexadiene and naphthalene, since these all represent structural fragments of fullerene cages.

For a given adsorbate, the calculations allow us to discriminate between stable, metastable and unstable binding configurations. Furthermore, the numerical methods may be used to investigate the electronic structure of molecular adsorbates. Such information can be compared with data from synchrotron and STM experiments, making it possible to ascertain which binding geometries a partiular adsorbate prefers.

FULLERENES AS LIGANDS

Fullerene is a ligand in organometallic chemistry. The [6, 6] double bond is electron-deficient and forms metallic bonds with $\eta = 2$ hapticity Fig. (4.6). C_{60} fullerene reacts with tungsten hexacarbonyl $W(CO)_6$ to the $(\eta^2\text{-}C_{60})W(CO)_5$ complex in a hexane solution in direct sunlight.

MULTISTEP FULLERENE SYNTHESIS

Although the procedure for the synthesis of the C_{60} fullerene is well established (generation of a large current between two nearby graphite electrodes in an inert atmosphere) a 2002 study described an organic synthesis of the compound starting from simple organic compounds.

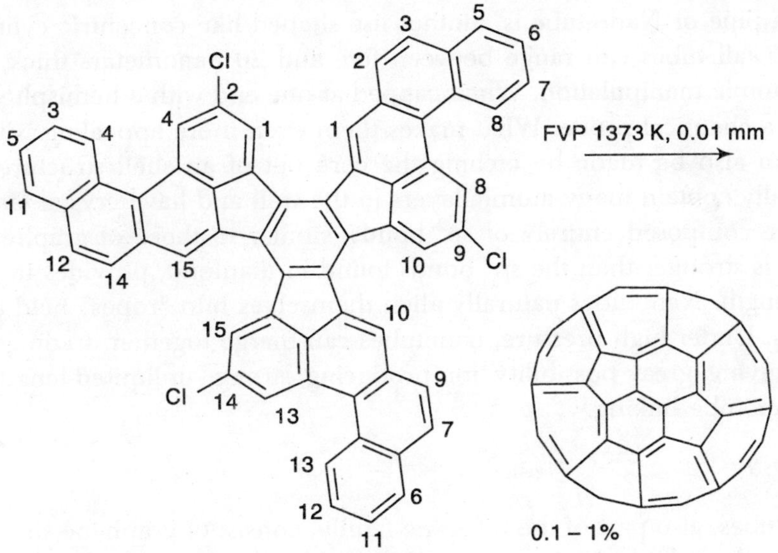

Fig. 4.6

In the final step a large polycyclic aromatic hydrocarbon consisting of 13 hexagons and three pentagons is submitted to flash vacuum pyrolysis at 1373 K, and 0.01 mm pressure. The three carbon chlorine bonds serve as free radical incubators and the ball is stiched up in a no-doubt complex series of radical reactions. The chemical yield is low: 0.1 to 1 %.

Nanotubes

Nanotubes are members of the fullerene structural family, Fig. 4.3(b)which also includes buckyballs. Whereas buckyballs are spherical in shape, a nanotube is cylindrical, with at least one end typically capped with a hemisphere of the buckyball structure. Their name is derived from their size since the diameter of a nanotube is on the order of a few nanometers (approximately 50,000 times smaller than the width of a human hair), while they can be up to several millimeters in length.

Certain compounds are capable of forming nanotubes where the tube consists of round shell of a single layer of atoms in a cylindrical lattice. Carbon nanotubes is the most famous example, but also other materials can form nanotubes such as Boron nitride, Molybdenum sulfide and others.

Their unique molecular structure results in unique macroscopic properties, including high tensile strength, high electrical conductivity, high resistance to heat, and relative chemical inactivity as it is round with no exposed atoms that can be easily displaced, such as in Benzene.

To make nanotubes, the team vaporized carbon with an electric current. The vapor condenses to form a sooty gob, rich in nanotubes. The workers mine the clump with cellophane tape, and then, holding a glue-dipped conventional tip, they lightly touch it to the wad of nanotube bundles and gingerly pull one out.

Improving the scanning-force microscope (SFM) tips (by replacing) in order to detect and nudge molecules by Hongjee Dai & Richard E. Smally of Rice University by fashioning some fullernene into a pipe or Nano-tube is another use shaped like concentric cylinders of chicken wire, these multiwall tubes can range between five and 20 nanometers thick, thus facilitating more accurate atomic manipulation. When capped at one end with a hemispheric fullerene, the tip can serve as a chemical probe. What makes them even more appealing is their durability.

Nanotubes can also be made by etching the core out of an shell structured rod, but such tubes will normally contain many atomic layers in the wall and have crystal facets on the sides.

Nanotubes are composed entirely of sp^2 bonds, similar to those of graphite. This bonding structure, which is stronger than the sp^3 bonds found in diamond, provides the molecules with their unique strength. Nanotubes naturally align themselves into "ropes" held together by Van der Waals forces. Under high pressure, nanotubes can merge together, trading some sp^2 bonds for sp^3 bonds, giving great possibility for producing strong, unlimited-length wires through high-pressure nanotube linking.

Nanotube Chemistry

Carbon nanotubes, also part of the fullerene family, consist of graphene sheets rolled up into a cylinder. Unlike the spherical fullerenes made up of hexagons and pentagons, nanotubes only have hexagons present but in terms of reactivity both systems have much in common. Due to electrostatic forces nanotubes have a nasty tendency to cluster together in large bundles and many potential applications require an exfoliation process. One way to do this is by chemical surface modification.

A useful tool for the analysis of derivatised nanotubes is Raman spectroscopy which shows a **G-band** (G for graphite) for the native nanotubes at 1580 cm^{-1} and a **D-band** (D for defect) at 1280 cm^{-1} when the graphite lattice is disrupted with conversion of sp^2 to sp^3 hybridized carbon. The ratio of both peaks I_D/I_G is taken as a measure of functionalization. Other tools are UV spectroscopy where pristine nanaotubes show distinct van Hove singularities where functionalized tubes do not and simple TGA analysis.

In one type of chemical modification an aniline is oxidized to a diazonium intermediate which after expulsion of nitrogen forms a covalent bond as an aryl radical Fig. (4.7):

Also known are protocols for Diels-Alder reactions, one assisted by chromium hexacarbonyl and high pressure.

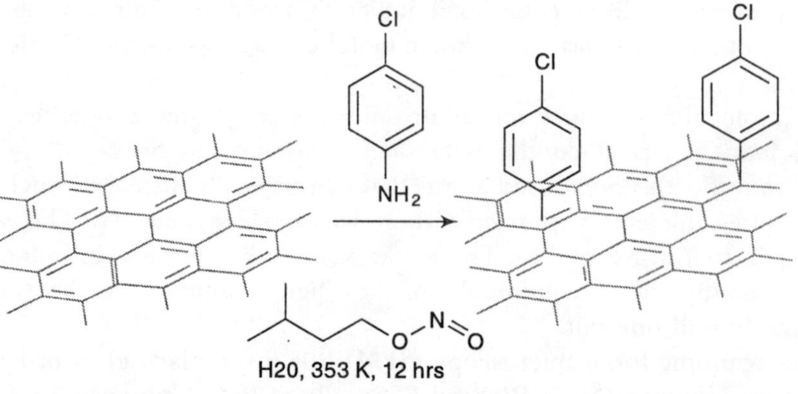

Cl
NH2
Cl
Cl
H20, 353 K, 12 hrs

Fig. 4.7

CARBON NANOTUBES

CARBON NANOTUBE BASICS

In beginning two allotropes of carbon were known as Graphite and Diamond. Graphite is formed from carbon atoms arranged in a honey comb pattern. These honey comb layers are stacked one above the other. A single sheet of graphite is very stable strong, and flexible. Since a single sheet is so stable by itself, it binds only weakly to the neighbouring sheets. Graphite consists of sp^2 hybridised carbons in hexagonal sheets which can slip over each other. Diamond consists of sp^2 hybridised carbons which form a tetragonal lattice.

There is another way of arranging the sheets which is even stronger. This is accomplished by wrapping the honeycomb pattern back on top of itself and joining the edges to form a cylinder. The cylinder can be tens of microns long with end "capped" with half a fullerene molecule a semi sphere cap is added to the end, the resulting structure is a tube of graphite, called a carbon nanotube.

Nanotubes are derived from a graphene carbon lattice (i.e. the same as in graphite) and were discovered in 1991 by the Japanese electron microscopist Sumio Iijima who was studying the material deposited on the cathode during the arc-evaporation synthesis of fluerenes.

These carbon nanotubes were found to have between two and over 50 watts, can be many micrometers in length.

Carbon nanotubes (CNTs) are allotropes of carbon Fig. 5.1 A carbon nanotube is a one-atom thick sheet of graphite (called graphene) rolled up into a seamless cylinder with diameter of the order of a nanometer. Carbon nanotubes generally have a length to diameter ratio of about 1000 so they can be considered as nearly one-dimensional structures. Carbon nanotubes (CNTs) are fullerene-related structures that consists of graphene cylinders closed at either end with caps containing pentagonal rings. One can get both the ends closed, or one end open, or even both ends open.

C-atoms placed in hexagons and pentagons form the end cap structures that twelve pentagons are needed in order to obtain a closed cage structure which consists of only pentagons and hexagons. The hexagonal structure of the 2-dimensional graphene sheet is due to the sp^2

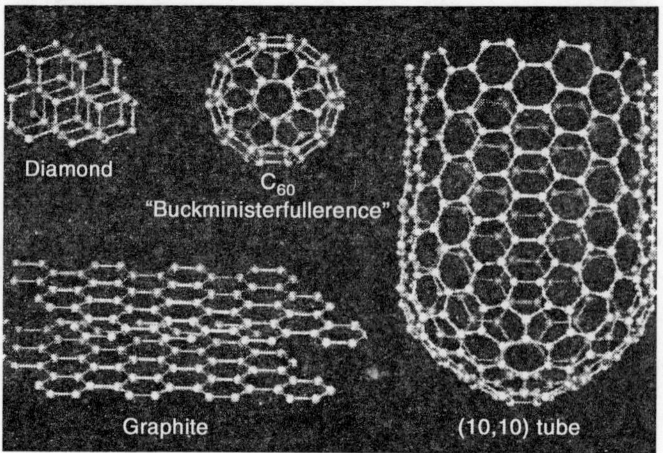

Fig. 5.1 *Various forms of Carbon*

hybridization of the carbon atoms, which causes three directional, in-place σ bonds separated by an angle of 120 degrees.

The combination of a pentagon and five surrounding hexagons results in the desired curvature of the surface to enclose a volume. A second rule is the isolated pentagon rule that states that the distance between pentagons on the fullerene shell is maximised in order to obtain a minimal local curvature and surface stress, resulting in a more stable structure. The smallest stable structure that can be made this way is C_{60} the one just larger is C_{70} and so on. Another property is that all fullerenes are composed of an even number of C-atoms because adding one hexagon to an existing structure means adding two C-atoms.

The other structure of which a SWNT is composed is a cylinder. It is generated when a graphene sheet of a certain size that is wrapped in a certain direction. As the result is cylinder symmetric we can only roll in a discreet set or directions In order to form a closed cylinder. Two atoms in the graphene sheet are chosen, one of which servers the role as origin. The sheet is rolled until the two atoms coincide. The vector pointing from the first atom towards the other is called the chiral vector and its and its length is equal to the circumference of the nanotube. The direction of the nanotube axis is perpendicular to the chiral vector.

Electronic Structure

A free carbon atom has the electronic structure $1s^2 2s^2 2p^2$. In order to form covalent bonds, one of the 2s electrons is promoted to 2p. In graphite, one carbon atom forms a strong σ bond called as sp^2 with three other adjacent atoms in a plane. The others in the p orbital have a weak π bond at 90° to this plane. This weak bond gives its semi-metal characteristics. The overlap of π orbitals on adjacent atoms in a given plane provides the electron bond network which gives graphite its relatively has electrical conductivity. In high-quality synthetic graphite the stacking sequence is generally *ABAB*, with an interlayer spacing approximately 0.334 nm.

The unit cell in this structure contains four atoms. In less perfect graphites, the interplanar spacing is found to be significantly larger than the value for single crystal graphite. In diamond which has a tetrahedral structure, each carbon atom is joined to four neighbours and the bonding here is sp^3. As diamond is less stable than graphite, it is converted to graphite at a temperature of 1973 K at normal pressure.

As illustrated in Fig. 5.2., there are two possible high symmetry structure for carbon nanotubes, known as 'zig-zag; and 'armchair'. In practice. it is believed that most carbon nanotubes do not have these highly symmetric forms but have structures called 'chiral' in which the honeycomb-sharped hexagons are arranged helically around the tube axis. The simplest way of specifying the structure of an individual tube is in terms of a vector *C*, joining two equivalent joints on the original graphene lattice, The cylinder is produced by rolling up the sheet such that the two end-points of the vector are superimposed, Fig. 5.3.

Each pair of integers (n, m) represents a possible tube structure in this image. Thus the vector C can be expressed as

$$C = na_1 + ma_2$$

where a_1 and a_2 are the unit cell base vectors of the graphene sheet, and $n \geq m$. It can be seen that $m = 0$ for all zig-zag tubes, while $n = m$ for all armchair tubes. All other tubes are 'chiral'. In the case of the two 'archetypal' nanotubes which can be capped by one half of a C_{60} molecule, the zig-zag tube is represented by the integers (9,0) Fig. 5.4(b) while the armchair tube is denoted by (5,5) Fig. 5.4(a).

Since the magnitudes of a_1 and a_2 are equally 0.246 nm, the magnitude of C in nanometers is $0.246\ (n^2 + nm + m^2)^{1/2}$, and the diameter d_t is given by

$$d = 0.246(n^2 + nm + m^2)^{1/2}/\pi \text{ (nm)}$$

The chiral angle, π, is given by

$$\theta = \sin^{-1}\{3^{1/2}m/2(n^2 + nm + m^2)^{1/2}\}$$

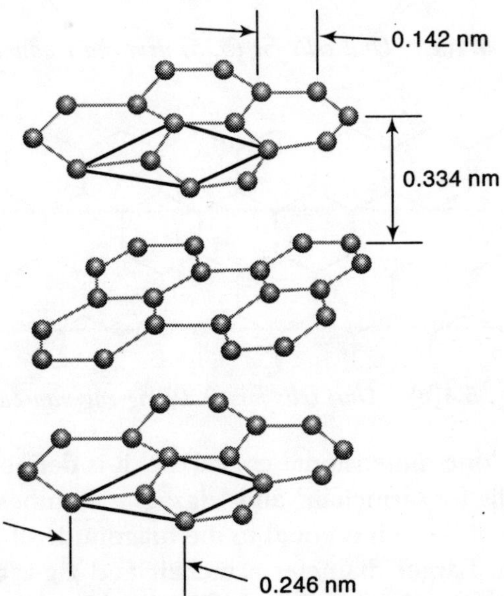

Fig. 5.2 *Structure of hexagonal graphite showing the unit cell*

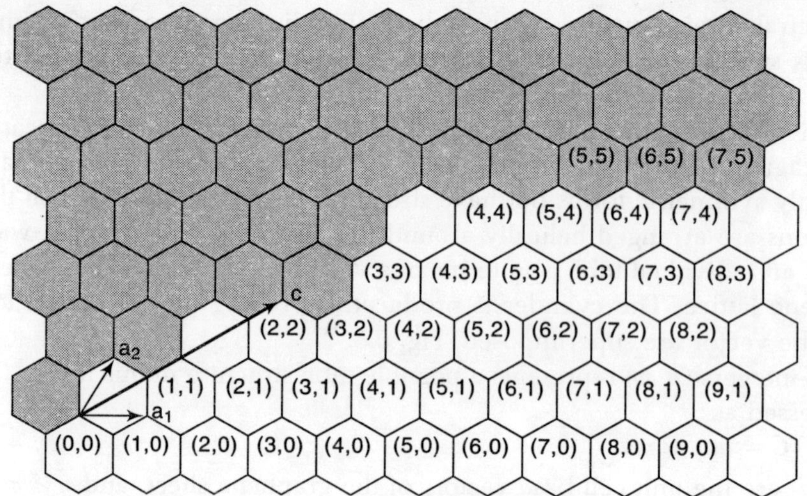

Fig. 5.3 *Graphites layer with atoms labelled using (n, m) notation*

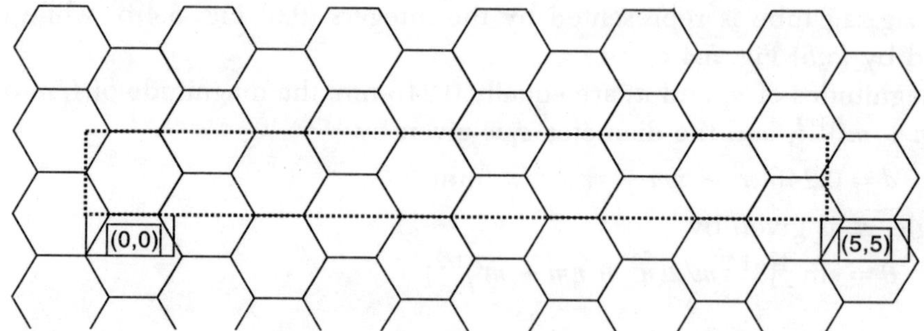

Fig. 5.4(a) *Unit cells for (5, 5) arm chair nanotube*

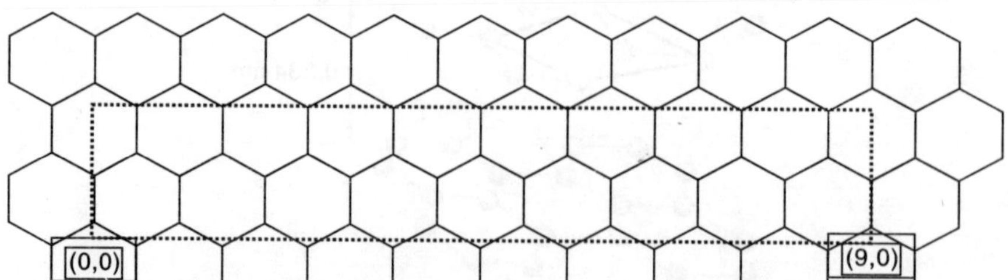

Fig. 5.4(b) *Unit cells for (9, 0) zig-zag nanotube*

Consider a nanotube as a 'one-dimensional crystal' i.e. it is defined as a translational unit cell along the tube axis. Unit cells for 'armchair' and 'zigzag' nanotubes are shown in Fig. 5.4. For the armchair tube, the width of the cell is equal to the magnitude of a, while for the zig-zag tube the width of the cell is $3^{1/2}$a. Larger diameter armchair and zig-zag nanotubes have unit cells which are simply longer versions of these. For chiral nanotubes, the lower symmetry results in larger unit cells. A simple method of constructing these cells is described in Fig. 5.5. This involves drawing a straight line through the origin *O* of the wedge normal to *C*, and extending this line until it passes exactly through an equivalent lattice point as is illustrated for the case of a (6.3) nanotube Fig. 5.5. The length of the unit cell in the tube axis direction is the magnitude of the vector *T*. For nanotubes in the diameter range ~ 2–30 nm, the unit cells can be very large. For example, the tube denoted (80, 67), which has a diameter of approximately 10 nm has a unit cell 54.3 nm in length containing 64,996 atoms.

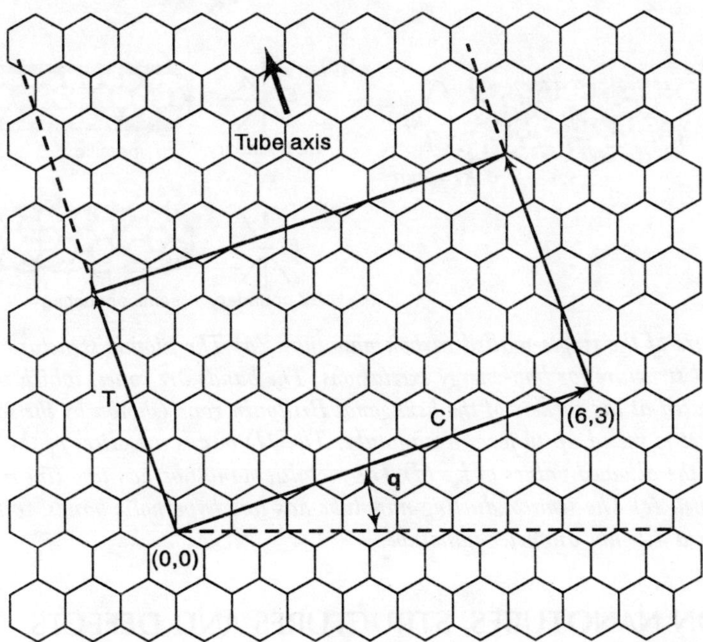

Fig. 5.5 *Construction of the unit cell for a (6, 3) nanotube*

Band Structure

Graphite is the starting point for considering the electronic structure of the carbon nanotube. Graphite is a solid form of carbon in which the atoms are covalently bonded in a hexagonal network to form 2D planes, called graphene, with the planes weakly bonded together through the van der Waals interaction. Figure 5.6(a) shows the graphene atomic structure. K_x and K_y are the wave vectors *x* and *y* directions. The low-energy band structure of graphene is unusual [see Fig. 5.6(b)]; the bands are cones, meeting in points at the Fermi surface at the *K* points in the Brillouin zone. When the graphene sheet is rolled up to form a nanotube, the circumfrential wavevector is quantized, such that $c \cdot k_r = 2\pi n$ (figure 5.6(c), where *C* is the circumference and *n* is an integer. In most directions in *k*-space, graphene is a semiconductor only for spccial directions [e.g., K_y" in Fig. 5.6 (b)] is graphene metallic.

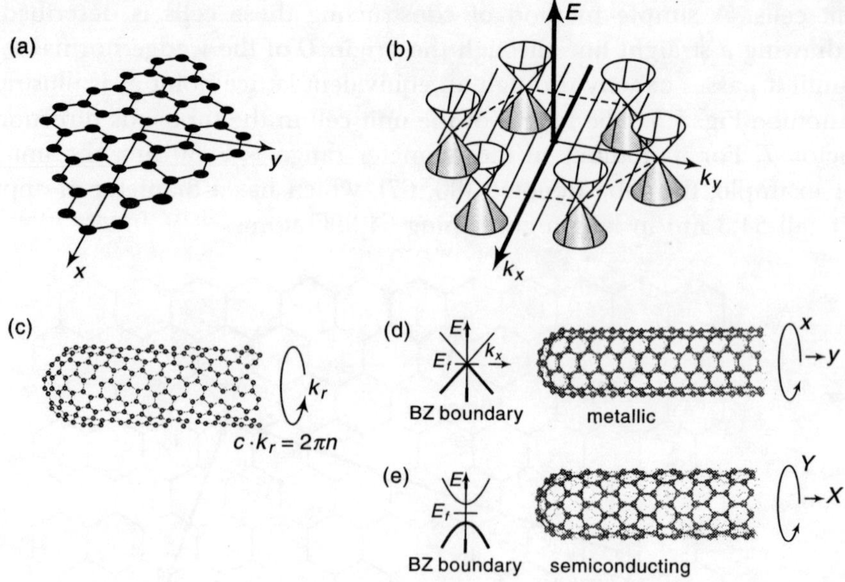

Fig. 5.6 *Band structure of the single-walled carbon nanotube. (a) The atomic structure of the graphene sheet. (b) The band structure for low-energy excitations. The bands are cones, which meet the Fermi surface at points located at the corners of the hexagonal Brillouin zone (shown by the solid hexagon). (c) The graphene sheet is rolled up to form a nanotube. The 1D bands are slices of the 2D band structure in (b) taken at the allowed values of k_r. (d) The metallic nanotube has two linear hands which cross at the Fermi level. (e) The semiconducting nanotube has two hyperholic bands, with an energy gap $E_g \approx$ 600 meV for a 1.5-nm diameter nanotube.*

TYPES OF CARBON NANOTUBES, STRUCTURES AND DEFECTS

Many exotic structures of fullerenes exist: regular spheres, cones, tubes and also more complicated and strange shapes. Below is described some of the most important and best-known structures.

There are three main types of nanotubes: *single-walled nanotubes (SWNTs) and multi-walled nanotubes (MWNTs) and circular nanotubes.*

Carbon (C) Nanotubes, are black nano scale cylindrical tubes of graphite having numerous applications as the stiffest and strongest known fibers and because of their unique electrical properties. They are in single walled, double walled and multi-walled forms.

The simplest type of carbon nanotube consists of just one layer of graphene rolled up in the form of a seamless cylinder, known as a single-walled carbon nanotube (SWCNT) with a typical diameter of just a few nanometers. Larger diameter nanotube structures are nanotube ropes, consisting of many individual, parallel nanotubes closed-packed into a hexagonal lattice, and multi-walled carbon nanotubes (MWCNTs) consisting of several concentric cylinders nested within each other.

SINGLE WALLED CARBON NANOTUBE

Conceptually, an SWNT is formed by cutting a strip of the graphene sheet and rolling it into a tube. The lattice vector on the graphene sheet that connects the two points that will roll into each other is called the circumferential vector *C*, and completely defines the type of SWNT, its diameter and chirality. SWNTs are then denoted by their circumferential vector in terms of the graphene lattice vectors $\boldsymbol{a}_1$ and $\boldsymbol{a}_2$; thus, an SWNT with a circumferential vector $n_1\boldsymbol{a}_1 + n_2\boldsymbol{a}_2$ would be denoted (n_1, n_2). (Two non-chiral types of tubes exist and are given special names in the literature: (n_1, n_2) tubes are called armchair and $(n, 0)$ tubes are called zig-zag, after the shapes of the bands of carbon atoms which encircle each nanotube.)

The electronic structure of an SWNT is approximated by simply taking the electronic structure of the graphene sheet and quantizing it in the circumferential direction such that $c \cdot k_r = 2\pi n$, where n is an integer.

This condition is equivalent to stating that the electronic wavefunction on the tube be single-valued around the circumference. Thus the graphene sheet is sliced at regular intervals in k-space [see Fig.5.6(b)]. The result is a 1D dispersion relation with many sub-bands corresponding to the various slices.

The bands of the graphene sheet only cross the Fermi surface at the *K* point. This means only those tubes whose band structure slices cross the *K* point will be metallic. For a tube (n_1, n_2) this can be expressed by the condition: $n_1 - n_2 = 3q$ where q is an integer. A more careful treatment which takes into account the curvature of the graphene sheet finds that only the tubes with $n_1 = n_2$ with be metallic, the other tubes satisfying $n_1 - n_2 = 3q$, with non-zero q will have a small gap on the order of milli electron-Volts, considerably smaller than room temperature.

The low-energy band structure of metallic and semiconducting SWNTs is depicted in Fig. 5.6 (d) and (e). Metallic SWNTs have two linear bands which cross at the Fermi level. Semiconducting SWNTs have two hyperbolic bands, separated by an energy gap, with $E_g \approx 900$ meV/*d*, where *d* is the diameter in nanometers. In both cases, the next closest bands are on order 500 meV away for typical diameter SWNTs; thus SWNTs are extremely 1 D at room temperature. where the thermal energy is 25meV.

This result is quite striking. It is an extreme example of a size effect in small structures, and manifests because of the unusual nature of the graphene Fermi surface–metallic for some directions in *k*-spaee, and semiconducting in others. This result has been verified by simultaneous measurement of the geometry and electronic structure of individual SWNTs by scanning tunneling microscope (STM) In these experiments, the atomic structure of individual SWNTs was measured topographically by the STM and then the STM tip was placed at a fixed distance above the SWNT to perform tunneling spectroscopy measurements of the band structure.

SWNTs fall into three broad classes: metallic SWNTs with $n_1 = n_2$, small-gap semiconducting SWNTs with $n_1 - n_2 = 3q$ but $q \leq 0$, and moderate-gap semiconducting SWNTs, with $n_1 - n_2 \leq 3q$. Furthermore, the typical energy gaps of the semiconducting SWNTs are found to depend almost entirely on diameter *d*, with the moderate semiconducting gaps scaling as $1/d$ and the small semiconducting gaps scaling as $1/d^2$. In fact. all the features in the density of states are determined to good approximation by diameter alone. Since the energy gaps in small-gap semiconducting SWNTs are smaller than room temperature for $d > 1.4$ nm. SWNTs may be produced with narrow diameter distributions, so the problem of obtaining an electronically uniform material is reduced to the separation of SWNTs into two classes, metallic (including small ban-gap semiconductor) and large-gap semiconductor.

SWNTs may be metallic or semiconducting, depending very sensitively on how the graphene sheet is wrapped up to form the nanotube. Depending on the chirality (the chiral angle between hexagons and the tube (axis), SWNTs can be either metals or semiconductors, with band gaps that are relatively large.

At room temperature small band-gap semiconductor SWNTs are treated as metallic. For example, SWNTs of diameter 1.5 nm will be either metallic, or semiconducting with a bandgap $E_g \approx 0.5$eV. Nanotubes are highly ID: a 1.5-nm SWNT has a sub-band spacing of order 1 eV, much greater than the room temperature thermal energy of 25 meV.

Because of the high stiffness and small number of 1 D modes of the nanotube (low density of states) nanotubes are relatively immune to the Peierls distortion, which destroys metallicity. The Peierls transition is estimated to occur only well below room temperature in metallic SWNTs.

Fabrication

The first attempts to attach electrodes to individual SWNTs were made using material derived from the laser ablation process. (See chapter on synthesis of CNTs)

These SWNTs are suspended in DCE via ultrasonication and then deposited on SiO_2-capped silicon substrates by spinning or simply applying a drop of suspension to the substrate for a short time and then washing off the excess solution with 2-propanol.

Two schemes for contacting SWNTs with electrodes are investigated: *deposition of SWNTs on top of prepatterned electrodes ("tube-on-top") and evaporating metal electrodes on top of SWNTs after deposition ("metal-on-op")*. Figure 5.7 shows AFM images of examples of each top of device, and illustrates each schematically. The prototypical SWNT device has two electrodes attached to

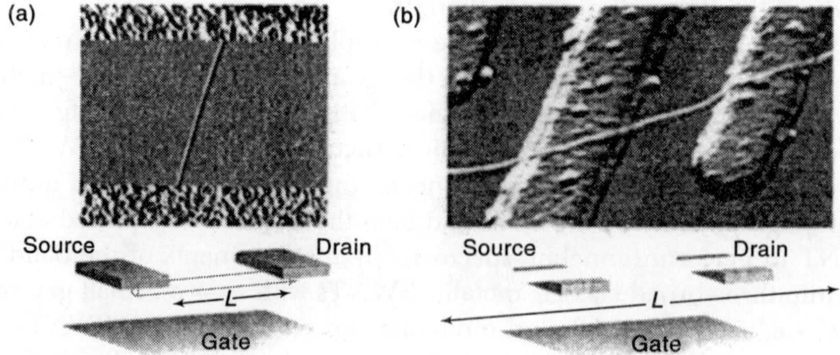

Fig. 5.7 *An AFM image or a "metal-on-top" device (a). The rough areas at top and bottom are Cr/Au electrodcs, and the narrow line is the nanotube, lying on an SiO_2-capped Si substrate. The device is shown below. A "tube-on-top" device is shown in (b). Here, a nanotube (thin line) is deposited from solution over pre-patterned Pt leads on SiO_2-capped Si. The metal leads are spaced approximately 140 nm apart. The lower panel shows the device.*

the SWNT ("sourcc" and "drain") and a third "gate" electrode (either the conducting silicon substrate beneath the oxide. or a nearby metallic electrode).

Tube-on-top devices are fabricated by prepallerning gold or platinum electrodes and then depositing SWNTs from suspension by spin coating or immersion in solution followed by rinsing. In general *tube-on-ton devices are found to have higher contact resistances than metal-on-top devices.* It is speculated that this is due to the sharp bending of the nanotube over the contacts which open an electronic energy gap in a nominally metallic SWNT. The contact resistance of tube-on-top devices is improved by planarizing the contacts before deposition. Metal-on-top devices are fabricated by first deposition SWNTs on sub-strates, and then patterning electrodes via a lift of process, which is a technique, to contact specific individual SWNT, that are located with AFM. First a pattern of metal (Cr/Au) alignment marks is defined on the subtrate using conventional electron-beam lithography and lift-off. An AFM is then used to image the SWNTs relative to the alignment marks. Resist is again spun on the substrate, and a second electron-beam lithography step established electrical contacts to the desired nanotubes.

To pattern the catalyst for CVD growth of SWNTs to fabricate devices with high yield, Islands of catalyst are patterned on substrates with separations of a few micrometers using electron beam lithography and lift-off. After CVD growth of SWNTs, a second lithography steps is used to cover the areas containing the catalyst islands with metal electrodes. Often SWNTs span the gaps between catalyst islands, thus are in contact to the electrodes.

MULTIWALL CARBON NANOTUBES (MWNT)

It is also possible to have multi-walled carbon nanotubes., These have two or more more layers and are about 2 to 200 nm diameter and inter layer spacings is 0.34 nm. (MWNT) consists of multiple layers of graphite rolled in on themselves to form a tube shape.)

They are constructed by rolling several cylinders of graphene sheets each cylinder with a slightly larger circumference then the first. Then each tube is placed into the tube that is slightly, created a multi-layered tube as each tube has a smaller version of itself hidden inside.

With widths in the range of 1nm, these cylinders of a rolled-up graphite layers can reach lengths of micrometers. Mechanically, these tubes are among the strongest materials known. If only one could grow them to macroscopic lengths and wind them to a string, it would be stronger than a steel rope.

There are two models which can be used to describe the structures of multiwalled nanotubes. In the first model sheets of graphite are arranged in concentric cylinders, eg a (0, 8) single-walled nanotubes (SWNT) within a larger (0, 10) single-walled nanotube (SWNT) within a larger (0, 10) single-walled nanotube. In the second model, a single sheet of graphite is rolled in around itself, resembling a scroll of parchment of a rolled up newspaper. The interlayer distance is close to the distance between graphene layers in graphite. The special place of Double-walled Carbon Nanotubes (DWNT) must be emphasized here because they combine very similar morphology and properties as compared to SWNT, while improving significantly their chemical resistance. This is especially important when functionalisation is required (this means grafting of chemical functions at the surface of the nanotubes) to add new properties to the CNT. In the case of SWNT, covalent functionalisation will break some C═C double bonds, leaving "holes" in the structure on the nanotube and thus modifying both its mechanical and electrical properties. In the case of DWNT, only the outer wall is modified.

Multi Walled Nanotubes (MWNT) can be considered as a collection of concentric SWNTs with different diameters. The length and diameter of these structures differ a lot from those of SWNTs and, of course, their properties are also very different. (Fig. 5.8)

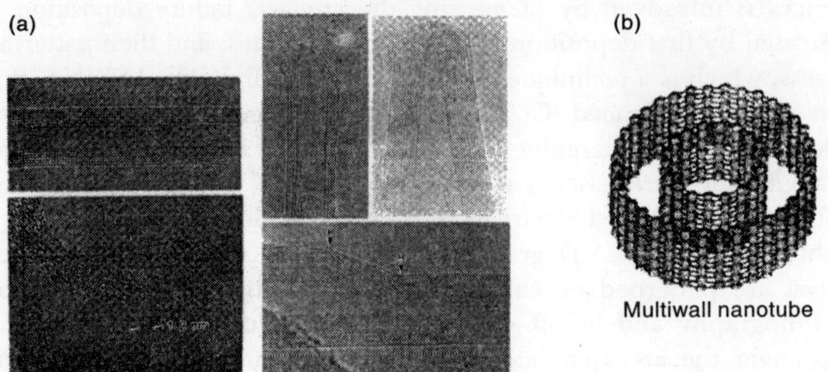

Fig. 5.8 *(a) Different structures of MWNTs. Top-left: cross-section of a MWNT the different walls are obvious, they are separated by 0.34 nm. Rotation around the symmetry axis gives us the MWNT. Top-right: Symmetrical or non-symmetrical cone shaped end caps of MWNTs. Bottom-left: A SWNT with a diameter of 1, 2 nm and a bundle of SWNTs covered with amorphous carbon. Bottom-right: A MWNT with defects.*

In Figure 5.8 carbon cones are also shown. It can be considered as a gradual transition from a large diameter to a smaller one without defects in the wall of the cone but with fewer pentagons in the end cap.

MWCNTs are composed of a number of SWCNTs in a coaxial geometry. Each nested shell has a diameter of $d = \sqrt{3}\, a_{C-C}(m^2 + n^2 + mn)^{1/2}/\pi$ where a_{C-C} is the length of the carbon-carbon bond which is 1.42 Å. The difference in diameters of the individual shell means that their chiralities are different, and adjacent shell are therefore in general non-commensurate, which causes only a weak intershell interaction.

The intershell spacing in MWCNTs is ~ 0.34 nm – quite close to the interlayer spacing in turbostratic graphite. The picture Fig. 5.8(a) is a representation of a multi-walled carbon nanotube.

Defects in CNTs

Introduction of defects can also result in various new structures such as Y-branches T-branches or SWNT junctions. (Figure 5.9) Under certain circumstances, these defects can be introduced in a 'controlled' way.

These defects result in special structures that will have other, but even more interesting, properties than their original forms. In the absence of external strain, carbon nanotubes are always straight unless carbon rings having a number of carbons defiant. From six (pentagons, heptagons, octagons, etc.) are present in the hexagonal network. The electronic properties of

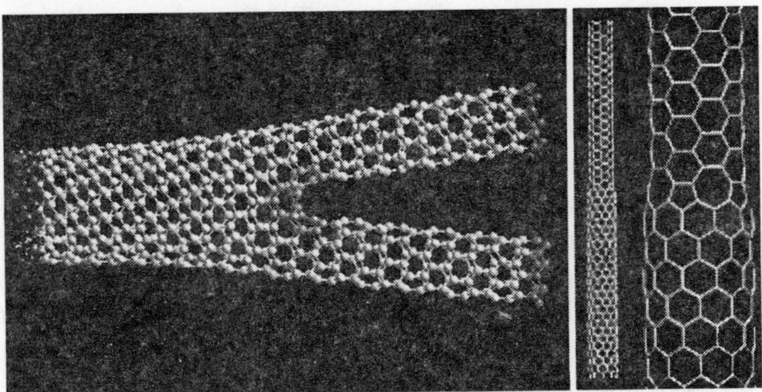

Fig. 5.9 *Left: A Y-branch, the defects are marked. Right: A transition from a metallic to a semi-conducting SWNT. The change is made by insertion of pentagons and heptagons*

perfect MWNTs are rather similar to those of perfect SWNTs, because the coupling between the cylinders is weak in MWNTs.

Circular Nanotubes: These are single wall nanotubes with their ends joined together.

Ropes of Carbon Nanotubes

The individual SWNTs packed into a close-packed triangular lattice with a lattice constant of ~ 17 A the density, lattice parameters, and interlayer, spacing of the ropes is dependent on the chirality of the tubes in the matrice (10, 10). Armchair tubes have a lattice parameter of 16.78 Å and a density of 1.34 gm cm^{-3}. Zigzag tubes of the chirality (17, 0) have a lattice parameter 16.52 Å and a density of 1.34 gm cm^{-3}. Matrice (12, 6) chiral SWNT's has a lattice parameter of 16.52 Å and a density of 1.40 gm cm^{-3}. The space between the tubes is also dependent on chirality. Armchair tubes have a spacing of 3.38 A, zigzag tubes have a spacing of 3.41 Å and (2*n*, *n*) chiral tubes have a interlayer spacing value of 3.39 Å. Compare these values to the spacing between the layers of graphite sheets, and the spacing between the variant walls of a MWNT, both about 3.4 Å; the lattice parameter in CNT ropes (bundled nanotubes) is $d + 0.34$ nm, where d is the tube diameter.

Fig. 5.10 *Nanotube Rope*

CELLULAR STRUCTURE OF CNT

Geometry plays a fundamental role in the evolution of a class of nonequilibrium systems called *cellular structures.* The evolution of stable cellular structures is characterized by universal or system-independent statistical distributions, which possess *scaling properties.* These systems play important roles in various technological applications. The uses of foams range from transport of granular media in pipes to fire suppression and explosion attenuation. *Polycrystalline thin films are used in electronic devices such as MOSFET transistors, magnetic storage media* and *conductors.* The technological performance of these materials depends on the *characteristics of their inherent cellular structure.* For example, the strength of a polycrystalline metal under stress depends on the average grain size. The phenomenon of *electromigration*, which occurs along grain boundaries in thin film conductors, is responsible for their electronic noise and eventual failure. Mean times to failures have been shown to depend on mean grain size and, surprisingly, on grain geometry. Controlled cellular structure formation in novel materials is, however, inherently interesting. The cellular structures out of vertical carbon nanotube arrays on silicon that can be floated off the substrate are also formed.

Carbon, besides being the basis of all living matter, contributes some of the most extraordinary properties to artificial materials. For example, ***a ring of n odd-numbered*** sp^2 ***bonded carbon atoms*** would be an ***insulator*** when n is small, say 5, 7 etc, a ***metal*** when n is large, > 13; and a ***semiconductor*** for intermediate numbers. These [$2n$] trannulenes still remain "theoretically interesting" molecules, but their cylindrical homologues, carbon nanotubes (CNT) (Fig. 5.11), have been synthesized as either concentric cylinders called multiwalled carbon nanotubes (MWNT) or single cylinders called single-walled carbon nanotubes (SWNT). The metallic carbon nanotubes transport electric current ballistically (they do not dissipate heat). They can also function as the active semiconductor in nanoscale devices, all as a result of their unique topologically controlled electronic properties.

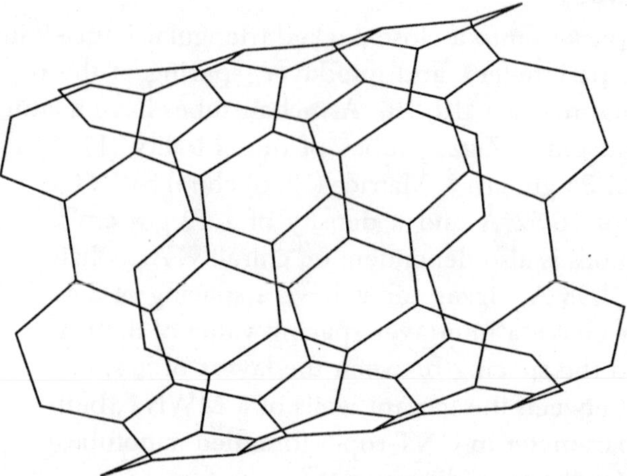

Fig. 5.11 *Schematic structure of a single-walled carbon nanotube*

Under suitable conditions, *MWNTs can be grown vertically in periodic arrays (Fig. 5.12) on various substrates.* The length *and diameter of the -nanotubes can be controlled The covalently bonded carbon structure gives rise to the high strength of the material* and *provides an extremely nyarophobic surface.* Most commonly, **chemical processing of CNTs** is carried out with the goal of overcoming the hydrophobicity of the surface, to render the material compatiable with solution state manipulations the behaviour of water and other polar solvents, confined within the internanotube hydrophobic channels, can be controlled.

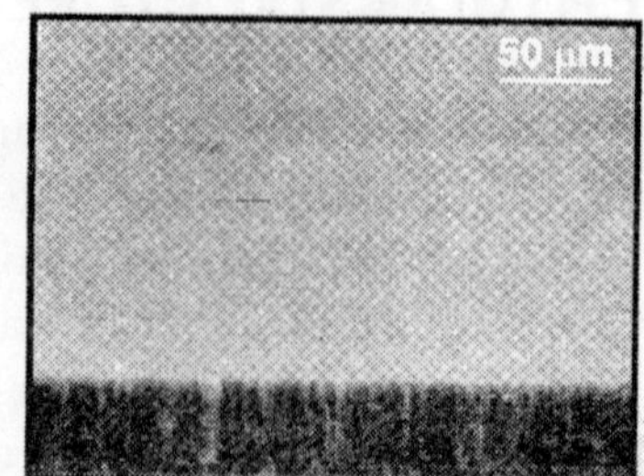

Fig. 5.12 *Growth of MWNT on a substrate*

The oxidation of the nanotubes in the arrays by using the oxygen plasma is assumed to result in the removal of amorphous carbonaceous materials and perhaps an oxidation of the tip of the nanotubes without affecting the integrity of the side walls. This chemical modification would create a hydrophilic environment at the top surface with a hydrophobic inner nanoscopic cavity. The solvent molecules trapped within nanoscopic space can form thin sheets, where the transport properties are highly collective. As a result, the entire sheet can slip back and forth under thermal agitation. Thus, when water is allowed to evaporate slowly, the capillary action of the nanotube channels forces them to collapse and the motion of the water sheet give rise to the grain boundaries. It is observed that freeze-drying of the wet nanotube membranes do not form the cellular structures, thus showing that the motion of the solvent sheets is necessary to form the boundaries.

The formation of the polygonal structures (Fig. 5.13), in contrast to crack patterns that are terminated by preexisting cracks (which are observed), indicates that the energy associated with the edges influences the topological properties of the edge network. The system tends to minimize its total energy by minimizing the total length of the edge network. It can do so only in a way compatible with local connectivity constraints as best illustrated by considering four points on the plane that one is asked to join by lines, so that the overall length of the latter is minimal. The rate of change of area of a cell thus formed to the number of its sides can be related by the von Neumann theorem. It governs the continuous evolution that drives a cellular structure toward topological changes and consequently to changes in the average characteristic scale. That the cells form independent of each other and continuously evolve to minimize energy is supported by the observation of multiple layers of the boundaries thus establishing that the average cell width changes inversely with the rate of evaporation of the solvent.

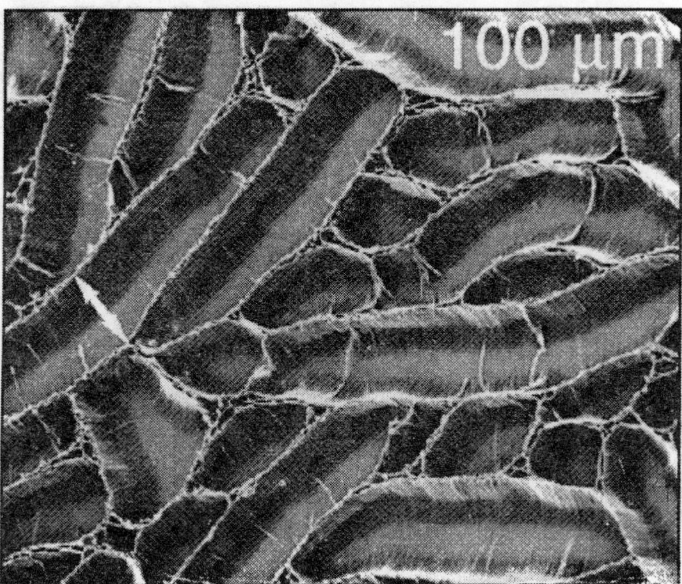

Fig. 5.13 *Topology of cellular structure formed from carbon nanotubes on Si.*

The observed alignment of the nanotubes at the cell boundaries is remarkable (Fig. 5.13). It indicates that the boundary formation process does not involve any exchange of water from within the membrane, during the evaporation process. This confinement would give rise to a membrane potential, much like in biological cells. Further shrinking the internanotube space from 50 nm to 1 nm would give rise to cellular structures controlled by the molecular properties of water. It is well known that in biological cells, the structure of water confined in nanoscopic space is very different from that in the bulk. The observation of the cell alignment in nanotubes grown on varying widths (Fig. 5.14) also indicates that the solvent enters the nanotube array only from the top, hydrophilic surface, thus leaving the side walls intact.

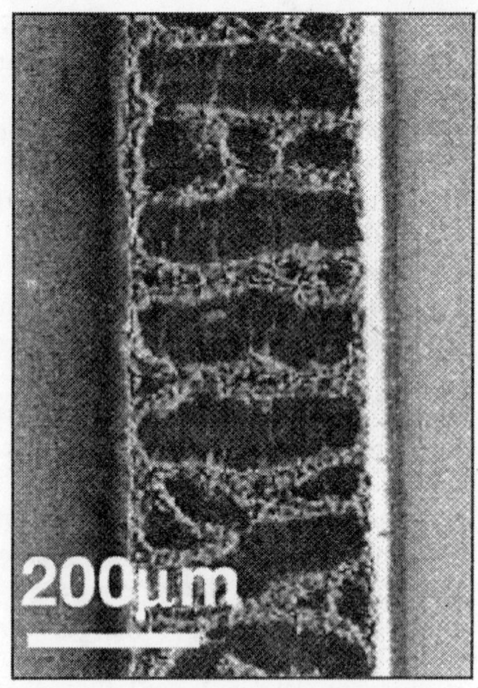

Fig. 5.14 *Alignment of the cells within a column of carbon nanotube array*

Finally, when the solvent was compartmentalized within a nanotube array (Fig. 5.15), the evolved structure was very similar to that simulated for the time averaged osmotically driven flow of water through semipermeable membranes of CNTs is Clearly, the solvent "slip" along the hydrophobic pore walls of the

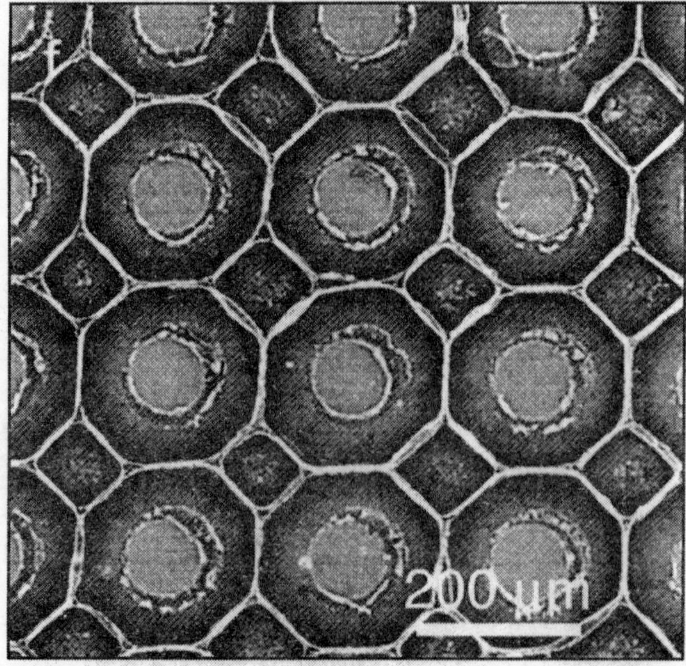

Fig. 5.15 *Cellular pattern formed by solvent migration through the nanoscopic pores in a carbon nanotube*

CNTs is controlled by fluid-wall interactions. Such slip-flow behaviour can be vital for generating the high-throughput rates required in nanofluidic devices.

Further refinement of the observed phenomenon can lead to the synthesis of living membranes, where concentration gradients will govern the directionality and flow of fluids. The observation of the cellular structures in this case was facilitated by the alignment and pinning of the nanotubes to the substrate, thus partially restricting two degrees of freedom. It is intriguing to consider the reverse situation, where the motion of nanoscopically confined fluids is allowed to reorganize a bulk anisotropic sample of carbon nanotubes.

BUCKY PAPERS (CARBON NANOTUBE (CNT) BASES FILMS)

CNT bucky papers (Fig. 5.16) are fabricated using a conventional filtration approach in the absence of polymer matrix or any other binder by filtering nanotube dispersion in aqueous or organic media and then peeling off the nanotube paper from the filter. SWNT films (bucky papers) have been processed from oleum, DMF, and nitric acid, and typical properties are given in Table 5.1.

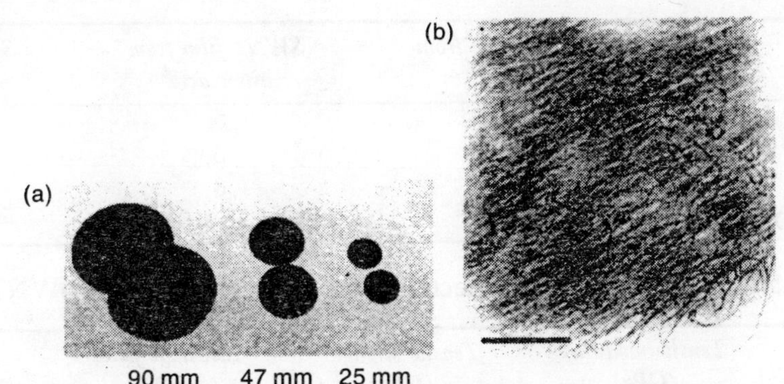

Fig. 5.16 (a) *SWNT bucky paper and (b) TEM image of magnetically aligned SWNT film, showing an area near the torn sample edge. The linear texture, running from the lower left to upper right of the image, is attributed to aligned ropes lying on average parallel to the applied magnetic field (scale bar 0.5 μm).*

Aligned nanotube films have been processed using high magnetic fields. In all cases of making nanotube films, the nanotubes are initially dispersed in liquid media. Typical films are made by depositing suspension of liquid media onto a filter in which nanotube bundles are randomly oriented in a plane. A preferential orientation of the nanotubes is achieved by using a high magnetic field. Drying the nanotube films while being subjected to magnetic field results in denser bucky papers due to the higher packing ratio of oriented nanotubes (Fig. 5.16(b)). Porosity of aligned nanotube films can range from about 30% to 90%, depending on the processing conditions. The measured anisotropic electrical and thermal transport properties of aligned nanotubes are higher along the alignment axis than in the transverse direction. As compared to typical bucky papers, the magnetically aligned nanotubes show significantly improved mechanical properties in the aligned direction due to dense packing of nanotubes.

Mechanical properties of SWNT bucky papers can be enhanced by reinforcing them with polymers such as PVA, PVP, and PS. Polymer-assisted reinforcement can be achieved by simply soaking the nanotube films in polymer solutions. *The bucky paper density increased after soaking in the polymer solution, and the SEM showed that polymer fills the pores between the SWNT ropes.* Mechanical properties of PAN/SWNT composite films are also substantially higher than either the PAN film or the SWNT bucky paper.

Mechanical properties of SWNT products are improved by covalent crosslinking saw by oxidation, the attachment of such functional groups as carboxylic acid, quinine, phenol, ester, amide, and switterions is found on the surfaces of oxidized SWNTs and poly-cyclic auromatic amorphous carbon.

Also the nitric acid treated bucky paper shows higher mechanical properties (Table 5.2) than the bucky paper processed from oleum. This may be due to crosslinking of the oxidized SWNTs in nitric acid. However, electrical conductivity decreases with nitric acid treatment, which is a result of the introduction of defects along the SWNTs wall due to chemical functionalization.

TABLE 5.1 Typical properties of the SWNT bucky papers

Property	*SWNT film from oelum*[a]	*SWNT film from nitric acid*[b]	*SWNT film from DMF*[c]
Tensile strength (MPa)	30	74	23
Tensile modulus (GPa)	8	5	3.0
Elongation at break (%)	0.5	3	2.4
Electrical conductivity (S/m)	1.3×10^5	1.2×10^4	3×10^4

TABLE 5.2 Mechanical and electrical properties of various SWNT films

Sample	*Tensile modulus (GPa)*	*Tensile strength (MPa)*	*Strain to failure (%)*	*Conductivity (S/m)*
Control SWNT film[a]	0.8	10	5.6	3.0 × 104
3M HN_3) SWNT film	1.4	16	1.4	2.3 × 104
6M HN_3) SWNT film	2.9	71	3.4	2.4 × 104
10M HN_3) SWNT film	5.0	74	3.0	1.2 × 104

[a]Control SWNT film is prepared from aqueous solution without nitric acid treatment.

DWNT *bucky paper* has been prepared using a two-step purification process–hydrochloric acid (18% by weight HCl at 373 K for 10 hr) treatment, which is used to remove iron catalyst and supporting materials, followed by oxidation in air at 773 K for 30 min to remove amorphous carbon. The resulting DWNT bucky paper shows high-purity DWNT structures and sufficient flexibility to be folded (Fig. 5.17).

The CNT/polymer films are also a developed using preformed SWNT bucky paper and resin infiltration. SWNT films are prepared by dispersing the SWNTs in water-based suspension with the aid of surfactant and sonication, followed by fitration. Subsequently, a diluted epoxy resin is infiltrated through the bucky paper in the thickness direction. As a result, CNT/

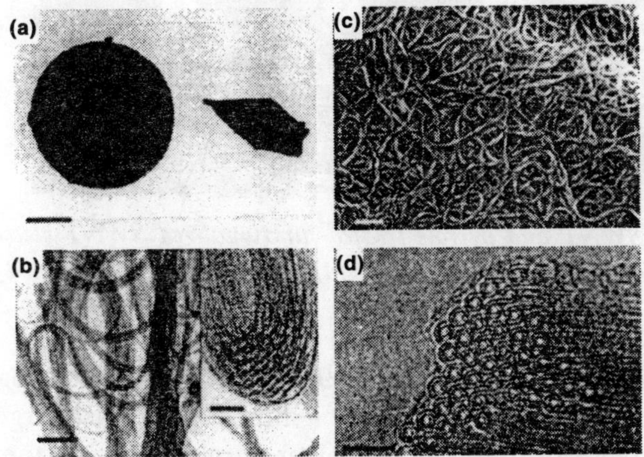

Fig. 5.17 *Production of pure and clean DWNTs in high yield. (a) Photograph of DWNT buckypaper; the paper (left) is tough and flexible enough to be folded into an origami plane (right). (b) SEM; and (c) low-magnification TEM of DWNT paper, showing bundles of carbon nanotubes (insert, a single bundle). Impurities are notably absent in these images. (d) Typical high-resolution TEM image of a different bundle, showing the perfect hexagonal packing structure in both cross-section and side view. Scale bars: (a) 1 cm; (b) 300 nm; (c) 150 nm; inset, 5 nm; (d) 5 nm*

polymer composites composed of multilayered bucky papers with high SWNT loading (up to 39 wt%) are produced by hot press molding prepreg bucky papers. The mechanical properties of resin impregnated bucky papers are higher than directly mixed nanocomposites.

CNT-based nano-structured materials are of tremendous interest in wide variety of energy storage applications such as membrane support catalyst for the electrocatalytic reduction of oxygen, storage of hydrogen, storage of lithium, and supercapacitors. In particular, CNTs are being widely investigated as the basis for electrochemical supercapacitor electrodes due to their unique properties and structure, including highly porous nanostructure, large surface area, high conductivity, and chemical stability. The basic operating mechanism of a supercapacitor involves two electrodes in contact with an electrolyte solution. Capacitance values from 20 to 180 Fg^{-1} have been reported, which depend on CNT purity and electrolyte, as well as specimen treatment conditions such as CO_2 physical activation, KOH chemical activation, or exposure to nitric acid, fluorine, or ammonia plasma. Supercapacitors have been fabricated with bucky papers consisting of catalytically grown nanotubes and SWNT/activated carbon. Physical or chemical activation of PAN or its copolymers is generally used for producing activated carbon with high specific surface area and porosity, and it is expected that similar activation treatment of SWNT/PAN composite films can also be used for developing SWNT/activated carbon composite films.

Supercapacitor electrodes based on pyrrole treated functionalized SWNTs were developed (Fig. 5.18). High values of capacitance (350 Fg^{-1}), power density (4.8 kW Kg^{-1}), and energy density (3.3 kJ Kg^{-1}) were obtained in 6 M KOH. The capacitance was almost seven times that of the control bucky paper. Specific capacitance of these materials has a strong dependence on the macropore surface area. The double layer capacity of pyrrole treated-functionalized SWNT

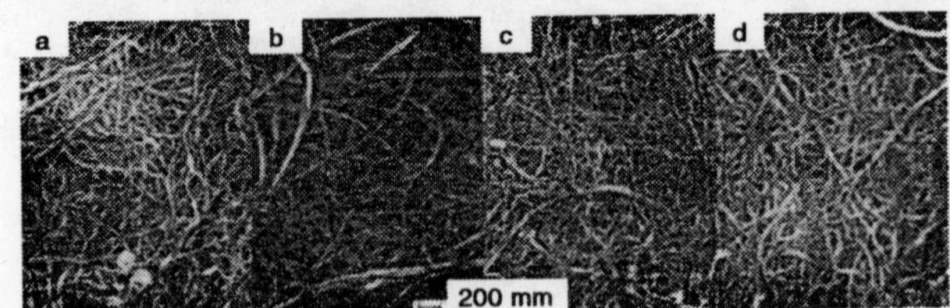

Fig. 5.18 *SEM images of SWNT and pyrrole treated-functionalized SWNT electrode surface. (a) Untreated HiPco SWNTs bucky paper, (b) 2; (c) 3; (d) 4, Numbers are explained in Table 4*

electrodes is 154 μF cm^{-2} based on the BET (Brunauer-Emmett-Teller) model and even higher based on the DFT model (Table 5.3). Figure 5.19 shows the specific capacitance of bucky paper and pyrrole treated-functionalized SWNTs as a function of discharge current density. The specific capacitances decreased rapidly at the lower current density, then tended to level off at higher current density in all the pyr-role treated samples. This decrease at the pyrrole treated SWNTs may be caused by some pores not being accessible efficiently at higher current density and also caused by larger internal resistance of the pyrrole moieties in comparison to the bucky paper.

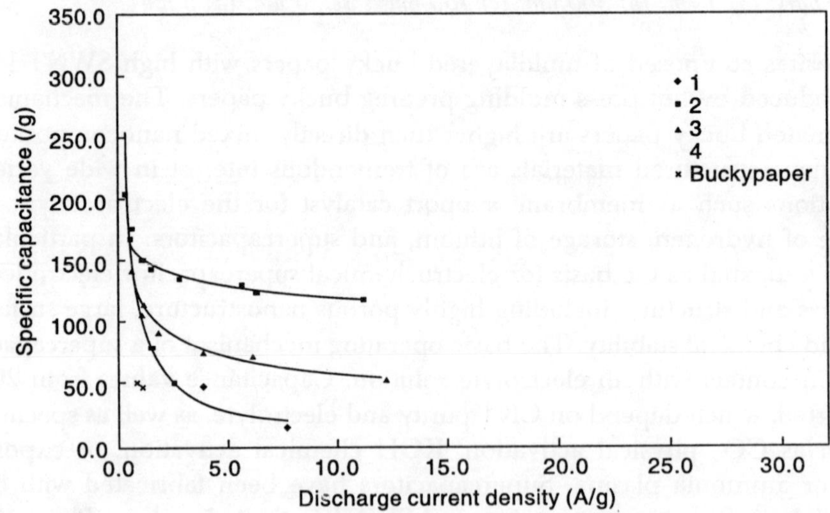

Fig. 5.19 *Specific capacitance of bucky paper and pyrrole treated SWNT electrodes as a function of discharge current density*

Recently, it is reported that some ionic liquids exhibit strong interactions with nanotubes and forms gel after being ground together. In addition, ionic liquid can be used as electrolyte in electrochemical storage applications. When CNT bucky papers are immersed in an ionic liquid, the mechanical properties of bucky papers decreased significantly because of strong interactions between ionic liquid and nanotubes.

TABLE 5.3 Capacitance, surface area, and pore volume of bucky paper and pyrrole treated SWNT electrodes.

		Untreated SWNT	*1*	*2*	*3*	*4*
Capacitance (Fg^{-1})[a]	0.1 V	55	176	151	114	305
	0.7 V	41	28	41	36	45
	mean value	49	102	93	80	183
BET surface area (m^2g^{-1})	546	148	205	113	198	
Total pore volume (cm^3g^{-1})	0.76	0.47	0.47	0.34	0.51	
DFT surface area (m^2g^{-1})	total	297	95	122	66	125
	micro[b]	167	12	40	32	24
	meso	128	27	46	27	25
	macro	2	56	36	7	76
DFT pore volume (cm^3g^{-1})	total	0.69	0.28	0.32	0.30	0.25
	micro	0.11	0.01	0.03	0.02	0.02
	meso	0.52	0.21	0.25	0.22	0.18
	macro	0.06	0.06	0.05	0.06	0.05
Cdl (μFcm^{-2})	BET	10	119	74	101	154
(0.1 V capacitance)	DFT-total	19	186	124	173	243
	DFT-meso & macro	43	214	183	330	302

1. Oxidatively *purified SWNT* (p-SWNT) heavily functionalized with arylsulfonic acid moieties (mainly individual tubes) and then treated with pyrrole.
2. *Raw SWNT* functionalized with arylsulfonic acid moieties and then treated with pyrrole.
3. *p-SWNT* functionalized with arylsulfonic acid moieties and then treated with pyrrole; approximately I in 33 carbons funclionalized by TGA.
4. *p-SWNT* functionalized with arylsulfonic acid moieties and then treated with pyrrole; approximately I in 54 carbons functionalized by TGA.

[a]1 mA discharge current.

[b]Micro-, meso-, and macropore refer to the sizes of <2. 2–50, and >50 nm, respectively.

Due to the exceptional thermal, mechanical, and chemical stability and high surface area, CNT-based membranes may compete with ceramic- and polymer-based separation membranes, The ability to fabricate robust membrane-like structures with CNTs can lead to a range of applications in separation industry, particularly highlighting the selective adsorption properties of nanotube surfaces. In order to perform various separation applications, appropriate large-scale structures need to be designed and built with nanoscale units. Though many ideas for separation technology application have been proposed, it is still a challenge to build macroscopic structures due to the difficulties associated with controlling the geometries and density for specific applications.

Transparent electrical conductors can serve as a critical component in video displays, image recorders, solar cells, lasers, optical communication devices, and solid state lighting. Ultra-thin, transparent, optically homogeneous, and electrically conducting films of pure SWNTs [174] have been prepared that were transferred to various substrates for electrical coupling in photonic devices (Fig. 5.20).

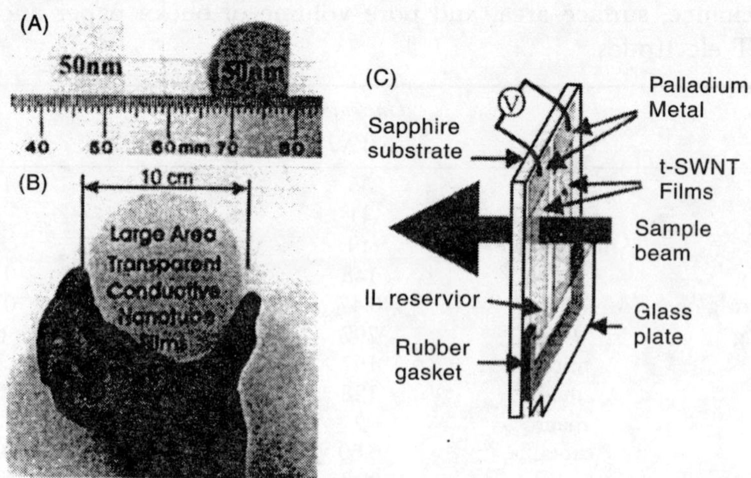

Fig. 5.20 *Transparent SWNT films. (A) Films of the indicated thickness on quartz substrates. (B) Large, 80-nm-thick film on a sapphire substrate 10 cm in diameter. The text in (A) and (B) lies behind the films. (C) Optical analog to the electrolyte-gated NFET (O-NFET).*

Hydrogen adsorption capacity of different types of activated carbon, as well as MWNTs, and SWNTs, depends on their pore structures. In general, adsorption ability is governed by porosity, pore structure, and external conditions, and it can be correlated to BET surface area. Among various types of carbon-based materials, the highest adsorption capacity (1.58 wt% at 77 K, 1.15 wt% at 87 K, and 0.02 wt% at 298 K) at atmospheric pressure has been observed in chemically activated SWNT samples.

The study of CNT actuators using SWNT sheets, low voltage artificial muscles are drawing attention for a range of industrial applications. Electromechanical actuators, often called artificial muscles, can be useful in robotics, e.g., switches, valves, or adjusting links, as well as in biomedical applications, such as microsurgical devices, artificial limbs or even, in the future, implants like artificial ocular muscles or hearts. The main requirements of these devices are, light weight, low maintenance voltage, large displacements, large force, fast response, and long cycle life etc.. The actuation properties are explained by a double layer charge injection which produces dimensional changes in covalently bonded direction for conjugated materials such as the nanotubes.

NANOTUBE MANIPULATION

In order to make devices out of nanotubes, it is important to be able to manipulate the nanotubes in a controlled way. A nanotube's position, shape and orientation, as well as cutting it is done by using an atomic force microscope (AFM). Firstly, The AFM is used in "non-contact" mode to obtain an image of the nanotube by scanning the AFM tip, shown in the schematic on the left, just above the surface. (Fig. 5.21(a). The AFM tip is then brought down to the surface and is used like a tiny plow to move the nanotube Fig. 5.21(b). Because of the strong

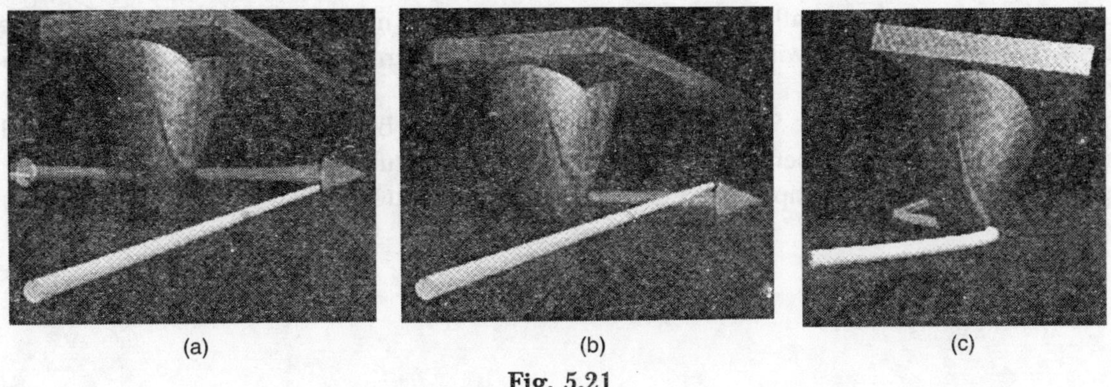
(a) (b) (c)

Fig. 5.21

interaction between the nanotube and the surface via van der Waals forces, the bent nanotube stays where it has been placed and maintains its shape, rather than snapping back to its preferred straight configuration Fig. 5.21(c).

Below is a series of AFM images (Fig. 5.22) illustrating the stepwise positioning or a nanotube across an insulating barrier so as to be able to pass current through it.

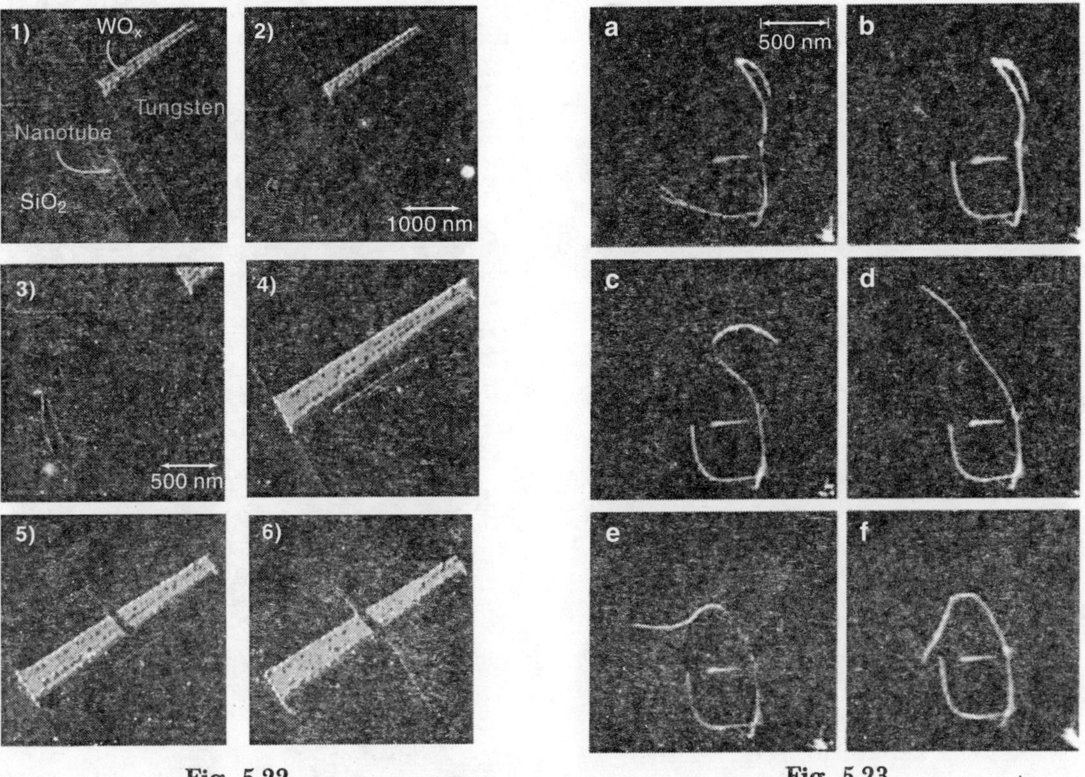

Fig. 5.22 Fig. 5.23

A single nanotube originally on an insulating substrate is manipulated in a number of steps onto a tungsten film thin wire, and finally is stretched across an insulating tungsten oxide barrier.

Another example is (Fig. 5.23), of how a nanotube can be manipulated to form complex shapes: the 6 frames are a series of AFM images of a nanotube on a silicon substrate. Not all steps are shown. The AFM tip is used to create the Greek letter "theta" from a 2.5 micron long nanotube.

SYNTHESIS OF CNTs

CNTs are stable molecules whose synthesis is highly exothermic. In this section, different techniques for nanotube synthesis are explained. First, the growth mechanism is explained, as it is general for all techniques, but typical conditions are stated at the sections of all the different techniques. The largest interest is in the newest methods for each technique.

Carbon nanotubes are generally produced by three main techniques, ***arc discharge, laser and chemical vapour deposition***. Growth of CNTs by arc discharge and laser ablation involve the addition of carbon dimers (C_2) to the CNT edges.

In the arc discharge, a vapour is created by an arc discharge between two carbon electrodes with or without catalyst. Nanotubes self-assemble from the resulting carbon vapour. In the laser ablation technique, a high-power laser beam impinges on a volume of carbon-containing feedstock gas (methane or carbon monoxide). Laser ablation produces a small amount of clean nanotubes, whereas arc discharge methods generally produce large quantities of impure material.

Chemical vapour deposition (CVD) results in MWNTs or poor quality SWNTs. The SWNTs produced with CVD have a large diameter range, (which can be poorly controlled).

GROWTH MECHANISM

Nanotubes are formed by gas-phase processes which start with a source of carbon, evaporated from a surface due to heating by the laser pulses or through heating by electric arcs or solar radiative fluxes. Carbon nanotube formation from evaporated carbon requires a metal catalyst such as nickel, cobalt, iron, yttrium, etc. It is found that bimetallic catalysts are more productive than single metals. Various configurations of laser-ablation experiments range from the dual laser pulsed system to continuous lasers.

Each technique has certain aspects in common and they have led to various proposed mechanisms for the formation of single-wall carbon nanotubes. The principal differences are is the time scales on which nanotubes form and the state of particles from which nanotubes grow. Thus in the growth mechanism for CNTs, more than one mechanism might be operative during their formation. One of such mechanisms consists of three steps. First a precursor to the formation of nanotubes and fullerenes; C_2, is formed on the surface of the metal catalyst particle Fig. 6.1(b). From this metastable carbide particle, a rodlike carbon is formed rapidly. Secondly there is a slow graphitisation of its wall.

The exact atmospheric conditions depend on the technique used. A simple growth process of CNTs, depicted in Fig. 6.1(a), involves diffusion of carbon atoms from a support surface into the transition metal catalytic surface, followed by extrusion from the catalysis surface.

Following are the more mechanisms for growth of carbon nanotubes.

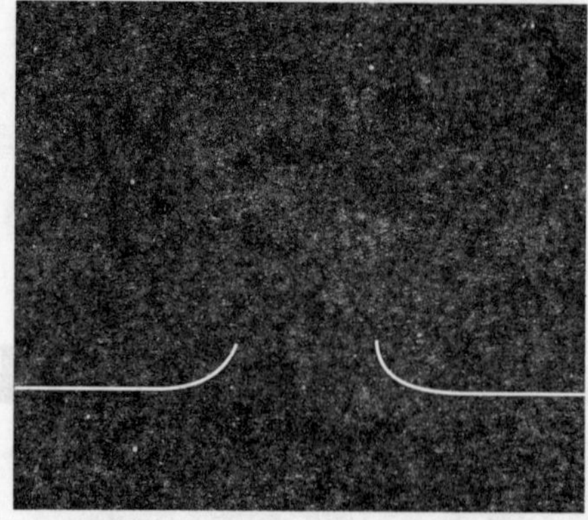

Fig. 6.1(a) *Sketch of the carbon nanotube growth process*

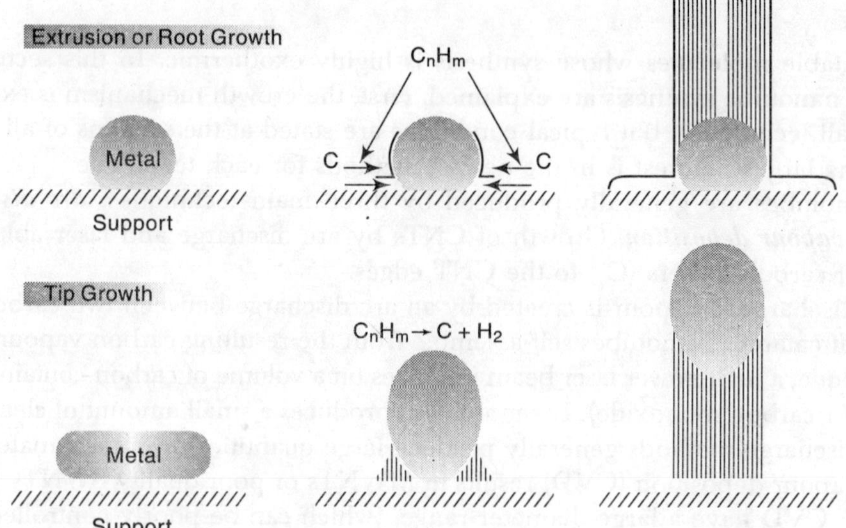

Fig. 6.1(b) *Visualisation of a possible carbon nanotube growth mechanism.*

The Scooter Mechanism

In this mechanism, atoms of the catalysts Ni and Co are envisioned to attach to a fullerene or a graphene sheet. The catalyst atoms prevent the closing of the fullerene as additional carbons attach and lead to the formation of tubes. A single atom or a few atoms circulate (scoot) around the open end of the tube and adsorb small molecules of carbon and convert them into a tabular graphene-like sheet. The tube grows until too many catalyst atoms aggregate on the end of the nanotube. The large particles either detach or become over-coated with sufficient carbon to poison the catalysis. This allows the tube to terminate. with a fullerene-like tip or with a catalyst particle. No mention of the time scale for nanotube formation is given for this model. The 'scooter' concept now is followed by a mechanism in which SWNTs grow out of catalyst particles saturated with carbon after they have had time to aggregate. Since it is found that bimetallic catalysts improved SWNT yield, it is suggested that carbon mobility increased with two catalysts metals as compared to only one.

Zhangest investigated the production of SWNTs in a nitrogen atmosphere, containing the product with that produced in argon in similar conditions. Their findings indicated that the nanotubes were created in the high-temperature zone close to the target, in a region where nitrogen was essentially excluded by the ablation products. Later, after mixing of the plume with background nitrogen, atmosphere carbons were formed that had nitrogen inclusions.

Reaction Mechanism

For the formation mechanism of carbon nanotubes, it is believed that the hydrocarbon molecules are decomposed on the free Ni-metal surface producing carbon atoms with simultaneous evolution of molecular hydrogen. The carbon atoms dissolve and diffuse through the bulk of the Ni catalyst and precipitate as graphite at the other side of the Ni particle.

A more detailed model of nucleation and growth of carbon nanofibers carbon nanotubes from methane by decomposition on supported Ni catalysts is based on magnetometric. The carbon atoms deposited form methane decomposition on the-surface of the Ni particle quickly dissolve into the bulk and form a substoichiometric nickel carbide at the early stage. This carbide phase then decomposes into metallic nickel particles and outside graphite layers which encapsulate the host nickel particle. The metal particle is then squeezed out by strong pressure buildup due to the formation of graphite layers at the internal surface of the graphite envelope. This is possible because of the liquid-like behavior of the metal due to melting point depression from the dissolution of carbon atoms that remained in the bulk, The fresh surface of the squeezed-out Ni particle would then further decompose methane and continue the growth of a carbon nanotube. Depending on conditions, the process can happen again, resulting in pulsed growth of zigzag nanotubes or, if the deposition of carbon atoms is not fast enough compared to the dissolution into the bulk Ni, the graphitization of carbon atoms can be continuous resulting in relatively straight nanofibers.

The model, based on methane decomposition, supported by TEM images for cyclopentadiene and benzene decomposition. TEM images of the catalytic growth of the carbon tubes formed from the decomposition of cyclopentadiene at the early stages. Many Ni particles just about to start growing their carbon fibers on a thick GCF used as the support. The high-resolution TEM images of the infant carbon tubes in the circles in Figure 6.2(c) are shown in Figures 6.2(a) and (b). The Tem image of Figure 6.2(b) also shows two Ni particles covered with 10 to 20 carbon layers and a young carbon tube with a Ni Particle at the tip. The carbon layers at the bases of the three Ni-carbon agglomerates in Figure 6.2(b) are all parallel to the graphite layers of the GCF support. The shape of the Ni particle at the tip of the short tube has become conical or is being tapered to the backward direction of tube growth. The state of a more developed carbon tube is shown in Figure 6.2(a). The size of this tube is roughly 2.5 times thicker and 1.5 times taller than that in Figure 6.2(b). The carbon layers near the root of the tube are also parallel to those of the GCF. The carbon layers near the center of the tube are not easily recognized, implying the presence of hollow structures along the center of the tube. The Ni particle at the tip of the tube in Figure 6.2(a) seems to turn its direction of growth to the left.

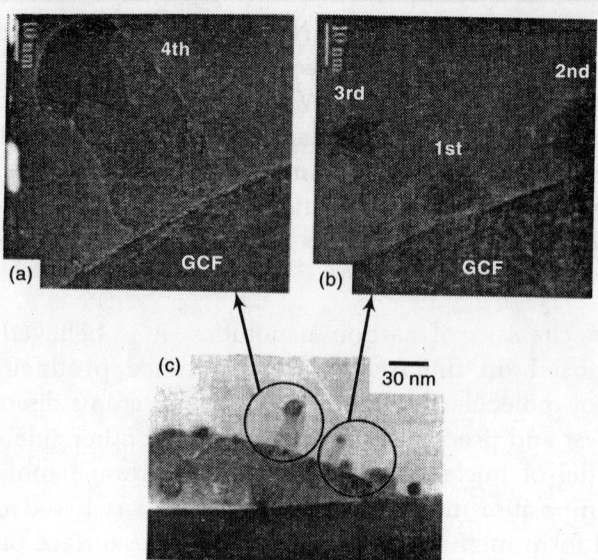

Fig. 6.2 *TEM images of child carbon nanotubes form cyclopentadiene on Ni (10 wt%) GCF at the very initial stage (C/Ni = 20).*

Continuous Wave Mechanism

A continuous-wave (CW) carbon dioxide laser was used to produce SWNTs by Maser et al. They heated a graphite rod containing Ni/Y and Ni/Co catalysts in an organ atmosphere sphere. They found that Ni/Y (4.2/1 at. %) yielded the most nanotubes, followed by the Ni/Y (2/0.5 at. %) mixture, then Ni/Co (2/2 at. %).

Because they found that the SWNTs were deposited very close to the target, they concluded that SWNTs are formed very rapidly in the gas phase, either in the hot plasma plume or very close to it, where the temperature is still high enough to support their growth. They suggested that the temperature of the hot region around the laser-ablation spot on the target is high enough to assure SWNT growth without an external oven. Since external gas flow rates are very low around the target, the time scale is long for nanotube formation. It is interesting to note that the CW laser ablation is similar to arc ablation in that the background gas and catalyst mix is found to be best in the arc process. This shows that heat transfer and diffusion dominates continuous processes as compared with pulsed processes.

Carbon–Co/Ni—mechanism

Using a 20-ms-duration pulsed CO_2 laser in single-pulse mode to ablate graphite containing 0.6 at. % Co/Ni catalysts, the effect of oven temperature on nanotube growth on investigation shows that larger diameter nanotubes are produced at higher temperature.

A yield greater than 60% at an oven temperature of 1473 K is found. Using high-speed video and emission spectroscopy, shows that carbonaceous materials with SWNTs are visible about 3 ms after the beginning of the laser pulse. Blackbody emission from clusters is observed for more than one second. Which shows that SWNTs grow from molten carbon–metal particles via supersaturation at ablation temperatures, followed by segregation at about 1593 K; so the temperature of the SWNT growth zone may be higher then 1473 K. This is based on C-Ni and C-Co eutectics. A continuous supply of hot carbon clusters and the maintenance of a hot SWNT growth zone during the 20-ms laser pulse are important.

There are several theories on the exact growth mechanism for nanotubes. One theory postulates that metal catalyst particles are floating or are supported on graphite or another substrate. It presumes that the catalyst particles are spherical or pear-shaped, in which case the deposition will take place on only one half of the surface (this is the lower curvature side for the pear shaped particles). The carbon diffuses along the concentration gradient and precipitates on the opposite half, around and below the bisecting diameter. However, it does not precipitate from the apex of the hemisphere, which accounts for the hollow core that is characteristic of these filaments. For supported metals, filaments can form either by 'extrusion (also known as base growth)' in which the nanotube grows upwards from the metal particles that remain attached to the substrate, or the particles detach and move at the head of the growing nanotube, labelled 'tip-growth'. The parameters controlling the growth are: nature of feed gas and its composition; flow rate, temperature, type of catalyst; Catalyst preparation technique and substrate material. Depending on the size of the catalyst particles, SWNT or MWNT are grown. In arc discharge, if no catalyst is present in the graphite, MWNT will be grown on the C_2-particles that are formed in the plasma.

SYNTHESIS METHODS

Following main six methods for synthesis of carbon nanotubes are discussed in detail in this Chapter.

(A) Arc discharge
(B) Laser vaporization
(C) Plasma enhanced chemical vapor deposition
(D) Thermal chemical vapor deposition
(E) Vapor phase growth
(F) Electrolysis/Flame synthesis

(A) ARC DISCHARGE

The carbon arc discharge method, initially used for producing C_{60} fullerenes, is the most common and perhaps easiest way to produce carbon nanotubes as it is rather simple to undertake. However, it is a technique that produces a mixture of components and requires separating nanotubes from the soot and the catalytic metals present in the crude procuct.

This method creates nanotubes through arc-vaporisation of two carbon rods placed end to end in the vessel and separated by approximately 1 mm, in an enclosure that is usually filled with inert gas (helium, argon) at low pressure (between 50 and 700 mm Hg) The two graphite rods are used as the cathode and anode, between which arcing occurs when DC voltage power is supplied. A large quantity of electrons from the arc-discharge move to the anode and collide into the anodic rod. Carbon clusters from the anodic graphite rod caused by the collision, are cooled to low temperature and condensed on the surface of the cathodic graphite rod. The graphite deposits condensed on the cathode contain both carbon nanotubes, nanoparticles and clusters. Fig. 6.3 shows a schematic diagram of arc-discharge apparatus for the synthesis of carbon nanotubes. This apparatus is connected to both a vaccum line with a diffusion pump, and to a helium supply. The electrodes are two graphite rods, usually of high purity. Typically, the anode is a long rod approximately 6 mm in diameter and the cathode a much shorter rod 9 mm in diameter. Efficient cooling of the cathode is essential in producing good quality nanotutes. The position of the anode is made adjustable from outside the chamber, so that a constant gap can be maintained during arc-evaporation. It is also possible to create nanotubes with the arc method in liquid nitrogen. A direct current of 50 to 100 A driven by approximately 20 V to 40 V, creates a high temperature discharge between the two electrodes. The discharge vaporises one of the carbon rods and forms a small rod shaped deposit on the other rod (hot plasma). The deposits are formed on the short black negative electrode while the long positive electrode is consumed. As the smaller rod is consumed, the rods are kept at a constant distance from each other (~ 1 nm). Producing nanotubes in high yield depends on the uniformity of the plasma arc and the temperature of the deposit form on the carbon electrode. The best rate of rod deposit is few nm per minute. The diameter of nanotube ranges between 2 and 20 nm and length is several micrometers.

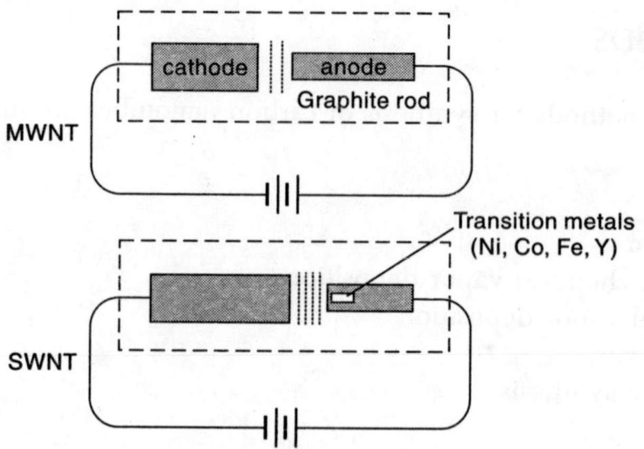

Fig. 6.3 *Schematic diagram of arc-discharge apparatus*

Carbon nanotubes synthesized by arc-discharge normally have multi-walled structures. After holes in the graphite rods are bored they are filled with appropriately proportional composites of graphite powder and catalytic Co. Ni, Fe, and Y powder. The pressure in the evaporation chamber and the current are the most important factors in producing a good yield of high quality carbon nanotubes. Increase in the yield is evident as the pressure is increased, but there is a fall in total yield at too high pressure. In addition, the current should be kept as low as possible. consistent with maintaining a stable plasma. Fig. 6.4(a) shows graphite deposits synthesized by a arc-discharge apparatus. In practice, carbon nanotubes are found in the core of the graphite deposits. Fig. 6.4(b) are SEM images of graphite deposits and multi walled carbon nanotubes in the deposits synthesized by arc-discharge method in the center. Multiwalled

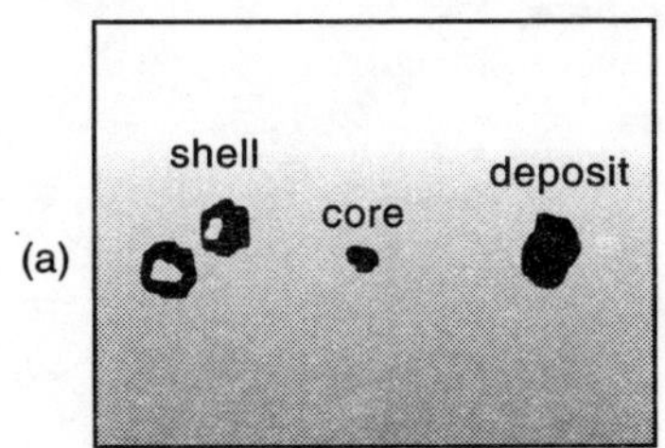

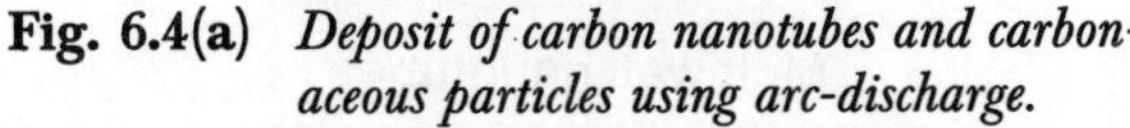
Fig. 6.4(a) *Deposit of carbon nanotubes and carbonaceous particles using arc-discharge.*

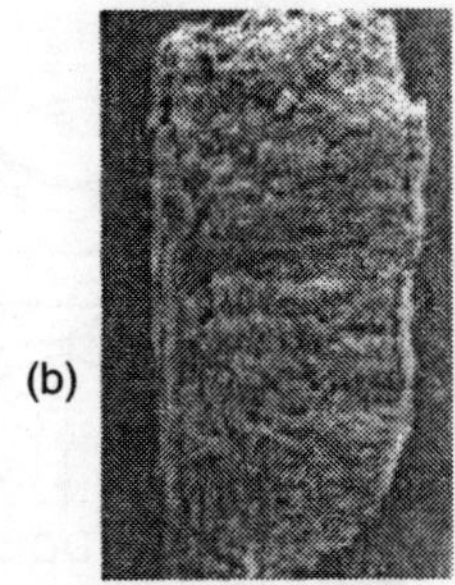

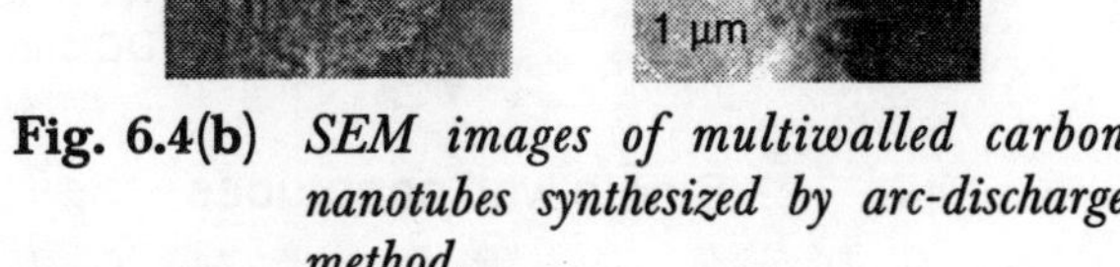
Fig. 6.4(b) *SEM images of multiwalled carbon nanotubes synthesized by arc-discharge method.*

carbon nanotubes are covered with a large amount of carbon deposits or carbon particles and partially exposed. In order to study basic properties. structures. and applications of carbon nanotubes. carbon particles stuck to carbon nanotubes should be purified.

Fig. 6.5 Shows TEM images of synthesized multiwalled carbon nanotubes. In these images, carbon nanotubes with the diameters ranging several nanometers are hollow in the core. Graphene sheets are aligned along either side of carbon nanotubes at a uniform distance of about 0.34 nm. The numbers of graphene sheets in these images are 5, 2, and 8 with multiwalled structures.

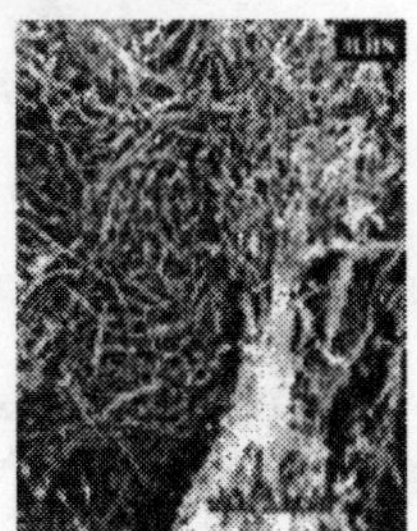
Fig. 6.5 *TEM image of multiwealled carbon nanotubes synthesized by arc-discharge method.*

Insight in the growth mechanism is increasing and measurements have shown that different diameter distributions have been found depending on the mixture of helium and argon. These mixtures have different diffusions coefficients and thermal conductivities. These properties affect the speed with which the carbon and catalyst molecules diffuse and cool., affecting nanotube diameter in the arc process. This implies that single-layer tubules nucleate and grow on metal particles in different sizes depending on the quenching rate in the plasma and it suggests that temperature and carbon and metal catalyst densities affect the diameter distribution ofnanotubes. By this method only 30% yield is obtained and contains a few structural defects. Tubes tend to be short with random sizes and directions.

Depending on the exact technique, it is possible to selectively grow SWNTs or MWNTs, which is shown in Figure 6.6. Two distinct methods of synthesis can be performed with the arc discharge apparatus.

Fig. 6.7 is a SEM image of carbon nanotubes. Here carbon nanotubes are densely entangled. As observed, the SWNTs exist as clustered ropes, having diameter upto several nm.

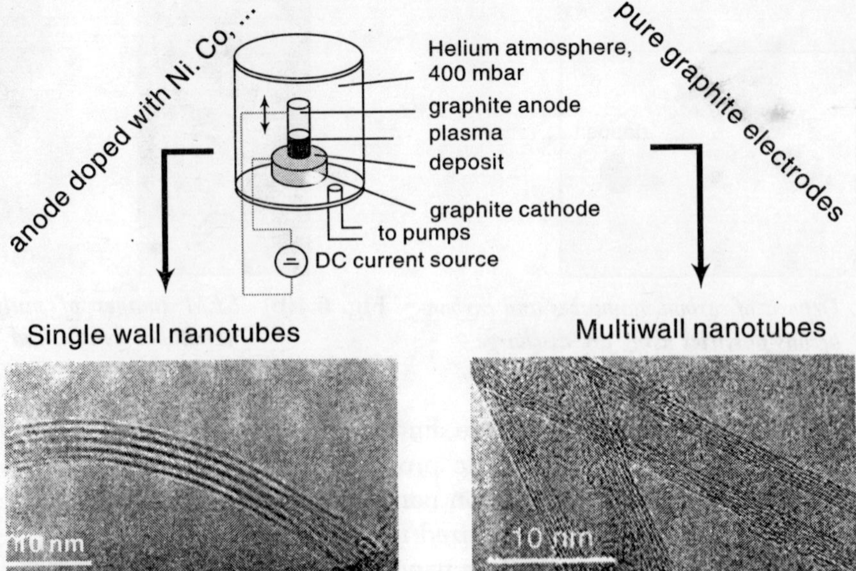

Fig. 6.6 *Experimental set-up of an arc discharge apparatus.*

Synthesis of SWNT by arc Discharge Method

SWNT can be synthesized by arc discharge on the cathode.

For synthesis of SWNTs, the anode has to be doped with metal catalyst, such as Fe, Co, Ni, Y or Mo. A lot of elements and mixtures of elements on testing give different results, even if the same elements are used. This is due to different experimental conditions. The quantity and quality of the nanotubes obtained depend on various parameters such as the metal concentration, inert gas pressure, kind of gas, the current and system geometry. Usually the diameters is in the range of 1.2 to 1.4 nm.

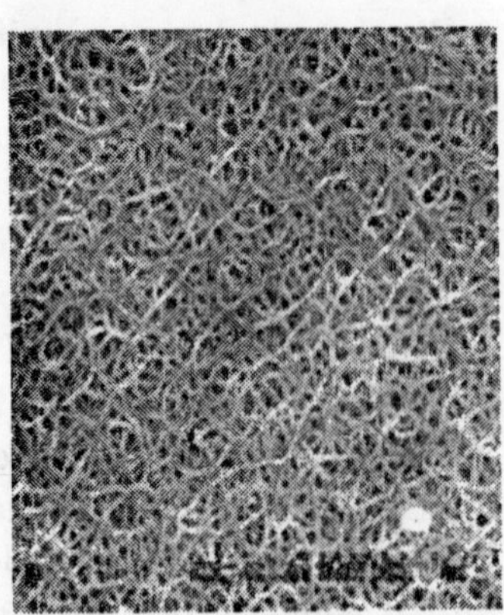

Fig. 6.7 *SEM image of carbon nanotubes synthesized by arc-discharge method*

Following are the useful processes of arcs discharge for improving the quality.

(a) Inert Gas

The most common problems with SWNT synthesis are that the product contains a lot of metal catalyst, SWNTs have defects and purification is hard to perform. On the other hand, an advantage is that the diameter can slightly be controlled by changing thermal transfer and diffusion, and, hence condensation of atomic carbon and metals between the plasma and the vicinity of the cathode can control nanotube diameter in the arc process. This was achieved in an experiment in which different mixtures of inert gases were used.

It appears that argon, with a lower thermal conductivity and diffusion coefficient, gives SWNTs with a smaller diameter of approximately 1.2 nm. The average nanotube diameter shows a 0.2 nm diameter decrease per 10 % increase in argon-helium ratio, when nickel/yttrium is used (C/Ni/Y is 94.8:4.2: 1) as catalyst.

(b) Optical Plasma Control

Optical plasma control the plasma temperature distribution is an intimate component of SWCNTs synthesis. For the arc process. The parameter controls species production and condensation, hence nanotube growth and morphology. To control this parameter the majority of laboratory scale arc reactors use a water cooling system. Plasma temperature control is accomplished by one of the following methods: (1) changing the plasma conditions by varying the electric current and/or 2) varying the anode to cathode distance (ACD) to maintain a constant voltage. These methods do not allow active control on the plasma temperature, independently of plasma parameters. An alternative approach consists of varying the conductance of the gap between the electrodes, using an argon-helium mixture as the inert background gas. But, the yield of nanotube is found to be much higher with helium than with argon. To improve nanotube production independent of the inert gas, an optical control technique based on the shape of the visible plasma zone is used. For a given inert gas composition, adjusting the gap difference, a strong visible vortices at the edge of the cathode is obtained. These vortices ensure a maximal mass flux of carbon and catalyst species from the plasma zone to the cold cathode region, and thus enhance the nanotube growth.

The structure of SWNTs is analyzed using Raman spectroscopy to determine diameter distribution; and Brunauer-Emmett-Teller (BET) surface analysis is done to estimate the surface area of produced nanotubes.

With a nickel and yttrium catalyst (C/Ni/Y is 94.8:4.2:1) the optimum nanotube yield was found at a pressure of 660 mbar for pure helium and 100 mbar for pure argon. The nanotube diameter ranges from 1.27 to 1.37 nanometre.

Experiment

SWCNTs are prepared by conventional *electric arc discharge method.* The reactor consists of a water-cooled reaction chamber, in which has been introduced an anode, made of pure graphite: carbon rod, 6 mm in diameter and 150 mm length. The cathode is also made in high polished graphite. Nickel and yttrium, ware introduced as catalysts in a drilled hole inside anode, of 3 mm diameter and 100 mm length. The anode, including the outer graphite shell, consists of 94.8 % atom, carbon 4.2% atom. nickel and 1 % atom, yttrium. A set of six experiments are done with different argon-helium mixtures (used as inert gases) under controlled pressure. Optimum nanotube yield is found at a pressure of 660 mmHg for pure helium and 100 mmHg for pure argon. Varying the percentage of helium from 0 to 100 (with a step of 20%), the pressure is arbitrarily kept between 100 mm to 660 mm. The thermal conductivity of the mixture depends on the relative concentration of the two gases. The thermal conductivity of argon is about eight times smaller than of the helium. So, varying the relative concentration of the two gases influences the temperature gradient between the plasma and the cathode. This affects the condensation of atomic carbon and metal catalysts, hence the nanotube diameter distribution.

The electric parameters of all the experiments are maintained at a direct current of 100 A to create the arc discharge. In the beginning the plasma is ignited by contact between the electrodes, which increases the temperature of the contact point until evaporation of the anode material. When the discharge runs, the plasma is observed through the quartz window of the reaction chamber, and a magnified image of the luminescent plasma is projected on a screen, using lens. The anode is moved back until vortices appear between the growing deposit on the cathode and the burning anode. The gap distance between the electrodes is maintained by moving the anode toward the cathode. The high temperature near the anode and the high energy densities in the plasma ensure a total vaporization of the anode material, while the water cooled cathode leads to high quench rates and to nanotube formation. The experimental results indicate that the carbon nanotubes are generated on the front.end surface of the deposit. The quench process is complex and uncontrolled hence, various products are obtained eg: soot on the reactor walls, web-like structures between the cathode and the chamber walls, a deposit on the cathode's edge, and a robber-like collaret around the deposit.

Pictures of the plasma are taken (through the video) and the distance between the electrodes in every moment is determined. A data acquisition system, composed of a power and harmonics clamp and a multimeter connected to a PC, allows recording the variations respectively of the intensity of the current, and of the difference of the potential. Nanotube diameter is determined using Raman spectroscopy using a spectrophotometer. Spectra is recorded at room temperature in ambient air with an argon laser at 514.5 nm. Surface analyses is accomplished with a BET instrument using nitrogen gas; in order to determine the BET surface areas as well as the pore size distributions. The method consists of one-hour gas evacuation at 393 K and of determination of the pore volume introducing helium gas, followed by adsorption of nitrogen at temperature of liquid nitrogen (77 K).

CHARACTERISTICS

I. Plasma Characteristics

By performing set of six experiments with different argon-helium mixtures as above, the plasma characteristics are reported only of pure argon plasma but are reproducible for all the experiments. The conditions of the synthesis of nanotubes namely *ACD* were modified in order to improve the yield of nanotubes. This parameter was not maintained constant but varied to establish a visible vortices around the cathode. These vortices are visible probably because of the presence of small particles of carbon acting as markers in the stream.

The shapes & colour of the Plasma

During all the experiments, two operative regimes concerning the shape and the color of the plasma, are distinguished qualitatively.

1. The first is the stable regime, during which the arc flame is uniform, in a *blue-green color*, and does not change its shape and brightness (figure 6.8(a); (b)). We adjust manually the distance between electrodes, until the appearance of the vortices (figure 6.8(c); (d)). when argon as inert gas is used the optimal distance between the electrodes, (encouraging the formation of vortices is between 15 to 17 mm.
2. The second regime is not desirable, the vortices do not appear; the colour of the arc flame is blue-purple and the brightness is weaker (figure 6.9(a); (b)). This regime is unfavourable

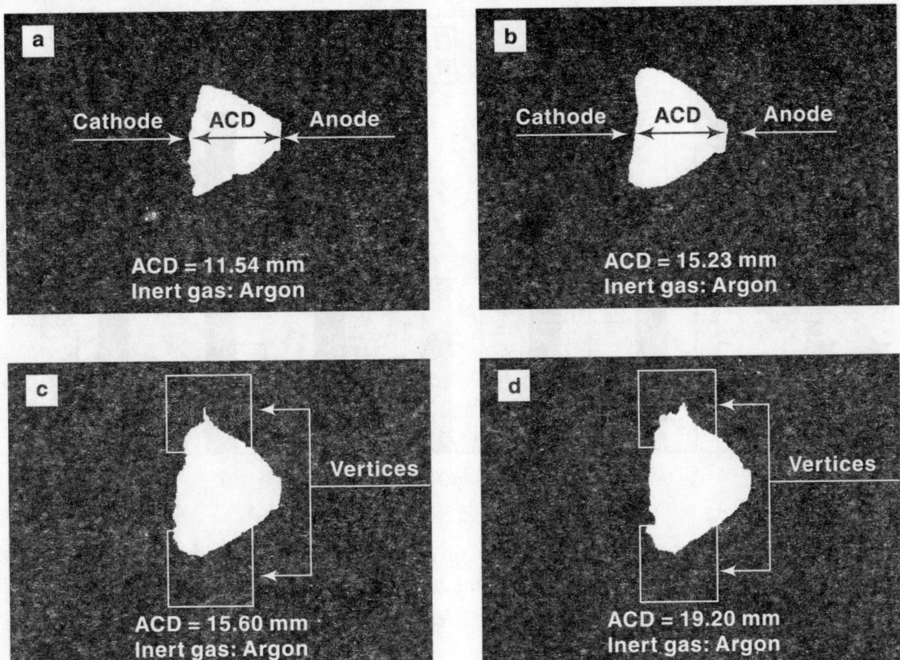

Fig. 6.8 *Development of vortices in argon plasma (Pressure = 100 mbar, Current = 100 A). The streamlines are curved outside the plane of symmetry.*

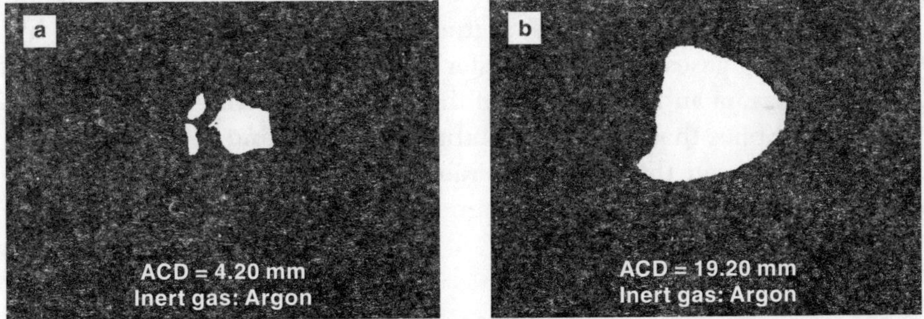

Fig. 6.9 *Pictures of the non-desirable regime. The vortices do not appear*

for the synthesis of nanotubes. It appears when the anode is very near (ACD = 1 to 7 mm) or when it is very far from the cathode. (ACD greater than 18 mm). On moving away the anode from the cathode it disappears.

The yield of Nanotubes

The effect of the control of the ACD on the *yield of nanotubes* is shown in reported on figure 6.10. In this figure optimal ACD vortices show improved yield of collaret, compared with ACD giving a constant voltage (U = 40 V). For some mixtures namely 0% and 40% argon, the deposition of webs on the quartz window makes ACD control difficult and explains the decrease of the mass of collaret. For pure argon the yield is improved by a factor 3.5.

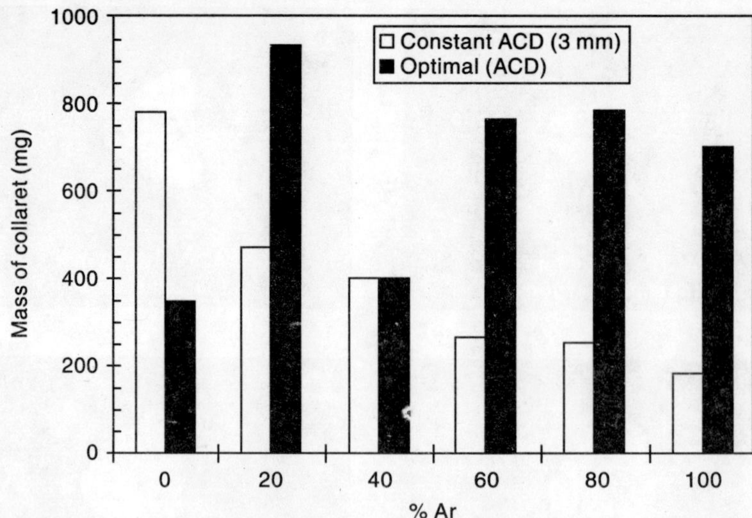

Fig. 6.10 *Effect of the anode to cathode distance on collaret mass collected in the cathode using following parameters: burned anode length 100 mm and current 1 = 100 A. ACD was maintained constant at 3 mm. ACD was varied to obtain vortices.*

The first experimental results demonstrate the efficiency of ACD control technique for different inert atmospheres. This new design provides a means to improve nanotube yield and also control gradient temperature in the arc, independently of plasma conditions *i.e.* current and voltage. It also provides the opportunity to control the morphology of nanotubes. Indeed, since the thermal conductivity of the argon-helium mixture is affected by the relative concentration of the two gases, thermal transfer and hence condensation of atomic carbon and metals between the plasma and the vicinity of the cathode can control nanotube diameter in arc process. This result implies that single-layer tubules nucleate and grow on a metal particle with different sizes depending on the quenching rate in the plasma and suggests that temperature and carbon and metal densities affect the diameter distribution.

II. Microscopic Characteristics

Nanotube Diameter

For each sample, spectra on at least ten different points is recorded to get the best view of all the sample. Two main regions of the Raman spectrum can be attributed to SWCNTs: (1) the high frequency (1500–1600 cm^{-1}) and (2) the low frequency (100–250 cm^{-1}) ranges. In figure 6.12(a), the overall spectrum and in figure 6.12(b), the low frequency range has been scaled up to better observe the breathing modes of nanotubes.

The relative intensity of these bands changes as the inert gas varies from pure helium to pure argon. To determine exactly the variation in the diameter distribution, several lorentzian shapes are fitted (one shape per band) to the whole group of peaks assigned to the breathing mode. Thus the proportion of each diameter for all recorded spectral is obtained. Calculation of average diameter distribution and an average diameter from all spectra obtained on the same

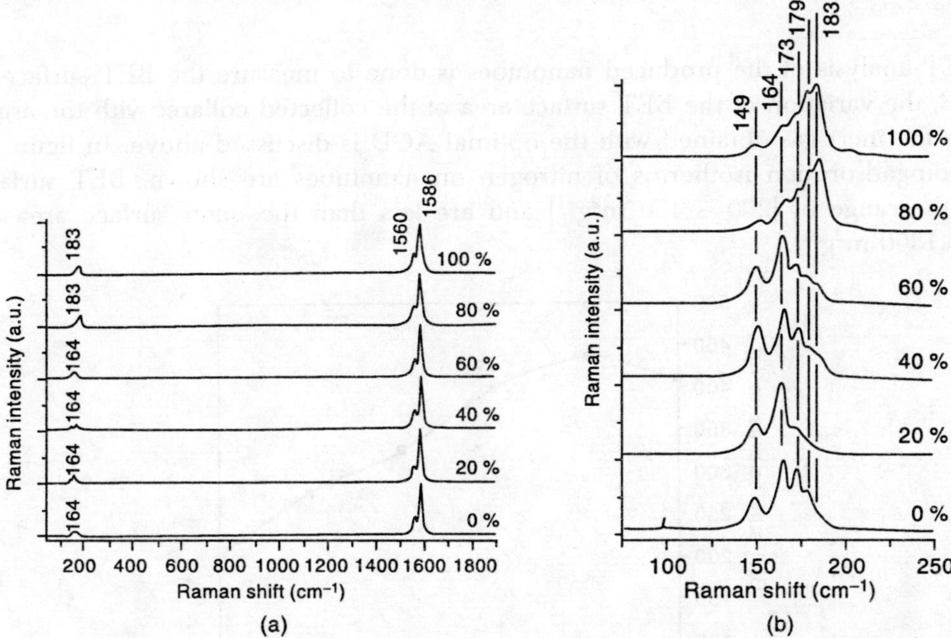

Fig. 6.11 *(a) Overall Raman spectrum and (b) Radial breathing mode of the SWNT produced with various concentration of Ar in the inert atmosphere. The percentage of Ar is indicated on the rigth side.*

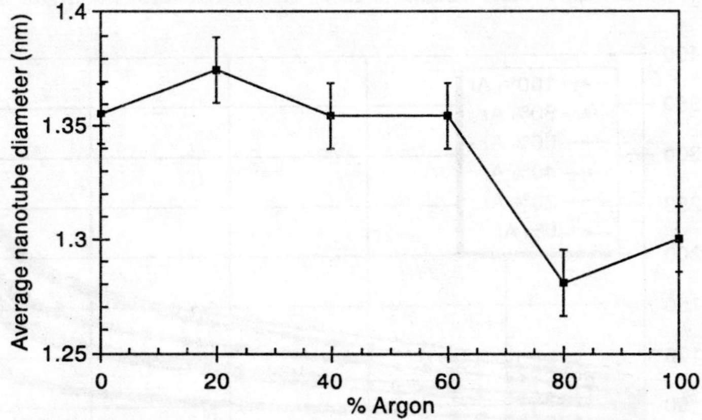

Fig. 6.12 *Variation of the average nanotube diameter as determined by Raman spectroscopy with the argon percentage in the arc chamber obtained with 1 = 100 A and optimal ACD giving vortices around the cathode.*

sampleis done; assuming the diameter distribution and average diameter depend on the inert atmosphere condition. This dependence is given in the figure 6.12. Varying the gas from helium to argon changes average diameter to smaller values.

Nanotube diameters range from ~ 0.04 to > 3 mm for SWNTs and from ~ 1.4 to atleast 100 nm for MWNTs.

BET Surface Area

The BET analysis of the produced nanotubes is done to measure the BET surface area. In figure 6.13, the variation of the BET surface area of the collected collaret with the argon mole fraction in the inert gas obtained with the optimal ACD is discussed above. In figure 6.14 the correspondingadsorption isotherms of nitrogen on nanotubes are shown. BET surface areas varies in the range of [200 – 450 m^2g^{-1}] and are less than the outer surface area of single nanotube (1300 m^2g^{-1}).

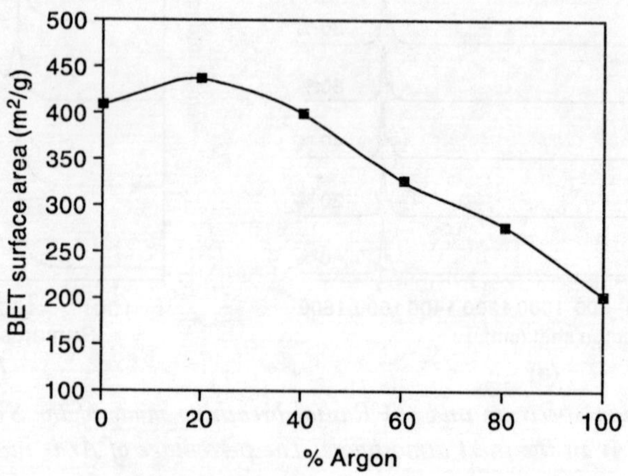

Fig. 6.13 *Variation of the BET surface area of the collaret with the argon mole fraction*

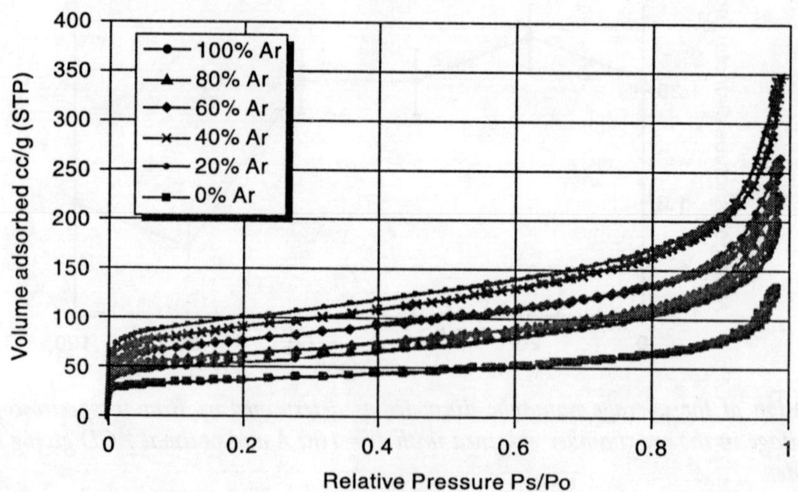

Fig. 6.14 *Adsorption isotherms of nitrogen at 77 K on as prepared nanotubes at different argon mole fractions.*

Thus, the measured BET surface area is attributed to the outer surface of a nanotubes organized in *bundles, or "ropes" of SWCNTs.* Hence, the parameter controlling BET surface area must be the rope diameter. But the relationship between individual nanotube diameter and rope diameter is not yet fully understood; Hence the macroscopic surface area of nanotubes is controlled by the argon-helium mixture gases. The maximal value of BET surface area, theoretically, never exceeds 350 m^2g^{-1}, Further High Resolution Transmission Electron Microscopy (HRTEM) observations of the samples are needed, to determine the number of individual nanotubes per rope. Here, the maximal value of BET surface area is found with mixture of 80% helium and 20% argon, and it decreases with increasing the argon percentage. In the most favorable case the BET surface is 440 m^2g^{-1}, which is bigger than the reported values (285 m^2g^{-1}).

Catalyst

By using the catalyst which is used for chemical vapour deposition (CVD) method, the same catalyst on using in arc discharge method yield the same result in nanotubes. But there is a correlation of diameter of SWNTs synthesized by CVD and arc discharge.

The diameter can be lowered to a range of 0.6–1.2 nm with arc-discharge by using a mixture of Co and Mo in high concentrations as catalyst. These diameters are considerably smaller than 1.2–1.4 nm, which is the usual size obtained from arc discharge.

Improvement of Oxidation Resistance

There is also progress in developing methods to improve the oxidation resistance of the SWNTs, which is a consequence of the defects present in nanotubes. A strong oxidation resistance is needed for the nanotubes if they have to be used for applications such as field emission displays. A modified arc-discharge method using a bowl-like cathode (see Figure 6.15), decreases the defects and gives cleaner nanotubes, and thus improves the oxidation resistance.

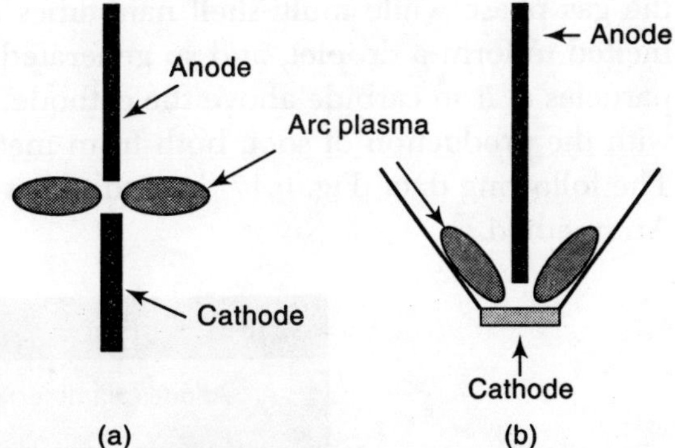

Fig. 6.15 **Schematic drawings of the electrode set-ups for (a) the conventional and (b) the new arc discharge electrodes.**

The Raman spectrum of the newly synthesised nanotubes show that the nanotubes formed are cleaner and less defective compared with those synthesised by conventional methods. The anode rod contained Ni and Y catalyst (C fNi/Y is 94.8:4.2:1).

Open Air Synthesis with Welding arc Torch

Only a couple of years ago, researchers discovered that it was possible to form MWNTs in open air. A welding arc torch was operated in open air and the process was shielded with an argon gas flow. The anode and cathode were made of graphite containing Ni and Y (Ni/Y is 4.2:1 at. %).

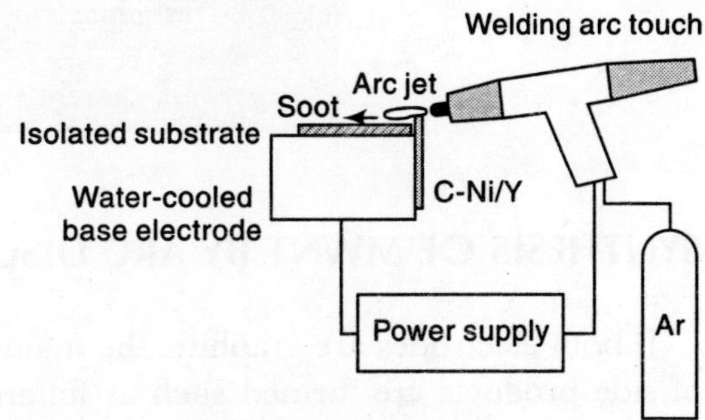

Fig. 6.16 **Experimental set-up of the torch arc method in open air.**

This method was modified for preparing SWNTS. A plate target made of graphite containing metal catalyst Ni and Y (Ni/Y is 3.6:0.8 at per cent), is fixed at the sidewall of a water-cooled, steel based electrode. The torch arc is aimed at the edge of the target and the soot is deposited on the substrate behind the target (see Figure 6.17). The arc is operated at a direct current of 100 A. Argon gas flows through the torch, enhancing the arc jet formation beyond the target.

In the soot, carbon nanohorns (CNHs) and bundles of SWNT with an average diameter of 1.32 nm are found. However, the yield is much lower than for the conventional low-pressure arc discharge method. There are two reasons for this fact. At first, because of the open air, the lighter soot will escape into the atmosphere. Secondly, the carbon vapour might be oxidised and emitted as carbon dioxide gas. In order to improve the yield in this method, contrivances for collecting soot and development of an appropriate target are required.

This method promises to be convenient and inexpensive once the conditions for higher yield are optimised. With a Ni/Y catalyst (Ni/Y is 3.6:0.8), SWNT bundles and CNHs are formed. In this case the SWNTs have a diameter of approximately 1.32 nm.

Single-shell Carbon Nanotubes of 1-nm Diameter

The synthesis of abundant single-shell tubes with diameters of about one nanometer. grow in the gas phase while multi-shell nanotubes are formed on the carbon cathode. The iron filings melted to form a droplet, and so generated iron vapor which cooled and condensed into small particles of iron carbide above the cathode. The vaporization of iron takes place simultaneously with the production of soot, both from methane and by evaporation form the carbon cathode. The following data (Fig. 6.17) is used to get 1 nm diameter of the single shell CNTs by Carbon-Arc method.

	Specification
Anode (diameter)	10 mm
Cathode (diameter) with a shallow dimple holding catalyst	20 mm
Catalyst	Fe, Ni or Co
Gas (pressure)	Methane (10 torr) + Ar (40 torr) or He (600 torr)
DC Current	200 A @ 20 V

Fig. 6.17

SYNTHESIS OF MWNT BY ARC DISCHARGE METHOD

If both electrodes are graphite, the main product will be MWNTs. But next to MWNTs, a lot of side products are formed such as fullerenes, amorphous carbon, and some graphite sheets. Purifying the MWNTs, means loss of structure and disorders of the walls. However scientist are developing ways to gain pure MWNTs in a large-scale process without purification.

Typical sizes for MWNTs are an inner diameter of 1-3 nm and an outer diameter of approximately 10 nm. Because no catalyst is involved in this process, there is no need for a heavy acidic purification step. MWNT, can be synthesised with a low amount of defects.

Synthesis in Liquid Nitrogen

A first, possibly economical route to highly crystalline MWNTs is the arc-discharge method in liquid nitrogen[14], with this route mass production is also possible. For this option low pressures and expensive inert gasses are not needed. The content of MWNTs can be as high as 70% of the reaction product The schematic is shown in Figure 6.18(a); (b).

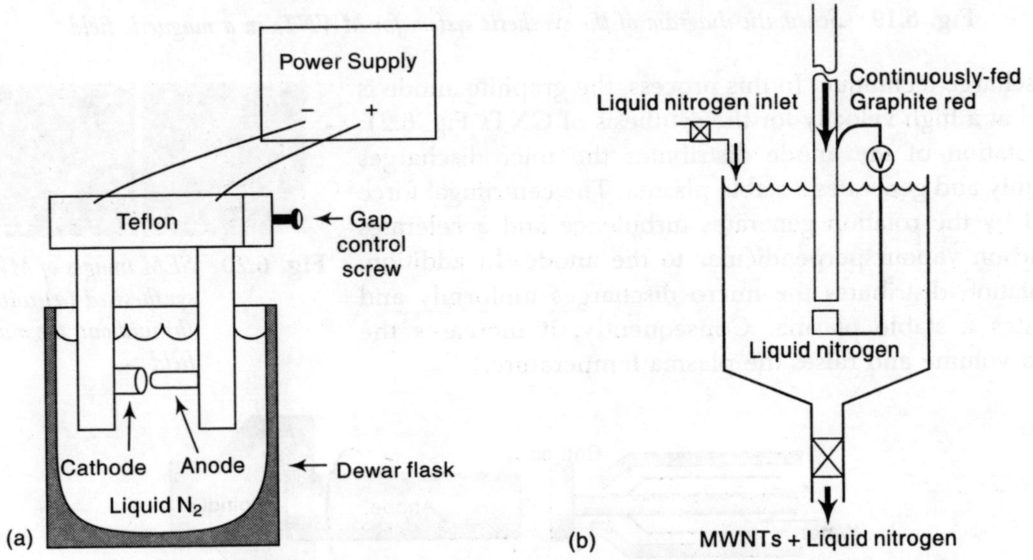

Fig. 6.18 *Schematic drawings of the arc discharge apparatus in liquid nitrogen*

Analysis with Auger-spectroscopy revealed that no nitrogen was incorporated in the MWNTs. There is a strong possibility that SWNTs can also be produced with the same apparatus under different conditions.

Magnetic Field Synthesis

Synthesis of MWNTs in a magnetic field (Fig. 6.19 and 6.20) gives defect-free and high purity MWNTs that can be applied as nanosized electric wires for device fabrication. In this case, the arc discharge synthesis is controlled by a magnetic field around the arc plasma.

Extremely pure graphite rods (purity > 99.999 %) are used as electrodes. Highly pure MWNTs (purity > 95 %) fire obtained without further purification, which disorders walls of MWNTs.

Plasma Rotating arc Discharge

A second possibly way of mass mass production of MWNTs is synthesis by plasma rotating

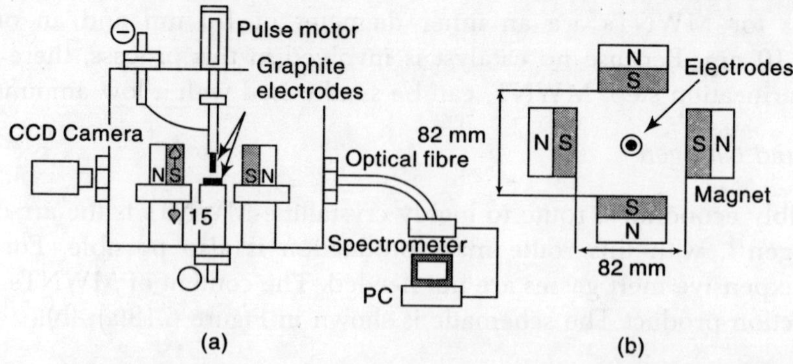

Fig. 6.19 *Schematic diagram of the synthesis system for MWNTs in a magnetic field*

arc discharge technique. In this process, the graphite anode is rotated at a high velocity for the synthesis of CNTs Fig. 6.21. The rotation of the anode distributes the microdischarges uniformly and generates a stable plasma. The centrifugal force caused by the rotation generates turbulence and accelerates the carbon vapour perpendicular to the anode. In addition, the rotation distributes the micro discharges uniformly and generates a stable plasma. Consequently, it increases the plasma volume and raises the plasma temperature.

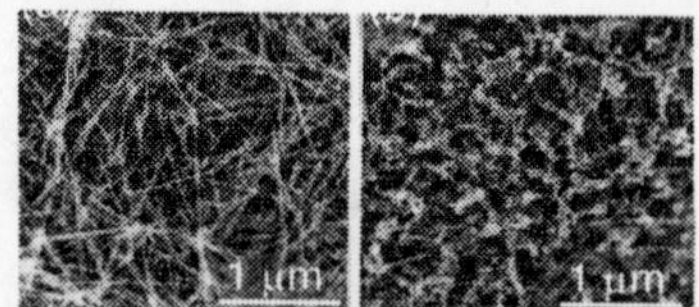

Fig. 6.20 *SEM images of MWNTs synthesised (a) with and (b) without, the magnetic field*

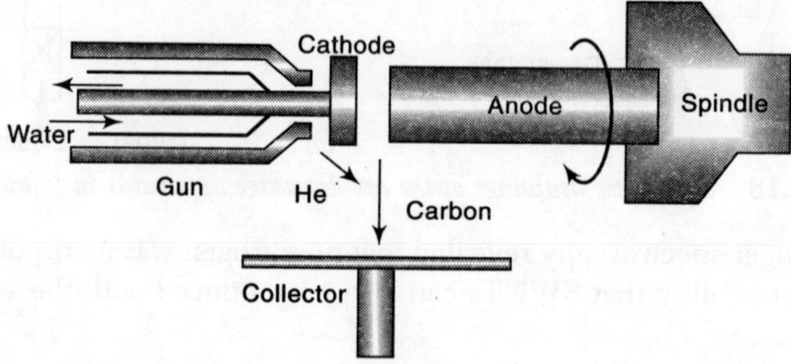

Fig. 6.21 *Schematic diagram of plasma rotating electrode system*

At a rotation speed of 5000 rpm a yield of 60% is found at a formation temperature of 1298 K without the use of a catalyst. The yield increases up to 90% after purification if the rotation speed is increased and the temperature is enlarged to 1150 dc.

(B) LASER ABLATION METHOD

Single-wall carbon nanotubes were first discovered in the soot of an arc process used for making fullerenes; arc production was improved with the use of suitable catalysts and gas media; but there has been a longstanding problem of making very pure SWNTs. In the meantime, the pulsed laser-ablation process for production of single-wall carbon nanotubes was developed.

The standard arc evaporation method produces only multi-walled tuubes. In the case of pure graphite electrodes, uniform SWNTs could be synthesized if a mixture of graphite with Co, Ni, Fe, or Y was used instead of pure graphite. It was found that the addition of metals such as cobalt to the graphite electrodes resulted in an extremely fine tube with single-layer walls. An alternative method of preparing single-walled nanotubes is by laser vaporization.

In 1995, Smalley's group are Rice University reported the synthesis of carbon nanotubes by laser vaporization. Fig. 6.22, is the laser vaporization apparatus by Smalley's group. A laser is used to vaporize a graphite target in an oven at 1473 K. Two Nd: YAG (Neodyaium: Yettrium–Aluminium–Garnet) pulsed lasers which impinge on a composite graphite and metal catalyst target. The lasers operate at 60 Hz pulse frequency. One laser operates at 1064 nm (IR) and the other at 532 nm (green). The current configuration uses pulse energies of about 0.3 Joules each with 10 nsec pulse widths. The lasers are controlled by a digital delay generator such that the time separation between the IR and green laser pulses is adjustable. For normal operation, the IR laser pulses follows the green pulse by 50 seconds. Two remotely controlled laser safety shutters are used on each beam to block them when requested. The beams then pass through beam expanding telescopes. The IR laser beam is then routed by reflection in a two 45° mirrors (coated for 1064 nm) after which it is transmitted through a 45° mirror used to route the green beam (coated for 532 nm). The two beams are adjusted to travel to uncoated right angle prism and impinge on the target as over lapping beams.

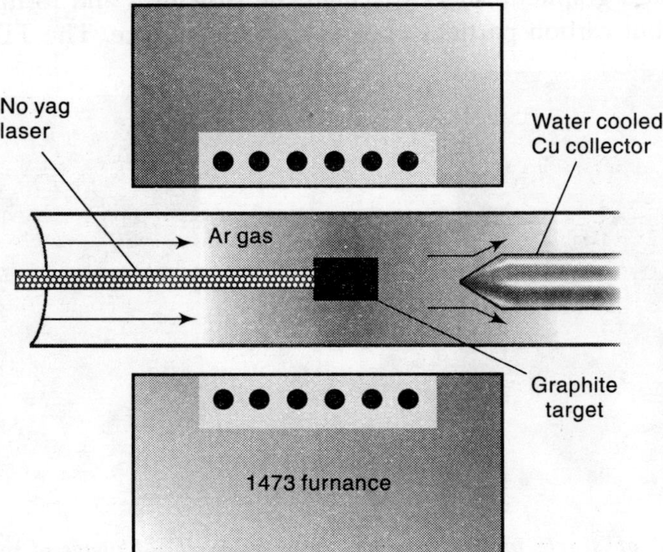

Fig. 6.22 *Schematic diagram of laser vaporization apparatus for the synthesis of multiwalled nanotubes*

Then helium or organ gas is filled to keep the pressure in the over at 67 Kpa. The laser pulses 50 ns apart maintain the growth conditions over a large volume and for a ling tie to provide more uiform vaporization of the carbon targets. Argon flows slowely (about 3 mm/s) down a 25 mm inner tube. A 12 mm diameter graphite target containing 1 at. % each of C and Ni is located at the end of the tube. The tube and the target are enclosed in 56 mm tube, all of which are placed in a tube furnace (oven).

A very hot vapour plume forms, then expands and cools rapidly. As the vaporised species cool, *small carbon molecules and atoms quickly condense to form* larger clusters possibly including fullerenes The catalysts also begin to condense, but more slowly at first, and attach to carbon clusters and prevent their closing into cage structures. Catalysts may even open cage structures when they attach to them. From these initial clusters, tubular molecules grow into single-wall carbon nanotubes until the catalyst particles become too large, or until conditions have cooled sufficiently that carbon no longer can diffuse through or over the surface of the catalyst particles. The collected material contains 40–70% nanotubes by weight. The rest is metal particles and graphite content.

It is also possible that the particles become that much coated with a carbon layer that they cannot absorb more and the nanotube stops growing. The SWNTs formed in this case are bundled together by van der Waals forces. Finally, flowing argon gas sweeps the nanotubes from the furnace to a water cooked collector outside the furnace.

The condensates obtained this way are mixed with carbon nanotubes and nanopartides. MWNT would be synthesized in the case of a pure graphite, but uniform SWNT could be synthesized if a graphite of a mixture of Co. Ni, Fe. Y were used instead of a pure graphite. SWNTs synthesized this way exist as 'ropes'. Fig. 6.23. and 6.24. show a SEM and a TEM image synthesized by Smalley's group using laser–vaporization. It is found from the SEM image that carbon nanotubes had curved shapes and carbon particles were stuck on the surface of the nanotubes. Fig. 15 shows that individual SWNTs with a uniform diameter of several nm are clustered evenly at a graphene sheet spacing. The structures and forms of graphene sheets are stable and dean, but carbon partides are stuck on the surface. The TEM image showing a

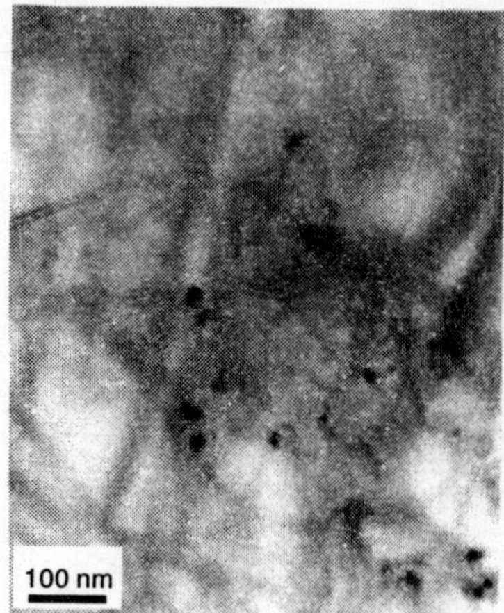

Fig. 6.23 *SEM image of single-walled carbon nanotubes synthesized by laser-vaporization method*

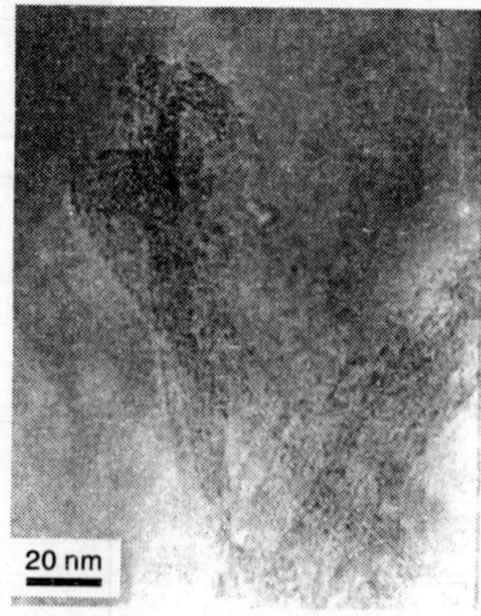

Fig. 6.24 *TEM image of rope for the single-walled nanotubes synthesized by laser-vaporization method*

cross-section of a carbon nanotube reveals that several SWNTs are clustered and typical 'ropes' shapes as said before. Laser vaporization are higher in yield than arc-discharge and can synthesize high quality SWNTs.

The purest SWNTs are produced by pulsed laser vaporization (PLV) of carbon containing metal catalysts, yield depends on the amount and type of catalysts, laser power and wavelength, temperature, pressure and type of buffer gad, and geometry of the fluid flow near the carbon target.

Nanotubes produced by laser ablation are purer (up to about 90% pure) than those produced in the are process.

There are some striking, but not exact similarities, in the comparison of the spectral emission of excited species in laser ablation of a composite graphite target with that of laser-irradiated C_{60} vapour. This suggests that fullerenes are also produced by laser ablation of catalyst-filled graphite, as is the case when no catalysts are included in the target. But, subsequent laser pulses excite fullerenes to emit C_2 that adsorbs on catalyst particles and feeds SWNT growth.

Laser ablation is almost similar to arc discharge, since the optimum background gas and catalyst mix is the same as in the arc discharge process. This might be due to very similar reaction conditions needed, and the reactions probably occur with the same mechanism. Thus, from the mechanisms of growth of single-wall carbon nanotube. (SWNT) formation in the pulsed laser-ablation process. It is concluded that. Laser pulses heat the target surface containing graphite and finely divided nickel and cobalt and vaporize them. A very hot vapor plume is formed that expands and cools rapidly. As the vaporized species cool, small carbon molecules and atoms quickly condense to form larger clusters, including possibly fullerenes. The catalysts also begin to condense, but more slowly at first, and attach to carbon clusters and prevent their closing into cage structures. Catalysts may also open cage structures when they attach to them. From initial clusters, tubular molecules grow into single-wall carbon nanotubes until the catalyst particles become too large, or until conditions have cooled sufficiently that carbon no longer can diffuse through or on the catalyst particles. It is also possible that the particles become so coated with a carbon layer that they no longer can absorb more and nanotube growth ceases.

By comparing the spectral emission of excited species in laser ablation of a composite graphite target with that of laser irradiated C_{60} vapor, women striking, but not exact similarities are there; that fullerenes are produced by laser ablation of catalyst-filled graphite, as is the case when no catalysts are included in the target. However, subsequent laser pulses excite fullerenes to emit C_2 that adsorbs on catalyst particles and feeds SWNT growth because the time scale for decay of C_2 Swan-band emission differs in the C_{60}-excitation experiments, compared with the present laser-ablation measurements. Repeated laser pulses vaporize some of the confined amorphous/graphitic carbon and possibly catalyst particles. Laser pulses excite any fullerenes which then decompose, leading to additional feedstock for SWNT growth. Because improved production and purity are observed when a 25-mm-diameter 'inner' tube is placed upstream of the target, we believe that the tube confines ablation products to the neighborhood of the laser beam. Without the confining inner tube. ablation products, consisting of graphitic carbon clusters and catalyst particles. escape from the influence of the laser pulse and its heating.

It may be possible to improve production and purity by reversing the flow of argon, thus maintaining the ablation products exposed to the laser beams longer and vaporizing more of the amorphous carbon and large catalyst particles. This would allow them to reform as nanotubes and reduce impurities. Smaller diameter inner flow tubes that would confine the ablation products totally within the laser beams may also improve the product.

Observations in the Experiment

Following observations are made during the experiment for Laser ablation method.

1. The Visible plume produced by the laser pulses is about 3 to 6 m m long and 6 mm in diameter. The plume temperature is estimated to be 3000–4000 K based on C_2 Swan-band spectra. As the plume expands, the temperature drops and its emission decreases significantly. This plume of ablation products expands away from the target and eventually forms a ring vortex that propagates upstream.
2. The transient intensity measurements of the plume, at several distances from the target, show that luminous ablation products propagate relatively slowly upstream.
3. By taking camera images on the flow in a single-pulse mode it is found that a ring vortex propogates very slowly (on order of milliseconds) far upstream from the target with continual pulses at 10 or 60 Hz, a quasi-steady-state condition develops in which ablation products containing vapors and condensed suspended particles propagate through the flowing argon and mix by diffusion and convestion eddies. The upstream of the target seem to have eddy-like motion seep in to the glow of laser-induced luminescene. (see Fig. 6.25).

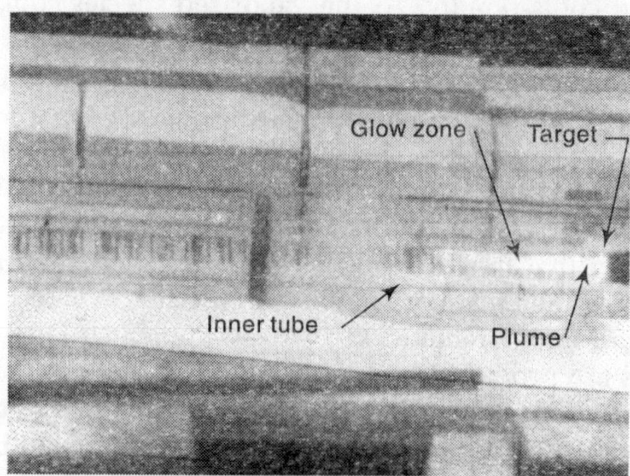

Fig. 6.25 ***Photograph of open laser-ablation oven showing laser plume and laser-induced fluroescene in the inner tube upstream of the target***

Note: (A notch filter is used to eliminate laser radiation from the exposure. It shows the target, the plume and the glow zone. The diameter of the glow zone is about the diameter of the laser beams).

4. Emission spectra taken in the bright plume zone at times as early as 1 μs are mainly due to C_2 Swan-band radiation. Somwwhat later, but still very early times (< 2 μs), C_3 near 400 nm. At much later time (greater than 50 μs), only blackbody emission. The C_2 Swan bands, persist for at least 50 μs, although the radiative lifetime of the upper state is only about 100 ns. This implies that the formation of excited C_2 must continue, for at least 50 μs after the laser pulse.

5. Exchange reactions could lead to formation of atomic carbon, which could recombine to form excited C_2, but these reactions are very fast under these conditions, small carbon molecules and fullerenes indicates that the principal production of C_2 may be form C_{60} photo-dissociation.
6. Trsnsient emissions from the plume zone (1.5 mm from the target) are obtained with a 10-nm bandpass filter centered at 510 nm (the $\Delta v = 0$ band of the C_2 Swan system).

   That the characteristic decay times are 1.2 μs and 95 μs, as shown in Table 6.1.

Table 6.1 Characteristic decay times of emission for C_{60} excitation and for laser ablation for SWNT production. Unless otherwise noted the wavelength range is limited by a 510±5-nm pass filter

Source conditions	*Time period, μs*	*Decay time, μs*	*Apparent radiating species*
Laser excitation of C_{60}	Very early 0.4–0.82	0.4 (no filter)	C_2 Swan bands and continum
Laser excitation of C_{60}	Early 0.7–2.6	0.3	C_2 Swan bands
Laser excitation of C_{60}	Late 4–14	4.6	C_2 Swan bands
Laser ablation for SWNT production, 12 mm from target	Very early 0.5–0.10	0.4 (no filter)	C_2 Swan bands and continuum
Laser ablation for SWNT production, 12 mm from target	Early 1 – 4.5	1.2	C_2 Swan bands
Laser ablation for SWNT production, 12 mm trom target	Mid 5–50	95	C_2 Swan bands, C_3 and continuum
Laser ablation for SWNT production, 12 mm from target	Late	Irregular	Blackbody continuum

Up to roughly 1 μs after the laser pulses, the spectra in the glow zone consist of the C_2 Swan system plus a little black-body or other continuum. Tills period is compared with the later period in which the C_2 Swan bands are absent, leaving only blackbody radiation that persists for many hundreds of microseconds. At a time as late as 100 μs the temperature of this blackbody radiation is about 2300 K, as determined from curve fits of the spectral intensity.

7. The quality of production of carbon nanotubes decreases when the inner quartz tube is replaced with a larger diameter tube or when the inner tube is removed. It appears that the inner quartz tube that is placed just in front of the target is important for production of good-quality SWNTs. In SEM images, shown in Fig. 4, raw product made with the inner tube is purer than that obtained without the inner tube.
8. Effect of background on nanotube production: It is found that argon and nitrogen gave similar results, but helium was found to yield poorer results. This suggests that thermal conductivity and, possibly, the speed of wave expansion plays a role in nanotube formation. Since these effects are dominant in the plume-expansion region and not in the area where the ablation products have cooled to the ambient oven temperature, we expect that

nanotube formation is dominant near the target and in the vicinity of laser-irradiated particles in the flow tube. The effect of the Oven is to maintain a base temperature such that there is greater target ablation, slower cooling of particles, and a higher vapor pressure of small carbon and catalyst clusters. Cooling of particles is mainly from evaporation at these conditions. Therefore, diffusion through the background gas would dominate the cooling process. Thus, particles would cool more slowly in a heavy molecular weight gas, such as argon or nitrogen.

SWNT VERSUS MWNT

The condensates obtained by laser ablation are contaminated with carbon nanotubes and carbon nanoparticles. In the case of pure graphite electrodes, MWNTs would be synthesized, but uniform SWNTs could be synthesised if a mixture of graphite with Co, Ni, Fe or Y was used instead of pure graphite. SWNTs synthesised this way exist as 'ropes', Laser vaporization using Co/Ni as catalyst, results in a higher yield for SWNT synthesis and the nanotubes have better properties and a narrower size distribution than SWNTs produced by arc-discharge. This is because Co/Ni catalyst prevents the ends of nanotubes from being "capped" during synthesis.

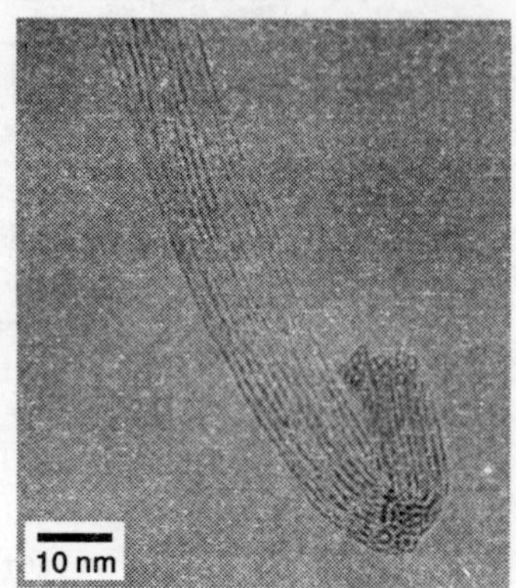

Fig. 6.26 *TEM images of a bundle of SWNTs catalysed by Ni/Y (2:0.5 at. %) mixture, produced with a continuous laser*

Nanotubes produced by laser ablation are purer (up to about 90 % purity) than those produced in the arc discharge process. The Ni/Y mixture catalyst (Ni/Y is 4.2/1) gave the best yield (Fig. 6.26).

The size of the SWNTs ranges from 1-2 nm, for example the Ni/Co catalyst with a pulsed laser at 1743 K gives SWNTs with a diameter of 1.3–1.4 nm[26] about 70–90% of the carbon target can be converted to SWNT. In case of a continuous laser at 1743 K and Ni/Y catalyst (Ni/Y is 2:0.5 at. %), SWNTs with an average diameter of 1.4 nm were formed with 20–30% yield. Primarily SWNTs, with a large diameter range can be controlled by varying the reaction temperature.

LARGE SCALE SYNTHESIS OF SWNT

The following two methods are used. By these methods. Scalping up is possible, but the technique is expensive due to the laser and the large amount of power required.

A—Ultra fast Pulses from a Free Electron Laser (FEL) Method

Usually the pulses in an Nd: Y AG system have width of approximately 10 ns, in this FEL system the pulse width is ~400 fs. The repetition rate of the pulse is enormously increased from

10 Hz to 75 MHz. To give the beam the same amount of energy as the pulse in an Nd: YAG system, the pulse has to be focused. The intensity of the laser bundle behind the lens reaches – 5×10^{11} W/cm^2, which is about 1000 times greater than in Nd: YAG systems.

A jet of preheated (1273 K) argon through a nozzle tip is situated close to the rotating graphite target, which contains the catalyst. The argon gas deflects the ablation plume approximately 90° away from the incident FEL beam direction, clearing away the carbon vapour from the region in front of the target. The produced SWNT soot, is collected in a cold finger. This process can be seen in Figure 6.27. The yield at this moment is 1,5 g/h, which is at 20 % of the maximum power of the not yet upgraded FEL. If the FEL is upgraded to full power and is working at 100 % power, a yield of 45 g/h could be reached since the yield was not limited by the laser power.

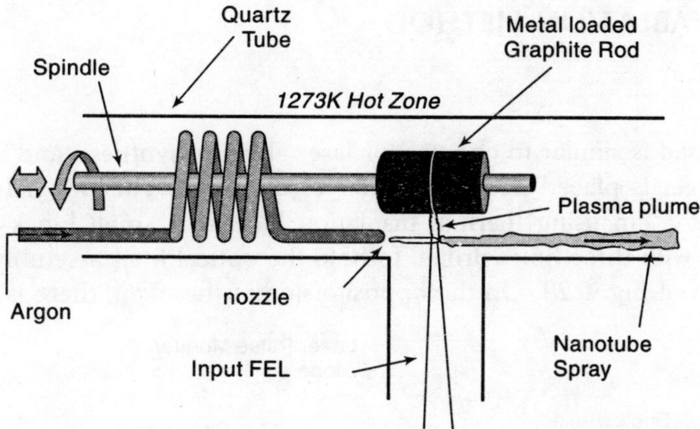

Fig. 6.27 *Schematic drawings of the ultra fast-pulsed laser ablation apparatus*

With this method the maximum achievable yield with the current lasers is 45 g/h, with a NiCo or NiY catalyst, in argon atmosphere at 1273 K and a wavelength of ~3000 nm. The SWNTs produced in bundles of 8-200 nm and a length of 5-20 microns has a diameter range 1-1.4 nm.

B—Continuous Wave Laser-powder Method

This is a highly productive laser-powder method of SWNT synthesis. It is based on the laser ablation of mixed graphite and metallic catalyst powders by a 2-kW continuous wave CO_2 laser in an argon or nitrogen stream. Due to the introduction of micron-size particle powders, thermal conductivity losses are significantly decreased compared with laser heating of the bulk solid targets in known laser techniques. As a result, more effective utilisation of the absorbed laser power for material evaporation is achieved. The set-up of the laser apparatus is shown in Figure 6.28(a).

The established yield of this technique is 5 g/h. A Ni/Co mixture (Ni/Co is 1:1) is used as catalyst, at a temperature 1373 K. In the soot a SWNT abundance of 20–40% is found with a mean diameter of 1.2–1.3 nm. An HRTEM-picture of this sample is shown in Figure 6.28(b).

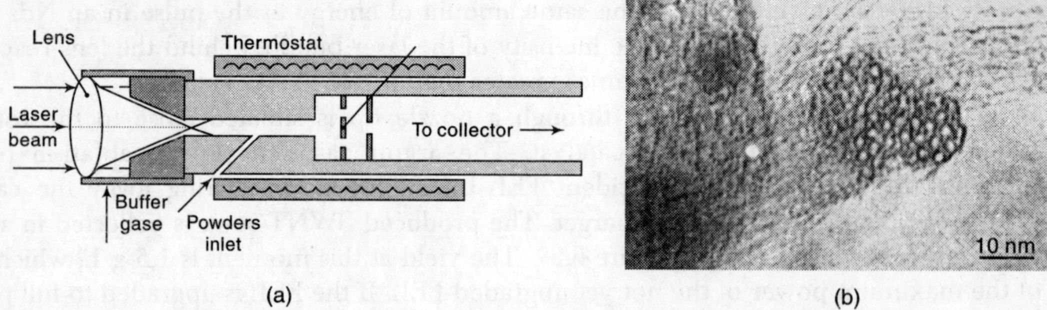

Fig. 6.28 *(Left) The principle scheme of the set-up for carbon SWNT production by continuous wave laser-powder method (Right) HRTEM of a SWNT-bundle cross-section*

OPTICAL LASER ABLATION METHOD

Experiment

The apparatus used is similar to one use for laser ablation synthesis and in similar conditions except that the target is placed near the center of the oven. The two halves of the oven are separated by about 3 cm using thermal insulation blocks. A small block is machined to the contour of the step with three holes drilled to hold the optical fiber assembly and is placed near the center of the oven Fig. 6.29. On the opposite side of the oven, there is a similar block and

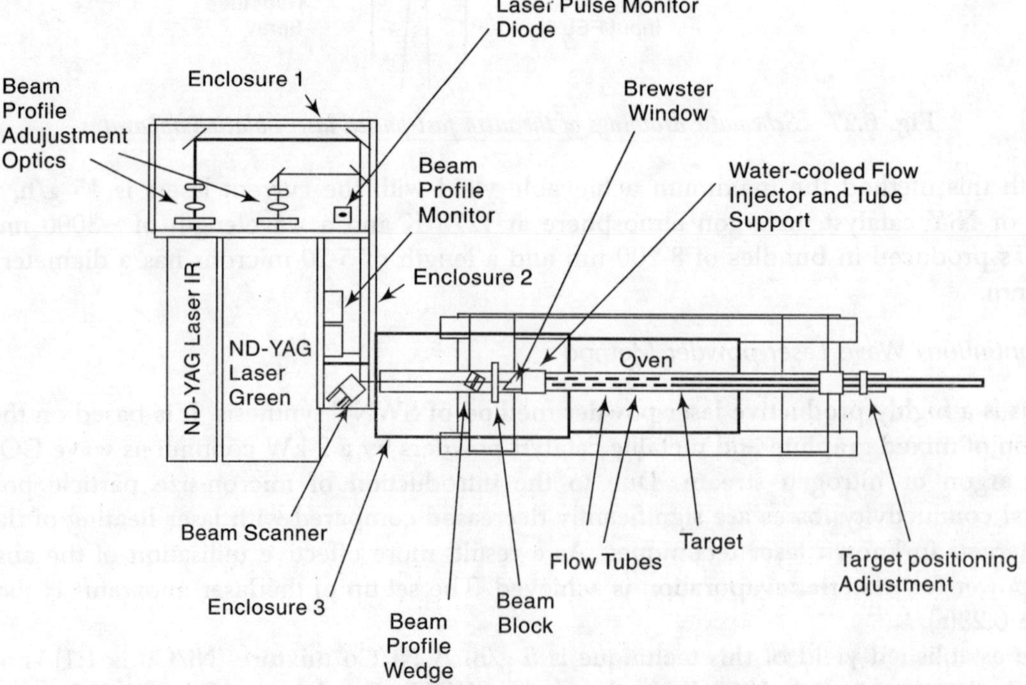

Fig. 6.29 *Layout of laser ablation process for carbon single wall nanotube production*

can be used for absorption measurements. The optical fiber assembly consists of a plano-convex lens (1.25 cm diameter and 2.5 cm focal length) which focuses the light from the central part of the ablation plane onto the end of a 800 micron quartz fiber. The other end of the fiber is connected to a fiber optic adapter attached to a spectrograqph (Fig. 6.30). Two detectors are used for recording the emission spectra: a gateable intensified CCD with 1024 × 256 pixels and a photomulplier tube. The. ICCD output is recorded using software. This allows background subtraction and adjustable gated detection (using a programmable pulse generator).

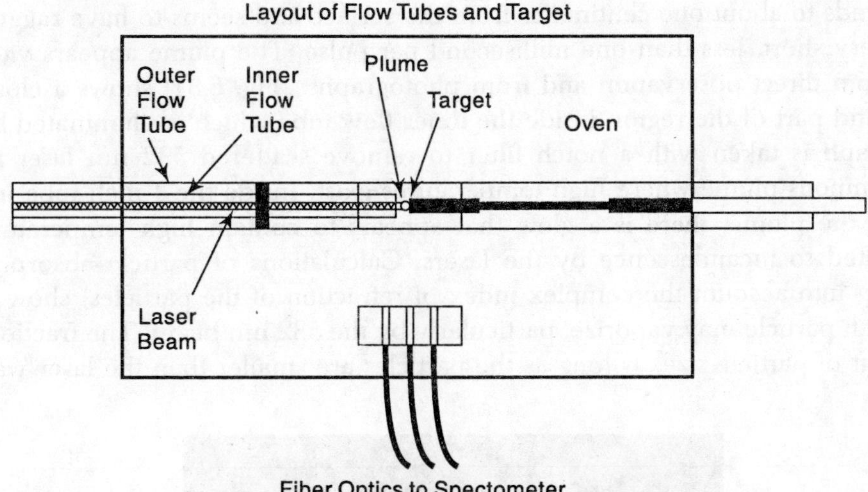

Fig. 6.30 *Configuration of flow tubes and target showing laser beams, plume and fiber optic light collectors*

Two background signals are obtained with the same number of scans (100) as in the measured spectra. One is obtained before and the other after the spectral measurements. These two background signals are averaged before subtracting them from the measured spectra. Most of the spectra are recorded with 25 micron slit widths using a 300 line/mm grating blazed at 500 nm. With a 300 line/mm grating, a coverage of about 350 mm in a single exposure of the ICCD is possible The resolution,taking into account the finite size of the intensifier channels, is about 1.9 on and the spacing between pixels is about 0.322 nm. All the spectra are intensity calibrated.

Wavelengths are calibrated using emission lines from low pressure mercury, neon, krypton and argon lamps.

Transient spectral data is collected by connecting the output of the R955 tube to a 32K channel transient digitizer which gives a minimum of 10 ns channel width or sampling time. The response time of the system can be as low as 3 ns. The time resolution for measurements is about 20 ns. All transient and steady state data are collected with stationary laser beams at the center of the target. Data collection is initiated 60 to 90 seconds after the start of the laser beams to allow for thermalization of the target surface.

Observations and Measurements

Three measurement techniques for determining properties of the laser ablation process for carbon nanotube production are

Photographic Observations

The first diagnostic technique, ***photography***, is used to obtain estimates of the size and location of the visible plume. The luminous plume is confined to a small region near the surface of the target. It extends to about one centimeter from the surface and seems to have ragged edges. Its duration is very short, less than one millisecond per pulse. The plume appears violet-white in color both from direct observation and from photographs. (Fig. 6.31) shows a close up of the laser plume and part of the region inside the inner flow tube which is illuminated by the laser. This photograph is taken with a notch filter to remove scattered 532 nm laser radiation. It shows the luminous plume where high temperatures exist. Inside the 1-inch tube in the region to the left of the plume, there is a glow that appears to contain high temperature particles, probably heated to incandescence by the lasers. Calculations of particle absorption of laser energy (taking into account the complex index of refraction of the particles) show that a large fraction of each particle may vaporize, particularly by the 532 nm beam. The fraction vaporized is independent of particle size, as long as the particles are smaller than the laser wavelength.

Fig. 6.31 *Photograph of reaction ozone inside 1-inch tube showing laser induced radiation. Photograph taken using notch filter to eliminate scattered laser light*

Emission Spectral Distributions

The second diagnostic technique is ***wavelength-resolved emission spectroscopy***. Spectra averaged over about 10 μS are obtained in the spectral range of about 320 to 630 nm (Fig. 6.32). These spectra are obtained at delay times of 0.4, 10, and 25 μs after the laser pulse. The predominant feature in these spectra is the C_2 Swan band emission system that results from the transition $d^3\Pi_g$-$a^3\Pi_u$. The lower $a^3\Pi_u$ state is not the ground state, but only 716 cm^{-1} above it. Fig. 6.32 shows spectra taken at various times. These spectra are normalized to the peak intensity at

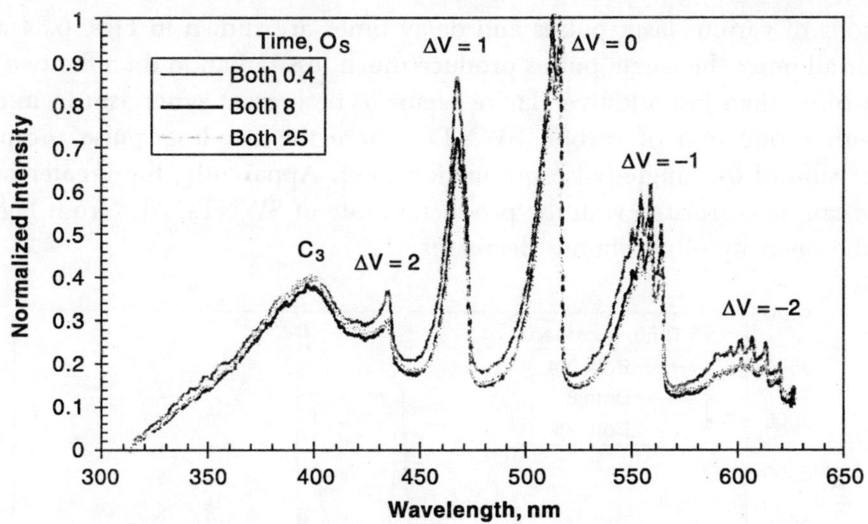

Fig. 6.32 *Normalized spectral distributions obtained in laser ablation plume at various times after laser pulses with identification of peaks associated with C_2 Swan bands and C_3*

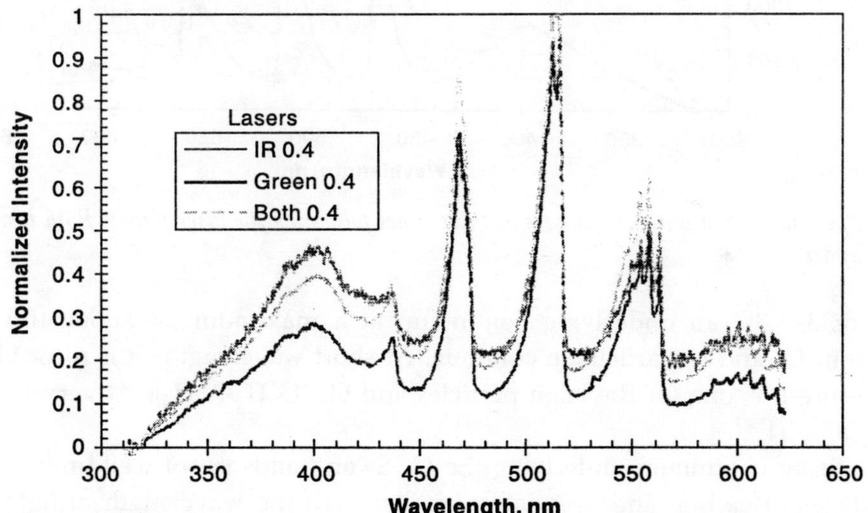

Fig. 6.33 *Normalized spectra taken 0.4 ms after laser pulses, with duration of 10 ms, showing C_2 Swan bands and underlying. Comparison is made for various laser pulses*

516 nm, which is the (v′, v″) = (0, 0) bandhead of the C_2 Swan band system. The peaks associated with various vibrational transitions, $\Delta v = v'-v''$, are indicated in Fig. 6.32, along with a tentative identification of C_3 emission. Fig. 6.33 shows similar spectra comparing that of single laser pulses, either "green" at 532 nm or "IR" at 1064 nm, with that produced by both the green and IR laser pulses in succession.

Comparisons of various laser pulses and delay times are shown in Figs. 6.34 and 6.35, it is found that for all times the single pulses produce much less radiation than the two *lasers together.* The effect is more than just additive. There seems to be a great synergism in intensity. This is consistent with production of carbon SWNTs, where the two-laser-pulse production rate is much greater sum of the single-pulse production rates. Apparently, the greater intensity of C_2 in the plume can be correlated with the production rate of SWNTs. Also from Fig. 6.35 as time progresses, the intensity of all plumes decreases.

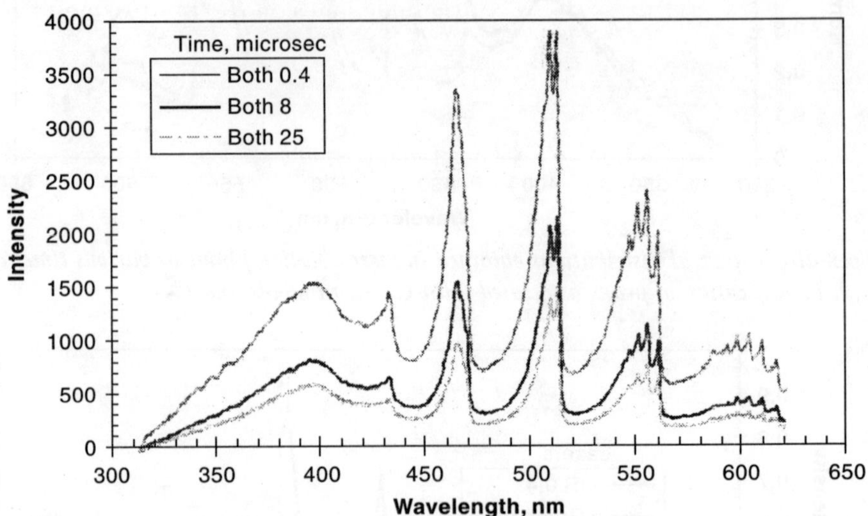

Fig. 6.34 *Spectrum taken with 10 μs gate at three times following the laser pulses. Both lasers are fixed 50 ns apart.*

In Figs 6.33–3.36, an underlying continuum at a maximum (at about 400 nm) is due to radiation from C_3 and to particulate continua. At short wavelengths it is most likely due to C_3. At low pressure it is due tot Rayleigh particles and C_2 $(C'\Pi_g\text{-}A^1\Pi_u)$. At very low pressure, it is due to C_3 $(\Sigma_g^+ - \Pi_u)$.

The significant continuum underlying the C_2 Swan bands is not well understood. It may be due to particles of carbon and/or catalyst smaller than the wavelength of light. However, the spectral distribution is not understood. From the black body and Rayleigh particle emission curves on calculating, there does not seem to be a temperature that would yield good agreement over the entire spectral range. As the laser ablates the target and forms a plasma plume, it is possible that electrons may be created that recombine to radiate free-bound continua. Calculations were made of $C^+\text{-}e^-$ continua for a range of temperatures. These calculations do not seem to support this hypothesis because spectral distribution is not consistent at low wavelengths. The measured spectra indicate a significant falloff at low wavelengths, with the intensity going to almost zero at about 310 nm. This falloff is not seen in the calculated spectra at any temperature.

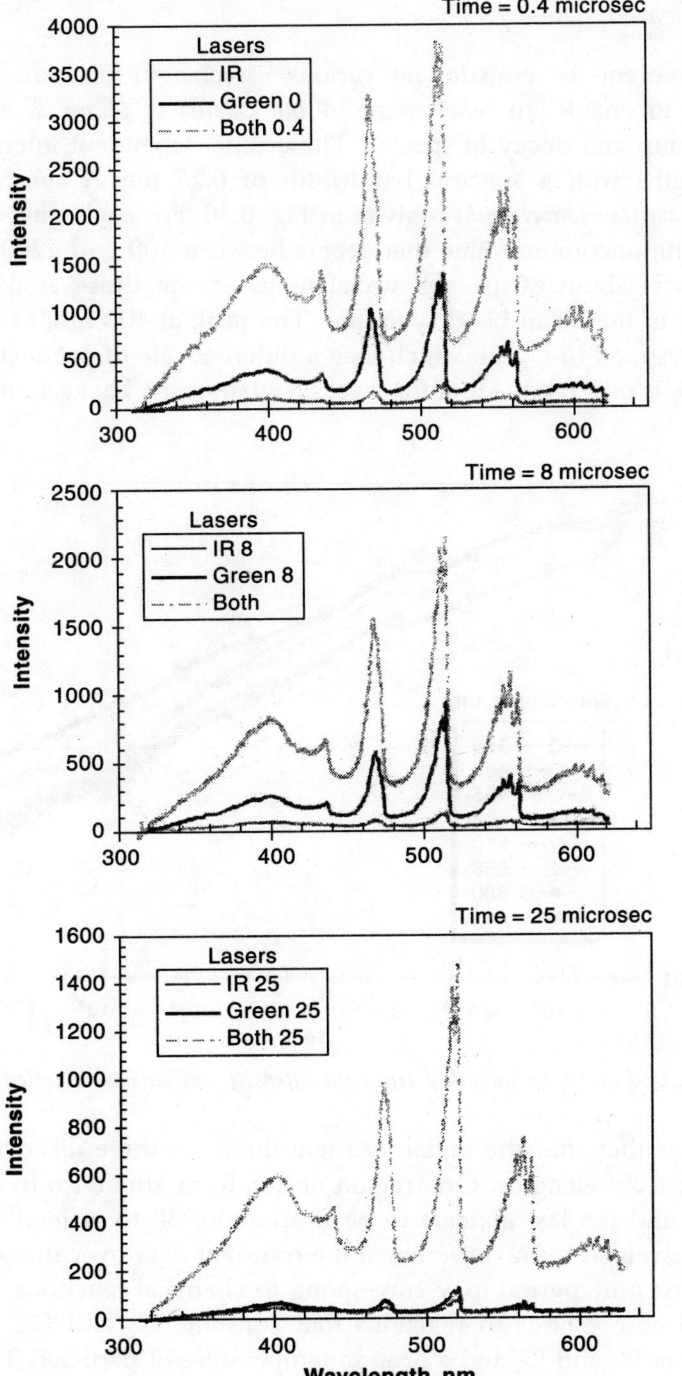

Fig. 6.35 *Un-normalized spectra taken 0.4, 8 and 25 μs after laser pulses for all laser pulses*

Transient Emission Measurements

Transient measurements of emission at various wavelength associated with peaks in the spectra were made to enable an assessment of *the kinetics of plume development* and to help understand the buildup and decay of species. These time dependent intensities were obtained at various wavelengths with a spectral bandwidth of 6.25 nm A summary of the *intensity measurement taken at various wavelengths* is given in Fig. 6.36. For each wavelength the intensities are normalized by the maximum value that occurs between 100 and 1200 ns. The duration of these measurements is about 80 μs. All wavelengths except those at 350 nm and 400 nm correspond to peaks in the Swan band spectrum. The peak at 400 nm, as well as the radiation at 350 nm may correspond to C_3, in. which case a different rate of production and decay of C_3 as compared with C_2 is observed. All of the traces contain some background from other species or particles.

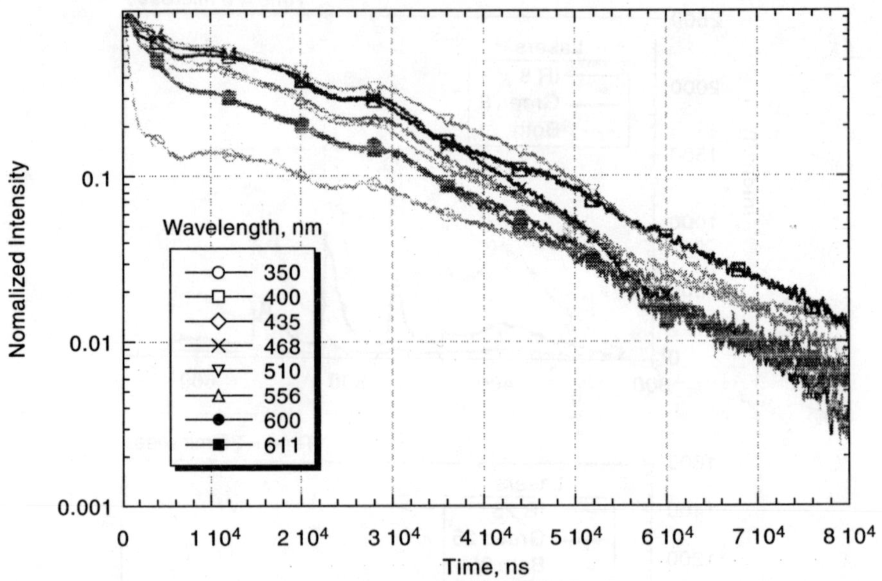

Fig. 6.36 *Normalized and then smoothed transient intensities at various wavelengths from 0 to 80 μs*

In Fig. 6.36 we see that after the initial transient there are three distinct regions of decay of the intensity for most wavelengths. One region occurs from about 1.6 to 6.0 μs, another from about 12 to 20 μS, and the last appears to be from about 30 to at least 80 μs. Characteristic decay times are determined from curve fits of the transient data over these times and are given m Table 6.2. The first time period may correspond to chemical reactions and to cooling of the plume. The second corresponds to reactions that consume C_2 and C_3. The last period is a combination of loss of C_2 and C_3 and a drop in temperature of particles. These decay times are much longer than the radiative lifetime of the Swan bands of 120 ns.

TABLE 6.2 Characteristic decay times in microseconds for three regions and eight wavelengths

Wavelength, nm	*1.6 to 6.0 μs* *Early Reaction Zone*	*12 to 20 μs* *Late Reaction and Cooling Zone*	*30 to 80 μs* *Particle Cooling Zone*
350	8.41	30.4	27.6
400	13.2	29.1	17
435	9.31	23.3	16.7
468	12.5	26.3	11.3
510	13.4	27.3	12.3
556	9.29	22.1	12.2
600	6.68	22.1	15.2
611	7.11	20.0	16.0

For the first 2.5 μs (see Fig. 6.37), 350 nm measurement shows high relative intensity spike very soon after the laser pulse. It is followed by a rise to a peak at 400 ns and then drops quickly to 20% of the peak by 1700 ns. This suggests that the radiation at 350 nm may be due to continuum from free-bound transitions early on, then C_3 or other species later.

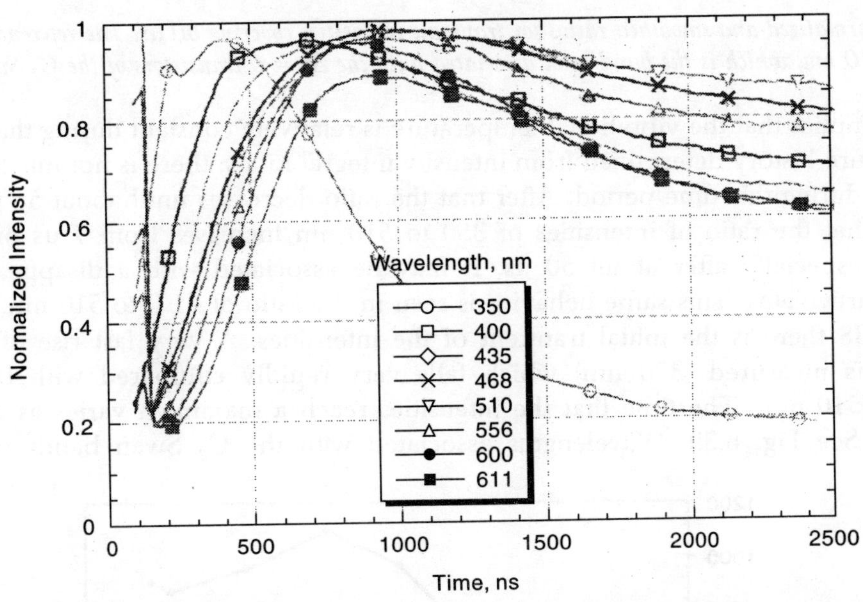

Fig. 6.37 *Normalized and then smoothed transient intensities at various wavelengths from 0.1 to 2.5 μs*

In Fig. 6.38 time dependent ratios of intensities taken at various wavelengths for the span of time up to 80 μs are there. Variations in decay rate (humps) of the intensities at various wavelength (Fig. 6.36) do not appear in the ratios (or are rather subdued). This implies that only one species is being observed and that an increase then a decrease of, C_2 (say) is not there. If the radiation are from different species this behavior would imply that they vary together in the same way; thus one depends on or is associated with the other. Also, since there are no humps in the ratio associated with various Δv bands of C_2 it is very likely that the temperature does not vary much. The ratio of 468 nm to 510 nm is almost constant from 4 μs until about

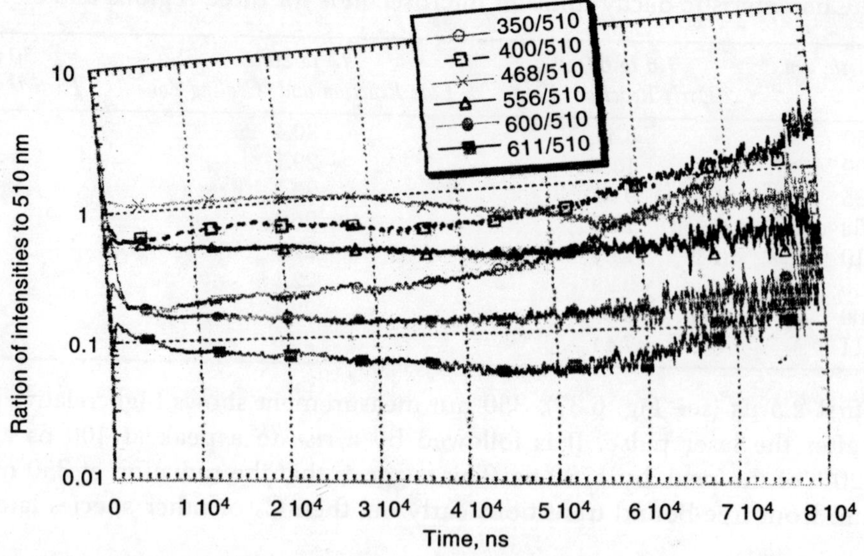

Fig. 6.38 *Normalized and smoothed ratios for transient intensities covering 80 μs. The reference wavelength is 510 nm, which is the band peak associated with the Δv = 0 transistors of the C_2 Swan band.*

20 μs. This implies that the vibrational temperature is relatively constant during that interval. In the temperature history determined from intensity integral ratios, there is not much variation in temperature during this time period. After that the ratio decreases until about 55 μs.

The fact that the ratio of intensities of 350 to 510 nm increases from 7 μs onward is not understood, especially after about 50 μs. It may be associated with a disappearance of C_2 relative to particulates. This same behavior is seen in the ratio of 61 0 to 510 nm.

In Fig. 6.38 there is the initial transient of the intensities. A very fast rise of the shortest wavelength is measured (350 nm) which falls very rapidly compared with the reference wavelength (510 nm). The time that the intensities reach a maximwn varies as a function of wavelength. See Fig. 6.39. Wavelengths associated with the C_2 Swan bands tend to peak

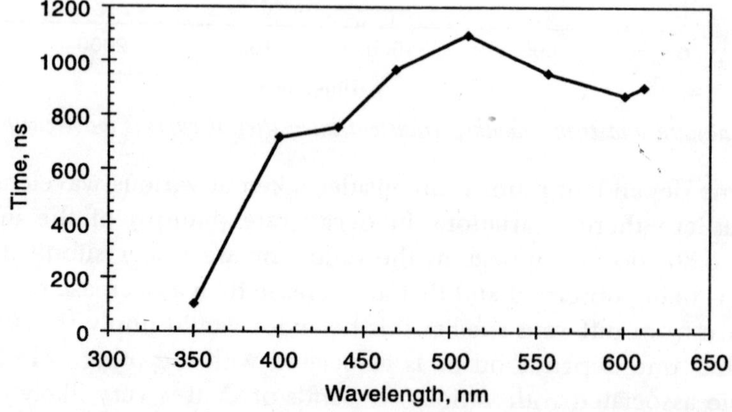

Fig. 6.39 *Time to peak intensity in the transient spectra of selected wavelengths*

around 800 to 1100 ns. At an estimated speed of sound of about 1500 ms^{-1} (based on a temperature of 5000 K) the distance the plume could travel would be about 1.5 mm. This is consistent with the location of the observation point from the target surface. The earlier time for the short wavelength (350 nm) may be due to a continuun excited by precursor (fast) electrons.

Estimates of vibrational temperatures are possible that show the relaxation of temperature with time. The ratio of vibrational peaks in the spectrum depends on the vibrational temperature. Theoretical spectra are calculated at various temperatures; and ratios of integrals over short intensity ranges are found as a function of temperature, in particular, the ratio of intensity integrals in the vicinity of the vibrational peaks (1,0) and (0,0) of the Cz Swan band ($d_3\Pi_g$-$a^3\Pi_u$). Integrated intensities over the wavelength ranges 465.04-475.87 nm and 509.87-519.07 nm, were determined, with an estimated underlying continuun subtracted. The ratios of measured intensities are then compared with integral ratios calculated as a function of temperature; and from that the temperatures are estimated. The first point covers a time interval from 0.4 μs to 10.4 μs. During that time there is a significant variation in the intensities. Therefore, the inferred temperature during that interval is an approximate average. The temperature histories are shown in Fig. 6.40. Note that the temperatures decrease slowly with time over the range covered, and that the temperature is highest with both laser pulses together. The temperatures for the green single pulse is higher than for the IR single pulse.

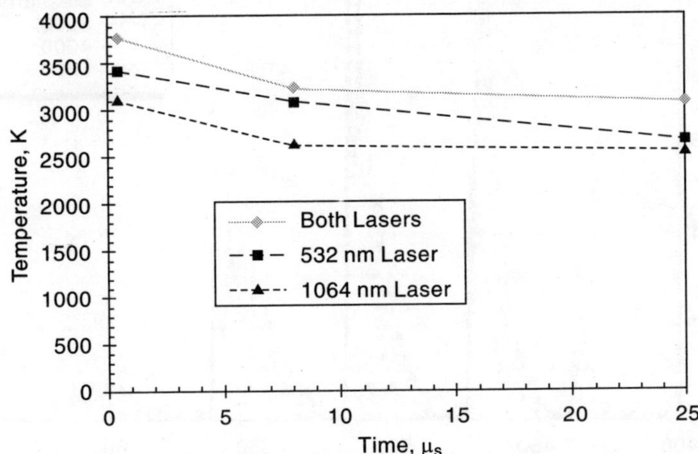

Fig. 6.40 *Temperatures in plume determined from C_2 Swan band intensity ratios of (1, 0)/(0, 0) bands*

The surface temperature of the target may be greater with two close pulses leading to additional ablation. Yudasaka, et al. made measurements of nanotube production in a similar laser ablation facility at various pressures from about 0.5 kPa to about 67 kPa. They found that higher pressures resulted in increased production of carbon single wall nanotubes. They proposed that this was due to greater evaporation of catalysts due to higher temperature surfaces at higher pressure. In the present experiments the two-laser pulse operation produces more SWNT; may be due to surface temperature. Also the C_2 Swan bands are much more intense; and therefore, the increased production to a greater amount of ablation of carbon is attributed.

Fig. 6.41 is a comparison of calculated spectra with measured spectra from which the strong continuum background is subtracted. Differences in the measured and calculated spectra at high vibrational/rotational energy levels toward the violet side of the band heads are taken. The continuum is estimated by choosing the points at the minima between peaks in the Swan bands. These points are then curve fit and interpolated values are subtracted from the measured spectra. It can be seen that the resultant spectrum measured at 400 ns after the laser pulses does not match calculated spectra over the entire range for any temperature. (The same disagreement was seen in the spectra for other delay times.) This indicates a significant amount of nonequilibrium in the vibrational states. This may be due to high level states being populated by recombination of carbon atoms (chemiluminescence) rather than by thermal excitation. Thus, there is a fair agreement at a temperature between 4000 K and 5000 K in the wavelength range from about 450 nm to 520 nm. This is at variance with the intensity ratio technique which showed that the temperature is closer to 3500 K. It is quite likely that the difference is due to dominance of the upper vibrational states by recombination that leads to a nonequilibrium vibrational population. This hypothesis is also supported by the nature of the distribution of intensities in the $\Delta v = -1$ system in the wavelength range 530 to 565 nm and in the $\Delta v = -2$ system in the wavelength range 580 to 620 nm.

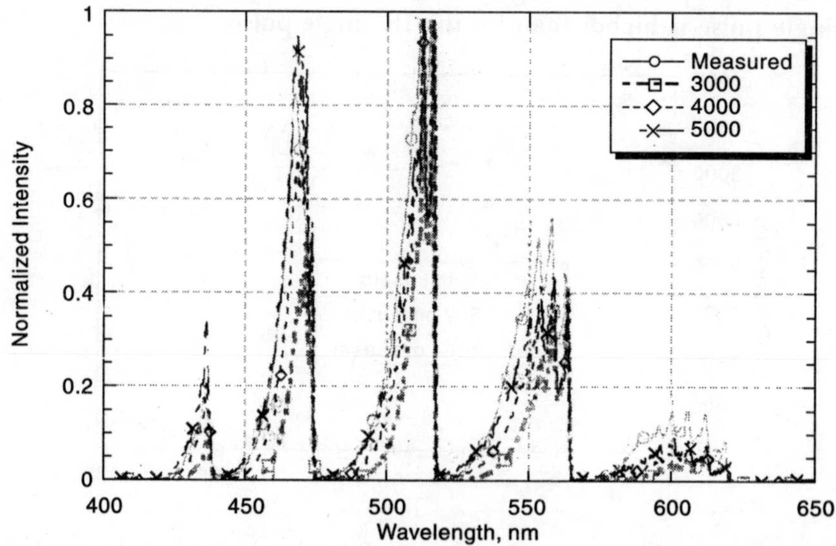

Fig. 6.41 *Measured spectrum with estimated continuum subtracted compared with calculated C_2 Swan bands at three temperatures. Measurement at 40 ns with duration of 10 μs with both laser pulses*

(C) CHEMICAL VAPOUR DEPOSITION

Chemical vapour deposition (CVD) synthesis is achieved by putting a carbon source in the gas phase and using an energy source, such as a plasma or a resistively heated coil, to transfer energy to a gaseous carbon molecule. Commonly used gaseous carbon sources include hydrocarbon gases, methane, carbon monoxide and acetylene. The energy source is used to "crack" the molecule into reactive atomic carbon. Then, the carbon diffuses towards the substrate, which is heated and coated with a catalyst (usually a first row transition metal such as Ni, Fe or Co) where it will bind the catalyst particles serve as seeds to nucleate the growth of nanotubes. Carbon nanotubes will be formed if the proper parameters are maintained. Excellent alignment, as well as positional control on nanometre scale, can be achieved by using CVD. Control over the diameter, as well as the growth rate of the nanotubes can also be maintained. The appropriate metal catalyst can preferentially grow single rather than multi-walled nanotubes Figs (6.42–6.43). Now, it is possible to fabricate high quality MWNTs and SWNTs directly onto substrates or in bulk as a raw material. A major advantage of CVD is that the CNTs can be used directly without further purification unless it is required to remove the catalyst particle.

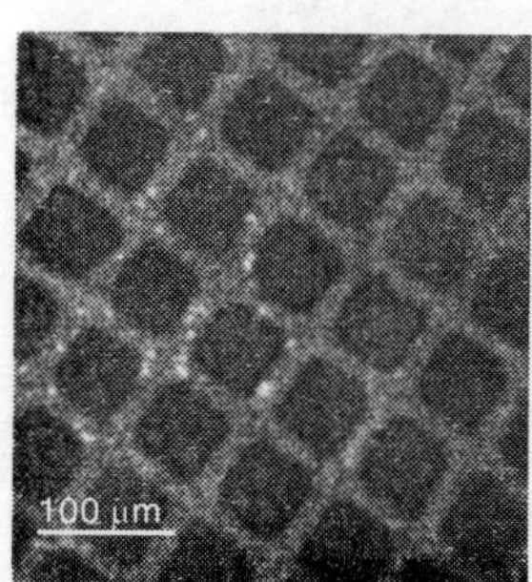

Fig. 6.42 *Nanotubes by CVD*

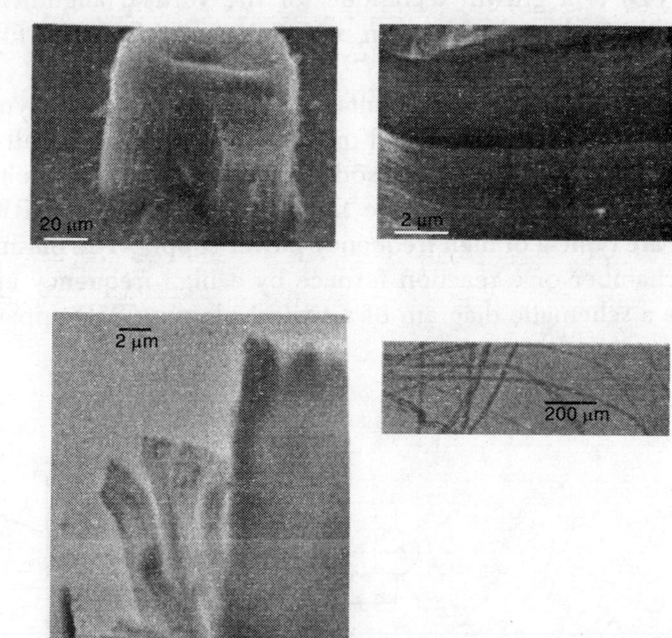

Fig. 6.43 *Multiwall carbon nanotubes grown by CVD*

The CVD methods can further be classified into:

(a) thermal CVD
(b) plasma enhanced CVD
(c) catalytic pyrolysis of hydrocarbon
  (I) vapor phase pyrolysis
  (II) liquid phase pyrolysis
(d) alcohol catalytic CVD
(e) high pressure disproportionation process
(f) aerosol-based pyrolysis
(g) nebulized spray
(h) solid state reaction

Basic Principles

CVD carbon nanotube synthesis is essentially a two-step process consisting of (1)catalyst preparation (2) synthesis of the nanotube.

The catalyst is generally prepared by sputtering a transition metal onto a substrate and then using either chemical etching or thermal annealing to induce catalyst particle nucleation. Thermal annealing results in cluster formation on the substrate, from which the nanotubes will grow. Ammonia may be used as the etchant. The temperatures for the synthesis of nanotubes by CVD are generally within the 923–1173 K range. Typical yields for CVD are approximately 30%.

Now, new different techniques for the carbon nanotubes synthesis with CVD have been developed, such as plasma enhanced CVD, thermal chemical CVD, alcohol catalytic CVD, vapour phase growth, aero gel-supported CVD and laser-assisted CVD. These different techniques are explained below.

Plasma enhanced CVD

(PECVD) is a growth technique for the vertical alignment of CNTs. The PECVD method generates a glow discharge in a chamber or a reaction furnace by a high frequency voltage applied to both electrodes.

Plasma CVD has an advantage of low temperature synthesis over thermal CVD. Carbon nanotubes can be synthesized on soda lime glass, especially for production of display devices, at less than 823 K, at which soda lime glass begins to melt. Generally, the power supplies for discharge of plasma CVD are DC and high frequency. RF(13.56 MHz) and microwave (2.47 GHz) are typical of high frequency power supply. The plasma method generates glow discharge in a chamber or a reaction furnace by a high frequency applied to both electrodes. Fig. 6.44 shows a schematic diagram of a typical plasma CVD apparatus with a parallel plate electrode

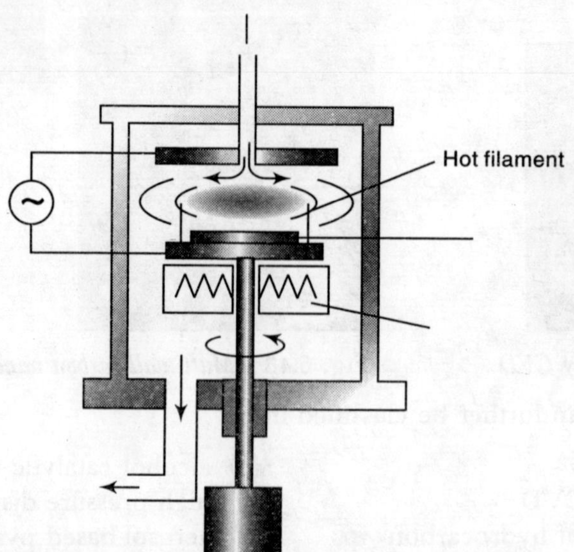

Fig. 6.44 *Schematic diagram of plasma CVD apparatus*

structure. A substrate is placed on the grounded electrode. In order to form a uniform film, the reaction gas is supplied from the opposite plate. *C_2H_2, CH_4 C_2H_4, C_2H_6, CO gases are typically used to synthesize carbon nanotubes.* Catalytic metal, such as Fe, Ni, and Co are used on a Si, SiO_2, or glass substrate using thermal CVD or sputtering. The deposited metal on substrate can be etched using ammonia or H_2 gas. After nanoscopic fine metal particles are formed, carbon nanotubes are grown on the metal particles on the substrate by glow discharge generated from high frequency power, Reaction gas, such as C_2H_2, CH_4, C_2H_4, C_2H_6, CO is supplied to the chamber during the discharge. The SEM images in Fig. 6.45. and 6.46 show carbon nanotubes grown on glass substrate by RF high frequency power using a mixture gas of C_2H_2 and NH_3. Fig. 6.45 shows SEM images of carbon nanotubes grown by plasma CVD method and cut with a razor. The SEM images show that uniform size carbon nanotubes were aligned vertically on the glass substrate without substantial carbon particles. Fig. 6.46 shows magnified SEM images of carbon nanotubes vertically aligned with uniform diameters. Catalytic metal cluster existing the list of carbon nanotubes. In this case. the diameter and density of carbon nanotubes depend

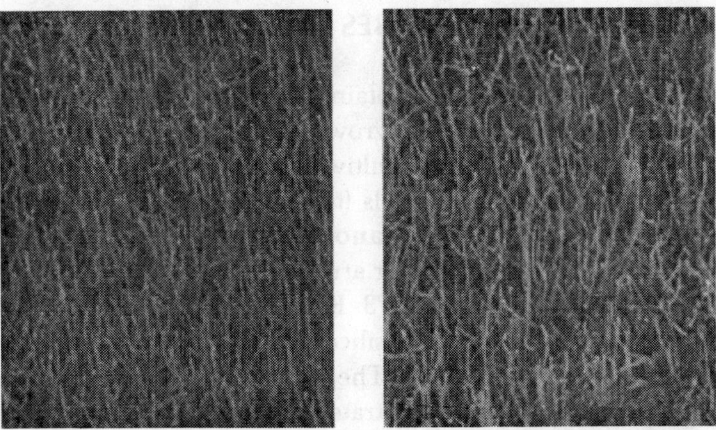

Fig. 6.45 *SEM images of carbon nanotubes grown by plasma CVD method*

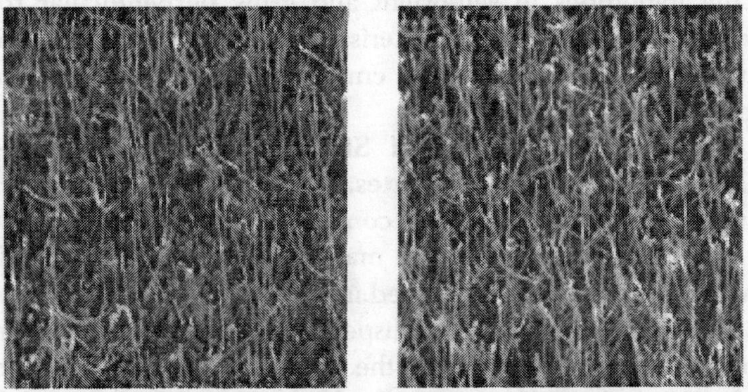

Fig. 6.46 *Magnified SEM images of carbon nanotubes grown by plasma CVD method*

greatly on the magnitude and density of catalytic metal particles. Fig. 6.47 is a TEM image of carbon nanotube grown on glass substrate, using Ni as a catalytic metal. As carbon nanotubes grown by plasma CVD are synthesized at low temperature, graphene sheets are known to be a little vermicular and structurally MWNT, even though the inside hollow of the nanotube looks dim.

The catalyst has a strong effect on the nanotube diameter, growth rate, wall thickness, morpholo microstructure. Ni seems to be the most suitable pure-metal catalyst for the growth of aligned walled carbon nanotubes (MWNTs). The diameter of the MWNTs is approximately 15 nm. The highest yield of carbon nanotubes achieved was about 50% and was obtained at relatively low temperatures below 603 K.

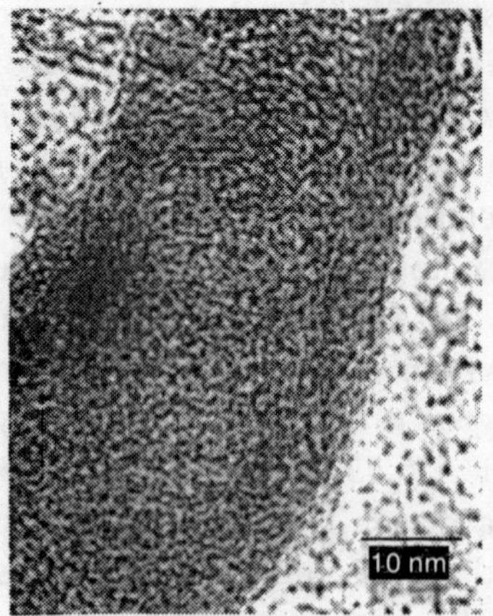

Fig. 6.47 *TEM image of carbon nanotube grown by plasma CVD method.*

PATTERNED GROWTH OF NANOTUBES BY CVD

A patterned growth approach is made to obtain organized nanotube structures. The idea is to position catalyst in arrayed fashions for the growth of nanotubes from specific catalytic sites on surfaces. Such patterned growth for both multiwalled and single-walled nanotubes is done by, self-assembly and active electric field controls (to manipulate the orientation of nanotubes).

(a) Ordered Arrays of Multiwalled Nanotubes–Self-Assembly by Intratube van der Waals Interactions. The synthesis of regular arrays of ordered towers consisting of multiwalled nanotubes (MWNTs) by CVD growth (973 K as described earlier, carbon source, C_2H_4; alumina-supported iron catalyst) on porous silicon or silicon substrates are patterned with iron particles in square regions (Figure 6.48(a)). The nanotubes within each tower are well aligned along the direction perpendicular to the substrate surface. The alignment is a result of nanotubes grown from closely spaced catalyst particles in each square self-assembling into rigid bundle structures due to strong intratube van der Waals binding interactions. The rigidity of the assembly allows the nanotubes to self-orient and grow perpendicular to the substrate. The arrayed nanotubes exhibit excellent characteristics in electron field emission, opening up the possibility of spatially defined massive field emitter arrays derived by simple chemical routes for flat panel display applications.

(b) Ordered Networks of Suspended Single-Walled Nanotubes–Self-Assembly by van der Waals Interactions with Substrates. Suspended SWNT networks with well-defined orientations on substrates is synthesized. It contained lithographically patterned silicon posts. Contact printing is used to transfer catalyst materials onto the tops of pillars selectively, and CVD (1173 K; carbon source, CH_4; supported iron catalyst derived by a template method from liquid precursors) on the substrates leads to suspended SWNTs forming nearly ordered networks with the nanotube orientations directed by the pattern of the silicon posts (Figure 6.48(b–d)). The mechanism for nanotube self-orienting in this case is due to van del' Waals interactions between nanotubes and the silicon posts. The nanotubes grown from the posts float and wave

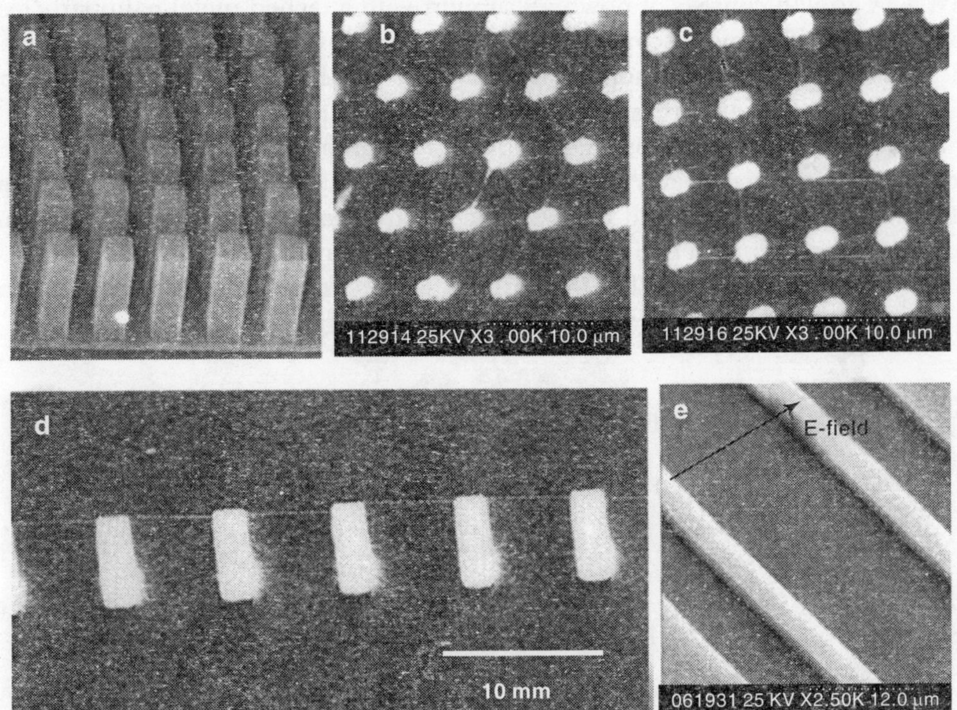

Fig. 6.48 *Ordered carbon nanotube structures obtained by direct chemical vapor deposition synthesis. (a) An SEM image of self-oriented MWNT arrays. Each tower-like structure is formed by many closely packed multiwalled nanotubes. Nanotubes in each tower are oriented perpendicular to the substrate. (b) SEM top view of a hexagonal network of SWNTs (line-like structures) suspended on top of silicon posts (bright dots). (c) SEM top view of a square network of suspended SWNTs. (d) Side view of a suspended SWNT power line on silicon posts (bright). (e) SWNTs suspended by silicon structures (bright regions). The nanotubes are aligned along the electric field direction.*

in the gas until they are fixed by nearby posts. The suspended nanotube networks are difficult to obtain otherwise by, e.g., postgrowth assembly with arc- or laser-grown materials and are useful for building novel nanoelectromechanicalcal devices (NEMs).

(c) Electric-Field-Directed Nanotube Growth. Electric fields are applied to actively control the growth directions of SWNTs. Highly aligned suspended SWNTs are grown under electric fields on the order of ~1 V/µm (Figure 6.48(e)). The alignment effect originates from the high polarizability of SWNTs. The induced dipole moment for a ~10 µm SWNT in a 1 V/µm electric field is about 10^6 D, leading to a large aligning torque that directs the nanotube parallel to the electric field. Importantly, the electric field alignment effect is stable against thermal fluctuations at the growth temperature and against gas flow randomization effects. Manipulation by applied fields during nanotube synthesis is highly promising for further exploration of both suspended molecular wires and complex nanotube fabric structures on flat substrates.

(d) From Powdery Catalyst to Isolated Catalytic Nanoparticles–Toward Nanotube Structural Control and Microscopic Understanding of Growth. Supported catalyst typically used for SWNTs growth consists of powdery support with attached metal nanoparticles that are difficult to characterize by microscopy. As a result, the size of the catalytic nanoparticles is hard to control. Some isolated iron nanoparticles are used for the synthesis of SWNTs by CVD. Controllable numbers of Fe^{3+} are placed into the cores of apoferritin to afford stable solutions of artificial ferritin that can be subsequently deposited onto flat substrates. Calcination in air leads to discrete Fe_2O_3 nanoparticles with average diameter ~1.5 nm, as characterized by atomic force microscopy (AFM) and transmission electron microscopy (TEM). AFM and TEM have revealed successful growth of SWNTs from the isolated nanoparticles (Figure 6.49). Further, synthesis of SWNTs directly on TEM grids has allows to clearly reveal the nanoparticle-nanotube relationship (Figure 6.50(a-f)).

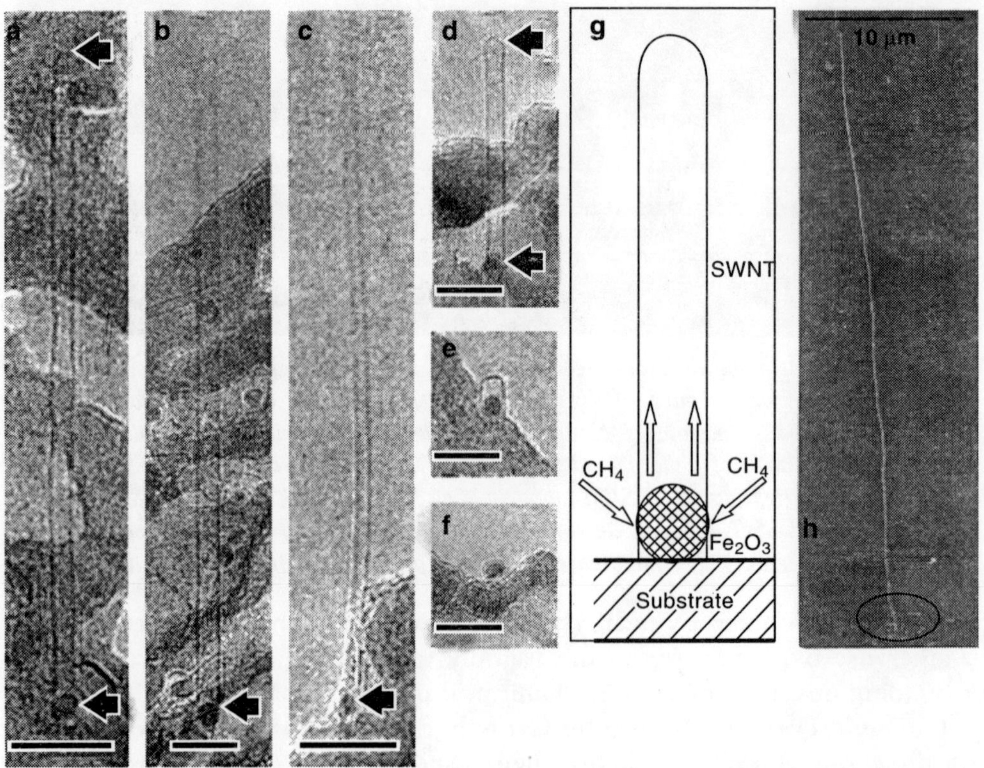

Fig. 6.49 *(a–f) TEM images of SWNTs grown from discrete nanoparticles, showing particle–nanotube relationships. Scale bars: 10 nm. (a–d) SWNTs grown from discrete nanoparticles (dark dots at the bottom of the images). The arrows point to the ends of the nanotubes. The ends extending out of the membrane in (b) and (c) are not imaged due to thermal vibration. The background roughness reflects the TEM grid morphology. (e) Image of an ultrashort (~4 nm) nanotube capsule grown from a ~2 nm nanoparticle. (f) Image of a nanoparticle surrounded by a single graphitic shell. (g) A schematic model for nanotube growth. (h) AFM image of a 50 mm long SWNT grown from nanoparticles patterned into the circled region*

The diameters of SWNTs are closely determined by the diameters of catalytic particles and TEM imaging of both ends of isolated SWNTs (Figure 6.48(a), (d), (e)) points to the so-called base-growth model of SWNTs in CVD process. That is, the catalytic nanoparticle remains on the supporting substrate, and a nanotube grows out from the particle with a dome-closed end (Figure 6.49(g)). The CVD reaction process then involves carbon atoms, decomposed from methane, sorbing into the anchored nanoparticle on the substrate to form a carbon-iron solid-state solution. As supersatu ration occurs, carbon atoms precipitate out from the particle, leading to the growth of a nanotube (Figure 6.49(g)). This growth model is in contrast with the tip-growth model, in which catalyst particles lift off from the substrate during growth.

Synthesis with controllable catalyst, although still in its early stage,[21-23] has led to important clues to the nanotube growth chemistry, mechanism, and structural control. Future work will include determining the nature of SWNTs (metal vs semiconductor) grown from individual particles, experimentally and theoretically investigating how a seed particle determines the nanotube chirality and obtaining highly monodispersed nanoparticles at the 1 or 2 nm level. A patterned growth is shown with ferritin particles placed in micrometer-scale squares on surfaces (Figure 3h). New lithographic approaches capable of 10 nm features are required to enable catalytic patterning of individual nanoparticles. Meeting these challenges will be extremely rewarding, leading to (1) rational routes to nanotube diameter and chirality control and (2) dense and ordered networks of nanotubes with predictable metallic or semiconducting electronic properties.

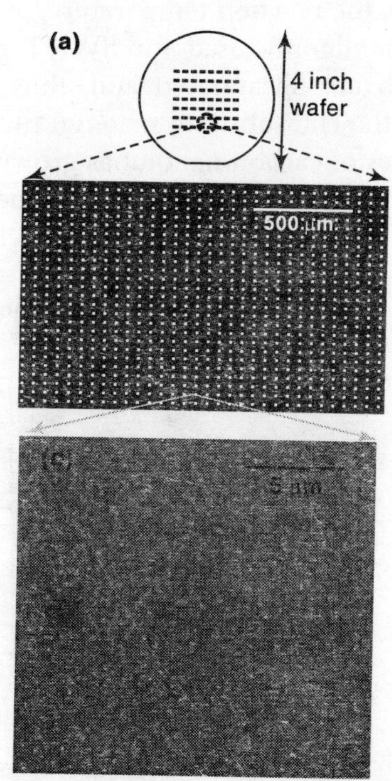

Fig. 6.50 *Wafer scale patterned nanotube synthesis. (a) An optical image showing patterned catalyst islands (bright dots). (b) AFM image of SWNTs grown and emanating from one of the catalyst islands on the wafer*

(e) Scalability of Patterned Growth. Assembly of nanotube arrays must be scaled-up on large substrates for various practical applicatins. To this end, patterned growth of SWNTs on full 4-in. SiO_2/Si wafers is explored and micrometer-sized catalyst islands with high uniformity over entire wafers by a photolithography approach is made (Figure 6.50(a)). Also investigated is, how the gas flow rates (CH_4 and H_2 co-flow) affect the growth of SWNTs. It is observed that, for relatively low CH_4 flow rates, the CVD process is ultrasensitive to the amount of H_2 co-flow, undergoing pyrolysis, growth, and inactive reaction regimes with increased H_2 addition.Thus growing of high-quality SWNTs (Figure 6.50(b)) from massive arrays of well-defined surface sites on full 4-in. wafers is achieved.

(D) THERMAL CHEMICAL VAPOR DEPOSITION

The synthesis method of carbon nanotubes using thermal chemical vapor deposition is as follows. Fe, Ni, Co, or alloy of the three catalytic metals is initially deposited on a substrate of silica, quartz, zeolite, or alumina, and then etched in dilute HF solution. After the substrate is etched in diuted HF solulion with distilled water, the specimen is placed in a quartz boat. The boat is positioned in a CVD reaction furnace, and nano-size fine catalytic metal particles are formed after an additional etching of the catalytic metal film using NH_3 gas at a temperature in 1073 to 1323 K. As carbon nanotubes are grown on these fine catalytic metal particles in CVD synthesis forming these fine catalytic metal particles is the most important process. Fig. 6.51. shows a schematic diagram of thermal CVD apparatus in the synthesis of carbon nanotubes. Reaction gas is supplied in one end of the apparatus, and gas outlet in the other. The nature and yield of the deposit obtained in the reaction is controlled by varying different parameters such as the nature of the catalytic metals and their supports, the hydrocarbon sources, the gas flows, the reaction temperature, and the reaction time, etc. Hydrocarbons such as CH_4 and CO are commonly used for SWNT growth, whereas hydrocarbons such as C_2H_2, C_2H_4, C_6H_6, which are unsaturated and, thus, have high carbon content, are typically used for MWNT growth. The substrate material may be silica, quartz, zeolite, or alumina. Fig. 6.52. shows SEM images of carbon nanotubes grown by thermal CVD method. In this figure, regularly patterned carbon nanotubes are shown to be vertically aligned. After Fe catalytic metal is patterned on a substrate, carbon nanotubes can be grown in accordance with the forms of catalytic metals.

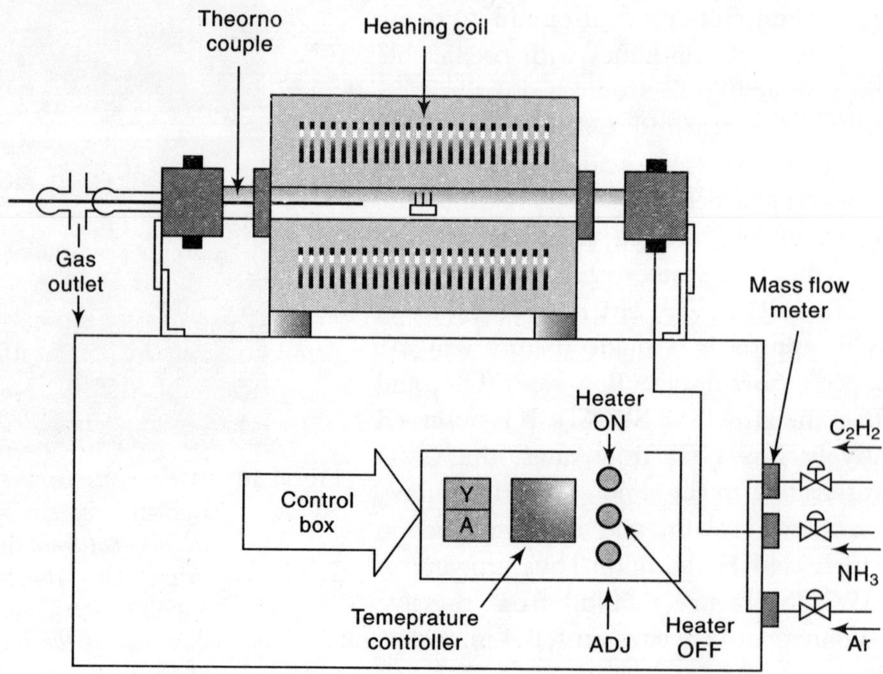

Fig. 6.51

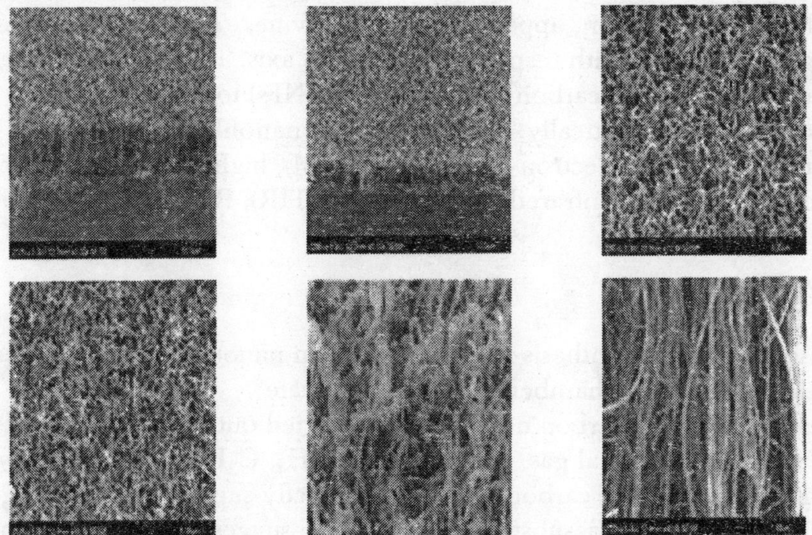

Fig. 6.52 *SEM images of carbon nanotubes grown by thermal CVD method*

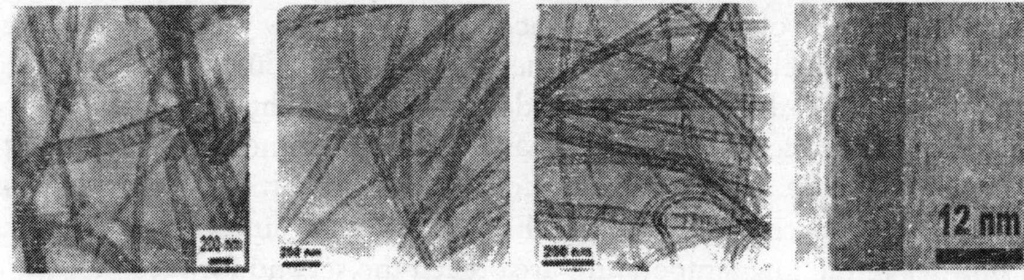

Fig. 6.53 *TEM images of carbon nanotubes grown by thermal CVD method*

Below is the magnified SEM image. It is known from the images that vertically aligned carbon nanotubes are bound together as bundles by Van der Waals force. Fig. 6.53 shows TEM images of carbon nanotubes grown by thermal CVD method. The diameter of each carbon nanotube is approximately 15 nm. It is hollow and a MWNT structure. composed of 10 graphene sheets. There are a graphene sheet structure with a little defects on the surface of the nanotube.

When growing carbon nanotubes on a Fe catalytic film by thermal CVD, the diameter range of the carbon nanotubes depends on the thickness of the catalytic film. By using a thickness of 13 nm, the diameter distribution lies between 30 and 40 nm. When a thickness of 27 nm is used, the diameter range is between 100 and 200 nm. The carbon nanotubes formed are multiwalled. Both methods thermal and plasma are usually used to grow SWNTs and MWNTs.

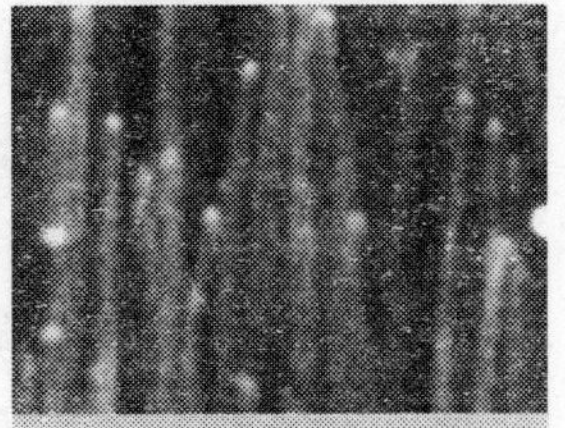

Fig. 6.54

Thermal CVD has allowed SWNT and MWNT growth on patterned substrates. In the ease of MWNTs, continued growth over 10 minutes yields nice towers of MWNTs. These towers consist of millions of MWNTs supporting each other by van der Walls force; individual nanotubes, in high magnification, appear to grow like vines. The nanostructures have interior walls exhibiting a small angle with respect to the central axis. This stacked-come shape material is preferably called multiwalled carbon nanofibers (MWNFs) to distinguish them from MWNTs. These are also known as vertically aligned carbon nanofibers (VACNFs). The nanotube characterization is done using electron microscopy (SEM), high resolution transmission electron microscopy; Fourier transform infrared spectroscopy (FTIR), Raman Spectroscopy and UV-Vis Spectroscopy.

Vapour Phase Growth

Vapour phase growth is a synthesis method of carbon nanotubes, directly supplying reaction gas and catalytic metal in the chamber without a substrate

Most synthesis methods of carbon nanotubes are carried out by deposition of catalytic metals on a substrate using conventional gas, such as C_2H_2, CH_4, C_2H_4, C_2H_6. However, vapor phase growth is a synthesis method of carbon nanotubes. directly supplying reaction gas and catalytic metal in the chamber without a substrate. It has been suggested as a good method for mass production. Fig. 6.55. shows a schematic diagram of vapor phase growth apparatus. A mass flow controller is placed in one corner and a boat for catalytic metal powder in the chamber. The chamber is composed of two stage furnaces. Relatively low temperature is kept in the first furnace while higher temperature is maintained in the hot second furnace where the synthesis occurs. Although hydrocarbon gas is not decomposed in the first furnace, relatively low temperature for vaporization of catalytic carbon is maintained. Although catalytic metals vaporized from powder are atomic. they are formed as fine particles down to several to tens of nm in the chamber amid collisions among the atoms. When fine catalytic particles vaporized from metal powder in the low temperature area reach the second furnace, decomposed carbons in the hot second area are adsorbed, diffused to catalytic metal particles, and synthesized as carbon nanotubes. The diameter of carbon nanotubes by using this method are in the range of 2-4 nm for SWNTs and between 70 to 100 nm for MWNTs.

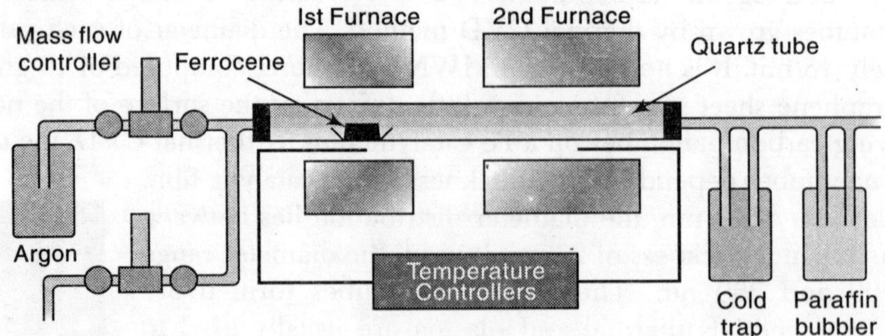

Fig. 6.55 *Schematic diagram of vapor phase growth apparatus*

Fig. 6.56 shows SEM images of carbon nanotubes grown by vapor phase growth method. It is found from the images that about 50 μm carbon nanotubes in length are densely synthesized without a substrate in the chamber and there were substantially no carbon particles. Fig. 6.57 shows TEM images of carbon nanotubes synthesized by vapor phase growth method. From the

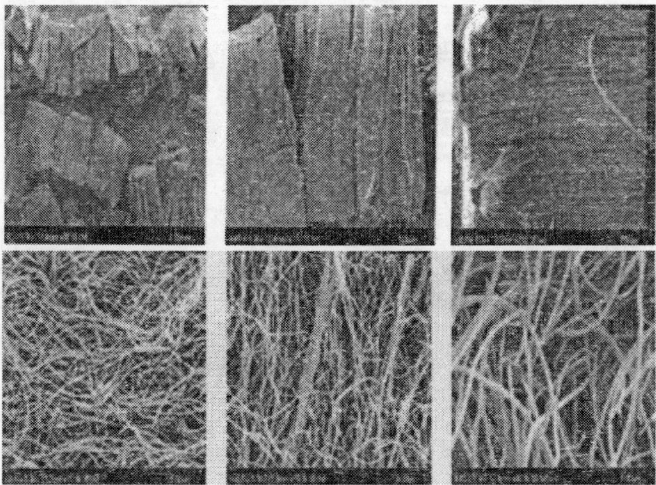

Fig. 6.56 *SEM images of carbon nanotubes grown by vapor phase growth method*

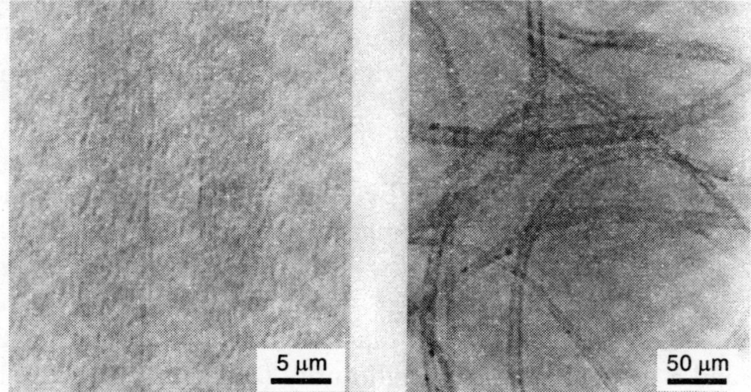

Fig. 6.57 *TEM images of carbon nanotubes grown by vapor phase growth method*

images, the tube is 30 nm in diameter and hollow in the middle. The dark area in the carbon nanotubes is where catalytic metal clusters (arrow 1) were adsorbed during synthesis, and carbon nanotubes were synthesized on catalytic metal clusters below (arrow 2). This indicates that carbon nanotubes can be grown without a substrate if there are catalytic metals.

Alcohol catalytic chemical vapour deposition

Alcohol catalytic CVD (ACCVD) is a technique that is being intensively developed for the possibility of large-scale production of high quality single wall nanotubes SWNTs at low cost. In this technique, evaporated alcohols, methanol and ethanol, are being utilised over iron and cobalt catalytic metal particles supported with zeolite. Generation is possible is possible at a relatively low minimum temperature of about 823 K. It seems that hydroxyl radicals, who come from reacting alcohol on catalytic metal particles, remove carbon atoms with dangling bonds, which are obstacles in creating high-purity SWNTs. The diameter of the SWNTs is about 1 nm. Figure 6.58 shows the ACCVD experimental apparatus.

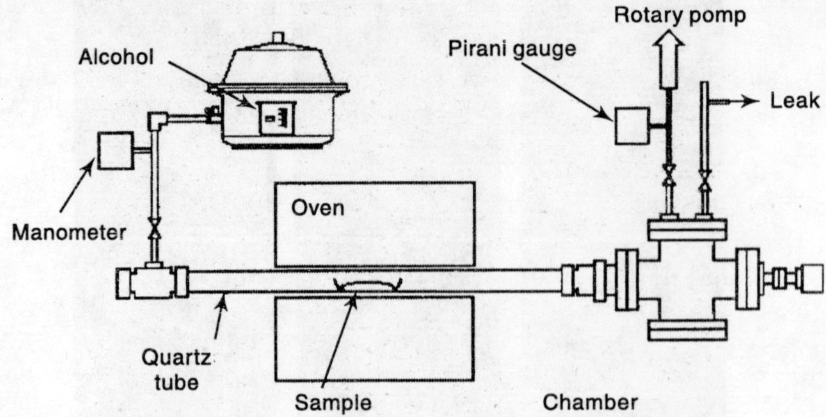

Fig. 6.58 *ACCVD experimental apparatus*

The lower reaction temperature and the high-purity features of this ACCVD technique guarantee an easy possibility to scale production up at low cost. Furthermore, the reaction temperature, which is lower than 873 K, ensures that this technique is easily applicable for the direct growth of SWNTs on semiconductor devices already patterned with aluminium.

Aero Gel-supported Chemical Vapour Deposition

In this method SWNTs are synthesised by disintegration of carbon monoxide on an aero gel-supported Fe/Mo catalyst. There are many important factors that affect the yield and quality of SWNTs, including the surface area of the supporting material, reaction temperature and feeding gas. Because of the high surface area, the porosity and ultra-light density of the aero gels, the productivity of the catalyst is much higher than in other methods. After a simple acidic treatment and a oxidation process high purity (>99%) SWNTs can be obtained.

When using CO as feeding gas the yield of the nanotubes is lower but the overall purity of the materials is very good. The diameter distribution of de nanotubes is between 1,0 nm and 1,5 nm. The optimal reaction temperature is approximately 1133 K.

Experiment

Catalyst Preparation

Alumina-aergel-supported Fe/Mo catalysts are prepared. 5.2 g ASB is dissolved in 35 ml hot ethanol. When the solution becomes homogeneous, 60 μl HNO_3 (69%) and 0.20 ml H_2O dilute in 25 ml ethanol are added to the mixture after it is refluxed for 2.5 h, 0.37 g $Fe_2 (SO_4)_3 \cdot 4H_2O$ and 0.10 g MoO_2 $(acac)_2$ dispersed in 15 ml ethanol are added to the clear sol. After another 1 h, the mixture is cooled to room temperature. A mixture of 1.0 ml ammonia and 1.0 ml H_2O plus 5 ml ethanol is then added to sol with vigorous stirring. A gel is formed within 30 s. It is left to age for about 12 h before being dried.

Drying of the gel is carried out under the supercritical condition of ethanol. Firstly, the gel (50–55 ml) is put into a 100 ml high-pressure container. The system is purged with N_2 for 15 min. Then a high pressure of N_2 (200 psi) is added to the container and the system to 533 K in 2.5 h. The system is held at this temperature for 20min.

The vessel is then depressurized slowly while the temperature is maintained at 533 K. finally, the aerogel is slowly cooled to room temperature with N_2 as purging gas. The powder obtained was calcined at 773 K in air for 1 h before being used for SWNT synthesis.

SWNT Synthesis

The method of synthesis has been described before. Briefly, an alumina boat is loaded with about 50 mg catalyst and put into a quartz tube. It is heated to 1133 K under an Ar flow (200 sccm). The Ar flow is shut off and CO (1000 sccm) is introduced when the temperature reaches 1133 K. After a certain reaction time (15 to 120 min), the CO flow is switched back to Ar while the system is cooled to room temperature. The yield is defined as the mass increase divided by the starting catalyst mass.

Sample Characterization

A 1.0 mg sample of SWNTs is sonicated in methanol and then a drop is placed onto a holey-carbon grid for transmission electron microscope observation. The content of carbon in the sample is measured by a TGA (Thermalgravimetric Analyzer) system under an airflow at a ramp rate of 5° K/min.

Feeding Gas

By using methane as a carbon source; a high yield of SWNTs is achieved. The purity of the raw products is very good when a thin layer of catalyst is used for growth. However, when a large amount of catalyst is used, more amorphous carbon in the product with methane as feeding gas is found, thus making the purification of the raw materials a challenge. Using carbon monoxide, the amount of amorphous carbon is significantly reduced. This difference is related to the thermo-equilibrium properties of the two chemical reactions:

$$CH_4 \leftrightarrow C + 2H_2 \qquad \Delta H° = +74.4 \text{ kJ mol}^{-1}$$

$$2CO \leftrightarrow C + CO_2 \qquad \Delta H° = -172 \text{ kJ mol}^{-1}$$

For CH_4, higher temperature favors carbon formation, while for CO, higher temperature favors carbon consumption. It is observed that the yield of nanotubes at higher temperature is lower when CO is used as feeding gas. When large amount of catalysts are used, the thermal exchange between the gas molecules and the surrounding environment is more efficient in the bulk of the catalyst compared with in the surface layer. Therefore, inside the bulk of the catalyst, the temperature of the gas molecules is higher. For methane, amorphous carbon is more likely to form, while for CO, the formation of carbon is not favorable. Experimentally, it is observed that when CO is used as feeding gas, only a thin surface layer of the catalyst (~ 1 mm in thickness) produces nanotubes; underneath, the catalyst remains unreacted.

When methane is used, more amorphous carbon is produced within the bulk of the catalyst. Although the yield of nanotubes in lower using CO as feeding gas, the overall purity of the materials is much better. In addition, the Raman spectrum also shows that the diameter distribution of the nanotubes made by the CO CVD method (1.0–1.5 nm) is narrower compared with the products made by methane CVD (0.9–2.7 nm).

Optimization of SWNT-synthesis Conditions

To optimize the growth conditions for nanotubes. It is found that the optimal reaction temperature is about 1133 K. When the reaction temperature is higher than 1153 K, the yield is lower. On the other hand, when the reaction temperature is lower than 1093 K, the product contains amorphous carbon.

The flow rate of CO also affects the experiment outcome. When it is higher than 1000 sccm, the product shows the best quality. If a lower flow rate of 500 sccm is applied, a small amount of multiwalled nanotubes (MWNTs) is observed together with SWNTs. The exact cause of the formation of MWNTs is still under investigation in out group.

Another important factor is the surface area of the catalysts. Simply evaporating the ethanol in the wet gel would cause the gel to shrink and significantly reduce the total surface area and pore volume. The resulting material is called a xerogel. A zerogel catalyst showed a weight gain of less than 20% with 60 min reaction time, while the aerogel catalysts gave a weight gain of around 80%. Typically the surface areas measured by the BET (Brunauer-Emmett-Teller) method on out aerogel and xerogel catalyst areas are 500–600 m^2/g and < 100 m^2/g, respectively. Therefore the surface area of the catalyst support is an important factor that affected the SWNT yield.

- *Aerosol-Based Synthesis Method* Aligned MWNTs can be prepared in high yield by pyrolysis of homogeneously dispersed aerosols generated from benzene/ferrocene solutions, at 1073 or 1223 K, using a compressed gas (Ar) driven atomizer. The aim of the atomization, using high Ar flow rates, is to produce minute hydrocarbon/catalyst droplets (aerosol), which homogenously fills the reactor volume, favoring the formation of evenly distributed metal clusters during pyrolysis Figure 6.59 shows the aerosol pyrolysis apparatus.

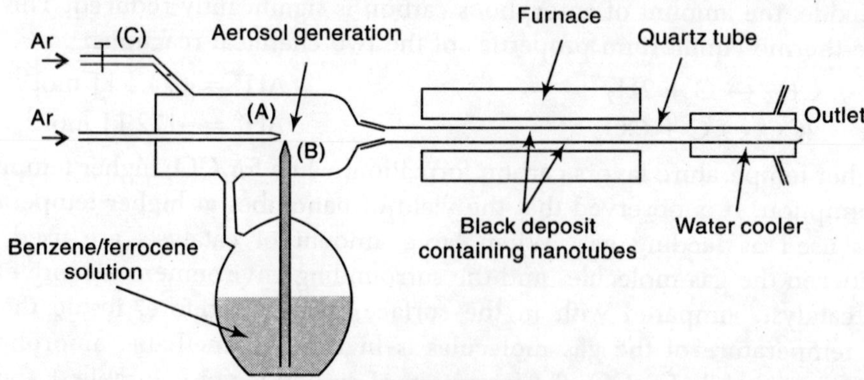

Fig. 6.59 *Aerosol pyrolysis apparatus*

- *Nebulized Spray Pyrolysis Method* The nebulized spray pyrolysis method can be conveniently used for the continuous production of MWNTs. Nebulized spray is a spray generated by an ultrasonic atomizer. MWNTs with fairly uniform diameters and aligned MWNT bundles were obtained by using a solution of ferrocene in hydrocarbon solvents (e.g., benzene, xylene, hexane, toluene) Figure 6.61 shows the schematic setup of the nebulized spray pyrolysis setup.

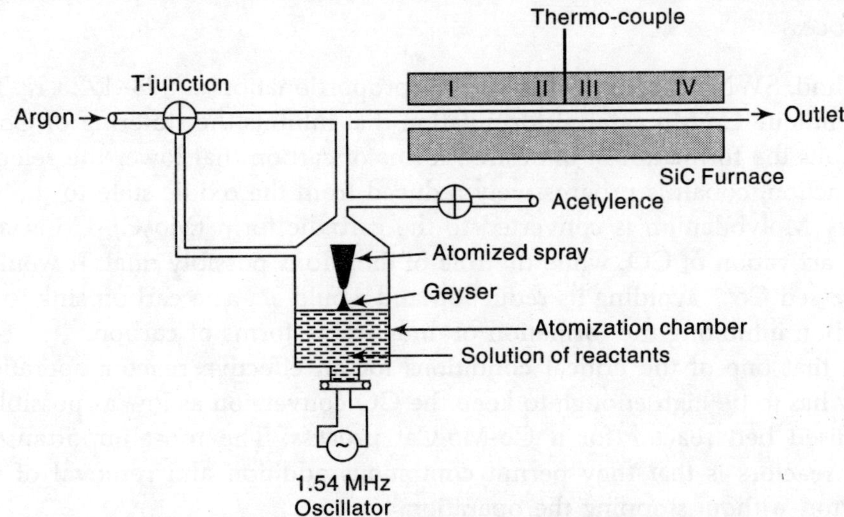

Fig. 6.60 *Schematic representation of the nebulized spray pyrolysis set-up*

(E) LASER-ASSISTED THERMAL CHEMICAL VAPOUR DEPOSITION

In laser-assisted thermal CVD (LCVD) a medium power, continuous wave CO_2 laser, which was perpendicularly directed onto a substrate, pyrolyses sensitised mixtures of $Fe(CO)_5$ vapour and acetylene in a flow reactor. The carbon nanotubes are formed by the catalysing action of the very small iron particles. Figure 6.61 shows the experimental set-up for laser-assisted CVD.

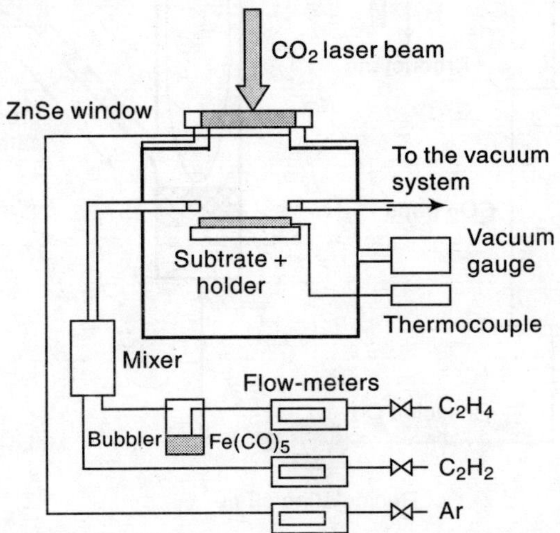

Fig. 6.61 *Experimental set-up for laser-assisted CVD*

By using a reactant gas mixture of iron pentacarbonyl vapour, ethylene and acetylene both single- and multi-walled carbon nanotubes are produced. Silica is used as substrate. The diameters of the SWNTs range from 0.7 to 2.5 nm. The diameter range of the MWNTs is 30 to 80 nm.

Co-Mo-Cat Process

In this method, SWNTs are grown by CO disproportionation at 973–1223 K. The technique is based on a unique Co-Mo catalyst formulation that inhibits the sintering of Co particles and therefore inhibits the formation of undesired forms of carbon that lower the selectivity. During the SWNT reaction, cobalt is progressively reduced from the oxidic state to the metallic form. Simultaneously Molybdenum is converted to the carbidic form (Mo_2C). Co acts as the active species in the activation of CO, while the role of the Mo is possibly dual. It would stabilise Co as a well-dispersed Co^{2+} avoiding its reduction and would act as a carbon sink to moderate the growth of carbon inhibiting the formation of undesirable forms of carbon.

It is found that one of the critical conditions for an effective reactor operation is that the space velocity has to be high enough to keep the CO conversion as low as possible. Figure 6.62 shows a fluidised bed reactor for a Co-Mo-Cat process. The most important advantage of fluidised bed reactors is that they permit continuous addition and removal of solid particles from the reactor, without stopping the operation.

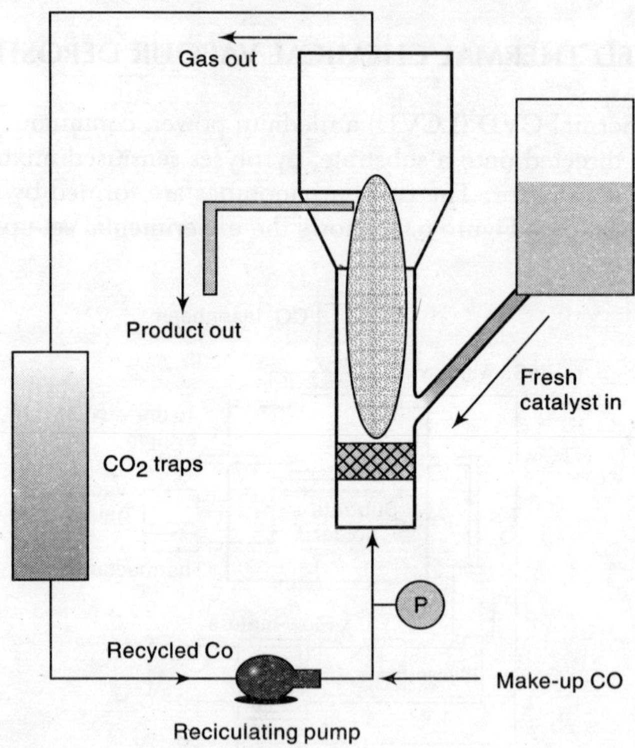

Fig. 6.62 *Schematic diagram of a Co-Mo-Cat apparatus*

This method can be scaled-up without losses in SWNT quality. By varying the operation conditions, SWNTs can be produced with different diameter ranges.

In Table 6.3 several diameter ranges at different temperatures are given.

Table 6.3 Diameter range *versus* temperature

Temperature K	*Diameter range*
1023 K	0,9 ± 0,05
1123 K	0,9 –1,25
1223 K	1,00 – 1,40

The Co-Mo-Cat catalyst has a high selectivity towards SWNTs, namely 80 – 90 %.

High Pressure CO Disproportionation Process

The High pressure CO disproportionation process (HIPCO) is a technique for catalytic production of SWNTs in a continuous-flow gas phase using CO as the carbon feedstock and $Fe(CO)_5$ as the ironcontaining catalyst precursor. SWNTs are produced by flowing CO, mixed with a small amount of $Fe(CO)_5$ through a heated reactor. Figure 6.63 shows the layout of CO flow-tube reactor. Size and diameter distribution of the nanotubes can be roughly selected by controlling the pressure of CO. This process is promising for bulk production of carbon nanotubes.

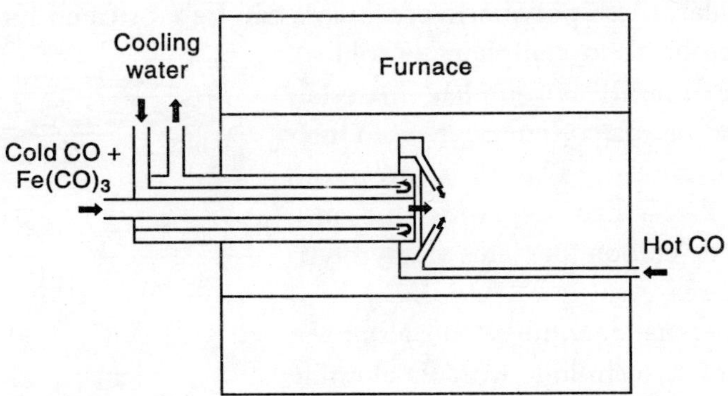

Fig. 6.63 *Layout of co flow-tube reactor*

Nanotubes as small as 0.7 nm in diameter, which are expected to be the smallest achievable chemically stable SWNTs, have been produced by this method. The average diameter of HiPco SWNTs is approximately 1.1 nm. The yield that could be achieved is approximately 70%. The highest yields and narrowest tubes can be produced at the highest accessible temperature and pressure. SWNT material with 97% purity can be produced at rates of up to 450 mgh^{-1} with this process.

Solid State Reaction

a. Electrolysis Multi-walled CNTs were produced by the electrolytic conversion of a graphite cathode in molten lithium chloride. Formation of CNTs in a condensed phase is possible by passing an electric current between carbon electrodes In addition to the electrolysis methods in fused alkali salts, a new electrochemical method for MWNTs from acetylene solution in liquid ammonia has been developed at 233 K. Single crystal silicon wafers (*n*-type Si (100) wafers) are used as electrodes and no metal catalysts are needed.

The chain radical reactions, which are involved in the formation of CNTs, are listed below.

$$2NH_3 \leftrightarrow [NH_4]^+ + NH_2^-$$
$$NH_3 + C_2H_2 \leftrightarrow [NH_4]^+ + C_2H^-$$
$$C_2H^- \rightarrow e + C_2H$$
$$NH_{2^-} \rightarrow e + NH_{2^-}$$
$$[NH_4]^+ + e \rightarrow NH_3 + H$$
$$C_2H_2 + H \rightarrow C_2H + H_2$$
$$C_2H + H \rightarrow C^* + H_2$$
$$C^* + C_2H \rightarrow C^* - C_2H$$
$$C^* - C_2H + H \rightarrow C^* - H_2$$

where C* indicates a carbon product from the synthesis (carbon nanotubes, or graphitic and turbostratic carbon or amoiphous carbon).

b. Pyrolysis Carbon materials can be converted from polymeric materials via simple thermal treatments. In particular, these polymeric precursors can be fabricated into various shapes with the assistance of porous solid templates. Zeolite-Y and mordenite were utilized as template materials for radical polymerization of acrylonitrile (within their open pores and channels). The thus prepared poly(acrylonitrile) (PAN) can be converted into graphite-like conducting carbon filaments via pyrolysis at elevated temperatures

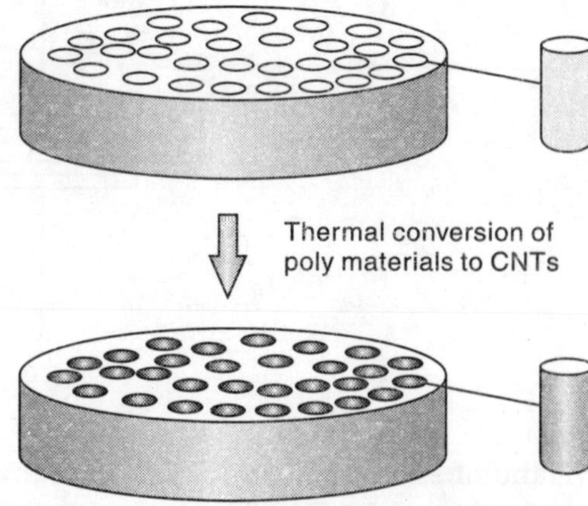

Fig. 6.64 *Schematic drawing of the porous inorganic template growth of CNTs with prefilled organic materials in one-dimensional channels and thermal conversion reactions*

Carbon 60(C_{60}, fullercne) nanotubes have also been prepared by the template technique with ID alumina channel with a density of 10^{11} channels cm^{-2}. See Figure 6.64.

The alumina template is repeatedly dipped into a C_{60} containing toluene solution. After several dipping-and-drying cycles, the C_{60} deposited templates are heated at 773 K for 5 hours in an argon-atmosphere and then cooled down to room temperature. The as-grown C_{60} nanotubes after removal of the template (with NaOH solutions) are well aligned (tube diameter 220-310 nm, and tube length 60 μm), showing a hexagonal long-range order in accordance with the original ID channel array. Table 6.4 gives. a summary of these three pyrolysis techniques.

Table 6.4 A summary of the major production methods and their efficiency

Method	*Arc discharge method*	*Chemical vapour deposition*	*Laser ablation (vaporization)*
Who	Ebbesen and Ajayan, NEC, Japan 1992	Endo, Shinshu University, Nagano, Japan	Smalley, Rice, 1995
How	Connect two graphite rods to a power supply, place them a few millimetres apart, and throw the switch. At 100 amps, carbon vaporises and forms a hot plasma.	Place substrate in oven, heat to 873 K, and slowly add a carbon-bearing gas such as methane. As gas decomposes it frees up carbon atoms, which recombine in the form of NTs	Blast graphite with intense laser pulses; use the laser pulses rather than electricity to generate carbon gas from which the NTs form; try various conditions until hit on one that produces prodigious amounts of SWNTs
Typical yield	30 to 90%	20 to 100%	Up to 70%
SWNT	Short tubes with diameters of 0.6-1.4 nm	Long tubes with diameters ranging from 0.60-4 nm	Long bundles of tubes (5-20 microns), with individual diameter from 1-2 nm.
MWNT	Short tubes with inner diameter of 1-3 nm and outer diameter of approximately 10 nm	Long tubes with diameter ranging from 10-240 nm	Not very much interest in this technique, as it is too expensive, but MWNT synthesis is possible.
Pro	Can easily produce SWNT, MWNTs. SWNTs have few structural defects; MWNTs without catalyst, not too expensive, open air synthesis possible	Easiest to scale up to industrial production; long length, simple process, SWNT diameter controllable, quite pure	Primarily SWNTs, with good diameter control and few defects. The reaction product is quite pure.
Con	Tubes tend to be short with random sizes and directions; often needs a lot of purification	NTs are usually MWNTs and often riddled with defects	Costly technique, because it requires expensive lasers and high power requirement, but is improving

The as-grown nanotubes thus prepared have many impurities in them. The following types of impurities are usually found in the tubes:

1. graphitic (wrapped up) sheets
2. amorphous carbon
3. metal catalyst
4. smaller fullerenes ($<C_{60}$)

These impurities can be removed using:

- oxidation
- acid treratment
- annealing

- ultrasonication
- magnetic purification
- micro filtration
- cutting
- functionalization
- chromatography

(F) FLAME SYNTHESIS

This method is based on the synthesis of SWNTs in a controlled flame environment, that produces the temperature, forms the carbon atoms from the inexpensive hydrocarbon fuels and forms small aerosol metal catalyst islands. SWNTs are grown on these metal islands in the same manner as in laser ablation and arc discharge.

These metal catalyst islands can be made in three ways. The metal catalyst (cobalt) can either be coated on a mesh, on which metal islands resembling droplets were formed by physical vapour deposition. These small islands become aerosol after exposure to a flame. The second way, is to create aerosol small metal particles by burning a filter paper that is rinsed with a metal-ion (e.g. iron nitrate) solution. The third way, is the thermal evaporating technique in which metal powder (e.g. Fe or Ni) is inserted in a trough and heated.

In a controlled way a fuel gas is partially burned to gain the right temperature of ~1073 K and the carbon atoms for SWNT production. On the small metal particles the SWNTs are than formed. As optimisation parameters the fuel gas composition, catalyst, catalyst carrier surface and temperature can be controlled. In the literature found, the yield, typical length and diameters are not stated.

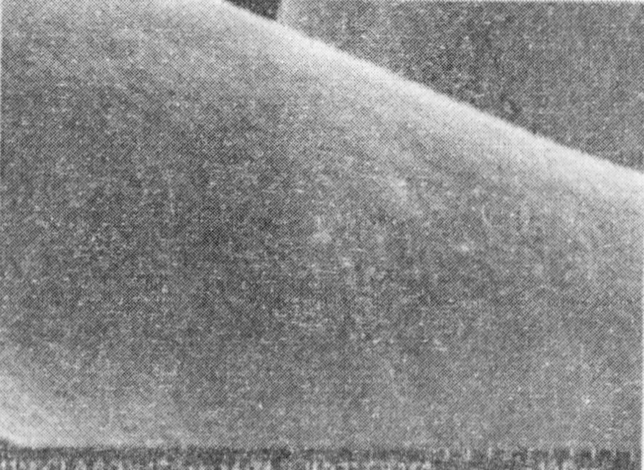

Fig. 6.65 *Meshes on which the metal catalyst is coated, used in flame synthesis*

(G) SOLAR SYNTHESIS

Another possibility is the solar route which is also used to produce single wall nanotubes by direct vaporization of a graphite-bimetal target under argon atmosphere with the highly concentrated solar radiation obtained with the solar furnaces.

Experimental Set-up

This method presents substantial advantages and small-scale experiments performed with a 2 kW solar furnace are well designed to perform accurate analysis of the structure evolution of produced materisl with synthesis parameters and studies of the SWNT growth mechanisms. By controlling most of the synthesis parameters, this small-scale experimental set-up limits the target size and yields a small amount of material due to the low power of the furnace and the small size of the focus area. Filled with a powdered mixture of graphite and catalysts and surrounded by a graphite pipe acting as a thermal screen. The vaporization occurs under an inert gas atmosphere (argon) which drives the vapors through the hot channel formed by the graphite pipe. By clear sky condition, the target temperature measured with a Heimann pyrometer (in the infra-red range) can be greater than 3000 K. When carbon nanotubes are produced large sheets (longer than 2 cm) of rubbery material can be collected into the graphite pipe Fig. 6.66. These samples generally contain in addition catalyst nanoparticles embedded in amorphous carbon, empty carbon vesicles and amorphous carbon. The more important parameter is the vaporization temperature.

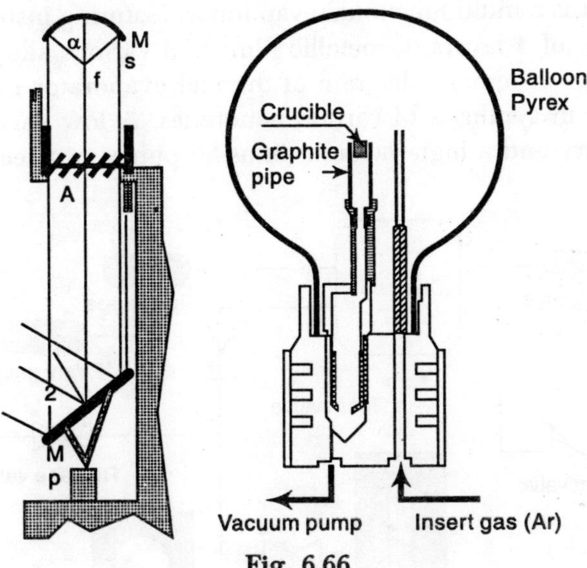

Fig. 6.66

Investigating the growth mechanisms of single wall carbon nanotubes (SWNTs) produced by the solar method as a function of the nature of catalysts and the temperature gradient in the condensing area. The Raman spectra clearly shows that the change of catalyst induce differences in the diameter of SWNT whereas TEM pictures enhance the change of both length and diameter of the bundles. The temperature at which SWNTs are formed is velated to the sublimation temperature of the target and to the eutectic temperature of the binary phase diagram.

Thus, Using an electron microscope, the nanotube material produced by either of these methods looks like a mat of carbon ropes. These ropes have diameters between 10 and 20 nm and lengths of up to 100 μm. When viewed through a transmission eledron microscope, each rope was found to consist of a bundle of single-wall carbon nanotubes aligned along a single direction. X-ray diffraction also shows that the diameters of the single-wall nanotubes have a narrow distribution with a strong peak at 138 ± 0.02 nm, very close to the diameter of an ideal (10, 10) nanotube. Also, the diameter distribution depends on the, composition of the catalyst, the growth temperature and some other growth conditions.

INSTRUMENTS USED FOR CNTs

(1) Evaporator

Evaporators are instruments to deposit thin film on a substrate, especially metal films. Evaporators are very important instruments in depositing catalytic metals on a substrate where carbon nanotubes are formed. The structure of an evaporator consists of a chamber, electrodes, a controller, and exhaust and cooling lines. The evaporator has the same structure and functions as conventional ones, but is a multi-functional evaporator. featuring installment of up to 4 boats which enables deposition of 4 layers of metallic films and various alloy films at a time under vacuum. Fig. 6.67 shows a schematic diagram of thermal evaporator in deposition of catalytic metal, which is essential in synthesis of carbon nanotubes. A low vacuum mechanical rotary pump is used as a primary and a high vacuum diffusion pump as a secondary. The valves for

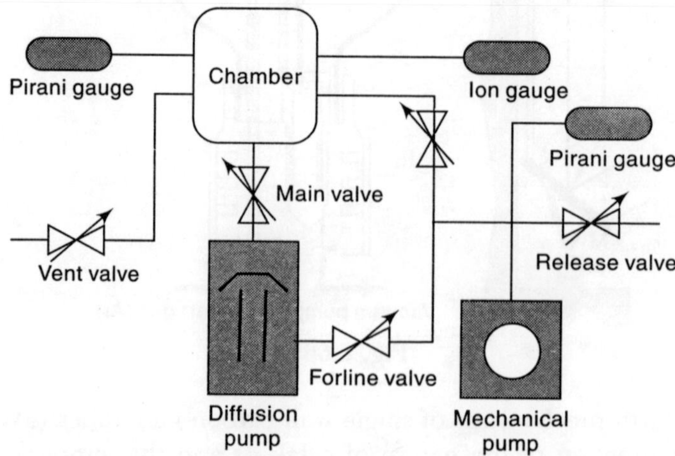

Fig. 6.67 *Schematic diagram of thermal evaporator*

the evaporator are pneumonic ones except release valves. Release valves are solenoid types for preventing oil of the rotary pump from flowing backward caused by an abrupt blackout or an operational error and do not function when normal.

Fig. 6.68 is a drawing of the internal structure of the thermal evaporator chamber. Stainless steel is used as material for the chamber. The chamber is a cylindrical shape with 400 mm in diameter and 530 mm in height. A pirani gauge and a ion gauge are installed in the chamber to measure pressure. A gate window for introducing deposition material and substrates into the inside of the chamber and a view window for watching reactions in the chamber is installed. As four electrodes for deposition are set up in the inside of the chamber, alloys can be deposited on substrates using various deposition materials simultaneously. Deposition boats can be positioned on the terminal of the electrode. The target substrate, on which deposition occurs is facing downward and hanging inside the chamber. Shutters are set up between electrode and target substrate to control thickness by manual operation and prevent deposition of impurities and unnecessary heat transfer. In addition, ports on the external wall of the chamber are positioned to facilitate setup of analysis and control instruments when necessary. The controller consists of a main power for the evaporator, power switches for a rotary and a diffusion pump, a power supply for the deposition electrode, a current controller, a vacuum gauge display of the chamber, on-off switches for valves, and warning systems for cooling water and air pressure. All the processes on actual deposition are controlled by the controller. When the warning system operates and power is cut off for unnecessary reasons, all the valves will be automatically shut off to prevent inflow of impurities in air or pump oil into the inside of chamber and keep all the exhaust lines but the chamber at atmospheric pressure, operating release valves at the same time. The operation procedures for deposition are as follows.

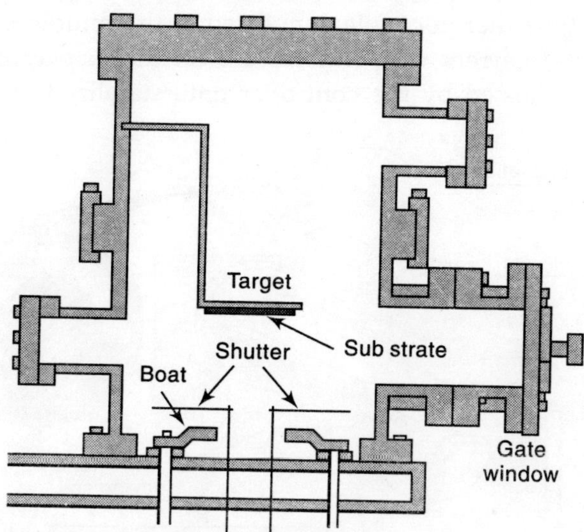

Fig. 6.68 *Internal structure of the thermal evaporator chamber.*

First, check cooling water and pressure. Pre-heat the rotary' pump for about 3 minutes to increase viscosity of oil and enhance efficiency. Open the fore-line valve to remove residual gas in the diffusion pump and close the valve again.

Now, open the roughing valve to maintain the chamber pressure at about 5×10^{-3}. Then open the fore-line valve gain, pre-heat the diffusion valve, increase the temperature to the boiling point of oil (taking 30-40 minutes), and open the main valve to keep the internal chamber pressure at 5×10^{-6} for deposition. When shutting down the diffusion pump. switch off the heater of the diffusion pump, wait for about 30-40 minutes to remove oil vapor, and then close the fore-line valve.

(2) Thermal Chemical Vapor Deposition Apparatus

A thermal chemical vapor deposition apparatus basically consists of a horizontal reaction tube, a heating element, a controller, and a flow meter and valves connected to each gas source. The shape and structure of the apparatus is similar to those of conventional chemical vapor deposition apparatuses, but cooling instruments around the injection and exhaust parts, and the length and shape of the resistance heater is set and optimized to control stable reactions in the chamber.

Fig. 6.69 shows a schematic diagram of thermal CVD apparatus for carbon nanotube growth. The horizontal reaction tube was made of quartz. The resistance heater is put in the straightly grooved area in the ceramic container so that the heating coil can go around the quartz tube. The horizontal tube goes through the heating coils, is fixed on the either support. An Q-ring comes between the reaction tube and the support and cooling lines are connected to each support to prevent overheating of the O-ring. An exhaust line and thermocouple are connected to the left terminal of the electric furnace and an injector and a flow meter to the right terminal. A window is installed to watch the inside of the reactor. P.I.D, programing, and auto tuning controls are possible and current control is manually done. Temperature change is recorded to the controller by the R-type thermocouple connected to the middle of the resistance heater in the reaction tube. Set and current temperatures are on the display of the controller and the current is automatically adjusted by the controller until stabilized. Even after the set value is

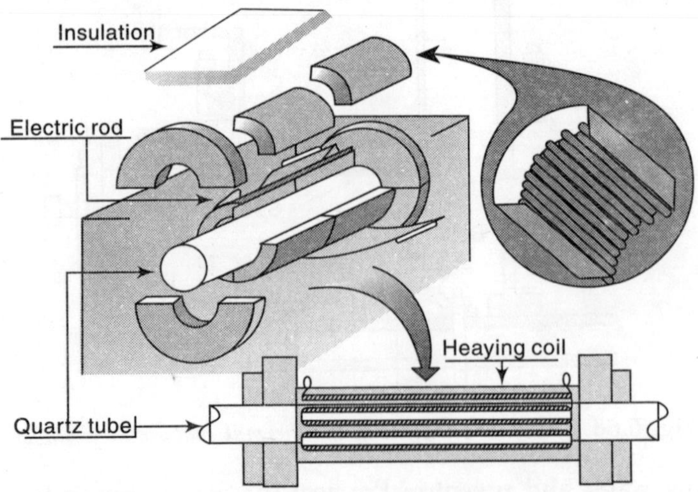

Fig. 6.69 *Schematic diagram of thermal CVD apparatus for carbon nanotube growth*

stabilized, heat loss caused by gas flow occurs. This loss is corrected by the controller. The error of the set temperature is within ±5° K. and the thermocouple is coated with ceramic to prevent cutting in high temperature and enhance mechanical strength. A flow meter is connected to the gas tank of Ar, C_2H_2, or NH_3 with teflon gas line and manual operation. It usually takes about four hours to cool the reaction tube to room temperature with an air cooling system.

About 3000 Å thick oxidized film is grown on a p-type (100) silicon with its resistivity 10-20. The oxidized film-coated wafer is sonicated in ethyl alcohol for about 5 minutes. After the ethyl alcohol on the wafer is dried with flowing nitrogen gas, the dried wafer is placed in the vacuum deposition apparatus for deposition of catalytic metal film. The substrate is 20 mm × 20 mm. Deposition is carried out under high vacuum of less than 2×10^{-6} mm Hg, with 140 A and 1.64 V. Deposition time is adjusted to several seconds and the thickness of catalytic metal film is approximately 50~100 nm. After the Fe gets deposited the substrate is roughly etched in approximately diluted HF solution with distilled water, the substrate in the Quartz tube is placed in the CVD reaction tube. Then the deposited face headed in the opposite direction to gas flow and downward. When the deposited substrate is positioned in the reaction tube. approximately 200 sccm Ar gas supplied to the inside of the reactor to prevent oxidation of the film until an approximate temperature for reaction is attained. When the set temperature is reached, Ar gas flow is stopped and NH_3 gas is flowed at 100-300 sccm for 20-40 minutes for etching of catalytic metal film. Carbon nanotubes were grown on the etched catalytic metal particles in the C_2H_2 atmosphere about 30-60 sccm for 10-20 minutes. When growth of carbon nanotubes was completed and temperature is down to 473 K. the substrate is retrieved from the tube. In order to study temperature effects on growth and structural characteristics of carbon nanotubes on a Fe-deposited silicon oxidized film substrate carbon nanotubes are synthesized at 1023, 1123 and 1223 K. After carbon nanotubes were synthesized SEM(JEOL, JSM-6400, 20 kV) was used to characterize the length, diameter, alignment. density, and uniformity of carbon nanotubes. In order to observe the structure and graphene sheets of tubes, TEM(Philips, CM20T, 200 kV) is used. In addition. measurement of the structure and crystallinity of carbon nanotubes is done using Micro-Raman(Ranishow, Micro-Raman 2000), which uses He-Ne laser at 632.8 nm wavelength. In order to observe the electric-electron emission and fluorescence pattern, carbon nanotubes grown on the substrate are positioned in the chamber under 5×10^{-6} mm Hg vacuum and its I-V is measured.

CATALYSTS FOR SINGLEWALL CARBON NANOTUBE SYNTHESIS

Single-wall carbon nanotubes (SWNTs) are usually produced with the help of metal catalysts that first break down various carbon species such as graphite, carbon clusters, amorphous carbon, or hydrocarbons and then release them in the form of SWNTs. These catalysts therefore must perform several functions in order to produce high quality, high yield, long SWNTs, and the finesse of these catalysts is controlled by the balance between those functions. Such a production is critical to future applications such as telecommunication, molecular electronics, sensing, and mechanical devices.

In general, multicomponent materials such as bimetallic nanoparticles (Bi-MNPs) are much better catalysts than those made of single component materials because the two metals in the Bi-MNPs eg. Ni/Co; can enhance certain functions by playing complementary catalytic roles. Furthermore, some functions can only be offered by Bi-MNPs because the two metals are structurally arranged in such specific manners that the combined properties would otherwise be impossible to access by monometallic nanoparticles (mono-MNPs).

Bimetallic catalysts have been proven superior with improved activity, selectivity, stability, and resistance to poisoning than their monometallic counterparts. For example, Bi-MNPs are preferred heterogeneous catalysts in petroleum reforming processes. Bi-MNPs of large diameters (>50 nm) have also been used to catalyze the growth of carbon filaments. Those catalysts have also been used in ammonia synthesis by barium promoted Ru catalysts supported on boron nitride. More recently, Bi-MNPs (e.g., Ni/Co) have been used to produce SWNTs, and it has been shown that the yield increases by many folds in comparison to that using their monometallic counterparts. In addition, Bi-MNPs have also been used to produce SWNTs of different diameters, cleanness, and lengths.

Bi-MNPs have superior catalytic abilities which can be attributed to the unique geometric and electronic structures they possess. The structural information obtained from the catalysts has enabled to understand the structure-function relationship associated with these nanoparticles. For instance, the catalytic mechanisms of mono-MNPs and Bi-MNPs used in both petroleum reforming processes and in multi-walled nanotube production are satisfactorily explained only after the structures of these catalysts is determined. In a Ru/Cu system, it is found that Cu atoms concentrate in the surface of small Ru/Cu Bi-MNPs (1 nm diameter). Such surface segregation helps regulate the rates of hydrogenolysis, which requires large patches of Ru atoms to function. In contrast, hydrogenolysis dominates in pure Ru nanoparticles, leading to undesirable results. Barium is used to, assist Ru in catalyzing ammonia synthesis. In this case, barium acts as a promoter, meaning that it alone possesses no activity. Reacting with oxygen, it forms an overlayer to enhance the catalytic ability of Ru MNPs. Another catalyst system, Pd-Pt bi-MNPs, which is more resistive to sulfur poisoning due to the special surface arrangement, has also been investigated. All these studies have clearly demonstrated that it is the special structure of those Bi-MNPs that leads to the novel physical and chemical properties, resulting in enhanced activities such as catalytic functions. Similarly, the structural factor may also be the key to understanding the catalytic growth mechanisms of SWNTs.

Transmission electron microscopy (TEM); x-ray photoelectron spectroscopy (XPS), and other methods are employed in the experimental investigations of both SWNTs and metal catalysts for understanding growth mechanisms. It has been shown experimentally that the overall compositions of MNPs can alter the yield and diameter of SWNTs. It has also been discovered that bi-MNPs are generally much better catalysts for SWNT production. For example, the presence of Ni/Co bi-MNPs can produce 100% SWNTs.

Theoretical studies have been performed on the growth of SWNTs as well. It has also been suggested that metal atoms may be responsible for the growth. Most other models, however, consider that the growth is the result of the interaction of metal nanoparticles and the carbon feedstock.

SYNTHETIC METHODS

Syntheses of mono-MNPs Co, and Bi-MNP Co/Ni and Co/Mo nanoparticles

1. *Co Nanoparticles.* The Co mono-MNPs were made using the high temperature decomposition method. Using this method, Bi-MNPs can also be made. Different sizes of Co mono-MNPs made, and these particles are used to synthesize carbon nanotubes. Three ratios of three surfactants, trioctylphosphine oxide (TOPO), oleic acid (OA) and trioctyl amine (TOA), are used in the nanoparticle synthesis to produce distinctive sizes of Co MNPs at 3, 5, and 12 nm, respectively. These nanoparticles are deposited on clean Si substrates, and the assembled samples were placed in a high temperature furnace. Methane, hydrogen, and argon gases are fed into the reactor at high temperatures ($>$ 1273 K). Nanotube growth using thes Co MNPs is conducted in the furnace at 1373 K.
2. *Co/Ni Nanoparticles.* The Ni/Co Bi-MNPs are contained in the product of materials produced in a laser vaporization chamber by vaporizing metal-graphite composite targets with 0.6 at. % of Co and 0.6 at. % of Ni. In the EXAFS measurements, micrograms of the nanotube-Bi-MNP samples are deposited onto tapes, which are then placed in the synchrotron x-ray beam. Highest yield of SWNTs is obtained (~ 100%). It indicates that these nanoparticles possessed certain properties that differed from that of a simple mixture of the two metals.
3. *Co/Mo and Other Nanoparticle Catalysts.* A brief description of Co/Mo(W) on SiO_2 system is given below. The preparation method used to produce other catalysts are provided below.

The Co-Mo catalysts are prepared by impregnatmg a Silica gel support with aqueous solutions of cobalt nitrate and/or ammonium heptamolybdate. The total metal content is kept at 6%. Calcined samples at 773 K are used for EXAFS/XANES measurements.

Mono-MNPs made during the nanotube synthesis are used as catalysts. For instance, Co, Ni, Fe, and other first row transition metals were used to grow SWNTs. Other metals such as Gd, Pt, Mo, Y, La, Sc, W, and other metals are also used as catalysts. However, pre-made catalysts deposited on substrates yield fewer nanotubes.

Mono-MNPs on support behave differently from those in space due to the influence of support-catalyst interaction on the growth. It is possible to obtain the desired structural information on those systems, even during the catalytic process. Nanotube synthesis using these nanoparticles is similar to those in gas phase if these nanoparticles are lifted off from the substrate by nanotubes.

Using the pre-made Co mono-MNPs described above, carbon nanotubes are produced. Figure 6.70 shows 12-nm Co nanoparticles. Normal EXAFS measurements are performed on these and two other sized nanoparticles, as shown in the right panel of Figure 6.70. The EXAFS data shows that large (12-nm) Co nanoparticles are partially oxidized while small ones (3 and 5 nm) are more heavily oxidized. From Figure 6.70 it is also clear that the average numbers of first neighbors for Co nanoparticles (~4) is much less than that in bulk Co.

A sample of the 3-nm Co nanoparticles that were deposited on a Si wafer before the nanotube growth, is shown in Figure 6.70. Although it is possible to observe the first coordination

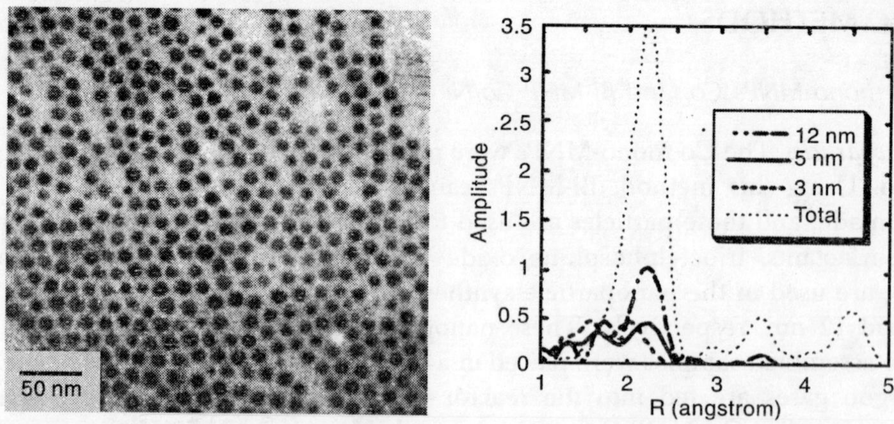

Fig. 6.70 *TEM image of 12-nm Co nanoparticles (left panel) and the EXAFS measurements on these particles (right panel). The total plot in the EXAFS is the theoretical model of ε-fcc of Co bulk. It is clear that the EXAFS signal of the Co particles, which is proponional to the number of nearest neighbors, is much smaller than that of the bulk*

shell, the signal-to-noise ratio has to be improved before a definite conclusion can be reached as to what the physical and chemical properties of these nanoparticles are example. reduction of the oxidized nanoparticles to their pure metallic forms before the measurement.

Effect of Catalyst on Nanotube Growth

For, the effect of different catalyst combinations on nanotube growth in a pulsed laser (YAG) ablation facility, a mechanism in which the laser heats the surface of the target, forming molten carbon and catalysts that are ejected from the surface is used. As the molten particles cool, dissolved carbon segregates from the metal and SWNT growth occurs. The yield depends on the laser power and on the amount of Ni and Co in the target. Their best yield was found to obtain with Ni/Co/C in the atomic ratio of 0.15/0.15/99.7.

Zhang and Iijima irradiated a target of pressed C_{60} fullerenes and Ni/Co catalyst powers within a tube furnace. They found that SWNTs are produced in low yield at 673 K, but almost none are found at room temperature. They compared this with a graphite/catalyst mix at 673 K and found almost no nanotubes. Carbon spices, mainly excited atoms and small clusters, first ablate from the target and form a Plasma Zone. The plasma cools by expansion and mixing with ambient argon until local thermodynamic equilibrium (LTE) is reached. SWNTs are formed at the stage where the plasma zone is hot and dense. The species ablated from the C_{60} and graphite targets differ due to the presence of pentagons in C_{60} and only hexagons in graphite. Di-atom production is greater in the C_{60} because of lower bonding energy. Thus, there is greater excess of thermal energy, hence a greater flux of growth species and a higher effect of catalysts that increases the growth rate of, SWNTs excited C_{60} vapor, by double-pulsed lasers and a long persisting emission of C_2 Swan bands. Because of the very short lifetime of C_2 Swan bands, this implies that excited C_2 is produced over a long period. Thus, photo-fragmentation of C_{60} results in production of electronically excited C_2. Fullerenes are products of laser ablation of graphite when no catalysts are present, and they are found with SWNTs when metal catalysts are present. Therefore, photo-fragmentation and decomposition of fullerenes and the 'shrink wrap' phenomenon (spontaneous decomposition of $C_n^* \rightarrow C_{n-1}^* + C_2$, $n \leq 58$) produces significant amounts of C_2 to serve as feedstock for SWNT growth.

The dynamics of the laser-ablation plume in an argon atmosphere is studied by recording plume images at various times from 100 μs to about 3 s; and the flow at various times and locations with LIF of Co, C, C_2, and C_3 and laser-induced luminescence (LIL) of particles. Spontaneous emission from the plume is observed at early times. It decreases until LIF/LIL dominates. There are no particles early in time, carbon and catalyst clusters form as the plume cools, long-persisting Co by LIF, from which one inferrs that catalyst clusters are slow to form (~ 2 ms at 1273 K).

CARBON NANOTUBES AND NICKEL NANOPARTICLES

Formate compounds (Fe, Ni or Co formate) are used as sources of catalysts for CVD synthesis. These compounds decompose rapidly at elevated temperatures, in an inert or reducing atmosphere to produce pure metals (releasing CO_2 and H_2). Carbon is removed by calcination at elevated temperatures (typically around 773 K) in air, to produce an oxide. Alternatively, oxides, or other inorganic salts (e.g. nitrates) that decompose to oxides, are used directly. The precursors decompose directly to form the required monodispersed metal nanoparticles. Certain combinations of precursor and support are effective at producing high yields of pure nanotubes. When the precursor molecules are dispersed into supporting materials like SiO_2, Al_2O_3 or CaSiOx powders, they are adsorbed onto the surfaces of the solid. At elevated temperatures, and just prior to the growth of CNTs, the precursor thermally decomposes and the metal nanoparticles are formed (1 – 3 nm). Such metal nanoparticles are chemically very active, thus serve as catalyst for the growth of CNTs.

Preparation of Ni Catalyst

Preparing the Ni catalyst is as follows: 500 mg SiO_2 powder is placed in a 100 ml beaker. To this add. 25 ml aqueous solution of $Ni(HCO_2)_2$ (1.08×10^{-2} M). The mixture is vigorously stirred at room temperature for 30 minutes, followed by further stirring at higher temperature (363 K) for about 20 hours to dry the sample. The metal loading in such a prepared sample is 3.2%, relative to the loaded silica. The sample is carefully grounded before introducing into the CVD furnace for the growth of CNTs. The CVD operation is normally performed at 873 – 1273 K, under argon atmosphere. Both SWNTs and MWNTs are produced. The quality of the nanotubes, involving the diameter, length, ratio of SWNTs over MWNTs and the amorphous carbon on tube surfaces, is controlled by controlling the catalyst loading, dispersion level in supports and the CVD growth parameters.

Production of High Purity Single-Walled Carbon Nanotubes

It is found that fumed silica powder yields the best results (pure SWNT with the highest yield). The silica support may be removed by hydrofluoric acid after the growth. At elevated temperatures, and just prior to tube growth. the catalyst is introduced into the growth furnace where thermal decomposition occurred rapidly to yield fresh Ni nanoparticles, signified by a blackening of the substrate. The subsequent CVD growth is performed using an Ar/CH_4 (1:1 by volume) atmosphere. Reaction temperatures are set between 873 to 1273 K, with an optimal temperature of 1133 K.

Carbon yields are measured by weighing the samples prior to and after the growth and also by thermal gravimetric analysis. In the former, yield is calculated on the basis of the following formula: yield (%) = $[(m_1 - m_0)/m_0] \times 100\%$; where m_1 is the mass of the catalyst/support sample after CVD growth, whilst m_0 is the mass of the sample after heating the sample under the same conditions but without methane. In the latter, the nanotube samples are heated under air to 1023 K. The yield is calculated by dividing the weight loss by the remaining sample weight. In both methods, the measured yields represent the carbon gains relative to the loaded supporting materials after CVD growth.

Figure 6.71 (a) and (b) show electron microscope images (SEM and HRTEM) of SWNTs grown by this method. The SEM picture indicates that the produced tubes are abundant. The HRTEM image shows that the tubes are pure single-walled, nearly monodispersed in diameter and almost no amorphous carbon on the tube surfaces. They are mainly in the form of small ropes or bundles although some individual tubes have also been found in other observations. Double wall nanotubes are rarely observed and thus considered in very low concentration in the samples.

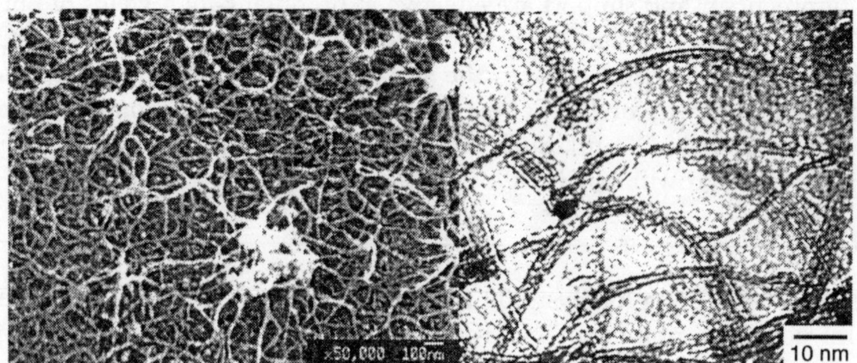

Fig. 6.71 *(a) SEM Image, the left one; (b) HRTEM image of SWNT from the as-prepared sample, grown by using Ar/CH_4 (1:1 in Volume) at 1173 K. SWCNTs are mainly seen in bundles and have diameters of around 1.0 nm*

Table 6.5 shows the measured yield data as a function of the Ni wt% loading. Figure 6.72 gives an example of the measured TGA profiles. It is seen that both analytical techniques result in consistent yield data. TGA shows that the sample begins to burn at about 788 K causing weight loss in the recorded profile, and also suggesting that relatively little amorphous carbon is present in the sample, which is in agreement with the HRTEM observations. The lower yields obtained by TGA reflects the existence of nickel oxide arising from thermal treatment in air. All wt% yields are relative to the weight of the silica support, however they can be converted into the nanotube to Ni yields, which we believe more clearly represents the productivity of the

Table 6.5 The measured yield of carbon nanotubes as a function of the Ni loading on Silica support. Both yield and metal loading are relative to the mass of the Silica support. The nanotubes are grown by using methane at 1173 K for 2.5 Hours

Sample No.	*Ni loading wt%*	*tube yield wt% (weighing method)*	*tube yield wt% (TGA method)*	*tube to Ni ratio (weighing method)*
1	0.1	34.6	23.2	346
2	0.8	47.9	3.8	59.9
3	3.0	58.3	–	19.4
4	5.4	60.7	57.3	11.2 (SWNTs & MWNTs)
5	6.8	61.2	54.0	9.0 (SWNTs & MWNTs)

catalyst. In these terms, for example, the 58.3 wt% yield is equivalent to a nanotube to nickel yield of ~1940 wt%. In other words, in this case, a gram of nickel metal may assist the growth of 19.4 grams of SWNT.

Comparing the yield measurements with the electron microscope observations for different samples, it is found that only pure SWNTs are produced for the catalyst samples with lower Ni loadings on the support (≤ 3 wt%, see Table 6.5). For the samples with higher metal loadings, some multi wall carbon nanotubes are observed along with the majority of SWNTs.

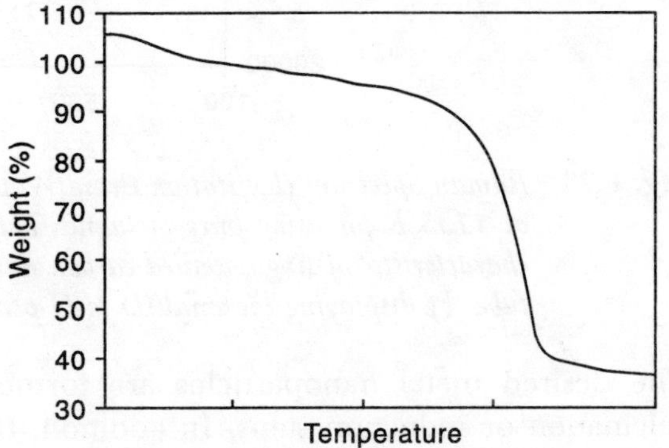

Fig. 6.72 *An example of the measured thermal gravimetric analysis profiles showing the weight loss of the nanotube/support sample by heating in air. The temperature increasing rate for the test is 10 Kmin^{-1}*

Raman spectra is collected from a range of samples. Figure 6.73 shows one of these spectra. The relatively broad D peak at 1293.3 cm^{-1} relates to the presence of defects. The G peak is strong and narrow. It shows the three peaks typically associated with single wall nanotubes at 1569, 1582 and 1595.6 cm^{-1}, relating to the in-plane vibrations of a highly curved graphene sheet. Ratios of the intensities of D to G peak can be used as an indicator of the extent of disorder within the nanotubes. The small ratio shown here, $I_D/I_G \approx 0.15$, indicates that the defect level in the atomic carbon structure is low. The radial breathing modes in the low frequency region show four components at 200.1, 212.5, 223.9 and 231.5 cm^{-1}, but with the 231.5 em^{-1} mode in dominance. The frequency of these modes is known to be inversely proportional to the diameter of the SWNT. Based on equation v (cm^{-1}) = 223.75/d, the tube diameter can be calculated as, d = 1.12, 1.05, 1.0 and 0.97 nm, respectively. The average diameter is then d = 1.04 nm, with a small diameter distribution (±0.07 nm). This result is consistent with the observation in the HRTEM, indicating a narrow size distribution in the produced SWNTs.

A straightforward method for the synthesis of high quality SWNTs *in high yields* is by using nickel formate. It acts as an ideal precursor for the formation of the Ni nanocatalysts particles.

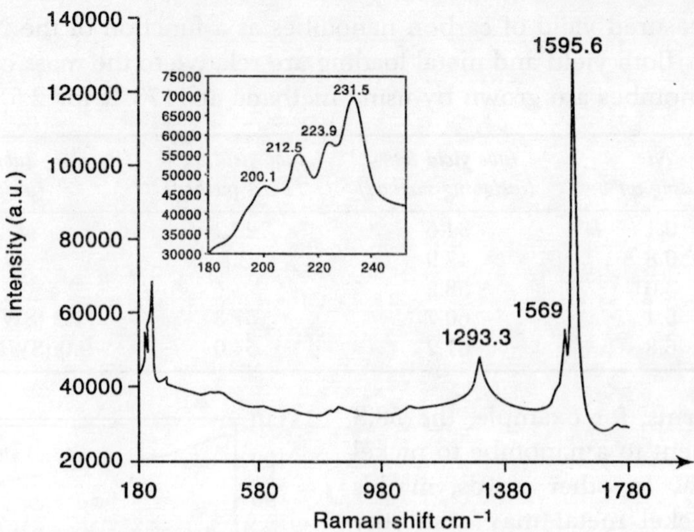

Fig. 6.73 *Raman Spectrum (Excitation Beam Wavelength: ~785 nm). Taken from a nanotubes sample grown at 1133 K for using pure methane. This spectrum shows peaks in the lower frequency region, the characteristic of single-walled carbon nanotubcs. It also indicates the low-level atomic defects in the tubes by displaying the small D to G peak intensity ratio*

The desired metal nanoparticles are formed directly from the formate, without the usual calcination or reduction steps. In addition, by optimising the experimental conditions, the Ni nanoparticles remain catalytically very active, having a small particle size and a narrow size distribution. These factors have enabled to produce high-quality SWNTs in high yields. In addition, we describe the use of Ni nanoparticles formed from its formate precursor as catalyst for the growth of multiwall carbon nanotubes (MWNTs) or a mixture of SWNTs and MWNTs by change of the carbon source in the CVD process and the substrates that the Ni nanoparticles are deposited on.

Nickel formate is such an ideal precursor for the formation of the desired Ni nanoparticles for the growth of SWNTs is because: in the decomposition mechanism; measurements by thermal analysis (TGA) have revealed that this compound dehydrates first at 433 K followed by a thermal decomposition at ~533 K. Unlike many other nickel salts that normally require H_2 as a reducing agent, nickel formate directly decompose to metallic nickel and gas species by a self-redox reaction.

$$Ni(HCOO)_2 \cdot 2H_2O \xrightarrow{433\,K} Ni(HCOO)_2 + 2H_2O$$

$$\downarrow 533\ K$$

$$Ni + H_2 + 2CO_2$$

In addition, the nickel particles could be obtained at a residence time as short as 0.1 seconds at 873 K. Comparing these results with our experimental conditions, particularly the CVD growth temperature (1133 K), we may assume that the formate decomposes instantane~usly in our case and a larger number of Ni nuclei are formed simultaneously. This type of decomposition mechanism favours the formation of small metal nanoparticles. On thermal decomposition of nickel formate dihydrate, by using QXAFS technique (quick scanning x-ray adsorption fine structure spectroscopy), It is concluded that the pyrolysis yields a finely divided nickel metal with mean particle diameter of about 1.2 nm. It is difficult to observe directly these very small Ni nanoparticles under the electron microscope after the tube growth presumably because the small metal particles have sintered in the porous silica and constitute only a tiny fraction of the overall yield. However, since the carbon nanotubes have a narrow size distribution with diameters that should match the size of the metal particles, therefore, the Ni nanoparticles must have a narrow size distribution. In addition, high purity of the formed nickel nanoparticles plays a vital role in their high catalytic activity. A catalyst sample heated to 1133 K in argon examined in with Raman spectroscopy, does not give a carbon signal peak in the spectra. This result suggests that the thermal decomposition of the formate at this temperature produces no amorphous carbon on the Ni surface.

Production of Multi-Wall Carbon Nanotubes or a Mixture of SWNTs and MWNTs

Multi-wall or a mixture of single wall and multi-wall carbon nanotubes are obtained by using acetylene rather than methane in an argon atmosphere. Prior to any run, a quartz reaction tube (14 mm inner diameter) is heated in argon and left for an hour for the temperature within the furnace. Once the temperature is equalibrated, an alumina boat containing about 10 mg of catalyst powder is placed in the centre of the furnace. Then, acetylene is added to the argon gas flow to grow the nanotubes. In some cases hydrogen is added to see its effect on the resultant product.

The temperature of the central furnace region is maintained between 873–1123 K. The resultant products are collected form the boat and then analysed. Waste gases are passed through paraffin oil and activated carbon balls to remove hydrocarbon and other toxic by-products.

In Figures 6.74 (a) and (b) show examples of SEM and HRTEM images of the carbon nanotubes grown from the nickel catalyst on CaSiOx support. Figure 6.75 shows an example of

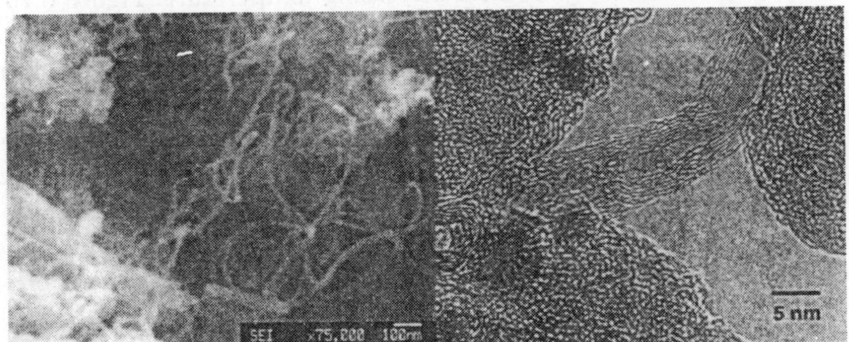

Fig. 6.74 *(a) SEM and (b) HRTEM Images of the MWNTs Produced by Using the Ni Catalyst on CaSiOx Support and C_2H_2 Gas at 923 K. The Mean Average Diameter of the MWNTs is about 10 nm*

the SEM images of the nanotubes produced from the silica (SiO_2) substrate. Comparing the tubes from the two substrates, it is noted that they are different in several ways. Firstly, the SiO_2 produces more straight tubes compared with CaSiOx, and the tubes seem to run or 'bridge' from one point of the support to another. Secondly, the yields of the tubes in the SiO_2 case are overall less than that of the CaSiOx. Furthennore, Laser Raman spectra (see below) indicate the tubes out of the CaSiOx support are mainly multi-walled, whilst the SiO_2 support favours the production of more single wall tubes.

Fig. 6.75 *An example of the SEM images of the carbon nanotubes produced from the Ni/SiO_2 catalyst by using C_2H_2 as carbon source in CVD process at 923 K*

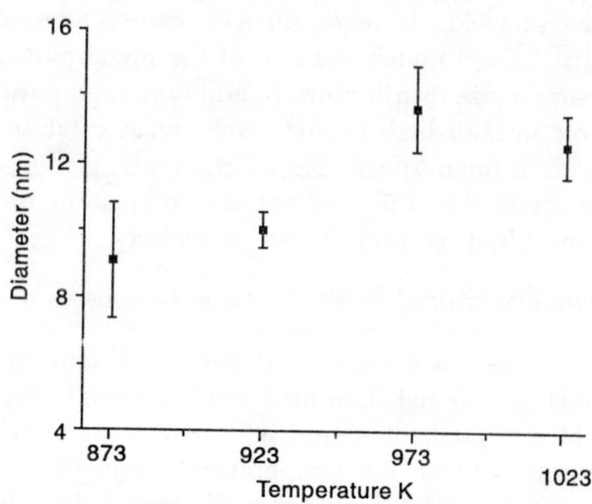

Fig. 6.76 *Average mean diameter of carbon nanotubes produced by using Ni/CaSiOx catalyst system under a flow rate of 220 ml/min of Ar:C_2H_2). The average outer diameter increases with increasing the temperature*

In both cases, the reaction temperature plays an important role in the nanotube yields and diameters. The yields are the highest at a temperature of about 923 K, and the average outer diameter of the tubes increases slightly with increasing temperature. Figure 6.76 shows a TEM analysis on the tube diameters grown at different temperatures. On the whole the outer diameter of the tubes are less than 15 nm, and the tubes have the narrowest diameter distribution at 923 K of about 1.0 run. The increase in the diameter may be explained by the effect of the mobility of the nickel nanoparticles on the catalyst supports. As the reaction temperature increases, the mobility on the support increases, favouring the formation of larger nickel particles by particle agglomeration, which would subsequently lead to larger diameter nanotubes.

Raman spectroscopic analyses were performed on various as-prepared nanotube samples using a 514 nm Argon excitation laser. Figure 6.77 shows a typical Raman spectra for the nanotubes prepared trom Ni/CaSiOx supports. The large D peak at 1350 cm^{-1} for the tubes grown at 973 K indicates the nanotubes consisting of lots defects. In comparison, the D peak for the tubes grown at 923 K is much smaller, suggesting a higher structure quality. The ratio of the D to G peak are given in Figure 6.78.

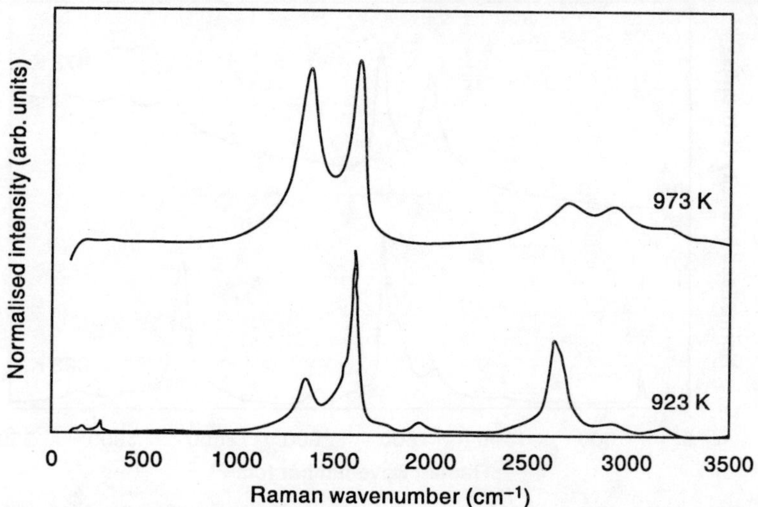

Fig. 6.77 *Raman spectra taken form nanotubes grown at 923-973 K. The breathing modes appeared at the lower profile are centred at 182 and 267 cm^{-1}, corresponding to a diameter of 1.23 and 0.84*

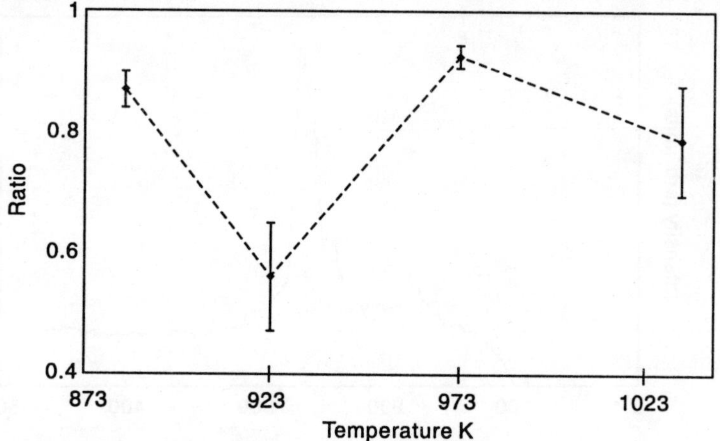

Fig. 6.78 *Average I_D/I_G ratios for nanotubes grown at different temperatures on calcium silicate supports*

The nanotubes grown at 923 K, seem to exhibit SWNT feature with the presence of breathing modes in the low wave number region (< 500 cm^{-1}). Here, the tube diameters can also be calculated in the range 0.84 to 1.23 nm. But according to the TEM and SEM analysis, the nanotubes should be primarily composed of MWNTs structure with a small percentage of SWNTs.

Figure 6.79 is shows an example of the Raman spectra on the carbon nanotubes, grown from Ni/SiO_2 catalyst system. The bottom spectrum, relating to the growth temperature of 923 K, clearly shows the presence of SWNTs. From the breathing frequencies (between 180 and 265 cm^{-1}, see (Figure 6.80), the single wall tube diameter can be estimated in the range of 0.84 and 1.24 nm.

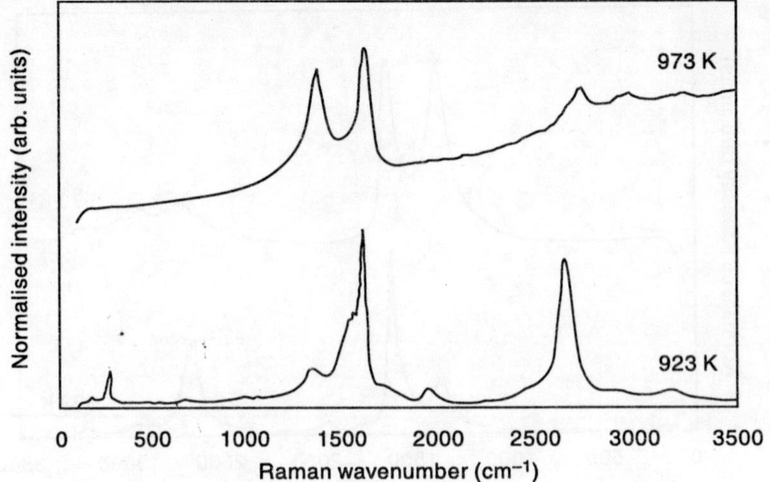

Fig. 6.79 *Raman Spectrum taken from the nanotubes grown from the Ni/SiO_2 catalyst system at 923-973 K. The bottom profile is typical for SWNTs. The peaks below 500 cm^{-1} are associated to the SWNTs, with a diameter range of 0.84 to 1.24 nm*

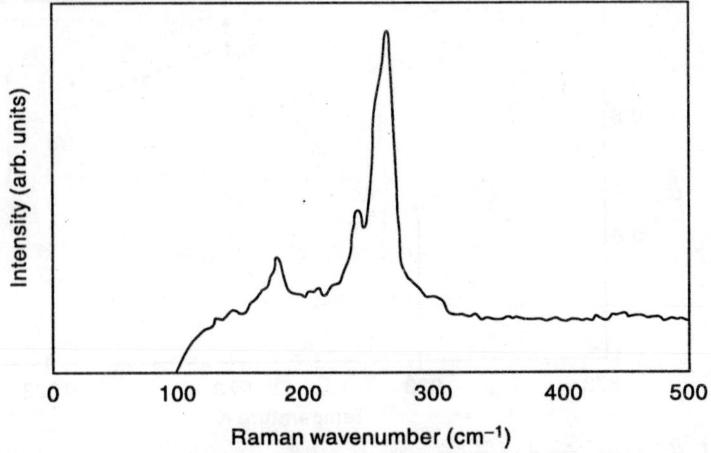

Fig. 6.80 *Breathing modes of the Carbon nanotubes grown from the Ni/SiO_2 catalyst system at 923 K. The peak are centred at 180, 241 and 265 em^{-1}, but with the 265 em^{-1} in dominance, suggesting that the corresponding SWNTs have diameters of 1.24, 0.93 and 0.84 nm and with the smallest diameter in dominance*

On examining the Raman spectra of the Ni/SiO_2 samples prior to the growth reaction, no Raman peaks appear in the breathing mode region, on the unreacted samples. In addition, the repeated reactions at 923 K indicate that the SWNTs feature are reproducible, suggesting the catalyst is good in catalysing the growth of SWNTs at such temperature and related growth conditions. The I_D/I_G ratios are also calculated based on the Raman data and they are presented in Figure 6.81. The trends are similar to the Ni/CaSiOx case, where the minimum point is at 923 K.

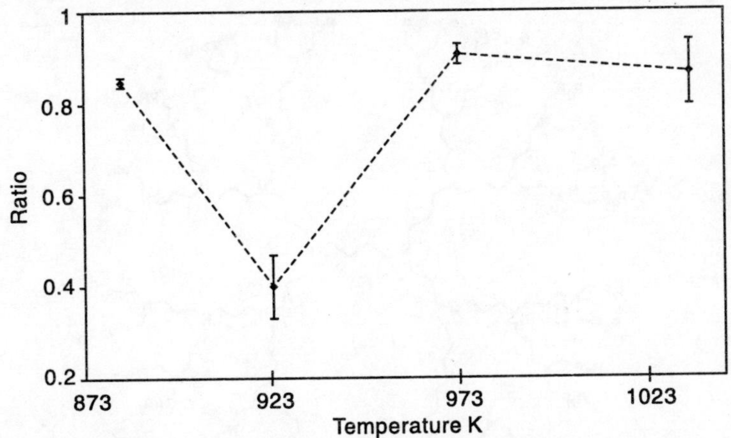

Fig. 6.81 *I_D/I_G Ratio for the nanotubes grown at different temperatures on Silica supports*

RUTHENIUM NANOPARTICLES AND CARBON NANOHORNS

It has been argued that the carbon nanotubes with the highest length/diameter ratio, hence the largest electrical field enhancement factor, are the best candidates in the carbon nanotube family for future emission devices. However, *short and stubby nanotubes* with intermediate diameters show the best field emission characteristics. Here, use of Ru/Au bimetallic cluster, [DAB-dendr-$[N(CH_2PPh_2)_2]_{16}(\mu{:}\eta^1\eta^1 - Au_2Ru_6C(CO)_{16})_{16}]$ (DAB-Ru96-Au32) is described as a molecular precursor to produce pure Ru nanoparticles as catalysts for growth of short, stubby and horn-like carbon nanotubes in a CVD pro:ess.

Synthesis of a Family of Nanoscale Ru-Containing Clusters of Clusters

The development of a dendrirner-based clusters of clusters as *quantum dots* provide unique advantages, as for example, in the preparation of monodispersed particles with well-defined molecular structures and surface functionalities.

A family of Ru-containing super clusters are assembled-around a molecular dendrimer template. High-resolution transmission electron microscopy (HRTEM), in conjunction with molecular modelling, using force-field calculations is employed to establish the morphology of these nanoparticles. In contrast to the 'magic' number synthesis of colloids wherein metal particles are stabilised by a 'passivating' external layer of organic ligands, a large internal organic scaffold [DAB-dendr-$[N(CH_2PPh_2)_2]_{16}]$ is used, to construct an array of sixteen clusters, thereby forming a nano-scale self-ordering molecule.

The compounds [DAB-dendr-$[N(CH_2PPh_2)_2]_{16}(\mu_3{:}\eta^1\eta^1\eta^1\text{-}Ru_5C(CO)_{12})_{16}]$ **1** and (DAB-dendr-$[N(CH_2PPh_2)_2]_{16}(\mu_3{:}\eta^1\eta^1\text{-}Au_2Ru_6C(CO)_{16})_{16}]$ **2** (Figure 6.82) were prepared by binding the 16[P-N-P] tridentate terminal functionalities of the organic third generation dendritic core to penta-ruthenium cluster unit $[Ru_5C(CO)_{12}]$ or di-gold hexa-ruthenium $[Au_2Ru_6C(CO)_{16}]$ cluster unit. Both **1** and **2** show a molecular multiple-layer structure (Figure 6.83). They may be viewed as a sphere consisting of a non-conducting organic core with a sheath of conducting metal particles

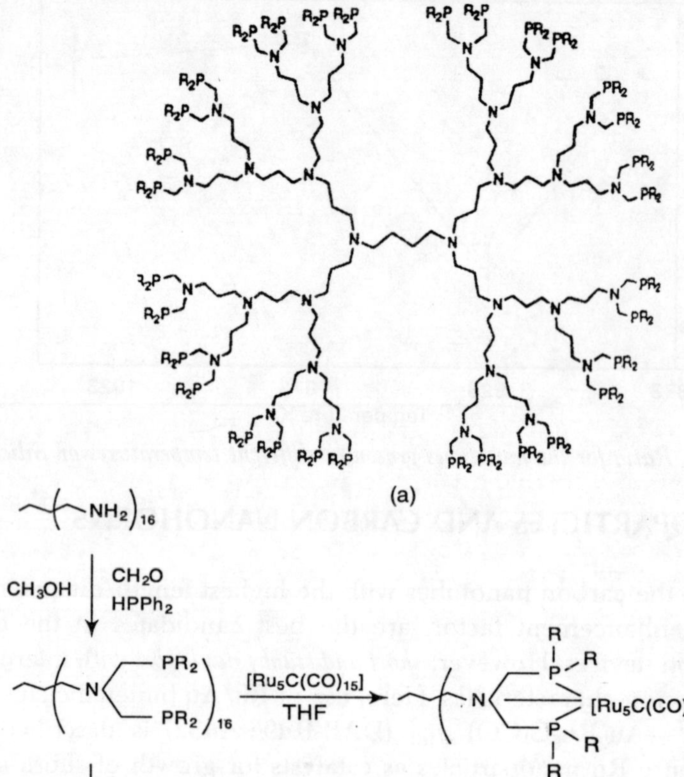

(a)

(b)

Fig. 6.82 *(a) Chemical formula of the Dendrimer Molecule Used for Assembling the Super Clusters of Clusters. (b) The Synthetic Route to Compounds **1** and **2**. $R = C_6H_5(Ph)$; $PPN = [P(Ph)_3 = N = P(Ph)_3]^+$*

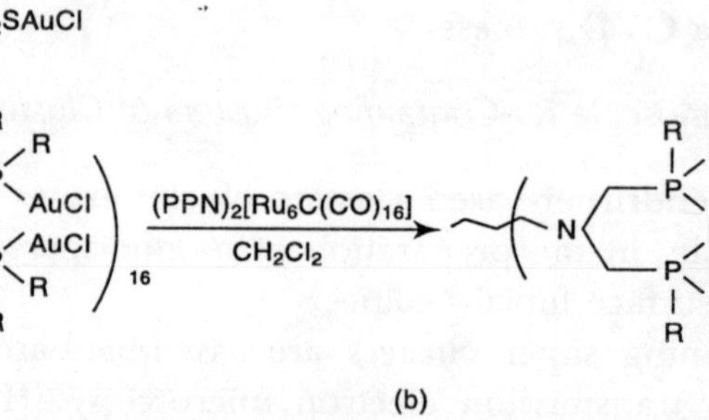

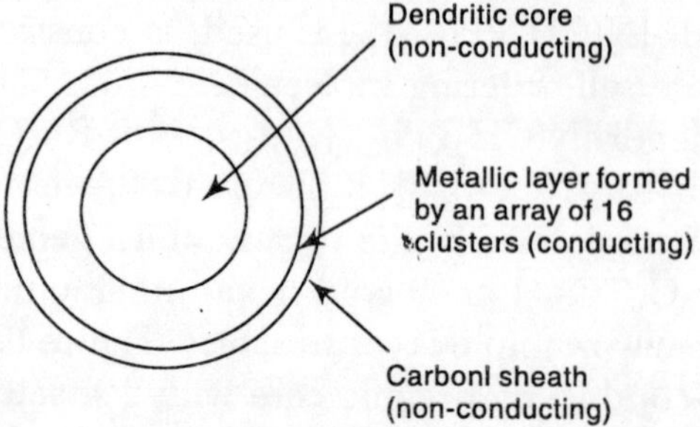

Fig. 6.83 *Representation of the Molecular 'Multi-Layer' Structure of Compounds **1** and **2***

over the surface, which in turn is covered by a further outer layer of insulating carbonyl ligands. For conduction to occur, an electron introduced into the molecule would first have to tunnel through the carbonyl sheath to the 'metallic layer', move through the metallic layer to an exit point and then tunnel again out of the molecule.

In addition to compounds **1** and **2**, the respective dimeric componds $[(CH_2)_4N(CH_2PPh_2)_2[Ru_5C(CO)_{12}]_2$ **3**, $[(CH_2)_4N(CH_2PPh_2)_2[Au_2Ru_6C(CO)_{16}]]_2$ **4** are prepared and characterised and a model compound $(PhCH_2N(CH_2PPh_2)_2[Au_2Ru_6C(CO)_{16}]]$ **5** in order to establish the bonding modes between the dendrimer and clusters. Crystals of **3** suitable for single crystal X-ray crystallography were grown from a chloroform solution by slow evaporation. Determined molecular structure is shown in Figure 6.84, along with selected bond lengths. Two penta-ruthenium clusters, related by a centre of inversion, are linked by the bis-tridentate ligand. The amino-nitrogen and the two phosphorus atoms of each terminal unit coordinate to three basal ruthenium atoms of the cluster, displacing three axial carbonyl ligands from the parent cluster. The pyramidal geometry of the pentaruthenium carbidocluster is retained. Single crystals of **5** were obtained from dichloromethane by addition of hexane. The same bonding mode is therefore anticipated in the dendritic compound.

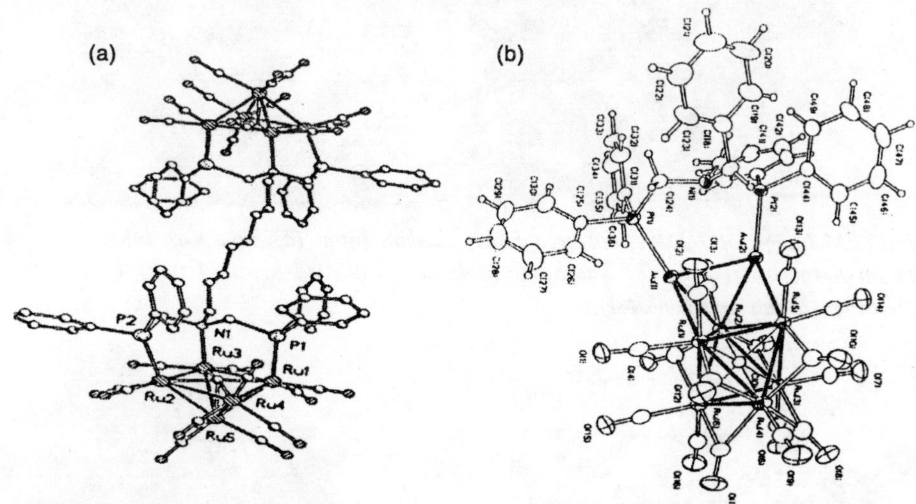

Fig. 6.84 *The Solid State Molecular Structure of the Model Compound* ***3****. Key Bond Lengths (Å): P(1)-Ru(5) 2.294(2), N(2)-Ru(1) 2.211(5), P(2)-Ru(2), 2.301(2), Ru(5)-Ru(1) 2.9218(8), Ru(1)-Ru(2) 28127(7), Ru(2)-Ru(4) 2.8966(8), Ru(4)-Ru(5) 2.8617(7)*

Compounds **1** and **3** display almost identical IR spectra in the CO region, strongly suggesting the coordinated clusters adopt a similar bonding mode. Evidence for the coordination of all the tridentate P atoms to clusters was provided by the observed single peak in the P NMR spectrum. Elemental analyses for both **1** and **2** and TGA measurement for 1 are in good agreement with those calculated for each dendrimer molecule containing sixteen $[Ru_5C(CO)_{12}]$ or $[Au_2Ru_6C(CO)_{16}]$ cluster units.

Initial attempts to record HRTEM images of **1** deposited on a carbon film, revealed that the material was highly perturbed by the electron beam, probably because of CO ejection causing movement. Despite this, good HRTEM images (Figures 6.85(a), (b)) were recorded by careful control of the beam current and exposure time. In Figure a single approximately cubic molecule is shown sat on the carbon surface of dimensions 46 Å × 46 Å. A second image was taken some 10 seconds later (Figure 6.85(b)) and shows the molecule to have collapsed inwardly to give a smaller denser particle. This may be explained by decarbonylation and agglomeration of the carbonyl clusters covering the perimeter of the dendrimer as we have observed previously. In direct contrast, Figure 6.86 shows an essentially spherical single molecule (*ca.* 8nm in diameter) of 2 supported on the carbon film. The large contrast observed is consistent with the large metal content, and may be compared to the molecular model (Figure 6.87) generated from Chern 3D based on modified MM2 parameters. These giant molecules of 2 show no further change to the electron beam even after several minutes of intense irradiation.

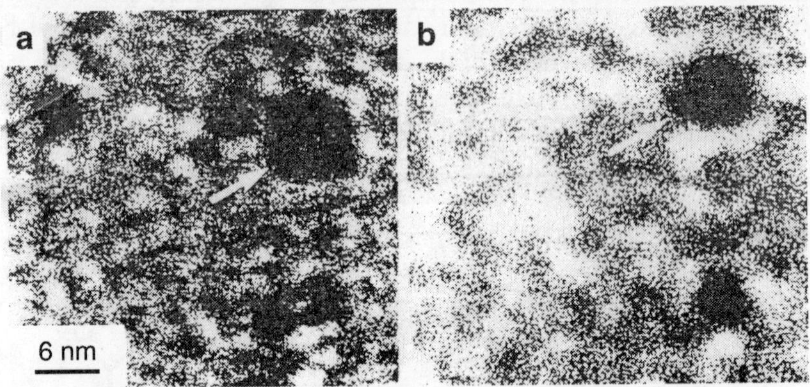

Fig. 6.85 *HRTEM Images of 1 Deposited on a Holey Carbon Film: (a) Shows an Intact Molecule (Indicated by an Arrow) at t ≈ 0. (b) The Same Molecule (Indicated by an Arrow) 10 Seconds Later After Decarbonylation and Collapse*

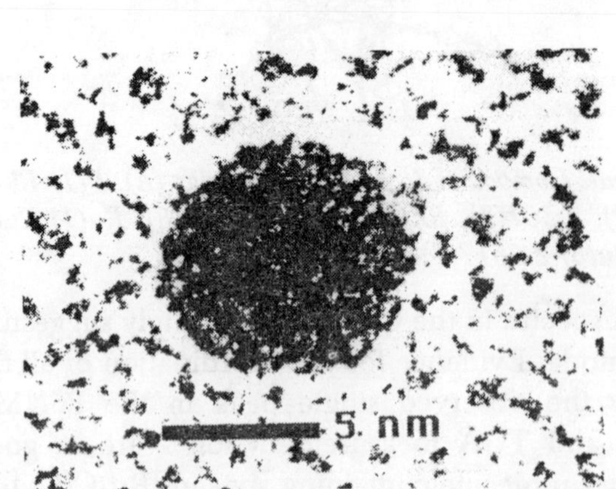

Fig. 6.86 *HRTEM Images of a Single Molecule of 2 Deposited on a Holey Carbon Film*

Fig. 6.87 *Chem 3D Model of 2 Minimised Using Modified MM2 Parameters and Molecular Dynamics*

Carbon Nanohorns (CNHs) Grown from Ruthenium Nanoparticles)

Compound 2 has been successfully used as a molecular precursor to produce pure ruthenium nanoparticles (seeds) as catalysts for the growth of carbon nanohorns (CNHs). For the growth, a TEM grid coated with thin carbon film was used as a substrate. It was first treated with iso-propanol to produce plenty of -OH groups on the surface, followed by the deposition of the precursor cluster onto the substrate from a dichloromethane solution. Growth of the CNHs was performed by CVD method. The catalyst was first annealed at 973 K under ammonia (3 Torr). Subsequently acetylene was added (C_2H_2 : NH_3 = 75 : 200 sccu) as the precursor of carbon for the vapour deposition.

In Figure 6.88(a) is shown a TEM picture of the CNHs grown in this way. The CNHs are short and hollow, with a ruthenium particle sitting at the top. The cavity diameter is dependent on the size of the encapsulated metal particle. This type of structure suggests a "tip growth mechanism" Figure 6.88(b) shows that the CNHs are not uniform along their central axis. The lower part of the tube body has larger diameter, suggesting that a possible growth mechanism may first involve the core formation of the CNHs, which then acts as substrate for subsequent thickening by deposition of secondary carbon atoms. Figure 6.89 is a HRTEM picture for several other CNHs. It can be seen that the body of the CNHs does not correspond to a well-ordered graphitic structure suchasnonnal multi-wall carbon nanotubes, nor of amorphous carbon. Instead they appear to possess an intermediate structure falling something between these two. Some well-layered graphitic regions are observed around the metal particles at the top thus, giving an insight into the activity of the catalytic metal particle in the growing process.

Both Figures 6.88 and 6.89 indicate that the metal nanoparticles have a wide size distribution. The smallest metal particle shown in Figure 6.89(a) is about 2 nm in diameter. As the size of a single precursor cluster is about 8 nm in diameter, it is assumed that the 2 nm metal particle is possibly formed from the collapse of a single DAB-Ru96-Au32 molecule after burning off the organic dendritic core and the periphery CO ligands. The larger metal nanoparticles are obviously produced by the aggregation of smaller particles and their subsequent coalescence.

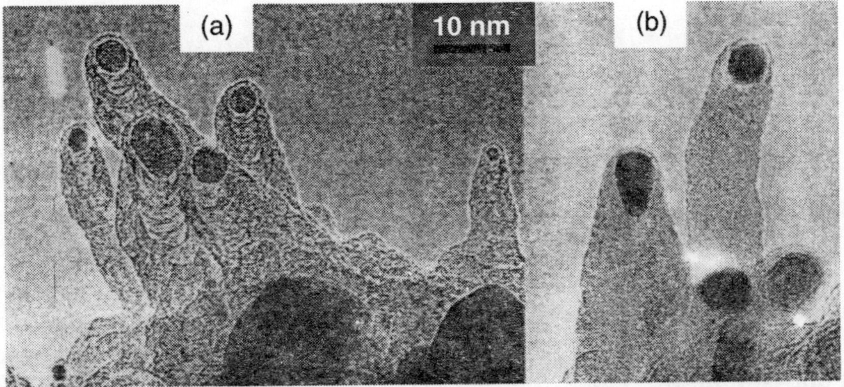

Fig. 6.88 *TEM Images Show That the CNHs Have a Hollow Core and a Partially Ordered Graphitic Body Catalytic Particles (2-10 nm dia.) Can Be Seen at the Top of the Nanohorns*

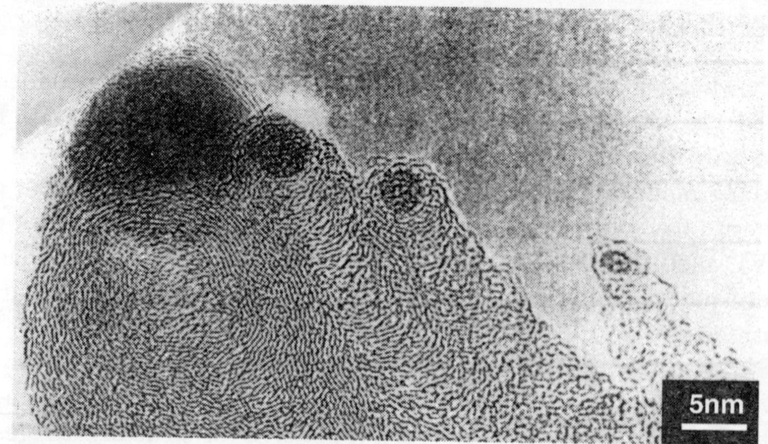

Fig. 6.89 *TEM Images Show Partially Ordered Graphitic Structure of the CNHs. This Image Also Shows a Large Dispersion in Size for the Ru Particles*

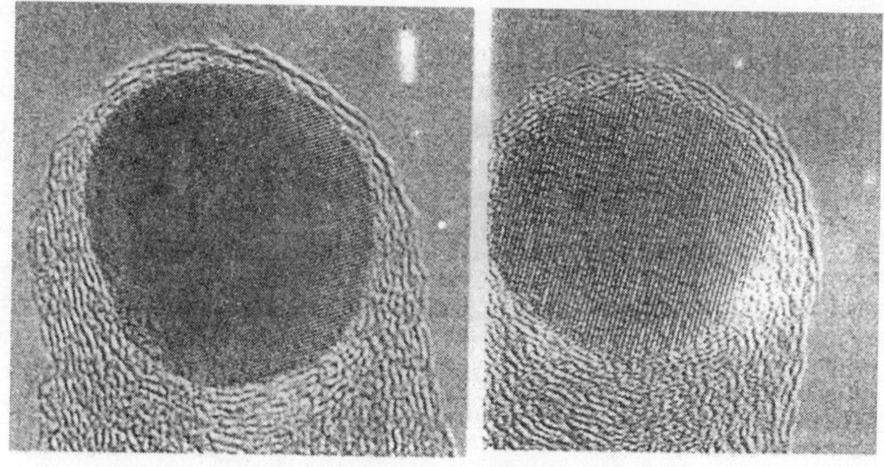

Fig. 6.90 *High-Resolutibn TEM Images of Ru Particles (Single Crystals) at the Tip of Two CNHs. The Left Particle Has a Lanice Spacing of 2.05 Å Which Corresponds to Ru (101) [2.05] Å], and the Right Particle Has a Measured Spacing of 2.14 Å Corresponding to Ru (002) [2.138 Å]. Scale Bar: 2 nm*

Figure 6.90 shows an example of the crystallinity of the metal nanoparticles. Measurements of the crystal lanice spacing for several similar particles but with different crystal planes and binges, confirm that they are pure ruthenium nanocrystals. This indicates that the thermal decomposition of the precursor results in the formation of Ru nanocrystals separating from the Au metal. This is consistent with the literature report that the Au and Ru do not alloy at elevated temperatures, and the catalytic effect described above is arising from Ru rather than Au or an alloy of the two.

Field emission measurements using the CNHs sample described above are taken (Figure 6.91). The tests are carried out in a vacuum of 10^{-7} torr by applying a positive voltage to a tungsten current probe, with tip diameter of about 10 μm and measuring the current at the cathode. The separation between anode and cathode is set to 10 μm. Current-field characteristics, recorded at different areas of the sample showed turn-on fields of the order of 5 V/μm, saturation current densities as high as 1 mA/cm^2. The threshold field calculated for the nanohoms, i.e. the field that corresponds to an emined current of 10^{-6} A, ranges between 6 and 10 V/μm in accordance with the values measured for multiwall carbon nanotube films.

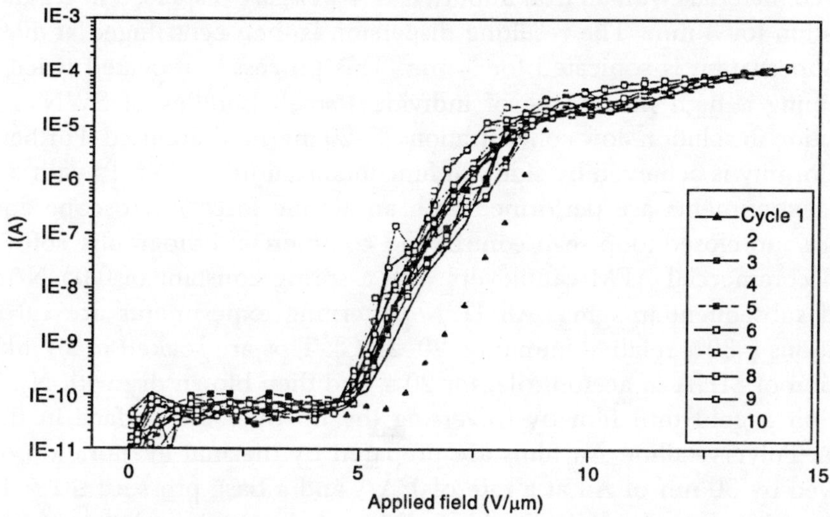

Fig. 6.91 *An Example of the I-V Characteristics Measured in a Vacuum of 10^{-7} mmHg by Applying a Positive Voltage to a Tungsten Current Probe*

SWNTS FROM SOLUTIONS

SWNTs have been successfully positioned along straight line features comprised of amine-terminated self-assembled monolayers (SAMs). Thus far, however, no one has demonstrated the ability to simultaneously control the position, shape, and linkage of SWNTs on the sub-μm scale.

To construct much more' sophisticated architectures from SWNTs including rings, electronic interconnects, and structured thin films.

SWNTs are attracted to the hydrophilic portions of a gold substrate patterned by dip pen nanolithography (DPN) and more specifically to the boundary between hydrophilic and hydrophobic SAM features made of 16-mercaptohexadecanoic acid (MHA) and 1-octadecanethiol (ODT), respectively. The process uses solvent as a preferred carrying media for the carboxylic acid-terminated features and allows one to control the manipulation and assembly (aligning, positioning, shaping, and linking) of SWNTs on the μm- to sub-100-nm length scale, and in some cases, over large areas by using microcontact printing (μCP) and parallel DPN.

The Experimental Method and Materials

Materials used are: MHA (90%), ODT (98%), MUO (97%), 1,2-dichlorobenzene (99%), and ethanol (200 proof), AUT (99%) Ti (99.7%) and Au (99.99%) wires PEF-SH is prepared as described.

SWNT FUNCTIONALIZATION AND SOLUTION PREPARATION

Purified HiPco materials with an iron impurity <1.4 wt%, are dispersed in 1,2-dichlorobenzene by mild sonication foı 3 min. The resulting dispersion is then centrifuged at 50,000 × g for 10 min, and the supernatant is sonicated for 3 min. This process is repeated twice, resulting in a solution containing a high percentage of individual/small bundles of SWNTs. To minimize SWNT aggregation in solution, low concentrations (5–20 mg/liter) are used. Further improvement in solution uniformity is achieved by sidewall functionalization of SWNTs with dodecyl groups.

DPN. DPN experiments are performed with an atomic force microscope equipped with a 100 μm scanner and closed-loop scan control and commercial lithography software.

Gold-coated commercial AFM cantilevers with a spring constant of 0.05 N/m are used for patterning and subsequent imaging. All DPN patterning experiments are carried out under ambient conditions (≈30% relative humidity, 20–24°C). Tips are soaked in an ink solution (e.g., saturated solution of MHA in acetonitrile) for 20 s, and then blown dry with N_2. MHA features are generated on a gold thin film by traversing the tip over the surface in the form of the desired pattern. Polycrystalline Au films are prepared by thermal evaporation of 10 nm of Ti on SiO_x followed by 30 nm of Au at a rate of 1 Å/s and a base pressure $\leq 1 \times 10^{-6}$ mmHg.

μCP. Stamps are fabricated and "inked" with 5 mM total alkanethiol solution by gently spreading a drop on the surface of the stamp with a Q-tip. After the stamp is dry, patterned structures are generated on the surface by bringing the stamp (by hand) into contact with a clean Au thin film for 10 s. Then, the substrate is rinsed with ethanol and dried with N_2. The regions surrounding the molecular features patterned by DPN or μCP are passivated with a monolayer of alkanethiol molecules (e.g., ODT) by immersing the substrate in a 1-mM ethanol solution for 10 min followed by rinsing with ethanol and water alternatively. Finally, the substrate is dried with N_2.

Monolayer Formation. SAMs of alkanethiol molecules are prepared on Au thin films by immersing the substrate in a 1-mM ethanol solution of the corresponding molecules for 1 h, followed by rinsing with ethanol and drying with N_2. Static contact angles are measured by using the half-angle technique.

SWNT Deposition. To a DPN-patterned, μCP-patterned, or monolayer-based substrate -is added a drop of 1,2-dichlorobenzene containing 5-20 mg/liter of SWNTs. The substrate is then tilted back and forth 5–10° to allow the drop to slowly roll through the patterned area five times. Subsequently, the substrate is rinsed gently with clean 1,2-dichlorobenzene to minimize nonspecific binding, and then let to dry in air.

AFM and Raman Characterization. Tapping mode AFM images are taken Raman images are obtained with a confocal Raman microscope, with a 633-nm excitation line.

Monte Carlo Simulations. In the Monte Carlo calculations only the first three carbon groups in the molecular SAMs are considered. CH_3CH_2COOH and $CH_3CH_2CH_3$ are used to represent MHAS and ODT, respectively.

Each SAM is constructed from an ensemble of 961 molecules in a hexagonal array, resulting in an overall simulation area of 13 × 11 nm, The optimized geometries of these molecular SAMs give an averaged intermolecular distance of 5.0 and 4.5 A for the COOH- and CH_3-SAMs, respectively.

By using the optimized geometries of these molecular SAMs and a 10-nm-long SWNT, made of 1,221 carbon atoms interaction energies, can be calculated. A 13-nm droplet, consisting of 9,948 1,2-dichlorobenaene molecules, is first optimized and sliced in half, which creates a flat surface comparable in size to the surface of the SAM. The SWNT is then put on this surface and the energy is minimized with the constraint that the SAMs, the solvent surface, and the nanotube are treated as rigid, i.e., the nanotube is only allowed to translate and rotate. The resulting interaction energy between nanotube and solvent is compared with those obtained with that between the nanotube and the SAM including the total interaction between the nanotube and its surroundings and the interaction between the solvent and the exposed part of the SWNT. Thus, this nanotube/exposed solvent interaction energy is same for all three surfaces.

To evaluate the prospect of using SAM boundaries for controlling the assembly of SWNTs, DPN is used to generate patterns of MHA consisting of lines, dots, rings, and even alphabetical letters on a gold substrate. Then, the exposed gold regions of the substrate are passivated with ODT. A drop of 1,2-dichlorobenzene containing SWNTs was then rolled over the patterned substrates (Fig. 6.92(a)). As 1,2-dichlorobenzene wets the MHA features but not the ODT

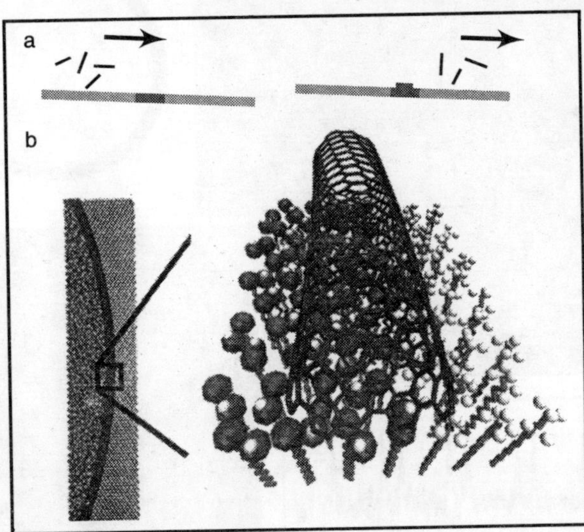

Fig. 6.92 *Schematic diagram of the directed assembly process. (a) Schematic illustrating the rolling of a drop of the SWNTs/1,2-dichlorobenzene solution on a two component surface, COOH-SAM and CH_3-SMA. (b) The SWNTs are selectively transported to the COOH-SAM and pinned at its boundary with the ODT SAM. Upon drying, the SWNT bends to precisely follow the molecular path of the patterned COOH-SAM*

passivated regions; so SWNTs are guided and localized on the hydrophilic regions of the substrate. As the solution containing the SWNTs evaporates, the nanotubes are attracted both to each other and the boundary between the ODT and MHA. This evaporation creates a high local concentration of the SWNTs at the SAM boundaries and almost exclusive assembly on the MHA features (Fig. 6.92(b)). There is a strong van der Waals attraction between the SWNTs and the carboxylic acid moieties of the MHA, which is apparent in molecular modeling studies (see below). Because the tubes are too long to assemble within an individual feature they are organized at the interface of the SAMs and bent along the perimeter of the features to maximize the overlap with the MHA feature and minimize the tension arising from nanotube bending. In the case of dots or rings, this assembly process results in circular structures and substantial bending of the SWNTs, the extent of which depends on the radius of curvature (Figs 6.93 and 6.94). Note that within these features the SWNTs follow the perimeter of the dot and form continuous architectures through intertube linking.

Circular structures made of SWNTs are unusual. These structures show interesting curvature-dependent magnetic and electronic properties. Rings of SWNTs, ≈ 500 nm in diameter, are first observed as a low-yield side product in nanotube synthesis. Recently, ring structures are produced in solution by an ultrasonication method and ring closure reactions. Such SWNT rings can easily be formed and positioned in an ordered array on a surface (Fig. 6.93(a)). For example, 1- to 3-μm-long nanotubes form ring structures on 170-nm-wide, 650-nm-diameter MHA ring

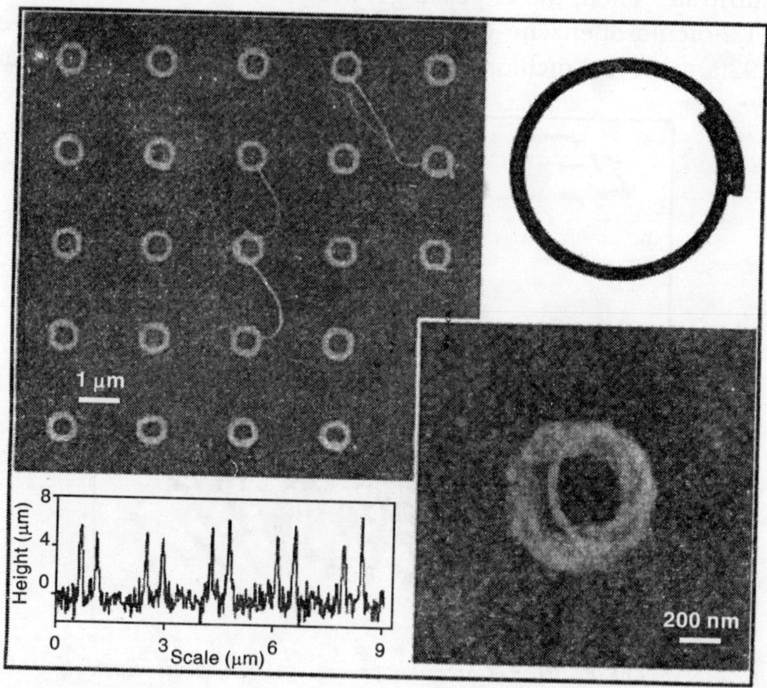

Fig. 6.93 *SWNTs assembled into rings and nano letters. (Left) AFM tapping mode topographic images (Upper) and height profiles (Lower) of SWNT rings in a 5 × 5 array are shown. (Right) A zoom-in view of one SWNT ring (Lower) and a molecular model of a coiled SWNT (Upper)*

features on an Au surface, with exposed Au passivated with ODT. The average height of each SWNT ring is 6 ± 2 nm, demonstrating that these are stacked intertwined structures. They are about five times thinner than the analogous structures formed by ultrasonication. By using shorter SWNTs (≈0.4–1.5 μm), they readily bend to form sub-μm-sized arcs. Even greater control is demonstrated by shaping SWNTs into nano letters with this approach.

It is found that the choice of affinity template SAM and passivating SAM pair is critical for achieving precision in the manipulation and assembly of SWNTs. When the passivation layer is 1-mercaptoundecanol (MUO) or 11-mercaptoundecyl-penta-ethyleneglycol (PEG-SH), the SWNTs assembled on the MHA features but not along the edges of such structures, as in the case of the OOT/MHA system (Fig. 6.94 (a)-(c)). Although dots of MUO, PEG-SR, and 11-amino-1-undecanethiol (AUT), where the gold substrate is passivated with ODT, all show affinities for the SWNTs, the interaction seems weaker as compared with MHA as evidenced by a lower density of SWNTs on such features (Fig. 6.94 (d)–(f)). MHA shows a higher tendency than AUT to assemble and surface-confine SWNTs.

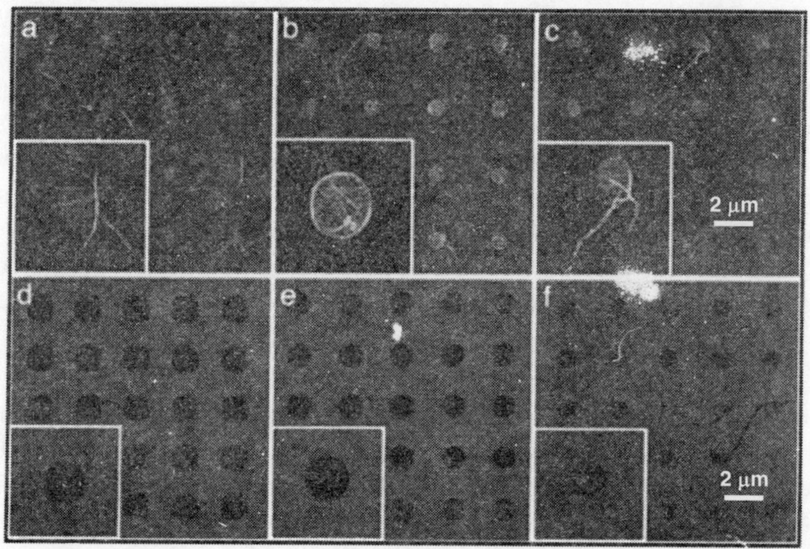

Fig. 6.94 *The assembly of SWNTs depends critically on the surface functional groups. (a-c) AFM tapping mode topographic images of a series of substrates with 1-μm MHA dot arrays passivated with MUO (a), ODT (b), and PEG-SH (c). These dots show that ODT is the superior passivation layer. (d) AFM phase images of a second series of 2-μm dots of MUO (d), AUT (e), and PEG-SH (f), passivated with ODT. None of these show the optimized SWNT assembly observed for the MHA/ODT system in b. (Insets) Shown are zoom-in images of a selected dot. All images were taken at a scan rate of 0.5 Hz. The height scale is 20 nm, and the phase lag is 10°*

Contact angle measurements show that 1,2-dichlorobenzene wets all but the OOT SAMs on gold. The static contact angle for 1,2-dichlorobenzene on OOT SAMs was determined to be 60 ± 2°, whereas the contact angles for the other SAMs were all <10°. Because the solvent wets the COOH-SAMs but not the CH_3-SAMs, no solvent remains on the CH_3-SAM as the substrate is pulled from the SWNT solution. In contrast, all of the other surfaces retain a thin liquid film, which results in nonuniform SWNT assembly. Compared with the SAMs of MUO, PEG-SH, and AUT, dot features of such materials with surrounding regions passivated with OOT show higher densities of SWNTs on the affinity template portions of the surface. These observations strongly suggest that the solvent/substrate interactions are in part responsible for localizing the SWNTs on the MHA patterns.

When a SWNT is driven close to the MHA pattern in the ODT/MHA system, Monte Carlo simulations show that van der Waals attractions between the MHA and SWNT provide the driving force for assembly. Using a parallel Monte Carlo program package and the Amber force field, we obtained the following relative interaction energies between a SWNT and its surroundings: –0.88 eV per nm SWNT for $E_{SWNT/solvent}$ (the interaction of a SWNT with the 1,2-dichlorobenzene solvent), –0.84 eV per nm SWNT for $E_{SWNT/CH3\text{-}SAM}$ (the interaction with a CH_3-SAM), and –1.05 eV per nm SWNT for $E_{SWNT/COOH\text{-}SAM}$ (the interaction with a COOH-SAM). This ordering, COOH-SAM >> solvent> CH_3-SAM, favors SWNT adsorption from solvent onto the COOH-SAM and not the CH_3-SAM.

Interestingly, the interaction of a SWNT with AUT, –0.87 eV per nm SWNT, is much weaker than that for MHA, manifesting the important role of van der Waals interactions in the assembly of SWNTs. In addition, if the tube is sufficiently long (length greater than the size of the MHA feature) and flexible, it will maximize its interaction with the COOH-SAM and minimize strain energy by aligning with the outer boundary of the COOH-SAM.

By balancing these interaction energies, it is possible to predict the size of a COOH-SAM required for trapping a SWNT of a given length. For a SWNT with a fraction χ of its length in van der Waals contact with the COOH-SAMs and the remainder, $1-\chi$, in contact with the CH_3-SAMs, the thermodynamic requirement for assembly is,

$$\chi \cdot E_{SWNT/COOH\text{-}SAM} + (1-\chi)E_{SWNT/CH3\text{-}SAM} \leq E_{SWNT/solvent}. \qquad (1)$$

Solving Eq. 1 gives $\chi \geq 19\%$. This χ value indicates that a SWNT can be stabilized on the surface even if only 19% of its length is in contact with the COOH-SAMs. In other words, a 19-nm-wide stripe of the COOH-SAM is sufficient for trapping a SWNT that is 100 nm in length, even in the extreme situation where the nanotube is perpendicular to the stripe. For a SWNT making a smaller angle with respect to the stripe, trapping requires an even smaller stripe width.

Based on the aforementioned observations, this approach appears ideal for the assembly of SWNTs in a dense array on surfaces. Indeed, Fig. 6.95(a) shows an array of parallel aligned SWNTs assembled on MHA patterns (1 μm × 130-nm lines at a line density of $5.0 \times 10^7/cm^2$) with nearly 100% occupancy. Atomic force microscopy (AFM) heights are consistent with individual SWNTs or small bundles on each feature. It is found that the line width of MHA required for the assembly is only ≈1/10 that of the NH_2-SAMs, resulting in a density > 10 times higher than achieved previously The decrease in required line width is essential for assembly of short SWNTs (10-50 nm) for high-performance field effect transistors.

As the size of the MHA pattern is increased, the same spot is often occupied by additional SWNTs, with the number being approximately proportional to the volume of the trapped solution. Because of the strong van der Waals attraction between SWNTs (–1.22 eV per nm

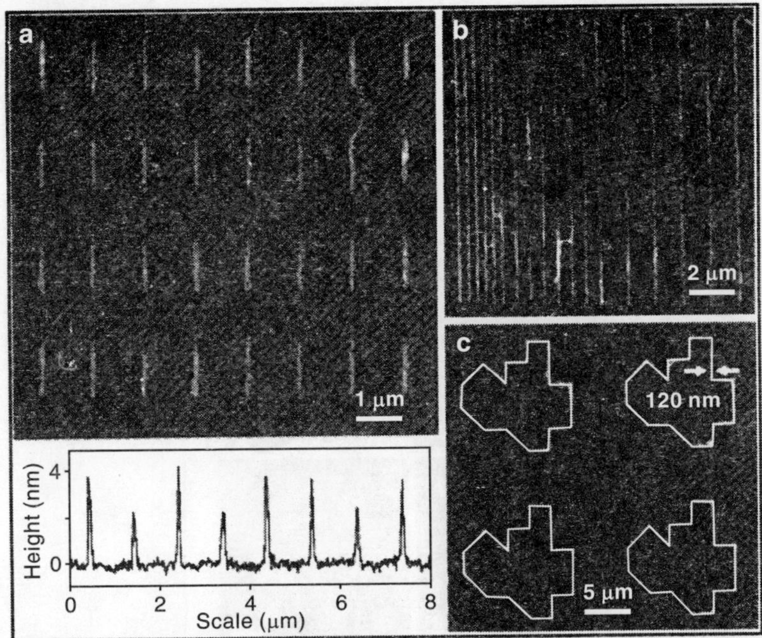

Fig. 6.95 *AFM tapping mode topographic images of selected SWNT arrays. (a) Parallel aligned SWNTs with a line density approaching 5.0 × 10⁷/cm². (b) Linked SWNTs following MHA lines (20 µm × 200 nm) spaced by 2 µm, 1 µm, and 600 nm. (c) Random line structure, showing the precise positioning, bending, and linking of SWNTs to a MHA affinity template. All images were taken at a scan rate of 0.5 Hz. The height scale is 20 nm*

contact), additional SWNTs can be deposited on the same site, intertwining with each other, and extending to bridge SWNTs on nearby MHA features. As a result, the spacing between SWNTs cannot be reduced to a distance much smaller than the tube length without effecting feature cross-linking or sacrificing SWNT feature occupancy. To Reduce the van der Waals attraction, we functionalized the SWNT sidewalls with dodecyl groups as described. The functionalization increased the solubility of SWNTs in 1,2-dichlorobenzene and resulted in more uniform assembly. Because the dodecyl groups can be thermally cleaved from the tubes at 473K, these groups act as nondestructive spacers for fine-tuning the interactions between the tubes. Alternatively, one can take advantage of van der Waals attraction to link tubes to make more sophisticated interlinked structures. For example, µm-long SWNTs were assembled into arrays of continuous, parallel aligned, sub-µm-spaced nanowires (Fig. 6.95(b)). Because of the unusual electrical properties of SWNTs, including an electrical conductivity rivaling copper and a current carrying capacity up to 10^9 A/cm^2 for metallic SWNTs , this technique enables SWNTs to be used as conductive interconnects in electronics. In another example, individual and bundled SWNTs were linked to follow an MHA path defined by DPN, exhibiting extensive flexibility that is favorable for making electronic interconnects (Fig. 6.95(c)).

This approach provides a simple route toward creating thin structured SWNT films that are currently unattainable. Fig. 6.96(a) demonstrates a "proof-of-concept" SWNT filtration membrane with a thickness of only 9 ± 2 nm. The SWNT composition has been characterized and mapped with a Raman confocal microscope (Fig. 6.96(b)). These membranes can be made uniformly over 1 × 2 cm by μCP (Fig. 6.96(c) and (d)), and their sizes appear to be limited only by the area and patterned structures defined by available lithography techniques. The thickness of these membranes can be controlled down to ≈5–10 nm, thereby promising an ultrahigh flux in filtration and structured, transparent conductors potentially useful in flexible displays.

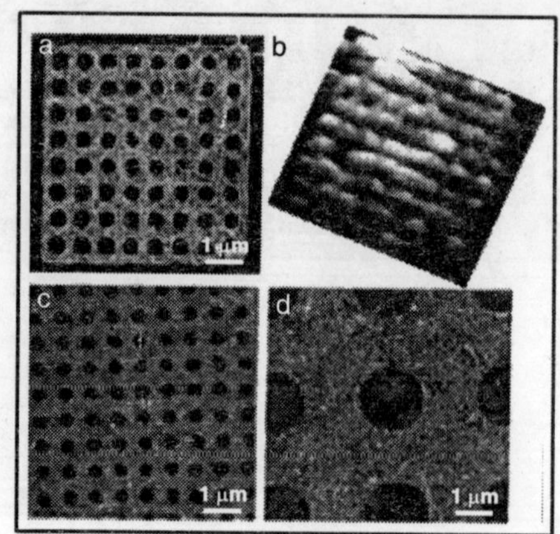

View larger version (160 K):
[in this window]
[in a new window]

Fig. 6.96 *Proof-of-concept SWNT filtration membranes. (a) AFM topographic image of a SWNT membrane with 600 ± 50-nm-diameter pores spanning an area of 10 ×10 μm. (b) Corresponding Raman image showing SWNT spectroscopic signatures. The false color topography represents the integrated Raman intensity over 1,320-1,620 cm^{-1}. (c and d) Representative AFM topographic images of a SWNT network with 1.6 ± 0.1-μm-diameter pores spanning an area of 1 × 2 cm. All AFM images were recorded at a scan rate of 0.5 Hz, and the height scales are 20 nm*

PURIFICATION OF CNTs

As the carbon nanotubes produced by arc and laser oven or any other process contain various impurities, a variety of procedures are used to purify carbon nanotubes. Also, the as-produced SWNT soot contains a lot of impurities. The main impurities in the soot are graphite (wrapped up) sheets, amorphous carbon, metal catalyst and the smaller fullerenes. These impurities interfere with most of the desired properties of the SWNTs. In order to understand the measurement better, the SWNT samples have to be as homogeneous as possible so it is preferred to obtain SWNTs as pure as possible. By using strong oxidation and acid refluxing techniques, the structure of the tubes is affected. Several purification techniques of the SWNT are there. Basically, these techniques are divided into two mainstreams, structure selective and size selective separations. The first one will separate the SWNTs from the impurities; the second one will give a more homogeneous diameter or size distribution.

CHARACTERIZATION TECHNIQUES

It is very important to monitor purity and kinds of impurities present at every stage of the purification process, examples are:

(a) In the arc discharge, a mixture of graphite powder and catalysts (typically 1:4 wt. % of Ni:Y) is vaporized in a direct current arc at 25-35 V and 100 A. The chamber is filled with helium at ~650 kPa, and the temperature within the electric arc can reach 6000 K. During the burn process, a substance transfer occurs between the electrodes and nanotubes, self-assembled from the vapor. In addition to nanotubes, carbon condenses into fullerenes, nanoparticles, and amorphous products. Organic impurities in the electrode material facilitate production of various hydrocarbons, which are revealed by NMR. The rate of the gas flow, size and orientation of the arc chamber, process scale, cooling system, the nature and purity of electrode materials, arc voltage, arc current etc. are other factors which affect the quantity, purity, and diameters of nanotubes produced. Localization of the resulting nanotube product in the arc chamber affects nanotube material properties as well. Nanotubes deposit in three distinct regions: collarette around cathode, webs that extend from collarette to chamber walls, and sooty deposit on the chamber walls. The best purity is obtained in the collarette material.

(b) In the laser oven, in addition to nanotubes, fullrenes and amorphous carbon products are formed. Also the presence of polyaromatic and hydrocarbons is revealed by NMR. EDS also shows small amounts of silicon. Various factors such as oven temperature, laser beam fluence flowgeometry and target composition, density and other factors affect the nanotube quantity, purity and lengths and diameters.

(c) SEM is the easiest tool. It provides medium-resolution images of the surface of nanotube material. It is possible to get a qualitative comparison of the purity of the samples, but it is sometimes not easy to distinguish nanotubes from the soot, especially when nanotubes are uniformly coated with impurities.

(d) As SEM is equipped with EDS. This allows seeing characteristic X-ray spectra of various elements and determining their atomic ratios in the sample. Unfortunately, concentrations of carbon are measured rather imprecisely.

Raman spectroscopy produces characteristic peaks from certain phonon modes in nanotubes. Low-frequency ("breathing") modes at 150-200 cm^{-1} are dependent on nanotube diameters, but only resonant tubes show up. It is impossible to get full information on the diameter distribution. The high-frequency (C-C stretch) mode at ~1585 cm^{-1} is sensitive to nanotube surroundings and can provide information.on nanotube chemistry, although interpretation is ambiguous. The disordered carbon peak at ~1340 cm^{-1} is related to amorphous impurities as well as to disorder in nanotube walls.

(e) TEM produces high-resolution images, but it is rather time-consuming. Also, can be looked at only a small fraction of the images (but not the whole sample). It is easy to distinguish nanotubes from impunties and qualitatively compare the purity of the samples. Nanotube diameter distributions can be obtained from TEM pictures, which require a specially prepared specimen and take a lot of time.

(f) XRD produces the lattice constant of a nanorope lattice. One can also see peaks from metal impurities and fullerenes. This is sensitive to the degree of crystallinity of the nanorope, which is affected by intercalation of impurities.

(g) NMR performed on solvent extracts of nanotube material. Any organic impurity produces proton peaks, whose positions are dependent on particular proton configuration; i.e., on the kind of steric group.

(h) TGA is simply a measurement of specimen mass as its temperature slowly increases up to 1073–1373 K. If done in inert gas (or vacuum), it shows that volatile compounds evaporate (certain hydrocarbons, fullerencs, etc.), leaving behinnd the fraction of non-volatile materials (typically nanotubes, metals, graphite, silicon, etc.) If done in air or oxygen, it shows temperatures at which nanotubes and impurities oxidize. The residual is then its ash content (typically metal oxides and silicon).

(i) HPLC is performed on solvent extracts or nanotubc material. It allows one to quantitatively determine the fraction of soluble fullerenes in the specimen and to determine other organic impurities, but poor performance of detector array in the ultraviolet region do not allow this. The HPLC apparatus at Rice University is equipped with a Cosmosil column for fullerene separation. There are reports that HPLC can also be used to separate nanotubes bylength if used with a size-exclusion column.

IDENTIFICATION OF IMPURITIES

EDS, SEM, and TEM have proved the existence of inorganic impurities. Annealed samples of laser-produced materials contain nickel, cobalt, and silicon. Nickel and yttrium found in arc material constitutes 10 to 20% by weight of the material total mass. Graphite-encapsulated catalyst metal (possibly carbide) particles are discovered and studied by means of SEM and TEM. EDS proves presence of metals; silicon and carbon in the encapsulated particles. In the EDS spectra taken over approximately 20 × 30 μm area on the specimens, reveal the presence and relative intensities pf Si, Co, and Ni peaks. XRD shows the presencec of metals as well as fullerenes.

TGA data in argon flow shows intense evaporation of water, volatile compounds and solvents at temperatures up to 473 K. Structures like –CH–, –HC=CH–, $(C_6H_x)_y$ evaporate in the temperature interval 473 to 523 K. Organic compounds evaporate at the temperature up to 723 K. Macromolecular compounds evaporate in the temperature interval up to 873 K. Fullerenes evaporate in the temperature interval 873 to 1023 to 1073 K. Decomposition of the material begins in the temperature interval 1023 to 1073 K or higher, and weight stabilizes in the interval 1073 to 1473 K.

TGA can also be performed in air (or oxygen) flow. Here, the temperature has to be raised to the point of complete oxidation of carbon to determine the elemental composition of ash, inorganic materials, and their weight fractions.

Organic compounds and fullerene are found by extracting raw materials with organic solvents: benzene, toluene o-xylene, 1, 2-dicblorobenzene, decaline, hexane, carbon disulfide, and other. An HPLC test of the solvent extract shows the presence of a full spectrum of fullerenes (C_{60}, C_{70}, C_{76}, C_{78}, C_{84}, C_{92}, C_{96}) and hydrocarbons. The amount of fullerenes extracted from raw nanotubes material by toluene and benzene slowly increases on a time scale of several hours before it saturates at as much as 9% by weight. SFM images before and after extracting are essentially the same, indicating that many insoluble impurities are left in the sample. After extraction sample typically loses 18 to 20% of its mass. This allows to estimate the amount of soluble hydrocarbons to be around 10% by weight.

NMR tests have determined the presence of H^+ in the extracted solutions. Localization of H^+ proves the presence of aromatic and linear hydrocarbons. Their weight fraction is 5 to 10% in accordance with the TGA data.

Raman peak in the area of 1340 em^{-1} shows presence of amorphous carbon typical laser material contains the following impurities by weight:

Fullerenes: 12 – 15%
Hydrocarbons: 12 – 15%
Metals: ~10%
Silicon: ~1 – 2%
Amorphous carbon: ~20%
Graphitic shells: ~5 – 10%
Nanotubes: 30 – 35%

PURIFICATION METHODS

Raw material contains nanotubes, fullerenes, catalytic metal particles, amorphous carbon, and various kinds of hydrocarbons and their compounds. The various methods of purification are oxidation, acid treatment, annealing, ultrasonication, micro filtration, ferromagnetic separation, cutting, functionalisation and chromatography. Most of the techniques used, are combined with other techniques.

Oxidation by acids is used to remove catalytic particles, amorphous carbon, and hydrocarbons. The following technique is tried on laser- and arc-produced material.

The soot is produced by a carbon-arc method using Ni and Fe as catalysts.

100 mg of the soot containing nanotubes is introduced into a flask with attachment together with 50 mL of distilled water and is heated at 373 K for 12 Hrs. The soot is disintegrated to submicron-size particles. The processed soot is filtered and dried at 333 K for 12 h. Then, fullerenes are washed out using toluene in a soxhlet. The residual soot. is heated at 743 K for 20 min. In the final step of the purification process, the soot is treated in 6 M hydrochloric acid and almost all the metal complexes are washed out.

There are many nanocapsules that have received damage on the graphitic layers. Especially, the graphitic layers of some empty capsules are completely destroyed.

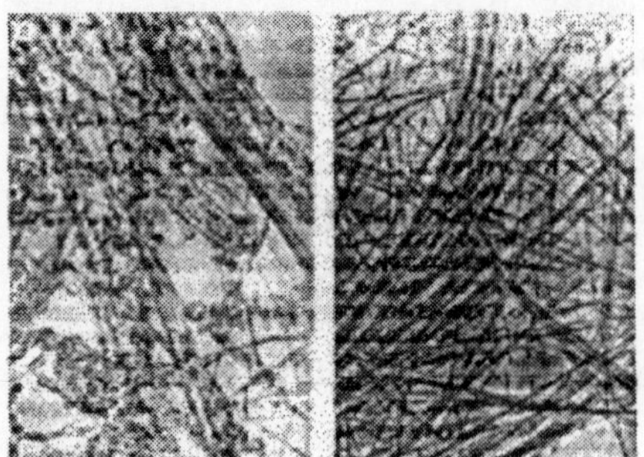

A typical nanotube sample produced by the carbon arc method (a) | A purified open nanotube sample (b)

The main idea of purification of nanotubes is separating nanoparticles from nanotubes and destroying nanotube caps (see adjoining figure). Inspired by the idea that nanotubes could be opened and thinned through oxidation; if the oxidation is allowed to proceed for long duration, all the caps and nanoparticles are eliminated. To remove all the particles from the samples, one has to oxidize more than 99% of the material, because the oxidative species are not able to reach evenly to all the parts of the sample as the ground-up nanotubes nanoparticles are intertwined. Chemical oxidation exhibits better yield and control as a purification technique.

Oxidation

Oxidation by acids is used to memove catalytic particles, amorphous carbon and hydrocarbons.

Oxidative treatment of SWNTs is a good way to remove carbon impurities on to clear the metal surface. Various oxidation methods used are:

(a) "Old Rice method". Oxidation by 5M nitric acid is performed over 24 hours at 369 K. The product is then neutralized by sodium hydroxide, suspended in 1 % Triton X-100 surfactant solution in water, and purified by cross-flow filtration. It is difficult to dry nanotubes afterward, since vacuum evaporation does not work well because of foaming under vacuum, and non-vacuum evaporation is very slow. Because oxidization and neutralization, insoluble hydroxides of Ni and Co precipitate with nanotubes, so heavy graphite particles, and organic Na salts which deposit into residue are difficult to remove. Also, 0.2 μm cross-flow filter cartridge is sufficiently permeable by nanotubes in Triton X-100; so, a significant fraction of nanotubes is lost with the waste. At the same time, many particles are larger than the pore size and stay with the purified material. Abundant quantities of nanotube are found in the waste, so it is not clear whether purified material is any purer than waste. This process is long and tedious, and results of cross-flow filtration are not reproducible.

The main disadvantage of oxidation by this method is that not only the impurities get oxidised but also SWNTs. These impurities get attached to the metal catalyst, which also acts as oxidising catalyst when the temperature is raised above 873 K.

SWNTs also get oxidised, even without a catalyst. This is possible in the case with thermal, fixed air and pure oxygen oxidations. These can easily oxidise all the components.

There are a couple of examples for clearing the metal surface, to prepare the sample for a metal removal step. The first one is (mild) oxidising in a wet environment with soluble oxidising agents, such as H_2O_2 and H_2SO_4. These will only oxidise the defects and will clear the surface of the metal.

Most commonly, the metal catalyst stays intact during these processes, but when oxygen is used in a wet atmosphere, the outer layer of the metal will be oxidised. Then, the density of this surface increases and the surface covering carbon layer ruptures. Not only the carbon impurities are oxidised but also the metal is partially oxidised and exposed.

Quite different from the oxidative technique is microwave heating. The microwaves heats up the metal and catalytically oxidises the carbon attached to the metal.

(b) "New Rice Method". It involves oxidization by 5M nitric acid for 6 hours, centrifugation to separate the liquid (decant) from the heavy part, washing by water, neutralization by NaOH, secondary oxidization by nitric acid for 6 hours, separation by contrifugation, secondary neutralization by NaOH, washing by methanol, and redisperslon in toluene. The filtered toluene solution has a characteristic color of fullerene mixture. HPLC test proves, the presence of C_{60}, C_{70} and higher fullerenes. Elemental analysis (EDS) of cleaned nanotubes shows a significant presence of catalytic metals, Na, and Si. TEM and SEM show encapsulated particles covered with amorphous carbon and nanoropes (i.e., nanotube bundles) covered by amorphous carbon.

It is possible that the nanotube surface becomes derivatized with products of nitric acid reaction with amorphous carbon, judging by its look on high-resolution SEM images. There is a lot of amorphous carbon left, which makes this technique a poor candidate. This technique (b) is better than the technique (a). Typical yield is 90 to 50%.

(c) Oxidation with Nitric Acid: HNO_3, is chosen because it is a strong oxidizer and is usually used for etching and oxidation of metals, but it is passive with respect to the surface of nickel and cobalt particles. HNO_3 reacts with non-metals and typical reducing agents. Since it is self-ionized in liquid form, .the reaction is weak and, with increase in temperature, the acid dissociates into NO and NO_2 while the encapsulated catalytic particles stay intact. This sample is subjected to four subsequent refluxes in nitric acid. SEM images after each reflux are taken and TEM images after the first and fourth reflux are taken.

Sample purity after the last reflux is not very good, there is still a significant quantity of amorphous carbon and nanoparticles. Raman spectra shows, that C-C stretch peak normally positioned at 1585 em^{-1} shifts to 1595 to 1605 cm^{-1} after the first reflux (which indicates a change, in the local environment) with products of nitric acid reaction with amorphous carbon. The disordered carbon peak at 1360 cm^{-1} increases significantly after the first reflux, also indicating possible intercalation and disorder in nanotube walls. C-C stretch peak goes back to 1590 cm^{-1} and disordered carbon peak disappears upon drying at 423 K in air, indicating that intercalation mostly goes away. Similar results are observed by X-ray. As-made material exhibits nanorope superlattice peaks, which are gone after reflux, indicating strong disorder and loss of crystallinity in nanoropes caused by intercalation. Upon drying nanoropes gain crystallinity back, indicating that intercalants are mostly removed. Impurities in electron microscopy images are also observed, this probably means that intercalants are not volatile enough to leave sample completely and rather redistribute on the surface of nanoropes. Typical yield is ~35%.

The activity of HNO_3 increases with the addition of HF, which helps to dissolve Si compounds with the formation of NO, H_2O, and H_2 (SiF_6). The sample is first refluxed in concentrated HF for 30 min, then HNO_3 is added and the sample is refluxed further for 45 min at 308 to 313 K. Gas bubble release (H_2 and SiF_6) is observed. This method has advantages for cleaning laser ablation material containing silicon. EDS spectra shows that silicon is essentially removed. Nevertheless, other amorphous carbon impurities still exist in the sample. Typical yield is 40 to 80%.

Following are some other oxidisers used for purification of nanotubes.

(d) Hydrochloric acid (HCl) is one of the strongest oxidizers. It reacts with metals with negative and normal potential with release of H_2. It also reacts with metal oxides and hydroxides and generates free acids from silicates. Samples look quite pure and the amorphous coating is mostly gone, The remaining coating swells with solution and is quite visible on SEM, but subsequent annealing demonstrates that the amount of coating is small. Typical yield is 70 to 90%.

(e) Sulfuric acid (H_2SO_4) is a strong acid but a weak oxidizer in diluted form. In concentrated form, it works as a strong oxidizer but passivates cobalt surfaces. Separation of nanotubes and the decant is possible with multi-step dissolution of acid by methanol in separation funnel. Acid has high viscosity and is difficult to wash from nanotubes, so this technique is not pursued further. Typical yield is 30 to 40%.

(f) Peroxidisulfuric acid ($H_2S_2O_8$) is a strong oxidizer that reacts with organic compounds to full decomposition. Reaction is performed at room temperature with rapid outgassing until outgassing stopped in about 1.5 hrs. The viscosity of the acid is very high and separation of solid and liquid phase is difficult. Therefore, the same method as with H_2SO_4 is used. Solid and liquid phases are divided in separating funnel into different layers by adding methanol to the acid. The pure solid phase product contains encapsulated particles, but it usually does not contain amorphous carbon. Since both $H_2SO_4$and $H_2S_2O_8$ catalyze the loss of water, there should not be any hydroxyl groups left. Typical yield is 25 to 45%.

3f. Hydrogen peroxide (H_2O_2) has oxidation-reduction properties, but the oxidizing function prevails because of the presence of O_2^-. The reaction is performed at 308 K for 3 hrs and is accompanied by rapid release of CO (bubbles) and H_2O. After oxidation by H_2O_2 the material looks very good on SEM images. The remaining amorphous coating swells with solution and is quite visible on SEM, but subsequent annealing demonstrates that the amount of coating is small. During the H_2O_2 oxidation of raw material, epoxy group formation on nanotube is possible. Typical yield 25 is 40%.

(g) Trifluoroacetic acid (CF_3COOH) is a strong acid that is used in organic synthesis as an oxidizer, a catalysis of electrophilic reactions, and a solvent of fluorine-containing compounds. The amount of silicon is distinctly lowered by the use of this acid. Also, the outer layer of the nanotubes looks smooth and clean. The remaining coating swells with solution and is quite visible on SEM, but subsequent annealing demonstrates that the amount of coating is small. Typical yield is 60 to 80%.

(h) Peroxyacetic acid (CH_3COOOH) is another strong oxidizer of organic compounds. The reaction is very active with the rapid outgassing (CO); and formation of epoxy compounds, polycycles, and ethers is possible. This acid helps to break up large molecules and modify nanotube surface. Typical yield is 80 to 90%.

Acid Treatment

Solvent extraction followed by acid reflux. This method of purification starts with extraction of fullerenes and hydrocarbons with organic solvents: toluene, benzene, acetone, butanol, propanol, methanol, chloroform, decaline, ethyl ether, hexane, O-xylene, 1, 2-dichlorobenzene, carbon disulfide, and some others.

SEM images of samples before and after extraction are taken. Since amorphous carbon and nanoparticles, which are visible on SEM, are unaffected by extraction the specimens. NMR spectra of extracts prove the presence of hydrocarbon, Typieally extraction removes 18 to 20% of the sample mass. out of which approximately half is soluble fullerenes and the rest are soluble hydrocarbons. There still may be some insoluble fullerenes and hydrocarbons left. Toluene (as a solvent), is recommended due to its lower toxicity. After extraction, the material is washed by methanol or acetone to remove the remains of the solvent used for extraction.

This is followed by oxidation by inorganic or organic acids. Inorganic acids used are: HNO_3, HCl, mixture of HF and HNO_3, mixture of H_2SO_4 and HN_3, $H_2S_2O_8$, H_2O_2. Organic acids CH_3COOH, CF_3COOH, and CH_3COOOH.

Acid oxidation is followed by washing of acid residue with hot water to a pH = 6 and then washing with methanol. Sometimes the neutralization is preformed with NaOH and/or NH_4OH.

When using a treatment in HNO_3, the acid only has an effect on the metal catalyst. It has no effect on the SWNTs and other carbon particles If a treatment in HCl is used, the acid has also a little effect on the SWNTs and other carbon particles. The mild acid treatment (4 M HCI reflux) is basically the same as the HNO_3 reflux, but here the metal has to be totally exposed to the acid to solvate it.

High-resolution SEM microscopy shows that outer layer of nanoropes looks different after oxidation with different acids. Nevertheless, there is always a layer of the products of acid reaction with amorphous carbon and hydrocarbons present in the sample. These products are not completely soluble in the solvents used for washing after oxidation. Since these impurities no longer constitute non-volatile amorphous carbon but are chemical derivatives thereof, they may be volatile and can be evaporated away. In addition, some impurities can be evaporated away before acid reflux. Therefore, several techniques in which samples are subjected to high-temperature annealing in vacuum or argon flow before and/or after acid reflux are performed. Vacuum annealing of laser material is performed at 1273 K and 10 mmHg, whereas arc material was annealed at 1073 K and 20 mmHg. The typical weight loss was 60 to 90 %. Fullerenes and organic compounds condensed on cold surfaces at the exit of the tube furnace. SEM images show clean surfaces of nanoropes and clean encapsulated catalyst particles present inside aggregates. TEM images also show clean nanorope surfaces, and a small quantity of amorphous carbon present close to encapsulated catalyst particles. Ropes are well organized and seem to have a smooth surface and clear structure.

Annealing

Due to high temperatures (873–1873 K) the nanotubes will be rearranged and defects will be consumed. The high temperature also causes the graphitic carbon and the short fullerenes to pyrolyse. When using high temperature vacuum treatment (1873 K) the metal gets melted and can also be removed.

(a) Simple Vacuum Anhealing. Unpurified material is placed in the alumina boat in the vacuum furnace at 10 mmHg. Temperature is increased in 473 K steps to 1273 K over 20 hrs and then stayed at 1273 K for 4 hrs. There is definitely quite a lot of amorphous coating left on nanotube surface that is not volatile even at such high temperature. Therefore, this technique is hardly suitable as a stand-alone purification method. It has to be combined with other steps (acid oxidation, solvent extraction). After annealing, the sample is studied by TGA in air flow up to 1273 K. At these conditions, nanotubes as well as amorphous coatings and graphitic particles oxidize and only metal oxides are left. The remaining weight fraction was ~54% of 1:1 CoO:NiO, which corresponds to about 35% by weight of Co and Ni. Specimen oxidation occurs at two distinct temperatures: ~703 K and ~823 K.

(b) (Toluene extraction +HNO_3 reflux) is followed by annealing in argqn flow up to 1373 K. Annealing is performed in TGA, and the specimen mass is monitored as the temperature is increased in 473 K. The sample has lost at least 50% of its mass and looks very good on SEM and TEM images. Nanoropes are well-formed and crystalline. Amount of amorphous coating is minimal, but there are still some nanoparticles left in the sample. Typical yield is 80 to 90% after extraction and ~50% after reflux.

(c) (Toluene extraction +HCl reflux) is followed by annealing in vacuum(for 24 hrs)up to 1273 K at 10 mmHg as temperature is increased in 473 K. The sample looks very good on SEM images. After vacuum annealing, the sample is studied in TGA in air flow up to 1073 K. At tliis condition, nanotubes as well as the amorphous coating and graphitic particles oxidize. Air that are left are metal oxides. The mass left after burning all nanotubes and carbon-containing dirt is ~33%, which corresponds to ~25% metals by weight. Typical yield is 80 to 90% after extraction and ~70% after reflux.

(d) A set of samples is annealed in vacuum (for 24 hrs) up to 1273 K at 10 mmHg as the temperature is increased in 473 K. Afterward all these samples are refluxed in various acids to observe how particular acids affect nanotube surfaces after annealing. Acids tried include CH_3COOOH, HNO_3, H_2O_2, and $H_2S_2O_8$.

The surface looks quite different after reflux in different acids. Note that all images are taken at the same magnification, but the visible nanorope diameter is different, which means that the nanorope coating adsorbs hydroxyl groups from the acid and intake is different for different acids. Nanoropes look quite similar after peroxyacetic and persulphuric acids reflux. They have smooth surfaces and less swelling. After hydrogen peroxide reflux, the nanoropes swelled a lot more, and the surface is smooth and somewhat fuzzy. After nitric acid reflux, nanoropes do not swell much and the surface is grainy. These results clearly show that acid reflux results in modification of amorphous coating on nanoropes, and this modification varies significantly for different acids. It is also possible that nanotubes themselves are derivatized to a different degree, but we cannot confirm it experimentally.

It is interesting that the swelling of nanoropes is reversible. One of the samples is studied in TGA after reflux in H_2O_2. Annealing is performed in TGA in argon flow and the specimen mass is monitored as the temperature is increased in 473 K to 1373 K. Approximately 66% of the sample mass is left, indicating that there is only 34% by weight of volatile compounds. The sample looks very clean after annealing, and the swelling is gone.

Some of these samples are checked by Raman spectroscopy are samples refluxed in $CH_3COOOHH_2S_2O_8$, and H_2O_2 after TGA. These can be compared with the unpurified sample and one only annealed before reflux. The C-C stretch peak position is located at 1588 cm^{-1} in unpurified and annealed samples; yet it shifts significantly after refluxes, indicating a change in the local environment of nanoropes. This may include charge transfer from amorphous coating reacting with acids and/or direct nanotube derivatization. The shift is largest for CH_3COOOH reflux, somewhat smaller for $H_2S_2O_8$ reflux, and even smaller for H_2O_2 reflux + annealing, which may be due to parual removal of intcrcalants during annealing.

(e) A set of samples after vacuum annealing (24 hrs up to 1373 K at 10 mmHg as temperature is increased in 473 K steps) is dispersed in various organic solvents. Solvents tried are carbon disulfide, dimethylformamide, water, hexane, propanol, toluene, dimethylsulfoxide, cyclohexane. Nanotubes behave very differently in different solvents. Amorphous coatings on nanorope surfaces adsorb solvents and swell to different degrees, possibly forming solvates. Dimethilsulfoxide and cyclohexane cause the least swelling, whereas propanol and hexane swell more, and toluene and dimethylformamide swell the most. Carbon disulfide and water hardly produce any swelling. It looks like carbon disulfide causes coagulation of the amorphous coating on the nanorope surface. Water may introduce hydroxyl groups into the coating, which may adsorb or chemically react with it. Note that all SEM images arc again taken at the same magnification.

Ultrasonication

In this technique particles are separated due to ultrasonic vibrations. Agglomerates of different nanoparticles are forced to vibrate and so they become more dispersed. The separation of the particles is highly dependable on the surfactant, solvent and reagents used.

The solvent influences the stability of the dispersed tubes in the system. In poor solvents the SWNTs are more stable if they are still attached to the metal. But in some solvents, such as alcohols, monodispersed particles are relatively stable.

When an acid is used, the purity of the SWNTs depends on the exposure time. When the tubes are exposed to the acid for a short time, only the metal solvates, but for a longer exposure time, the tubes will also be chemically cut.

Magnetic Purification

In this method ferromagnetic (catalytic) particles are mechanically removed from their graphitic shells. The SWNTs suspension is mixed with inorganic nanoparticles (mainly ZrO_2 or $CaCO_3$) in an ultrasonic bath to remove the ferromagnetic particles. Then, the particles are trapped with permanent magnetic poles. After a subsequent chemical treatment, a high purity SWNT material is obtained. Figure 7.1 shows a schematic diagram of the apparatus for magnetic purification.

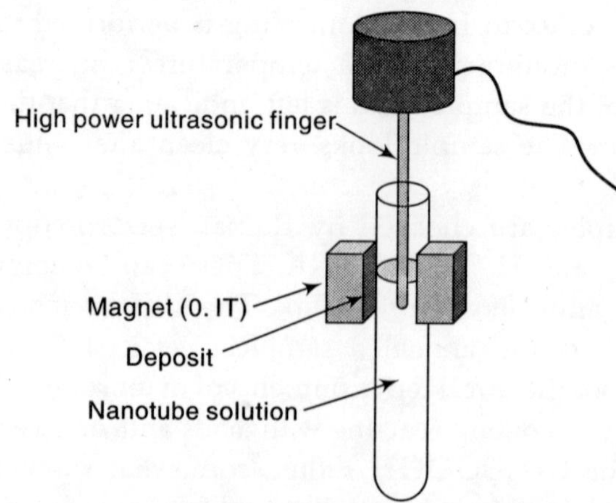

Fig. 7.1 *Schematic diagram of the apparatus for magnetic purification*

This process does not require large equipment and enables the production of laboratory-sized quantities of SWNTs containing no magnetic impurities.

Micro Filtration

Micro filtration is based on size or particle separation. SWNTs and a small amount of carbon nanoparticles are trapped in a filter. The other nanoparticles (catalyst metal, fullerenes and carbon nanoparticles) are passing through the filter. Figure 7.2 shows a schematic diagram of a micro filtration cell.

One way of separating fullerenes from the SWNTs by micro filtration is to soak the as-produced SWNTs first in a CS_2 solution. The CS_2 insolubles are then trapped in a filter. The fullerenes which are solvated in the CS_2, pass through the filter.

A special form of filtration is cross flow filtration. In cross flow filtration the membrane is a hollow fibre. The membrane is permeable to the solution. The filtrate is pumped down the bore of the fibre at some head pressure from a reservoir and the major fraction of the fast flowing solution which does not permeate out the sides of the fibre is fed back into the same reservoir to be cycled through the fibre repeatedly. A fast hydrodynamic flow down the fibre bore (cross flow) sweeps the membrane surface preventing the build-up of a filter cake.

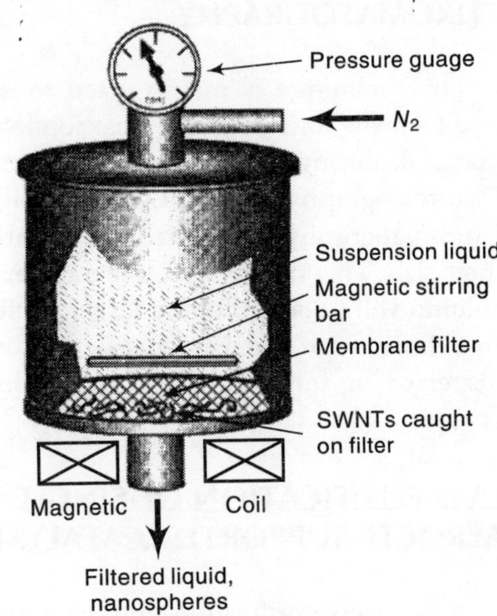

Fig. 7.2 **Schematic diagram of a micro filtration cell**

CUTTING

Cutting of the SWNTs can either be induced chemically, mechanically or as a combination of these.

SWNTs can be chemically cut by partially functionalising the tubes, for example with fluor. Then, the fluorated carbon will be driven off the sidewall with pyrolisation in the form of CF_4 or COF_2. This will leave behind the chemically cut nanotubes.

Mechanical cutting of the nanotubes can be induced by ball-milling. Here, the bonds will break due to the high friction between the nanoparticles and the nanotubes will be disordered.

A combination of mechanical and chemical cutting of the nanotubes, is ultrasonical induced cutting in an acid solution. In this way the ultrasonic vibration will give the nanotubes sufficient energy to leave the catalyst surface. Then, in combination with acid the nanotubes will rupture at the defect sites.

FUNCTIONALISATION

Functionalisation is based on making SWNTs more soluble than the impurities by attaching other groups to the tubes. Now it is easy to separate them from insoluble impurities, such as metal, with filtration.

Another functionalisation technique also leaves the SWNT structure intact and makes them soluble for chromatographic size separation.

For recovery of the purified SWNTs, the functional groups can be simply removed by thermal treatment, such as annealing.

CHROMATOGRAPHY

This technique is mainly used to separate small quantities of SWNTs into fractions with small length and diameter distribution. The SWNTs are run over a column with a porous material, through which the SWNTs will flow. The columns used are GPC (Gel Permeation Chromatography) and HPLC-SEC (High Performance Liquid Chromatography - Size Exclusion Chromatography) columns. The number of pores the SWNTs will flow through, depends on their size. This means that, the smaller the molecule, the longer the pathway to the end of the column will be and that the larger molecules will come off first. The pore size will control what size distribution can be separated. However, a problem is that the SWNTs have to be either dispersed or solvated. This can be done by ultrasonication or functionalisation with soluble groups.

CVD PURIFICATION OF SINGLE-WALLED CARBON NANOTUBES ON AEROGEL-SUPPORTED CATALYST

A two-step purification involving acid washing and oxidation in diluted air produced high-purity SWNTs.

Purification of SNWTs

The raw product contains the aerogel support, catalyst particles and a small amount of amorphous carbon (mostly as a coating on the catalyst nanoparticles) as important (Fig. 7.3a). A purification scheme was designed to remove all of these undesired impurities. In a typical purification experiment, about 100-mg raw material was mixed with 5-ml hydrofluoric acid (52%) and 15-ml H_2O to dissolve the aerogel support. It was stirred for 18 h and then centrifuged to remove HF acid. The SWNTs obtained still contained many metal nanoparticles, presumably because the nanoparticles are coated by carbon layers. To further purify the product, a two-stage gas-phase oxidation procedure is done. First, the sample is heated in a 15% air/Ar mixture at 673 K for 1 h. A wash in 5-ml concentrated HCl acid (37%) dissolved the metal oxides that formed during the oxidation process. In the second stage, the oxidation is carried out at 723 K for 1 h and then the sample was washed again in HCl acid. Such oxidation steps burned the carbon coating on the catalysts particles, making them exposed to acid attack, while leaving the nanotubes unaffected because of the higher stability of nanotubes against oxidation compared with amorphous carbon (Fig. 7.3b). The final purified product consists of mainly ($>$ 90%) SWNTs in bundle form (Fig. 7.3c). Thermogravimetric data shows that the purified SWNTs contain a negligible amount of metal oxide (Fig. 7.4).

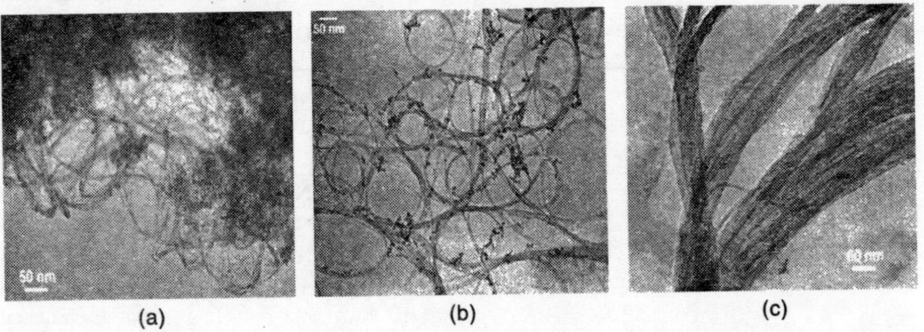

Fig. 7.3 *(a) TEM images of a raw product, (b) SWNTs after the second oxidization step and (c) purified SWNTs*

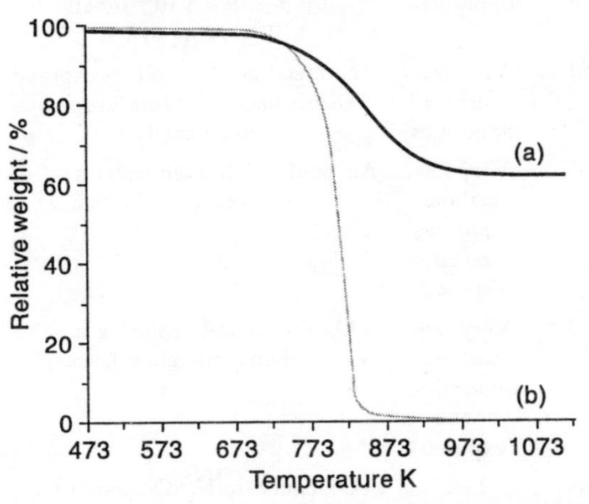

Fig. 7.4 *TGA plots of (a) raw materials (30-min runtime) and (b) purified SWNTs*

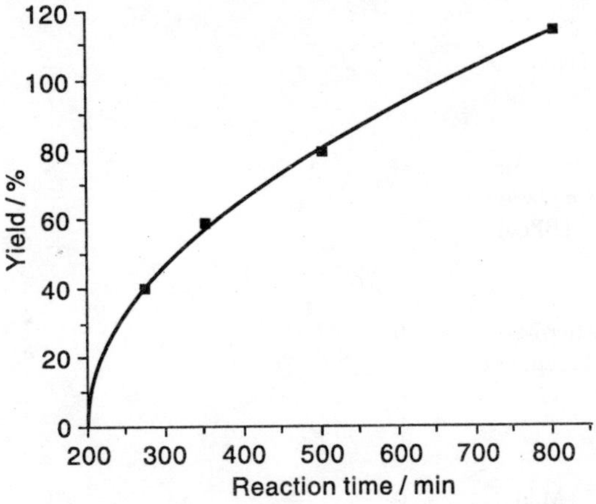

Fig. 7.5 *Weight gain of the catalyst as a function of reaction time.Conclusion and Comments*

Conclusion

The relationship between the SWNT yield and the reaction time is shown in Fig. 7.5. The fit gives a time dependence of $t^{1/2}$. The trend of this curve shows that the growth rate decreases with time. One obvious reason is that the gas-diffusion rate becomes lower with more SWNTs grown on the catalyst. This also explains why catalysts with high porosity like aerogel can greatly enhance the SWNT yield.

With the different techniques for purification as described above, different results are achieved. Care should be taken when the technique is chosen, as the effect on the entire sample will also depend on the composition and the amount of the sample. What is desired are techniques that only tear down the carbon impurities and the metals, without changing the nanotubes. But only few techniques are capable of this. Most techniques have a (sometimes devastating) influence on the SWNTs, such as cutting the SWNTs or oxidation reactions. Here extra care should be taken in adjusting the process variables such as temperature, scale and time. Table 7.1 and 7.2 on the next page, shows several techniques used in purification. How the techniques effect the SWNTs and other particles, is stated in this table. Also given is how these techniques effect the yield and purity.

TABLE 7.1 Several techniques for purification og cuts

Technique	*SWNT*	*Carbon particles, granhite*	*Short fullerenes C_{60}, C_{70}, etc*	*Metal catalyst*	*Yield*	*Purity*	*Short description of the technique and literature*
Thermal Oxidation	0	–	–	+	No Record	No Record	Heating from 623–873K. Metal is used as oxidation catalyst.
Wet Oxidation (H_2O_2)	0	0	0	+	No Record	No Record	Reflux in hydrogen peroxide. Process makes it easier to remove metal catalyst
Pure Oxygen Oxidation (Size Selective)	–	–	–	+	Poor	Very few carbon impurities	Preferentially bums SWNT with smaller diameter. $P = 1, 5 - 5.10^{-6}$ mbar; $T = 723 - 873$ K
Fixed Air Oxidation	0	–	–	+	Good	Very few carbon impurities	The metal catalyst will be exposed. Process makes it easier to remove metal catalyst
Wet Air Oxidation (HiPco)	0	–	–	+	Good	Very few carbon impurities, metal exposed	Air bubbled through water oxidises and exposes the metal.
Microwave Treatment	0	–	–	–	Very good	Very few carbon impurities, metal exposed	Oxidation and rupturing of the carbon passivation layer.
Mild Oxidation	0	–	–	+	Good	Less defects	Oxidation in H_2SO_4 and H_2O_2. Shortens the tubes and consume sidewall defects
HNO_3 Treatment	+	+	+	–	Good	Up to < 0.1 wt% metal	Reflux or sonication in nitric acid. Catalyst will solvate.
HCl Treatment	0	0	0	–	Good	Up to < 0.1 wt% metal	Reflux or sonication in hydrogloric acid. Catalyst will solvate.
Mild Acid Treatment	+	+	+	–	Very good	Up to < 0.2 wt% metal	Removes most of the metal, when the metal surface is totally exposed to the acid. (4 M HCl reflux)
Functionalisation (In diethylether)	+	+	+	+	SWNTs ~100%	~0,4% Metal	Functionalised with 1,3 dipolar cycloaddition and then solvated. Ref: [66]
Functionalisation (GPC)	+	+	+	+	SWNTs ~74%	Very pure	Functionalised with octadecyamine. Size separation by GPC.

+ = no effect on particle
0 = effects particle slightly
– = effects particle

TABLE 7.1 (*Contd*)

Technique	*SWNT*	*Carbon particles, granhite*	*Short fullerenes C_{60}, C_{70}, etc*	*Metal catalyst*	*Yield*	*Purity*	*Short description of the technique and literature*
Thennal Oxidation	0	–	–	+	No Record	No Record	Heating from 623–873 K. Metal is used as oxidation catalyst.
Wet Oxidation (H_2O_2)	0	0	0	+	No Record	No Record	Reflux in hydrogen peroxide. Process makes it easier to remove metal catalyst
Pure Oxygen Oxidation (Size Selective)	–	–	–	+	Poor	Very few carbon impurities	Preferentially bums SWNT with smaller diameter. P = 1, 5 – 5.10^{-6} mbar; T = 723 – 873 K
Fixed Air Oxidation	0	–	–	+	Good	Very few carbon impurities	The metal catalyst will be exposed. Process makes it easier to remove metal catalyst
Wet Air Oxidation (HiPco)	0	–	–	+	Good	Very few carbon impurities, metal exposed	Air bubbled through water oxidises and exposes the metal.
Microwave Treatment	0	–	–	–	Very good	Very few carbon impurities, metal exposed	Oxidation and rupturing of the carbon passivation layer.
Mild Oxidation	0	–	–	+	Good	Less defects	Oxidation in H_2SO_4 and H_2O_2. Shortens the tubes and consume sidewall defects
HNO_3 Treatment	+	+	+	–	Good	Up to < 0.1 wt% metal	Reflux or sonication in nitric acid. Catalyst will solvate.
HCl Treatment	0	0	0	–	Good	Up to < 0.1 wt% metal	Reflux or sonication in hydrogloric acid. Catalyst will solvate.
Mild Acid Treatment	+	+	+	–	Very good	Up to < 0.2 wt% metal	Removes most of the metal, when the metal surface is totally exposed to the acid. (4 M HCl reflux)
Functionalisation (In diethylether)	+	+	+	+	SWNTs ~100%	~0,4% Metal	Functionalised with 1,3 dipolar cycloaddition and then solvated.
Functionalisation (GPC)	+	+	+	+	SWNTs ~74%	Very pure	Functionalised with octadecyamine. Size separation by GPC.

+ = no effect on particle
0 = effects particle slightly
– = effects particle

Table 7.2 Several techniques for purification of SWNTs

Technique	*SWNT*	*Carbon particles, granhite*	*Short fullerenes C_{60}, C_{70}, etc*	*Metal catalyst*	*Yield*	*Purity*	*Short description of the technique and literature*
Magnetic Purification	+	+	+	–	Very good	Metal hardly detectable	Magnetic particles are trapped with pennanent magnetic pols.
Vacuum Annealing	+	–	–	+	Increases	Increases Less defects	Reammges the tubes and consumes defects. P = 5.10^{-6}; mbar; T = 873 K
High Temperature Annealing	+	–	–	+	Increases	Increases Less defects	Rearranges the tubes and consumes defects. Annealing in AT at T = up to 1073 K.
High Temperature Vacuum Treatment	+	–	–	–	Increases	Increases Less defects	Rearrangement of SWNT walls. removal of graphitic carbon and metals T = 1873 K; P = 10^{-3} Pa.
HPLC (Size Separation)	+	+	+	+	SWNTs 100%	Very pure	Size separation. SWNTs must be cut and purified in advance.
(Ultra)sonication (in alcohol)	+	+	+	+	SWNTs ~100%	Enhances purification techniques	Enlarges reactive surface, often in combination with other techniques. Suspends particles. Mass separation.
Ultrasonication (in acid)	–	+	+	–	decreases with time	First increases. then decreases	For short times. solvates metal. For longer times. also cuts the tubes.
Cutting through Fluorization	–	+	+	+	SWNTs ~55%	Shorter tubes, good purity	Cutting SWNTs with initial length of 1 um. Average length: less then 50 um.
Ball-milling or grinding	–	–	–	+	decreases with time	SWNTs shorten and more carbon impurities	The SWNTs are mechanically shortened and disordered carbon is fonned.
Filtration	+	+	+	+	Good	Enhances purification techniques	Size or particle separation. Often in combination with other techniques.
Cross Flow Filtration	+	+	+	+	Very good	Enhances purification techniques	Size or particle separation. with cross flow. Often in combination with other techniques. Contiunuous process.
CS_2 Filtration	+	+	–	+	SWNTs ~100%	No more fullerenes	Easy way to separates fullerenes without harming them.

+ = no effect on particle
0 = effects particle slightly
– = effects particle

Material produced by laser ablation contains fullerenes, and a higher fullerene concentration usually means cleaner raw material, which has m9fe nanotubes. The presence of silicon in the laser material was observed by elemental analysis (EDS). The arc material does not contain silicon, but the quality and quantity of nanotubes in the arc material is lower than in laser ablation material. It was observed that, in laser-produced nanotubes, purity correlates with silicon content. The smaller silicon content corresponds to lower purity. Silicon may, therefore, be a factor that affects the quality and quantity of nanotubes formed by either of the processes. But since no systematic studies were performed, this is only a hypothesis.

The results of purification show the following trends. Incorporation of dry steps (annealing) into purification methods generally produces better results. Hydrocarbons and fullerenes evaporate {lithe processes of high-temperature annealing, after which the outer layer of the nanoropes is much deaner and does not contain much amorphous material. Encapsulated catalyst particles are not covered by amorphous carbon, while a significant amount of amorphous carbon is found on the surface of nanoropes close to encapsulated particles, where the nanoropes interlace the most. This first step can be replaced with toluene extraction, which also removes soluble fullerenes and hydrocarbons. Nevertheless, annealing looks more promising as a first step, since it can also disrupt graphitic shells on metal particles and facilitate subsequent metal removal.

Subsequent oxidation by hydrochloric acid or hydrogen peroxide peroxide helps to remove some of the remaining amorphous coating and makes the rest more susceptible to vaporization, so that it can be removed in the second vacuum annealing. The most promising techniques, therefore, are (c) and (d) (H_2O_2 or HCl oxidation). (c) (HCl oxidation) generally produces higher yields.

Annealing as a last purification step also improves the crystallinity of nanoropes and helps to get rid of intercalants introduced into inter-tubular spaces by acid reflux. Nitric acid reflux on any step gives inferior results to H_2O_2 or HCl. All acid oxidation methods probably cause either dissolution of the nanotube walls or addition of different radicals and/or compounds to the surface of nanotubes. This results in radicals on nanorope surfaces, which are difficult to remove by vacuum annealing at temperatures up to 1273 K C. Nitric acid is probably the worst in this sense, giving nanorope coatings that are more difficult to remove.

PURIFICATION AND SIZE SELECTION OF CARBON NANOTUBES

A non-destructive method for the separation and size-selection of carbon nanotubes and nanoparticles produced by arc discharge is give here. A liquid-phase separation of nanotubes and nanoparticles is performed by filtering a kinetically stable colloidal dispersion consisting of the carbon material in a water/surfactant solution, allowing thus to extract the nanotubes from the suspension while leaving the nanoparticles is the filtrate. Further purification is accomplished by size-selection through controlled foculation of the dispersion. Final separation yields as high as 90% in weight, and are obtained without any damage to tube tips or tube walls following the separation process.

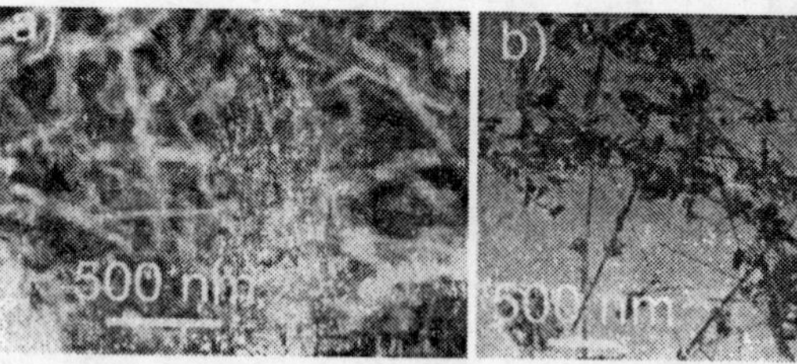

Fig. 7.6 *SEM (a) and low magnification (TEM (b) micrographs of the raw powder (after removal of the large graphitic particles by sedimentation and centrifugation). The tube proportion was typically of 28% in number and 61% in weight.*

Promising applications for carbon nanotubes range from nanoscale devices and materials to field emission and scanning probe microscopy (SPM). However, difficulties are encountered in the synthesis of high quality nanotubes. Carbon nanotubes are typically produced by arc discharge, with a wide distribution of lengths, helicities of the layers, and diameters. Furthermore, nanoparticles consisting of nested closed graphitic layers of polyhedral shape make up to one half in weight of the nanomaterial deposit. Fig. 7.6 displays typical scanning electron microscopy (SEM, Fig. 7.6(a) and transmission electron microscopy (TEM, Fig. 7.6(b) micrographs of raw arc-discharge deposit, which consists of particles and tubes that are loosely bound and attached together.

Up to now, the only available purification method was oxidation, either in gas-phase (by heating pristine or brominated deposite in air or flowing oxygen at ~973 K), or in liquid-phase. The basal planes of graphite are subject to oxidation only if defects (vacancies/dislocations) are present. The tips of the tubes and the particles, incorporating reactive five-membered carbon rings, have a higher curvature and higher strain and thus are preferentially attacked. For the nanoparticles, the reaction proceeds at a constant rate until complete oxidation. The tube ends get invariably opened or at least damaged during the purification process and the outer tube layers are removed during the treatment, presumably because of defects in the hexagonal network. That the tube tips have electronic properties, which differ from the cylindrical part of the nanotubes and they determine the field emission properties to a great extent; and the SPM capability of the tubes are also known. It is is so of great importance to obtain purified tubes with intact ends and non-damaged sidewalls.

A possibility to narrow the size distribution of nanotubes is to be separate them after their formation in the liquid-phase. Therefore, a stable colloidal suspension of well-seperated tubes and particles has to be obtained. None of the usually used solvents has actually proven successful to separate the nanotubes from the nanoparticles or even to maintain the nanomaterial in suspension for more than a few hours. Brombenzol, chloroform, and sulphuric acid are among the best suited solvents for forming a kinetically stable colloid with the nanotubes. Ethanol, acetone and methanol (as well as water) rank among the less suited solvents and are not able to maintain the material in colloidal suspension over more than a few hours.

The reason for this is the intrinsic instability of most of the colloidal suspensions which areas from the van der Waals attractive forces between colloidal particles. A solution of nanotubes and nanoparticles in a solvent can be viewed as a colloid, i.e., a system consisting of a dispersed phase (the nanotubes) distributed evenly throughout a dispersion medium (the solvent).

These forces overcome any other repulsive forces (e.g., electrostatic or steric), and make dispersed colloidal suspensions thermodynamically unstable. However, surface active agents (surfactants, polymers or other colloidal particles) can modify the particles-suspending medium interface and prevent aggregation over long time period (upto years). They provide an additional repulsive force (electrostatic and/or steric), lower the surface energy, and change some rheological surface properties which contribute to increasing the metastability of the colloidal suspension. In the starting carbon mixture needs to be dispersed, with sonication as an external mechanical energy which helps the particles to overcome the attractive Van der Waals force at contact.

In order to obtain metastable nanotube suspensions, use of surfactant is done because it is especially easy to wash away to get finally clean carbon nanotube surfaces. Three surfactant/solvent solutions are considered (a) water with an anionic surfactant (SDS, or sodium dodecyl sulfate, $Na^{+}CH_3(CH_2)_{11}(SO_4^-)$, (b) a 50% mixture of water and ethanol with a non ionic surfactant (Synperonic NP10), and (c) tetradecane with another non-ionic surfactant. Fig. 7.7 shows images taken with the phase contrast microscope (01 mg of raw material sonicated for 10 minutes in 1 ml of solvent) a few minutes after sonication. An optical phase contrast microscope permits to characterize directly the aggregation state of the suspension on the micron scale. In ethanol (and in methanol and acetone as well), particles are not dispersed and formed big aggregates of about 100 μm, which sedimented rapidly (Fig. 7.7(a)). Water-nanotube suspensions stabilized with SDS (sodium dodecyl sulfate) show dramatic change in their aggregation state (Fig. 7.7(b)). A few carbon aggregates are imaged by optical microscopy, meaning that most of the material had been dispersed. No sedimentation indicates the required metastability to start a purification process. The optimum is reached when the SDS concentration is slightly above the critical micellar concentration (CMC, Fig. 7.7(c).

The CMC for SDS in water is equal to ~2.5 g/l, at which surfaces are saturated by the surfactant and surfactant molecules start to self-aggregate into micelles. Below this concentration (0.4 CMC, Fig. 7.7(b), the total amount of suspended nanomaterial in comparison is small. Above this concentration (12 CMC, Fig. 7.7(d)), the tubes and particles tended to form larger aggregates. The aggregation is reversible in the sense that when the surfactant concentration is lowered down to the CMC, the aggregates spontaneously disappeared.

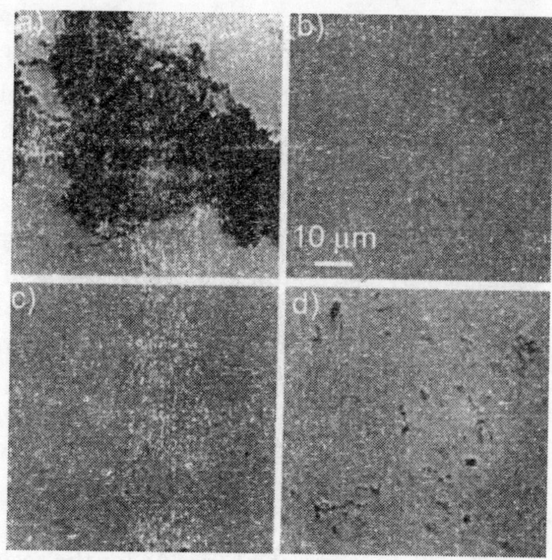

Fig. 7.7 *Optical-phase contrast microscope photograph of nanotubes i.e. (ii) chanol (b) distilled water, with SOS at 0.4 CMC, (c) distilled water with SDS at 2 CMC (d) distilled water with SDS at 12 CMC*

New, nanotube colloidal suspensions are stabilized in water with SDS at 2 CMC, and a part fo the enhanced dispersion is taken to perform a liquid-phase separation of nanotubes and nanoparticles. The suspension is filtered, in order to extract the larger particles (i.e., the tubes), while leaving the smaller constituents (i.e., the nanoparticles) in the filtrate. All filtration experiment a started with a suspension prepared from 500 ml distilled water, 2.5 g SDS and 50 mg of raw material sonicated during 15 minutes. Sedimentation and centrifugation (at 5000 rpm for 10 minutes) removes all graphitic particles larger than 500 nm from the solution.

The filtration appartus used is a funnel large enough to allow the sonication of the colloidal suspension during the filtration process with an ultrasonic probe. The probe is held about 5 mm above the filter, which is placed on a 47 mm diameter fritted glass support, and the filtration is aided by a water pump. Track-etched polycarbonate membranes yielded a significant separation of nanotubes and nanoparticles. Such membranes have vertical and well-separated pores with a well-defined size, with the advantage that the residue can be recovered simply by scraping the deposit off the filter. The best results a obtained with 0.4 μm pores and strong sonication.

To enhance the separation yield, successive filtrations to both residue and filtrate a carried out. The residue is redispersed in a fresh water-SDS solution and centrifuged to remove the debris picked up following the recuperation on the filter. Fig. 7.8 shows the material extracted after two successive filtrations of both filtrate (Fig. 7.8(a), SEM and Fig. 7.8(b), TEM) and deposit (Fig. 7.8(c), SEM and Fig. 7.8(d), TEM). For SEM and TEM analysis, all the materials as extracted by filtering the tube-or particle-rich colloid through a 0.02 μm pore size silica filter and washing the obtained film with distilled water to remove the remaining traces of surfactant. The filtrate shows a large proportion of nanoparticles, along with some short tubes (≤ 500 nm). The tube proportion amounted to 5% in number and 18% in weigh tin the filtrate (which is an overestimation due to the shorter-than-average tube length), whereas it reached typically 28% in number and 61% in weight in the raw deposit after sedimentation and centrifugation (see Fig. 7.6). The weight of a nanotube (and of a nanoparticle) can be roughly estimated by calculating the total surface of the graphitic planes of the tubes. The weight of an avcerage tube with a length of 1 μm, and inner cavity of 3 nm and a total diameter of 14 nm (as determined from a statistical analysis), amounts thus to 0.32 fg. Similarly, the weight of an average nanoparticle of 40 nm diameter and an inner cavity of 10 nm is 0.076 fg, i.e., ~25% of the weight of a tube. As for the residue, it was obviously enriched in nanotubes: only a few particles are still present in the powder. Analysis show that the tubes represent 61% of the material in number, and 87% in weight. Thus the filtration allows to obtain after two steps a separation of nearly 90% in weight of

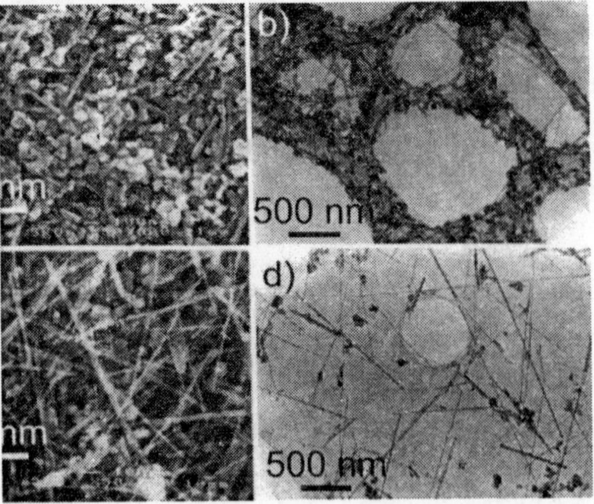

Fig. 7.8 *SEM (a, c) and low-magnification Tem (b, d) micrographs of purified nanoparticle (a, b) and nanotubes (c, d) after two filtration steps. The tube proportion amounted to 5% in number and 18% in weight and 61% in number and 87% in weight for purified nanoparticles and nanotubes, respectively. The magnifications are the same as in Fig. 1.*

nanotubes and nanoparticles, respectively. The loss of nanomaterial (sedimented and centrifuged dispersion) during two separation steps amounts to about 25% on the filters.

The above procedure is repeated until the desired purity is reached. Another way to further purify the nanotube suspension is based on the fact that the aggregates formed in the presence of a large excess of surfactant contain mainly large objects, while smaller objects remain dispersed. These aggregates sediment after a certain time, typically a few hours, leaving the smaller particles in suspension. We will see in the following that this reversible floculation is due to the depletion interaction induced by the surfactant micelles on large colloidal particles (see Fig. 7.9). For instant, this interaction induces a size-dependant floculation of droplets in emulsions, and is the basic principle of a size purification technique by fractionated crystallization.

Whether the depletion interaction can be induced for the foculation of nanotubes, by leaving carbon nanoparticles in dispersion. The depletion energy at contact of two nanotubes or particles can be evaluated. Let, SDS micelles be represented as spherical objects with an exclusion diameter, σ_1. They exert an overall osmotic pressure *P.* on other colloidal objects (nanotubes or nanoparticles). When the distance between two objects, *d*, becomes smaller than σ_1, the micelles are excluded from a volume V. This induces an uncompensated force from osmotic pressure that results in an attraction of the two particles (Fig. 7.9(a).

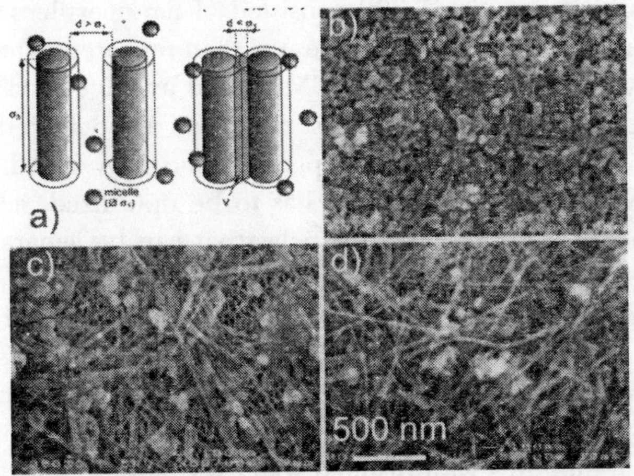

Fig. 7.9 *(a) Mechanism of aggregation of nanotubes in presence of surfactant micelles. Micelles are excluded from the depletion volume, V, between tubes separated by a distance, d, smaller than the micelle diameter σ_1, which induces an attractive force due to the uncompensated micelles osmotic pressure; (b) to (d) SEM micrographs of the iterative size-selection procedure started from a purified nanotube suspension: (b) material left in suspension after the first floculation step (12 CMC); material left in suspension (c) and aggregated (d) after the third floculation step (8 CMC and 6 CMC for the second and the third step, respectivel). The average tube length amounts to 0.6 μm (b) 1.3 μm (c) and 1.7 μm (d). The magnifications are the same as in Fig. 1.*

Assume that the micelles behave like an ideal gas, so that $P = nkT$ where n is the micelle concentration, *k* the Boltzmann constant and *T* the temperature. The depletion contact energy

u_c^s for two spherical carbon particles of diameter σ_2 ($\sigma_2 \gg \sigma_1$) is given by $u_c^s = -(1/4)\pi P\sigma_1^2$, whereas the depletion contact energy u_c^c for two cylindrical carbon nanotuibes of diameter σ_3 and contact length L is given by $u_c^c = \text{PL}[\sigma_3/2\sqrt{4\bar{\sigma}^2 - \sigma_3^2} + 2\bar{\sigma}^2 \text{ arc sin } \sigma_3/2\bar{\sigma}]$, where $\bar{\sigma} = (\sigma_1 + \sigma_3)/2$. By taking an average diameter of 40 nm for carbon particles, 14 nm and 1000 nm for nanotubes diameter and contact length (which is smaller then the total tube length due to the curvature of the tubes), respectively, and taking into account the thickness of the surfactant layer (~2 nm), it is found that $u_c^c = 60u_c^s$.

Usually a contact energy $U = 5$ kJ is sufficient to induce foculation two colloidal objects, which corresponds for the nanotubes objects, which corresponds for the nanotubes to 3 CMC (considering that a micelle is formed by ~70 SDS molecules, with a diameter of 4 nm). AT this concentration, carbon particles experience a depletion contact energy of about 0.1 kT, which is far too low to modify the dispersion state at usual colloidal particles volume fractions between 0.1 and 10%).

Preliminary experiments a done on purified nanotube suspensions (Fig. 7.8(c) and (d), and typical results are depicted in Fig. 7.9(b) to (d). Floculation is observed above ~12 CMC, which is comparable to the estimated value above for contact lengths smaller than 400 nm. The material remaining in suspension, Fig. 7.9(b) consisted of nanopartilces with a few short tubes (average length: ~600 nm). The above procedure was then repeated twice on the sediment with decreasing SDS concentration (8 CMC and 6 CMC), and average tube lengths of 1.1 μm, 1.3 μm, and 1.7 μm (±0.3 μm) were found. The last two steps are depicted in Fig. 7.9(c) and (d), which shows the material that remained in suspension and aggregated, respectively, after the third floculation step. Although this procedure has to be optimized, it should allow both the elimination from the suspension of all nanoparticles that can be separated from the tubes by sonication, and a size-selection in length of the nanotubes.

One point of concern with such separation methods to be remove throughly the surfactant after separation, as its presence may influence substantially the physical properties of the nanomaterial and induce unwanted chemical reactions such as oxidation. Energy-dispersive X-Ray analysis in the SEM on non-washed films (obtained from suspensions with 12 CMC SDS) showed sufficient to wash the films with distilled water for half an hour to remove these traces, indicating a SDS content of less than 0.1%. No surfactant traces are detected by electronic paramagnetic resonance measurements on washed films.

Finally, the structural quality of the tubes is evaluated by high resolution TEM The tubes show intact tips and walls, covered in some places with some amorphous material, which is quite common for non-oxidized arc-discharge tubes. This indicates that the SDS does not chemically react with the tubes, and that neither tips nor walls were damaged during the separation process.

OVERVIEW OF PURIFICATION TECHNIQUES

1. *Method for purifying single wall carbon nanotubes*

SWNTs, produced with arc discharge, Ni:Y:Ti:graphite = 2:2:2:94, are purified with two oxidation steps.

First step:

1. Thermal oxidation: heating from 623–873 K.
   Metal catalyst acts as oxidation catalyst for the graphite carbon layers.
2. Wet oxidation: reflux in hydrogen peroxide solution between 333–393 K.
   Metal catalyst surface is cleared, so the metal can be solvated in acid, the SWNTs remain suspended.
3. Wet oxidation: reflux in nitric acid.
   The metal catalyst is exposed to acid and solvated; the SWNTs remain suspended.

Second step:

4. Wet oxidation: reflux in hydrogen peroxide solution
   Metal catalyst acts as oxidation catalyst; both SWNT and carbon oxidise. Because there is only a small amount of catalyst left, this step does not affect the yield very much.
5. Wet oxidation: reflux in hydrochloric acid.
   The metal catalyst is exposed to acid and solvated; the SWNTs remain suspended.

A sonication step is performed between all steps. Smaller catalyst diameter and thinner carbon layers are beneficial for the method.

2. *Reduced diameter distribution of SWNTs by selective oxidation*

Both a chemical treatment in HNO_3 and oxidation in a reduced O_2 atmosphere lead to a selective burning of the narrower SWNTs in bulk samples and to a diameter distribution which is smaller by a factor two.

Reduced oxygen atmosphere

1. The sample is annealed in high vacuum (5×10^{-6} mm Hg) at 873 K in a quartz tube to remove the remaining fullerene particles and amorphous carbon.
2. Into the same quartz tube, pure oxygen is fed with a partial pressure of $1.5 - 5 \times 10^{-6}$ mbar, within a temperature range of 723 – 873 K. Total annealing time is 70 h. This step selectively burns SWNTs with smaller diameter.
3. The sample is annealed in high vacuum (1×10^{-7} mmHg) at 1023 K to remove the oxygen. In this method there is still metal catalyst left in the sample.

Chemical oxidation

1. The sample is annealed in high vacuum (5×10^{-6} mmHg) at 873 K to remove the remaining fullerene particles and amorphous carbon.
2. The material is then refluxed in 2 M HNO_3 for 43 h at room temperature. This selectively oxidises the narrower SWNTs and also removes the metal catalyst.

3. The solution is neutralised and filtrated 6 h using crossflow filtration.
4. The sample is annealed in high vacuum at 873 K for 0.5 h to remove the remaining solvent and acid.
5. The sample is annealed in high vacuum at 1073 K for 0.5 h to remove the remaining solvent and acid.

Regarding the metal content a purity of 99.9 % is reached.

The samples are checked with AAS (atomic absorption spectroscopy) and TEM pictures. This shows that the metal content in the samples is reduced from 31 wt% to 0.1 wt% and that the diameter distribution is narrowed. A narrower size distribution is created, with a larger average diameter.

3. *Plasma etching for purification and controlled opening of aligned carbon nanotubes*

For the usage of aligned MWNTs in FED display industry, a thin layer (ca. 0.1 – 0.5 μm) of amorphous carbon film has to be removed. Most of the techniques known are tedious, have low yields and damage the vertical alignment.

The unique layered structure of the amorphous carbon film formed on top of the aligned MWNTs, makes the radio frequency glow-discharge plasma etching a good method for purification.

1. The substrate is exposed to a water-plasma under vapour pressure in the range of 0.3 – 0.5 Torr. (80 min)
2. The power needed for the plasma is supplied by a radio frequency of 250 kHz
3. The plasma leads to a high concentration of atomic oxygen.
4. This oxygen will controllably oxidise a thin carbon layer on top of the vertically aligned tubes.
5. As a result purified aligned one-end opened nanotubes are obtained.

The result is confirmed by SEM and TEM images, which clearly show that the end-caps with the metal catalyst have been removed.

4. *Purification and characterisation of SWNTs*

This method has been developed for the removal of metal catalyst and carbon impurities from the laser grown SWNTs and initially cleaned with nitric acid. It describes two studies of a relatively high temperature gas phase oxidation process.

The first study:

1. Water reflux for removing aromatic carboxyl acids produced by the cleaning with nitric acid.
2. The sample is heated in a 5 % O_2/Ar, 1 atm mixture for 1 h. The temperature is varied from 573–773 K in four steps.
3. After each heating step, a sonication in concentrated HCl is followed.
4. Best steps are the oxidations at 573–773 K.

The second study was done with a two-step oxidation to produce the best results for metal removal with minimum loss of nanotubes:

1. Water reflux for removing aromatic carboxyl acids produced by the cleaning with nitric acid.
2. The sample is heated at 573 K in a 5 % O_2/ Ar, 1 atm mixture for 1 h.
3. After this heating step a sonication in concentrated HCl is performed.
4. The sample is heated at 773 K in a 5 % O_2/ Ar, 1 atm mixture for 1 h.
5. Another sonication in concentrated HCl is performed.
6. The final metal content is 0.1 % relative to the carbon content. In the last step, the oxidation is metal catalysed.
7. Finally, annealing at 1173 K follows.

The samples are checked with SEM and TEM images. TGA indicates the samples that are heated show the best temperature resistance. Also the effect of more metals can be checked with TGA. From the Raman spectra, it can be seen that the annealing step increases the intensity of the SWNT peak. Looking at the van Hove peaks gained with UV-vis-near-IR, it can be concluded that this oxidation removes smaller diameter SWNT preferentially. Also concluded is that the intensities of the van Hove peaks is increased.

5. *Purification of SWNTs by selective microwave heating of catalyst particles*

1. *Microwave treatment at 773 K for 20 min.*
   The sample (100 mg of arc-discharge SWNTs) is placed in a quartz tube that passes through a cavity. The microwaves couple to the residual metal catalyst, raising significantly the local temperature, leading to both the oxidation and rupturing of the carbon passivation layer over the catalyst particles.
2. *Mild acid treatment*
   With this protective carbon coating weakened or removed, a mild acid treatment (4 M HCl reflux for 6 h) is then sufficient to remove most of the metal in the sample, leaving the nanotubes in tact.

Optimal conditions: Flow (dry air) = 100 sccm
Frequency = 2,45 GHz
Power = 150 W
Temperature = 773 K

The entires process is monitored by TEM and HRTEM pictures. TGA shows three different peak structures that are obtained, which indicate that the metal catalyst partially oxidise the tubes, to form amorphous carbon and multi shell carbons. Raman spectrometry indicate that the microwave process gives a small broadening of the SWNT peak, the HCl treatment hardly damaged the SWNT and the HNO_3 reflux step gives a significant peak broadening.

Using microwave processing and a mild acid treatment it is possible to remove residual metal to a level lower than 0,2 wt%. The total yield for the two-step purification procedure is about 35 wt%.

6. *Length sorting cut SWNTs by high performance liquid chromatography*

Previously cut SWNTs by an acid base process are sorted by length by HPLC based, size exclusion chromatography (SEC). The SWNTs were grown by pulsed laser vaporisation. Purifying and cutting SWNTs

1. Wet oxidation with reflux in HNO_3
2. Crossflow filtration
3. First cutting was affected by 24 h ultrasonication in a 3: 1 mixture of HNO_3:H_2SO_4 at 318 K.
4. Next to further shorten the tubes and to consume sidewall defects, the SWNT reacted in a 4:1 mixture of H_2SO_4:H_2O_2 for 10 min.

HPLC

1. The cut SWNTs are run over a SEC-HPLC.
2. The biggest SWNTs leave the HPLC first (after ± 20 min).
3. At the end, when all the SWNTs have already past, the C_{60}, C_{70} etc, come out (after ± 30 min).

The results above are are also obtained from AFM (atomic force microscopy) pictures and the UV-vis measurements from the HPLC.

7. *Purification of SWNTs synthesised by the hydrogen arc-discharge method*

SWNTs synthesised with the hydrogen arc-discharge method have been purified with ultrasonication in alcohol, oxidation in fixed air and soaking in hydrochloric acid. The yield of the method is ~41% of the as-produced soot which contained ~55% SWNTs. The obtained purity is 96%.

1. Ultrasonication in alcohol for 30 min to cleave off impurities from the SWNTs.
2. Dried to air at room temperature.
3. The sample is oxidised in fixed air at 813–823 K for 40 min, which oxidises the carbon impurities.
4. The product is then soaked in hydrochloric acid.
5. Afterwards, the sample is washed with de-ionised water and dried in an oven at 423 K.

The TEM pictures show that almost no visible impurities remain in the sample. This is confirmed by TGA (thermo graphic analysis), which included TG (thermogravimetry), DTG (differential thermogravimetry) and DTA (differential thermal analysis). Raman spectrometry shows the removal of the carbon impurities peak and sharpening of the SWNT -peak.

The carbon impurities start to combust at 801 K and the SWNTs start to combust at 873 K. This is why a temperature of 823 K is chosen.

8. *High-quality SWNTs from arc-produced soot*

A three-step process, for purifying the raw soot, which contained 33 % metal and 67 % carbon is given below.

1. *Soft oxidation.* The as-produced soot (150 mg) is subjected to 393 K reflux in a 2,8 N HNO_3 solution (200 ml) for 6, 12 or 24 hours. The solution is filtered and washed with distilled water several times, followed by drying at 373 K for more than 10 hours. In this step an oxide layer is formed on the surface of the metal catalyst and damage to the amorphous carbon occurrs.
2. *Air oxidation.* The dried sample is heated in air at 828 K for ten min to remove the amorphous carbon.
3. *High-temperature vacuum treatment.* The sample was heat-treated at 1873 K at a pressure of 10^{-3} Pa for three hours. In this final step the graphitic carbon and metals are removed, the oxide layer decomposes and the rearrangement of SWNT walls occurred.

After the final step, about 20 % of the weight of the initial raw soot remained and the final product contained less than 1 % metal.

TEM images are analysed. TGA samples of the different steps have been made, which show the loss in metal and carbon. Raman spectra give information on the diameter size (RBM peaks) and the SWNT content (G/D values) by comparing peaks.

9. *High-yield purification process of SWNTs*

A two step process for the purification of as-produced SWNTs from the catalytic arc discharge process (Ni/Co/FeS = 1:1:1). The process involves rotating thermal oxidation in air and an acid treatment.

1. The as-produced SWNTs are grinded and heated at 743 K for 50 min while rotated at 30 rpm. This step removes the carbonaceous particles, weight was reduced to 40 wt% of the initial sample.
2. The oxidised powder is immersed and filtered in 6 M HCl to dissolve the metal catalyst. The yield was 70 wt%.
3. To unbundle the SWNTs, the sample is refluxed in 30 % HNO_3 for 4-6 hours.
4. The obtained suspension is filtered with a PTFE (poly-(tetrafluoroehtylene» membrane in de-ionised water.

The final yield is 25-30 wt% with a purity of ~96 %, containing 1 % metal atoms (initially 8%).

SEM and TEM images, Raman spectra and TGA of the process samples are taken. The Raman . spectra clearly show that the samples purity and defects content are improved by the various processes. The ratio of the peaks at 1600 cm^{-1} and 1300 cm^{-1} increases.

10. *Purification and characterisation of SWNTs obtained from the gas-phase decomposition of CO (HiPco process)*

1. Raw HiPco tubes compressed onto a dry filter paper (vacuum).
2. SWNTs are placed in a quartz tube furnace.
3. Mixture of O_2 in Ar is passed trough a water bubbler and run over the sample.

4. Nanotubes and a continuous flow of wet Ar and O_2 are heated at 498 K for 18 h. This heating step oxidises the metal. This metal oxide has a lower density, which causes the carbon shells to break open and exposes the metal. This could also be done at 598 K (1,5 h) or 698 K (1 h).
5. Sonication (15 min.) or prolonged stirring in concentrated HCl solution (overnight).
6. Filtered onto a 47 mm (1 mm pore size) Teflon membrane.
7. Washed several times with de-ionised water and methanol
8. Dried in a vacuum oven at 373 K (minimum of 2 h).
9. Annealing at 1073 K in Ar for 1 h.

From the TGA measurements it can be seen that the samples first gain weight, because of oxidation of the metal and then loose weight because of the (catalysed) carbon removal. UV-vis-near-IR indicate the preferential loss of smaller diameter during the cleaning process and give insight in the van Hove features during the process. With Raman spectra the loss of intensity due to the process and recovery due to annealing is shown. SEM en TEM images show the raw and the purified SWNTs. A large reduction of impurities can be seen. Also a comparison is made with laser-grown SWNTs by comparing the UV-vis-near-IR and Raman spectra.

The reference shows the different outcomes for the different temperatures. This method oxidises the metal and the carbon impurities. SWNTs with smaller diameters are oxidised preferentially.

11. *Ultrasonic reflux system for one-step purification of carbon nanostructures*

Here, the soot containing amorphous carbon, metal catalyst and carbon nanostructures is purified in one system.

1. Sonication, to solve soot in the first flask.
2. Solvated soot is run over a filtering membrane and removed from the sonication area, non-solvated soot remains in flask.
3. Solvated soot is further purified in second flask with a reflux column.
4. Oxidising gas is inserted to oxidise the amorphous carbon.
5. In the second flask acid removes metal catalyst.

12. *Purification of SWNTs by microfiltration*

This technique allows to separate the as-prepared SWNT mats into three separate fractions without the use of acid, heat, or oxidative treatment.

1. The sample (SWNTs prepared by pulsed laser ablation) is soaked into CS_2.
2. The CS_2 insolubles are then trapped in a filter.
3. Sonication of the insoluble solids in aqueous solution (0,1 % surfactant).
4. Micro filtration. Most of the carbon nano spheres (CNS, C_{60} and C_{70}) and metal nanoparticles are passing through the filter, while the SWNTs and a small amount of residual CNS and metal nanoparticles are trapped. The microfiltration process is repeated

for three cycles to mini mise the amount of residual CNS and metal nanoparticles trapped between the ropes of the SWNTs.

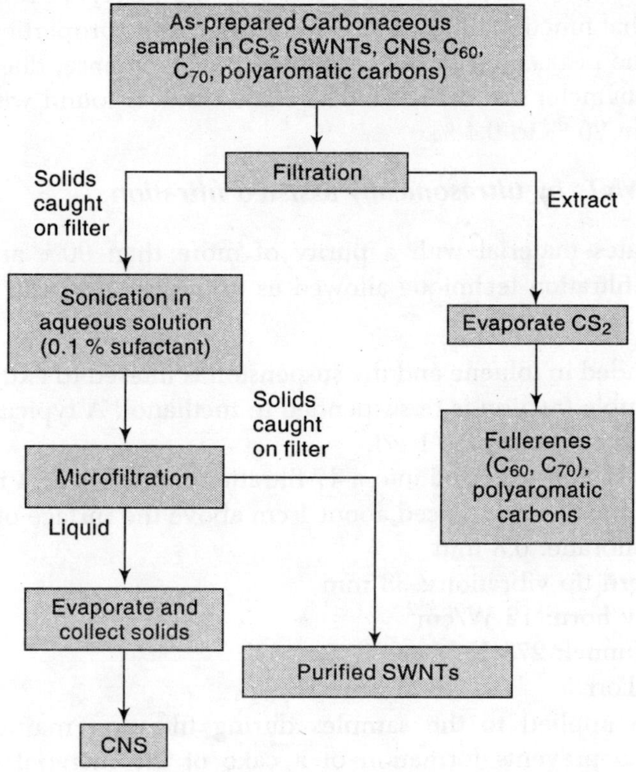

Fig. 7.10 *Scheme of the purification process*

HRSEM and TEM pictures gave proof of that what is stated above. The Raman spectra of the CS_2 extracts, the as-produced and the purified SWNTs show that C_{60} and C_{70} are removed.

The individual weight percentages of the separated fractions are 6, 10 and 84 wt% for the CS_2 extract, CNS and SWNTs respectively.

13. *Purification of HiPco carbon nanotubes via organic functionalisation*

1. Organic functionalisation of the as-produced nanotubes (p-SWNTs) based on 1,3-dipolar cycloaddition of azomethineylides in DMF suspension.
2. Purification of the soluble functionalised nanotubes (f-SWNTs). The purification is based on the improved solubility of the nanotubes and the insolubility of the metal particles. To further purify f-SWNTs, basically removing the amorphous carbon particles, a slow precipitation process is employed, occurring when diethyl ether is added to a chloroform solution of f-SWNTs (85 mg in 300 ml).
3. Removal of the functional groups and recovery of the purified nanotubes (r-SWNTs) by thermal treatment at 626 K, followed by annealing to 1173 K.
4. The metal content is reduced from 26 % to 0.4 %.

TEM images of the purified and as-produced nanotubes are taken. DSC (differential scanning calorimetry) shows that the purification process has few impacts on the structure of the SWNTs, as the curves are almost the same. The DSC curve of the f-SWNTs are quite different. NIR (near-infrared) shows that functionalisation altered the electronic properties of the tubes. Further could be seen that some peaks after purification became less intense, due to the fact that more SWNTs with smaller diameter are oxidised. The same result is found with Raman. The metal content is reduced from 26 % to 0.4 %.

14. *Purification of SWNTs by ultrasonically assisted filtration*

This process generates material with a purity of more than 90% and yields of 30–70%. Ultrasonically assisted filtration technique allowed us to purify up to 150 mg SWNT soot in 3-6 h.

1. The soot is suspended in toluene and the suspension is filtered to extract soluble fullerenes. The toluene-insoluble fraction is re-suspended in methanol. A typical concentration of the raw material in the suspension is 1 g/l.
2. The suspension is then transferred into a 47 filtration funnel. A 25,4 mm ultrasonic horn is inserted into the funnel and is placed about 1 cm above the surface of the filter membrane.
    - Pore size membrane: 0,8 mm
    - Amplitude horn tip vibrations: 33 mm
    - Power diss. by horn: 12 W/cm^2
    - Temperaturefunnel: 273 K
    - Pressure: 50 Torr

   Ultrasonication applied to the samples during filtration maintains the material in suspension. It also prevents formation of a cake of the material on the filter surface. Methanol is continuously added to the filtration funnel to maintain a constant filtration volume.
3. After filtration, the material is washed with 6 M sulphuric acid to remove traces of metal, introduced into the sample by the ultrasonic horn.

TEM and SEM images of the purified samples are made. They are characterised by Raman spectroscopy. The method decreases the amorphous carbon peak. TGA shows that the purified SWNTs have more defects. The 50 % oxidation point is reached 303 K earlier as for purification.

15. *Mechanical purification of SWNT bundles from catalytic particles*

In basic principle of this purification method the energy of elastic impact between encapsulated catalysts and small hard inorganic particles to eject the metal kernels and trap them with a strong magnet is used.

1. SWNTs are first suspended either in soap solution or in toluene and then are dispersed in various solvents such as toluene, N,N-dimethylformamide or 30 % nitric acid.
2. A powder of nanoparticles (ZrO_2, NH_4Cl, $CaCO_3$ or diamond), that is not soluble in the given medium, is then added to the suspension. This slurry is sonicated typically for 24

hours. The ultrasonic bath mechanically removes the ferromagnetic particles from their graphitic shells.
3. The magnetic particles are trapped with permanent magnetic poles and with a subsequent chemical treatment A high purity SWNT material is obtained.

The samples are checked with ESR (electron spin resonance) and TEM images, which prove that the main part of the metal catalyst is removed. This method is useful for lab-sized purifications.

16. *Cutting SWNTs by Fluorination*

This process describes the fluorination of purified HiPco SWNTs (~1.0 nm, no amorphous carbon, no metal), followed by pyrolysis of the partially fluorinated tubes (CF_x, $x < 0.2$). This gives short tubes with average lengths of < 50 nm. The initial length ~1 μm.

1. Purified SWNTs are fluorinated for 2 hours to a stoichiometry of CF_x, ($x < 0.2$).
2. The partially fluorinated SWNTs are pyrolised up to 1273 K (TGA control, 10 C/min) in an argon atmosphere.
3. The fluorine is driven off the sidewall of the tubes to form CF_4 or COF_2.
4. This leaves behind chemically cut SWNTs in a yield of 55 %, with average lengths of ~40 nm. The yield is based on the amount of SWNTs.

The length of the tubes is measured by AFM imaging. TGA, elemental analysis and IR show that during pyrolysis the fluorine is completely driven off the nanotube. The elements that appear due to pyrolysis are also shown in a derivative weight percentage. Raman spectra of the purified, fluorinated and cut SWNTs indicate that the defects in the carbonaceous species increase.

17. *Enhanced saturation lithium composition in ball-milled single-walled carbon nanotubes*

The nanotubes are grinded for several minutes up to 20 min, thus breaking and disordering them. This process is highly time dependent, as longer ball-milling tear down all the SWNT structures.

The TEM and SEM images and Raman spectra are taken. The tubes are shortened and Raman spectra indicate that more disordered carbon is produced.

18. *Chromatographic purification of soluble SWNTs*

SWNTs (1.4 nm and lengths of 100 – 300 nm) produced by arc-discharge, after several purification steps, are dissolved and run over a gel permeation chromatographic column:

1. Short SWNTs are functionalised with octadecylamine and solvated in THF.
2. The solution is run over a gel permeation chromatographic column, styragel HMW7.
3. Two main fractions are obtained.
4. The first fraction gives semi-conducting SWNTs.

5. The second fraction does not contain SWNTs, but contains nanoparticles and amorphous carbon.

AFM images have been made of the different fractions. The three HPLC fractions were analysed with continuous UV-vis measuring and a complete spectrum of each fraction was made. NIR spectra suggest that fraction 1 hardly contains metallic tubes. The yield was estimated with NIR. The Raman spectra. AFM and NIR show the purity of the fractions.

This method claims to separate and purify semi-conducting and metallic SWNTs. Finally, an estimate of 50% of the initial amount of SWNTs in the soot is recovered, with a high purity.

19. *Chromatographic purification and properties of soluble SWNTs*

SWNTs (1.4 nm and lengths of 100 – 300 nm) produced by arc-discharge, after several purification steps, are dissolved and are run over a gel permeation chromatographic column:

1. Short SWNTs are functionalised with octadecylarnine and solvated in THF.
2. The solution is run over a gel permeation chromatographic column, PLgel MIXED-A.
3. Three main fractions are obtained.
4. The first fraction gives serni-conducting SWNTs.
5. The second fraction gives traces of SWNTs and nanoparticles.
6. The third faction gives amorphous carbon.

AFM images are made from the different fractions. The three fractions are seen with continuous UV-vis measuring and a complete spectrum of each fraction is made. The yield is estimated with NIR. The Raman spectra, AFM and NIR show the purity of the fractions. The fluorescence en NIR spectra suggest that fraction I hardly contains metallic tubes.

This method separate and purify semi- and metal-conducting SWNTs. The estimated yield is a lot higher, 74 % of the initial amount of SWNTs in the soot with high purity.

PROPERTIES OF CNTs

The properties of carbon nanotubes are the characteristics of the material. As Carbon nanotubes have a structure of a sheet of graphite rolled into a cylinder, and are macromolecules they are unique for their size and shape. Electronic, molecular and structural properties of CNTs are determined to a large extent by them, nearly one dimensional structure. Various physical properties like diameter, length, twist, chirality and electrical, thermal, mechanical properties of carbon nanotubes are discussed below.

(A) MECHANICAL PROPERTIES

The properties of the carbon nanotubes are very intriguing and are yet to be fully understood because they display wide range in their electrical, thermal, and structural properties. These properties depend upon the lattice parameters density etc.

A table 8.1 of the experimental data pertaining to the lattice parameters, density and interlaying spacing for the different nanotube arrangements is given below:

TABLE 8.1 Lattice pattern, density and spacing

Lattice Parameter:		
	(10, 10) Armchair	16.78 A
	(17, 0) Zigzag	16.52 A
	(12, 6) Chiral	16.52 A
Density:		
	(10, 10) Armchair	1.33 g-cm^{-3}
	(17, 0) Zigzag	1.34 g-cm^{-3}
	(12, 6) Chiral	1.40 g-cm^{-3}
Interlayer Spacing:		
	(n, n) Armchair	3.38 A
	(n, 0) Zigzag	3.41 A
	(2n, n) Chiral	

Nanotubes have remarkable mechanical properties that can be exploited to strengthen materials or to act as "tips" in scanning probe microscopes. And because they are composed entirely of carbon, they have a low specific weight. Carbon fiber is already used to strengthen a wide range of materials, and the special properties of carbon nanotubes make them the ultimate high-strength fiber.

Molecular mechanics consists of simulating the behaviour of an object (e.g. a nanotube) by calculating the interaction of every atom with its neighbours. This powerful technique enables to calculate how tubes of different diameter are distorted when placed on a substrate. It requires for the tubes, energy to bend or deform, but in turn they can gain energy from the van-der-Waals interaction with other tubes or with the substrate they are deposited upon. Figure 8.1 shows the radial deformation of adsorbed single-wall carbon nanotubes calculated using molecular mechanics. The radial deformation increases as the tube diameter is increased. The system can lower its energy by increasing the area of contact between the tube and the substrate. (See also propeties of distorted CNTs)

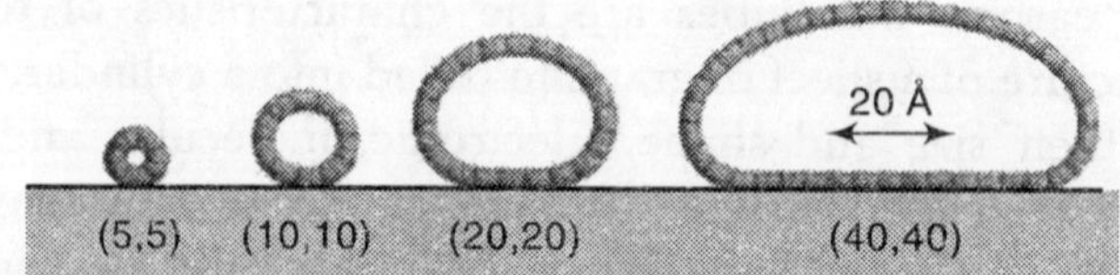

Fig. 8.1 *Radial deformation of adsorbed single-wall carbon nanotubes on graphite*

Below (in Figure 8.2) is a simulation of the situation where two adsorbed nanotubes cross each other at a right angle. The bottom nanotube is compressed by the top tube which is strongly attracted to the substrate. This one can compute the force exerted on the lower tube, and understand how nanotubes deform.

In a sheet of graphite each carbon atom is strongly bonded to three other atoms, which makes graphite very strong in certain directions. However, adjacent sheets are only weakly bound by van der Waals forces, so layers of graphite can be easily peeled apart. But it is not so easy to peel a carbon layer from a multiwall nanotube. Carbon fiber is used to strengthen a wide range of materials, and their special properties mean that they could be the ultimate high-strength fiber.

Figure 8.3 shows a microscopic view of a carbon nanotube being bent and stretched. The bending stiffness. of the carbon nanotube is an important physical property. It can be measured by placing the nanotubes across "nanopores" and using an atomic force microscope to bend them in the middle. Occasionally a nanotube will

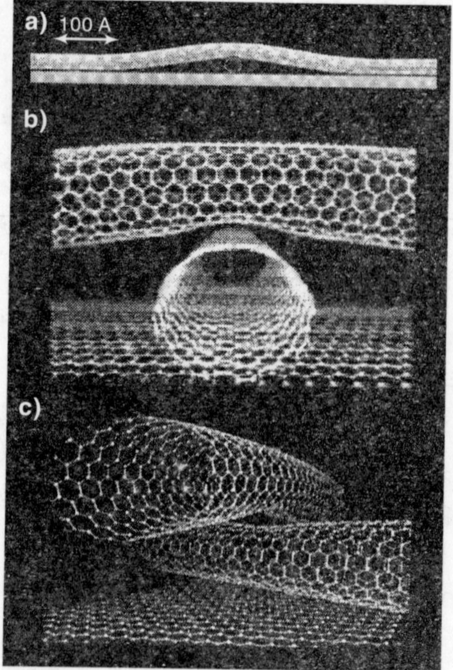

Fig. 8.2 *Simulation of compression of 2 carbon nanotubes*

span one of the pores and a microscope was used to measure how the deflection, which is inversely proportional to the Young's modulus, varied with the applied force. Whether multi-walled or single-walled, these tubes bend to surprisingly large angles, when pressed on the tip without causing any damage to the tip or the tube. They produce kinks or get deformed on the application of the force, but these deformations disappear completely when the force is removed and the tip of the nanotube or the whole tube recovers back to its original state. Taking the example of silicon carbide, which is another very strong material, and if the force is applied beyond its Euler's limit, the silicon carbide rods fractured at high loads whereas carbon nanotubes do not break. When a rod is compressed along its length, initially the rod remains straight, but as the compression increases, its starts bending at the value called Euler's limit. These deformations are permanent in silicon carbide or in other words, they are plastic deformations. Multi-walled carbon nanotubes did not break in contrast to silicon carbide rods.

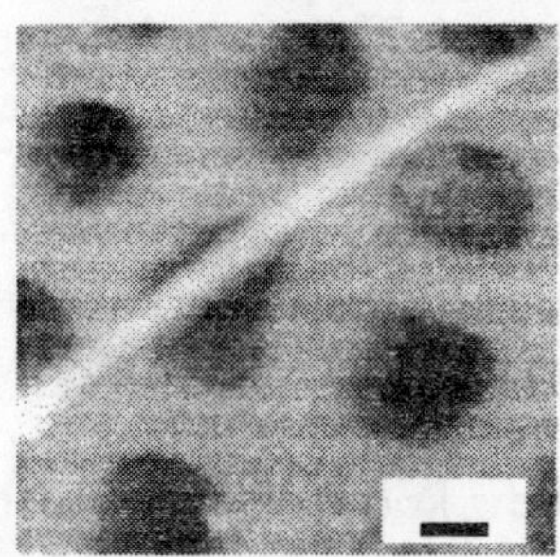

Fig. 8.3 *Microscopic view of a carbon nanotube being bent and stretched*

Young's Modulus

Carbon nanotubes have a very large Young modulus in their axial direction. The nanotube as a whole is very flexible because of the great length. The average modulus is found to be 1.8 Tpa. This is measured by first allowing a tube to stand freely and then taking a microscopic image of its tip, the modulus is calculated from the amount of blur seen in the photograph at different temperatures. It is found that a (10, 10) armchair tube had a modulus of 0.64 Tpa, a (17, 0) zigzag tube had a modulus of 0.648 Tpa, and a (12, 6) tube had a value of 0.674 Tpa.

The Young's Modulus is also dependent on the size and chirality of the SWNT, ranging from 1.22 Tpa for the (10, 0) and (6, 6) tubes to 1.26 Tpa for the large (20,0) SWNT.

SWNTs are 10 to 100 times stiffer than steel and are resistant to damage from physical forces. Pressing on the tip of the nanotube causes it to bend without damage to the tip (or the whole CNT). When the force is removed, the tip of the nanotube recovers to its original state.

The diameter of the nanotube is determining factor for it's elastic modulus. Different trends have been predicted for the dependence of Young's modulus on the diameter of the tube. The value of SWNTs does in fact depend on diameter; an individual tube had a modulus of about 1 Tpa while bundles (or ropes) of 15 to 20 nm in diameter have a modulus of about 100 Gpa. The small-diameter SWNTs are quite stiff and exceptionally strong, meaning they have a high Young's modulus and high tensile strength. They have a tendency to form bundles or ropes. Small diameter nanotube ropes extend elastically by ~5.8% before breaking, their strength being ~37 Gpa.

The Young's modulus of CNTs also depend on the-thickness of the walls of the nanotube. If the tube is considered to be a solid CNG cylinder, then it would have a lower Young's modulus but if it is hollow, the modulus is higher, and thinner the walls of the nanotube are, the higher the modulus will be. However, multi-walled CNTs modulus measurements do not depend on the diameter. The modulus depends on the disorder or defects in the multi-walled nanotubes. The disorder or defects in turn depend upon the method of preparation. MWNTs, grown by arc discharge have a modulus of about 1 TPa, whereas those grown by the catalytic decomposition of hydrocarbons have a smaller modulus. The bending stiffness (by bending the tubes in the middle), and the variation of deflection (with the applied force) are measured with AFM. This deflection is inversely proportional to Young's modulus which varies from 0.5 TPa to 5.5 TPa of the tube having a diameter of 1 nm.

It is now known that the Young's modulus of a carbon nanotube should approach a value of 1.28 terapascals. This is true both for multiwall and single-wall nanotubes because the modulus is mainly determined by the carbon-carbon bonds within the individual layers. Table 8.2 shows the relative Young's modulus for a multi and single wall nanotube (MWNT and SWNT) as well as the tensile strength and thermal properties (Table 8.3).

Table 8.2 List of the Young's modulus for a single and multi-walled nanotube

Young's Modulus (SWNT)	~1 Tpa
Young's Modulus (MWNT)	1.28 Tpa
Maximum Tensile Strength	~30 Gpa

Table 8.3 List of thermal properties associated With the mechanical properties

Thermal Conductivity	~2000 W/m/K
Phonon Mean Free Path	~100 nm
Relaxation Time	~10^{-11} s

Some measurements pertaining to the carbon nanotube are summarized in Table 8.4 given below. This gives a basic idea of the structure and currently measured properties for the nanotube.

Equilibrium Structure

Table 8.4 Structure data pertaining to a single-wall nanotube

Average Diameter of SWNTs	1.2–1.4 nm
Distance from opposite Carbon Atoms (Line 1)	2.83 A
Analogous Carbon Atom Separation (Line 2)	2.456 A
Parallel Carbon Bond Separation (Line 3)	2.45 A
Carbon Bond Length (Line 4)	1.42 A
C-C Tight Bonding Overlap Energy	~2.5 eV
Group Symmetry (10 10)	C_{5V}
Lattice: Bundles of Ropes of Nanotubes Triangular Lattice	(2D)
Lattice Constant	17 Å

Tensile Strength

The maximum tensile strength of carbon nanotubes can reach up to 300 Gpa. The tensile strength of multi-walled carbon nanotubes is measured by attaching the ends of a multiwall nanotube to a pair of AFM tips and stretch it until it breaks. It is found that the tensile strength is at least an order of magnitude lower than expected. The outermost layer has to take most of the stress because the AFM tips only make contact with the outside of the nanotube. It is found that the outermost tube ruptures first at the tensile limit, and that this is followed by a sudden elongation and that the ruptured outer tube then slide over the inner tubes. The tensile strength of this layer ranges from 11–63 Gpa for the set of 19 MWNTs that are loaded. A record strength and toughness of greater than 460 MPa of a yarn of carbon nanotubes is achieved by using a method for continuous spinning of composite fibers.

The fiber strength of 4.2 GPa and a modulus of 167 GPa have been achieved by incorporating SWNTs in a high-strength, high-modulus polymer; electrical and thermal conductivities are low because of limitations on nanotube content. Carbon nanotubes, apart from the mechanical strength, should be good thermal and electrical conductors too. If one tries to get good conduction, the mechanical properties have to be compromised and the mechanical strength becomes far lower than can be obtained using a polymer matrix. Though there has been good progress in synthesizing single-walled nanotubes (SWNTs) using wet spinning and making both multi-walled nanotubes (MWNTs) and SWNTs using dry spinning, the highest strength achieved with any of these spinning methods is about (an order of magnitude lower than the strength of individual SWNTs) 37 GPa. The challenge is to produce yarns that are at the same time strong, creep resistant, highly conducting, and reversibly deformable over relatively large mechanical stresses.

The high strength of carbon nanotubes makes them promising candidates in reinforcement applications, but there are many outstanding problems that must be overcome. First, the properties of the individual tubes must be optimized. Second, the tubes must be efficiently bonded to the material they are reinforcing (the matrix) so that they actually carry the loads. Third, the load must be distributed within the nanotube) itself to ensure that the outermost layer does not shear off.

Electrical Properties

Depending on their chiral vector, carbon nanotubes with a small diameter are either semi-conducting or metallic.

Electrical Conductivity

Based on different arrangements of Carbon nanotubes, the tube can possess different electrical properties which are the result of the electron moving differently in the tube depending on the carbon nanotube arrangement. Based on the structure of the nanotube, the, carbon is bonded to its neighbours by 3 valence electrons which leaves one conducting electron to move in the tube. The electron more or less travels a straight path and can only propagate along the axis of the tube and the carbon atoms must line up with the circumference of the tube. In general, if you subtract the integer value m from n and the result is divisible by 3, the nanotube is considered metallic, otherwise, the nanotube is a semi conductor.

Because of the near perfect orientation of a metallic carbon nanotube, they make perfect conductors for electrons. Because of the carbon nanotube's ability to change from conductor to semi-conductor, if a defect is introduced into the honeycomb arrangement, then a Shottky barrier is formed which is the basis for many electric devices.

Energy States

The graph (Figure 8.4) represents the energy states that are associated with a metallic nanotube conductor.

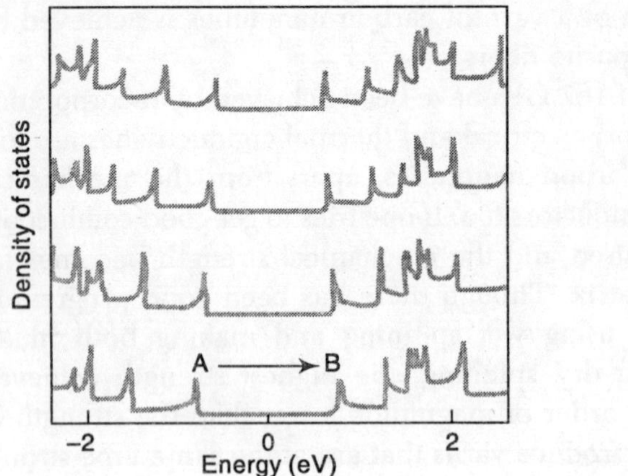

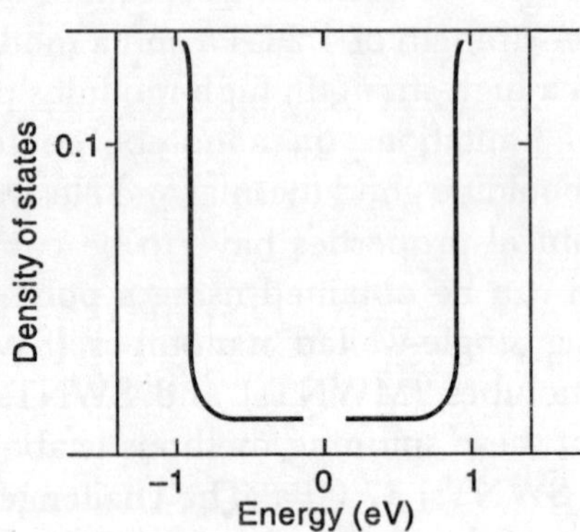

Fig. 8.4 *Graph of the energy states in metallic nanotubes*

The differences in conducting properties are caused by the molecular structure that results in a different band structure and thus a different band gap. The differences in conductivity can easily be derived from the graphene sheet properties. It is shown that a (*n, m*) nanotube is metallic when, $n = m$ or $(n - m) = 3i$, where i is an integer and n and m are defining the nanotube. The resistance to conduction is determined by quantum mechanical aspects and independent of the nanotube length.

Despite the fact that different choices in *n* and *m* determine the conductivity of the nanotube, the chemical bonds of the carbon atoms are exactly the same, which is surprising since a 2-D graphene sheet is a semi-conductor with a zero band gap. This means that the valence bond has the same energy as the bottom of the conduction band which means a zero band gap. The following table shows the band gaps for a conducting or semi-conducting carbon nanotube

Fundamental Gap:

For (*n, m*); *n-m* is divisible by 3 [Metallic]	0 eV
For (*n, m*); *n-m* is not divisible by 3 [Semi-Conducting]	~0.5 eV

At the Fermi Energy (the highest occupied energy level), the density of states is finite for a metallic tube (through very small), and zero for a semi-conducting tube. As energy is increased, sharp peaks in the density of states, called Van Hove singularities, appear and specific energy levels. The optical spectrum is given by:

$$x = E$$

where E is the energy difference between occupied and unoccupied states, especially near the peaks. The spectrum is dominated by the transitions between the Van Hove singularities.

Conductance Quantization in One Dimension

The starting point for the discussion of electronic transport properties of highly 1 D-SWNTs is the conductance.

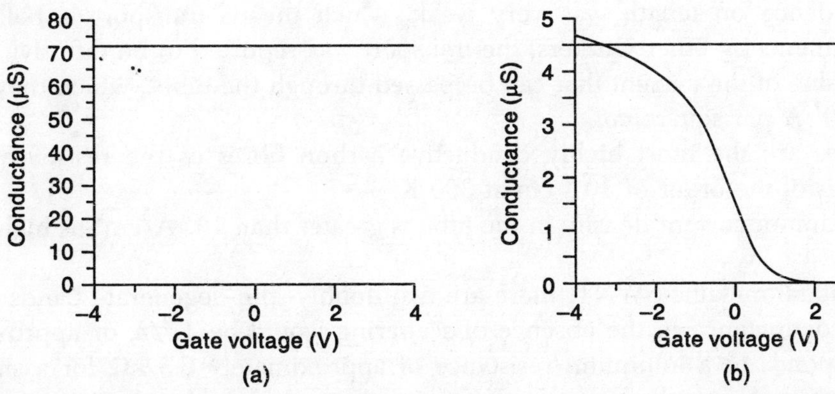

Fig. 8.5 *Thc conductance G as a function of gate voltage V_g for metallic and semiconducting SWNT devices at room temperature. The conductance of the metallic nanotube (a) is nearly independent of gate voltage, while the conductance of the semiconductaing nanotube (b) varies strongly with gate voltage, conducting at negative gate voltage, and becoming insulating at positive gate voltage. Both dcvices are "metal-on-top", fabricated using nanotubes prepared by CVD*

Landauer (1958), derived the conductance of an ideal (zero scattering. or ballistic) 1-D channel, and found that the conductance was finite, and quantized and a nanotube acts as a ballistic conductor. Nanotubes also have a constant resistivity, and a tolerance for very high current density of the nanotube is any where from 10-100 kohm and the resistance is around 12.9 kohm. The conductance $G = e^2/h$, for each mode of the channel, where e is the electronic charge. and h is Planck's constant

The general result for a 1-D channel with multiple modes and less than unity transmission probability is

$$G = \frac{e^2}{h} \sum_i T_i$$

where T_i is the transmission probability of the *i*th mode.

By contacting nanotube fibres with a Hg surface and measuring their conductance using SPM (Scanning Probing Microscopy), it is found that the nanotubes behave as ballastic conductor, with quantum behaviour. As additional nanotubes are touched to the Hg surface, the MWNTs conductance jumps by increments (Fig. 8.6) of 1 G_0, where, $G_0 = 2e^2/h$. The value of G_0 is 1/12.9 K^{-1}.

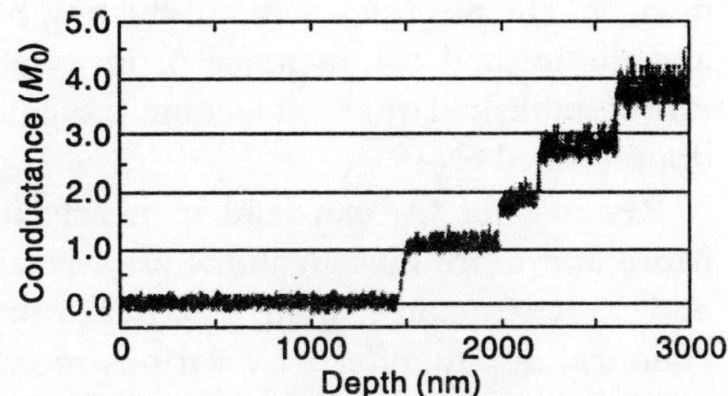

Fig. 8.6 *A carbon nanotube is a quantum conductor. The conductance rises by 1 G_0 as the depth increases*

In the case of carbon nanotubes, theory predicted that the conductance should be ballistic and the conductance should be quantized. Neither of these effects were observed experimentally until, in 1998, Frank and co-workers did an interesting experiment to measure the conductance of multi-walled carbon nanotubes. They found that all the tubes have nearly the same conductance and the dependence on length was very weak, which means transport is ballistic. But in previous experiments by other workers, the transport was reported to be diffusive. Frank et al. also gave the value of the current that can be passed through the tubes, which they reported to be as high as 10^7 A per square cm.

SWNTs ropes are the most highly conductive carbon fibres as the resistivity of metallic SWNTs, ropes is of the order of 10^{-4} cm at 300 K.

Also the maximum current density in the tube is greater than 10^7 A/cm^2 as high as $\approx 10^3$ A/cm^2.

For an individual metallic SWNT, there are two doubly-spin-degenerate bands at the Fermi level, and the conductance in the absence of scattering would be $4e^2/h$, or approximately 155 μS. This corresponds to a minimum resistance of approximately 6.5 kΩ for a single SWNT. This quantized resistance appears as a contact resistance–a drop in the electrochemical potential occurs at each contact to the SWNT. The two-terminal resistance of a nanotube device is

$$R = h/4e^2 + R_i + R_{c1} + R_{c2}$$

where $h/4e^2$ is the quantized contact resistance of the nanotube; R_i the intrinsic resistance from scattering processes within the tube arising from, for example, disorder or phonons. and the contact resistances R_{c1}, the resistances due to transport barriers formed at the metal electrode/nanotube junctions.

TRANSPORT PROPERTIES

It is revealed by transport measurements (Figure 8.7(a)) that the majority of individual SWNTs synthesized by CVD on supported catalyst are semiconducting in nature. This type of nanotube exhibits field effect transistor (FET) behavior at room temperature and has been intensively explored in recent years for nanoelectronics devices, including transistors and logic devices. The as-grown semiconducting SWNTs are hole-doped p-type FETs with hole depletion and diminished conductance (from typically 100 kΩ to 1 MΩ) by positive gate voltages. It is now established that molecular oxygen adsorbed on the nanotubes is responsible for hole doping of SWNTs.

Removal of O_2 can lead to nearly intlinsic semiconducting behavior (Figure 8.7(b) inset). More and more investigations are revealing that, although carbon nanotubes are highly robust and inert structures, their electrical properties are extremely sensitive to charge transfer and chemical doping effects by various molecules.

The second type of observed SWNTs synthesized by CVD appears to be quasi-metallic with small band gaps on the order of 10 meV. These nanotubes are not as sensitive as semiconducting SWNTs to electrostatic doping by gate voltages but exhibit a conductance dip associated with the small band gap. These nanotubes correspond to a class of non-armchair SWNTs, and the origin of the band gap is slight Sp^2 to Sp^3 hybridization due to the nonflat nature of the

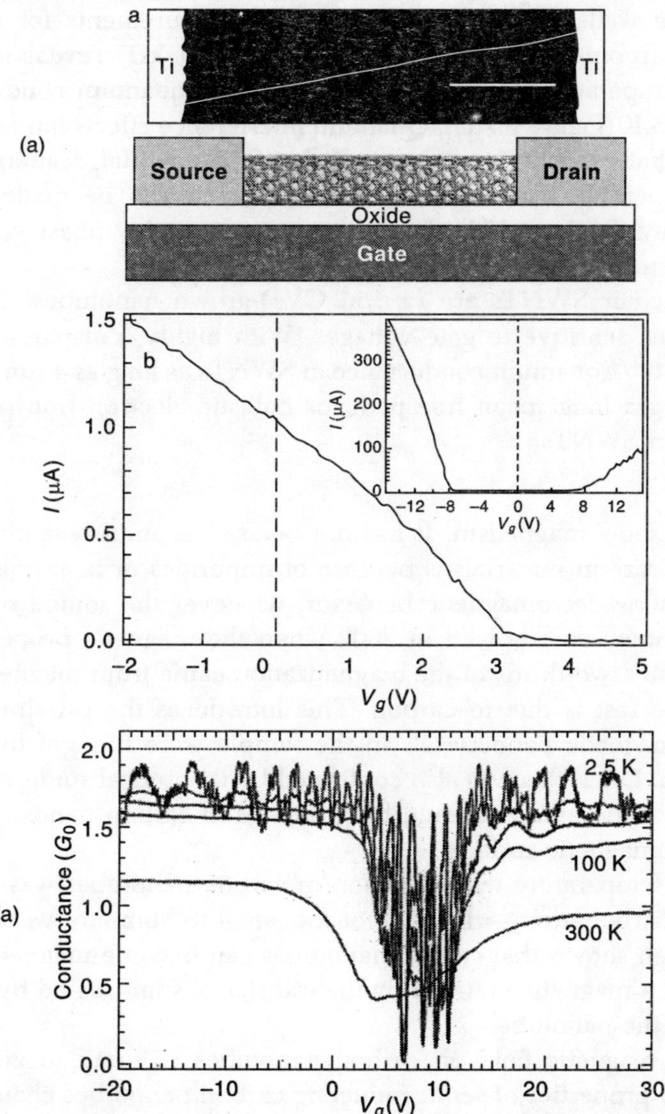

Fig. 8.7(a) *Electrical properties of individual nanotubes. (a) AFM image of a SWNT contacted by two Ti electrodes and a schematic cross-section view of the device. Scale bar: 1 mm. A gate voltage (V_g) can be applied to electrostatically couple to the nanotube and shift the Fermi level. (b) I–V_g curve for a typical as-made semiconducting SWNT showing p-type FET characteristics. Inset: The p-type FET evolves into a nearly intrinsic semiconductor after removal of surface-adsorbed oxygen. (c) Conductance G (in units of quantum conductance, $G_0 = 2e^2/h$) vs V_g (Fermi energy) for a quasi-metallic SWNT at various temperatures. The conductance approaches the theoretical limit $2G_0$ at low temperatures, with conductance fluctuations due to quantum resonance effects*

hexagons on the tube walls. Temperature-dependent measurements for some of the quasi-metallic-like SWNTs (room-temperature resistance ~10–20 kΩ) reveals increased electrical conductance at low temperatures, reaching the $4e^2/h = 2G_0$ quantum conductance (6.45 kΩ in resistance) limit at ~1.5 K (Figure 8.7(c)). Quantum interference effects have also been observed. The results suggest that (1) phonon is one of the funda-mental scattering mechanisms in SWNTs at room temperature, (2) excellent ohmic contacts can be made to nanotubes with transmission probability $T \approx 1$, and (3) electron transport is highly phase coherent and ballistic in nanotubes at low temperatures.

Truly metallic armchair SWNTs are rare in CVD-grown nanotubes. The conductance of these nanotubes is least sensitive to gate voltages. With highly transparent ohmic contacts to these nanotubes, near $4e^2/h$ quantum conductance in SWNTs as long as 4 μm at low temperatures is observed, suggesting a long mean free path for ballistic electron transport in high-quality/perfection CVD-grown SWNTs.

Magnetic Properties

Carbon nanotubes show magnetism. It has not been clear until recently whether the weak magnetism shown by carbon materials is because of impurities or is an intrinsic property. But carbon materials do show ferromagnetic behavior, whatever the source may be, though it is very weak. The dilemma was resolved in 2002 when the magnetic properties of a meteorite sample showed that only two-thirds of the magnetization came from magnetic minerals present in the sample and the rest is due to carbon. This introduces the proximity effect to carbon nanomaterials. Ferromagnetic nanocrystals in the sample were thought to induce a magnetic moment in the carbon. Later, this was also confirmed by theoretical studies where it was shown that a measurable magnetic moment could be produced in carbon nanostructures if they were placed close to a ferromagnetic surface.

The average room temperature magnetization in carbon nanotubes was calculated to be 0.1 Bohr magnetons per carbon atom, which is not too small to compare with iron, for which the value is 2.2. It has been shown that carbon nanotubes can become magnetized, when they are placed in contact with a magnetic material by the transfer of spin(carried by electrons) from the magnetic substrate to the nanotube.

Also, the effect of magnetic field on carbon nanotubes is found to go beyond proximity effects. Basic electrical properties of semiconducting carbon nanotubes change, when they were placed inside a magnetic field because "band gap" of semiconducting nanotubes narrowed down steadily in the presence of a strong magnetic force. Though, according to the quantum mechanical theory, nanotubes are crystals with well-defined lattice periodicity (Aharonov-Bohn effect). Here, solutions of nanotubes are placed inside a chamber containing very strong magnetic fields. Fields as high .as 45 Tesla is applied on them. Lasers are shined the light that was emitted and absorbed by the samples land to the conclusion. that at very high magnetic fields, the band gap disappears completely and the nanotubes can become a metal.

This discovery lead to novel new experiments on one-dimensional magneto-excitons, quantum pairings (quantum computing nonlinear optics) and quantum optics.

OPTICAL PROPERTIES

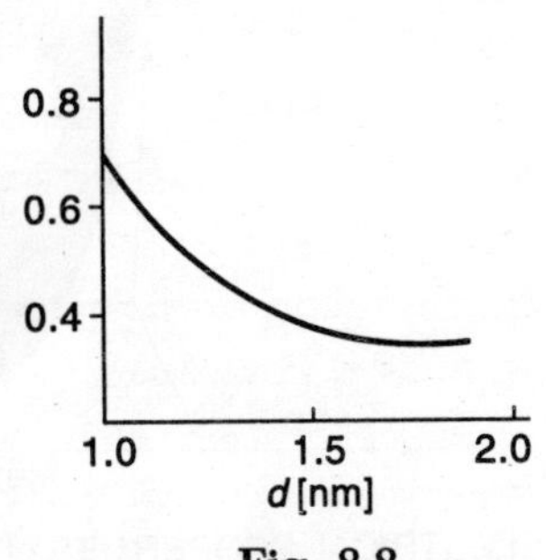

Fig. 8.8

Nanotubes of type $n - m = 3l$, where l is zero or any positive integer, are metallic and therefore conducting. The fundamental gap (HOMO-LUMO) would therefore be 0.0 eV. All other nanotubes, behave as a semi-conductor. The fundamental gap is a function of diameter, where the gap is in the order of about 0.5 eV. Data shows that the energy gap reflects, the graph (Fig. 8.8) at right. This graph can be modeled by the function:

$$E_{gap} = 2y_0 a_{cc} / d$$

Where y_0 is the C-C tight bonding overlap energy (2.7 0.1 eV); a_{cc} is the nearest neighbor C-C distance (0.142 nm), and d is the diameter. This shows that the fundamental gap ranged from around 0.4 eV – 0.7 eV, which is in good agreement with the values obtained from one-dimensional dispersement relations. Thus the fundamental gap of semiconducting nanotubes is determined by small variations of the diameter and bonding angle. A small gap exists at the Fermi level in metallic nanotubes, because of the bonding orbitals and anti-bonding orbitals mixing due to the curvarure in the graphene sheet of a SWNT. Also the conducting nanotubes shows E_{gap} to range from 1.7–2.0 eV.

Note: The optical activity of chiral nanotubes disappear if the nanotubes become larger.

ELECTRONIC PROPERTIES

The composition and geometry of carbon nanotubes engender a unique electronic complexity, due to their size, at the nanometer scale.

Whereas most electrical conductors can be classified as either metals or semiconductors, graphite is one of the rare materials known as a semi-metal. By combining graphite's semimetallic properties with the quantum rules of energy levels and electron waves, carbon nanotubes emerge a conductors.

For example one rule of the quantum world is that electrons behave like waves as well as particles, and electron waves can reinforce or cancel one another. Thus an electron spreading around a nanotube's circumference can completely cancel itself out; only electrons with just the right wavelength remain. Out of all the possible electron wavelengths, or quantum states, available in a flat graphite sheet, only a tiny subset is allowed when we roll that sheet into a nanotube. That subset depends on the circumference of the nanotube, and the nanotube twist.

In a graphite sheet, one particular electron state (Fermi point) Fig. 8.9 gives graphite almost all of its conductivity; none of the electrons in other states are free to move about. Only one third of all carbon nanotubes combine the right diameter and degree of twist to include Fermi point in their subset of allowed states. These nanotubes are truly metallic nanowires.

The remaining two thirds of nanotubes are semiconductors. That means that, like silicon, they do not pass current easily without an additional boost of energy. A burst voltage can knock electrons from valence states into conducting states where they can move about freely. The amount of energy needed depends on the separation between the two levels and is the so-

THE ELECTRICAL BEHAVIOR OF NANOTUBES

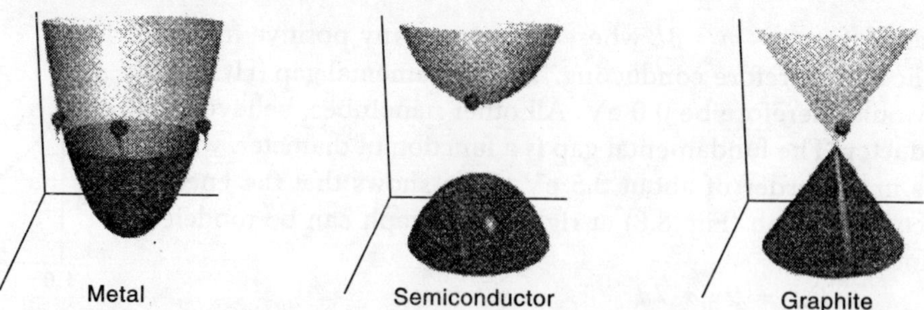

ELECTRICAL PROPERTIES of a material depend on the separation between the collection of energy states that are filled by electrons and the additional "conduction" states that are empty and available for electrons to hop into. Metals conduct electricity easily because there are so many electrons with easy access to adjacent conduction states. In semiconductors, electrons need an energy boost from light or an electrical field to jump the gap to the first available conduction state. The form of carbon known as graphite is a semi metal that just barely conducts, because without these external boosts, only a few electrons can access the narrow path to a conduction state.

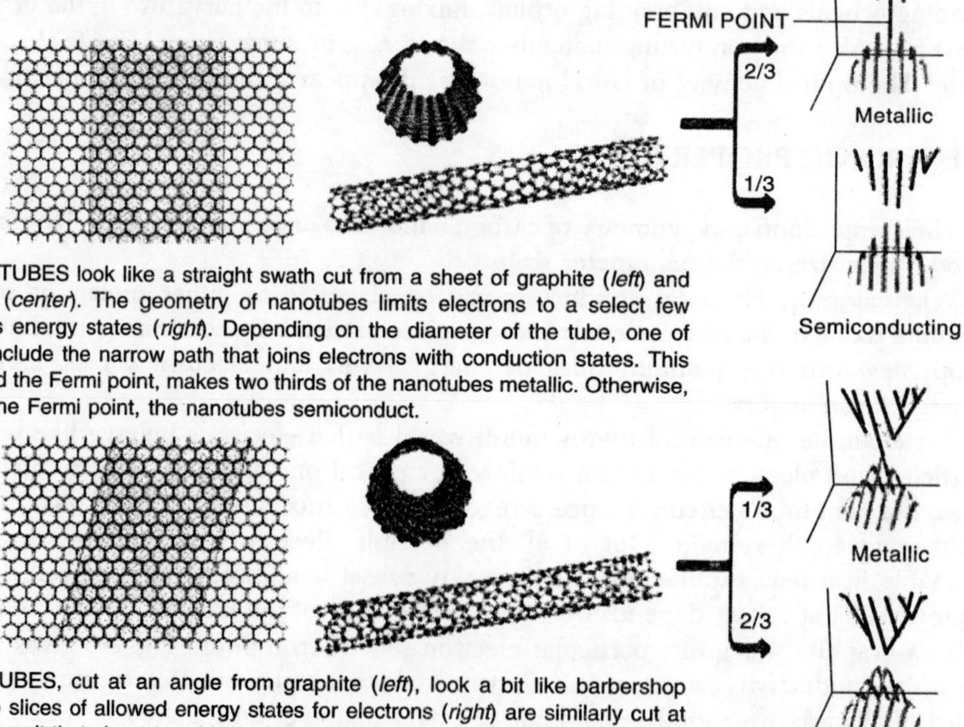

STRAIGHT NANOTUBES look like a straight swath cut from a sheet of graphite (*left*) and rolled into a tube (*center*). The geometry of nanotubes limits electrons to a select few slices of graphite's energy states (*right*). Depending on the diameter of the tube, one of these slices can include the narrow path that joins electrons with conduction states. This special point, called the Fermi point, makes two thirds of the nanotubes metallic. Otherwise, if the slices miss the Fermi point, the nanotubes semiconduct.

TWISTED NANOTUBES, cut at an angle from graphite (*left*), look a bit like barbershop poles (*center*). The slices of allowed energy states for electrons (*right*) are similarly cut at an angle, with the result that about two thirds of twisted tubes miss the Fermi point and are semiconductors.

Fig. 8.9

called band gap of a semiconductor that are useful in electronic circuits; and by having materials with different band gaps, a vast array of electronic devices are available today.

Carbon nanotubes don't all have the same band gap, because for every circumferences there is a unique set of allowed valences and conduction states. The smallest-diameter nanotubes have very few states that are spaced far apart in energy. As nanotube diameters increase, more and more states are allowed and the spacing between them shrinks. In this way, different-size nanotubes can have band gaps as low as zero (like a metal), as high as the band gap of silicon, and almost anywhere in between. No other known material can be so easily tuned.

The main factor responsible for determining the electronic properties in carbon nanotubes is their size. The diameter is in the nanometer range and length of the tube several microns. Whether they are metal, narrow gap semiconductors, wide gap semiconductors or insulators depends mainly on the diameter of the tube and its chirality.

Carbon nanotubes have been found to be good materials for field emission because of their small diameter. Field emission is a process by which a device emits electrons when an electric field or voltage is applied to it. Both open and closed end nanotubes can emit currents as high as 0.1 mA on the application of a few hundred volts. But it was astonishing to find that closed end nanotubes were much more efficient than the closed ones. It opens up new possibilities for these tubes to be used in applications such as electronics and photonic development. The explanation for light emission from nanotubes is related to its small diameter. Carbon nanotubes are like long hollow cylinders with diameters of 1–2 nm only. Thus, when an electric current is passed through a carbon nanotube, negatively charged electrons in the nanotube combine with positively charged 'holes' that are locations in the molecule where electrons are missing. When an electron fills a hole, it emits light. Infrared light was first found to be emitted by applying a voltage to nanotubes.

To improve the field emission properties, several kinds of doping and composites have been tried. Composite structures of single-walled carbon nanotubes and poly (3-octylthiophene) show very good emission properties. The composites of carbon nanotubes with polypyrrole have shown good charge storage capacities. The same composite also shows enhancement in conductivity by two orders of magnitude with an increasing weight percent of the nanotubes.

Another interesting study and experimental evidence for the ballistic conduction in carbon nanotubes was by cutting the nanotubes into small pieces so that the length of the tube becomes comparable to their diameter. It was found that in the band structure of metal carbon nanotubes, there are two distinct linear bands in contrast to single band in normal metals. One band is formed by bonding molecular orbitals and another one is formed by anti-bonding molecular orbitals, and the electrons move in opposite directions in these two bands. Thus it is very unlikely that the electron moving in one direction will change direction as then it has to move to another band i.e., another molecular orbital for changing direction. This keeps them moving in a single direction, and that means no scattering and, hence, ballistic nature.

Considering (6, 6) armchair nanotubes of different lengths. Find their bandgap values the DOS spectrum and binding energy of carbon atoms as a function of length of the tube segment. It is found that a band-gap for short nanotubes whose value decreases toward zero with increasing tube length. Cut the nanotubes into segments of a few tens of nanometers in length and use tunneling spectroscopy to obtained an evidence of increased band-gaps in these short tubes. Interestingly, the variation of the band-gap is not monotonic but shows strong oscillations as a function of tube length. While all computational techniques used predict these oscillations, the actual values of the band-gap vary strongly with the technique used. Computed DOS spectra and local density of states distributions show that the transition from a 0-D (quantum dot) to a 1-D (quantum wire) like spectrum takes place already in very short 5–10 nm tubes. The binding energy of the carbon atoms increases with increasing tube length.

Also, multi-walled carbon nanotubes showed no-electron-scattering behavior. The tubes carry such high current density that it heats the nanotube to its vaporization temperature. But electrons are thought to be strongly de-coupled from the lattice and, hence, the heat is not transferred to the lattice.

The σ bonds are mainly responsible for the mechanical properties, while the electronic properties are mainly determined by the π bands. By a tight-binding approach the band structure of these π bands can be calculated.

In a 2-D graphene sheet, the π-bonding and π^*-antibonding states become degenerate at the K-point of the hexagonal Brillouin zone, resulting in a zero band-gap semiconductor. In a nanotube, quantization of the wave function along its circumference restricts the allowed wave vectors to certain directions of the graphite Brillouin zone so that $C \cdot k = 2\pi j$, where j is an integer. If at least one of these wave vectors passes through the K-point, the tube is metallic; otherwise, it is a semiconductor with a finite band-gap. Thus, $(n, 0)$ zigzag tubes are expected to be metallic if $n/3$ is an integer and semiconducting otherwise. As the role-up vector C rotates away from the $(n, 0)$ direction, the resulting (n, m) tubes are chiral and are expected to be metallic if $(2n + m)/3$ is an integer. Otherwise, they are semiconductors with a gap $\propto 1/R$, where R is the tube radius. Finally, when C is rotated 30° away from the $(n, 0)$ direction $n = m$ and the resulting armchair tubes are metallic.

This gives us three types of band gaps in nanotubes. These are:

- Metallic: These are the armchair tubes where the small shift of the degenerate point away from the K point does not cause a band opening for symmetry reasons.
- Small-bandgap semiconducting: These are characterized by $n - m = 3j$ with j being an integer. Here, the wave vectors of the allowed states cross the K point, but due to the slight shift of the degenerate point a small gap will be present, the size of which is inversely proportional to the tube diameter squared with typical values between a few and a few tens meV.
- Semiconducting: In this case $n - m \neq 3j$. This causes a larger bandgap, the size of which is inversely proportional to the tube diameter: $E_g = k/d$ with experimental investigations suggesting a value of k of 0.7–0.8 eV/nm.

Typically the bandgap of the type 2 nanotubes is so small that they can be considered metallic at room temperature. Based on this it can be inferred that 1/3 of all tubes should behave metallic whereas the remaining 2/3 should be semiconducting. However, it should be noted that due to the inverse proportionality between the bandgap and the diameter of the semiconducting tubes, large-diameter tubes will tend to behave metallic at room temperature. This is especially important in regards to large-diameter MWCNTs.

Electronic Property (Single SWNTs)

To study the electronic properties of SWNTs, models for energy bands and Density of States (DOS) in SWNTs are made. These models have to be experimentally confirmed, so a way has to be found in which atomic structures and electron densities can be visualised. Scanning Tunnelling Microscopy (STM) can display atomic structures as well as measure the DOS.

As is shown in (Figure 8.10) and (Figure 8.11(A)), chirality is determined from STM measurement and the effect of nanotube chirality on electronic properties is studied. The spectroscopic image (Figure 8.11(B)) shows the density of states as measured by a STM. As these two images show, this technique can also be used to determine properties of intra-molecular SWNT junctions. In addition, it can also be used for measuring influence of symmetry, defects, doping, electronic contacts and so on.

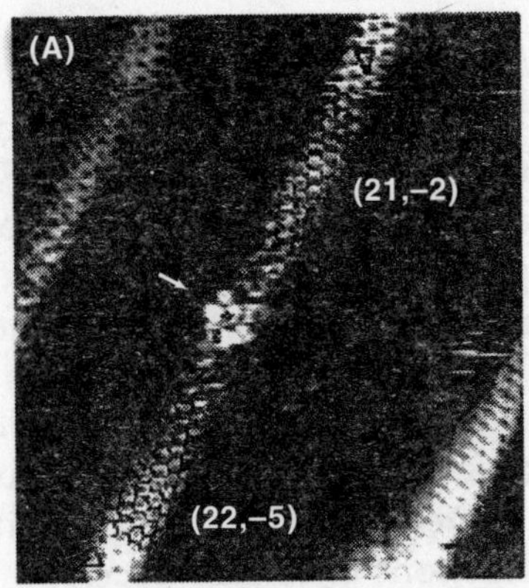

Fig. 8.10 *STM image of a SWNT. The chirality can be determined easily*

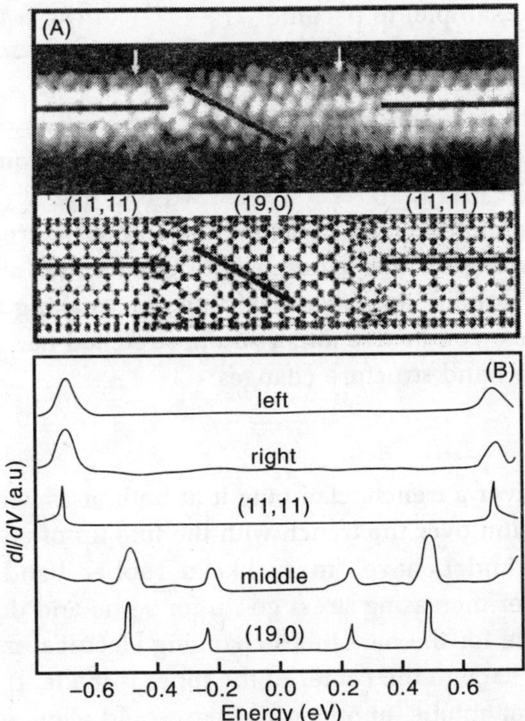

Fig. 8.11 *(A) STM atomic resolved image of intra molecular junction of SWNT. (B) STM image measurement representing the Density of States corresponding to lines in A.*

ELECTRONIC STRUCTURE

Bending of Nanotubes

Nanotubes look at the molecular level like rolled-up chicken wire Fig. 8.12. Carbon atoms are linked into hexagons, and the hexagonal pattern makes up the walls of a cylinder only nanometers wide. A nanotube's chiral angle (the angle between the axis of its hexagonal pattern and the axis of the tube) determines whether the tube is metallic or semiconducting.

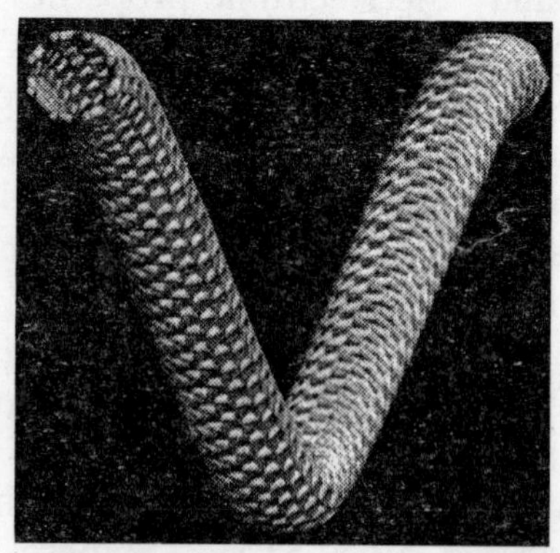

Fig. 8.12 ***Springy Nanotubes****. A carbon nanotube, when pressed in the middle, will stretch like a rubber band. According to experiments, the stretching modifies the quantum states available to electrons, which alters the nanotube's electrical resistance*

Metals conduct electricity easily because electrons have easy access to the quantum states that carry them long distances, states in the so-called conduction band. In semiconductors there is a band gap–an energy barrier that electrons must overcome to reach the conduction band. The electrons need ah extra push of energy (from heat or an electric field, for example) to become conducting.

Pressing on a carbon nanotube can change its electrical properties, Stretching a nanotube can either increase or decrease its conductivity–depending on the type of nanotube–by changing the structure of quantum states available to electrons.

Bending or stretching a nanotube could change its band structure, which would change its electrical properties. Depending on the nanotube's original chiral angle, the stretching could either induce a band gap or modify an already existing one, making the nanotube either more or less conductive become less conductive under strain, stretched nanotubes become both more and less conductive, and the band structure changes.

Experiment

Suspend each nanotube over a trench, clamping it at both ends with gola electrical contacts. Then press down on the region over the trench with the fine tip of an atomic force microscope (AFM). "They're springy," and behave "more like a rubber band than a steel rod." The nanotubes conductance under increasing stress go up for some and down for others. The band gap changes were responsible for the nanotubes changing electrical properties, using AFM tips to apply varying levels of voltage to the center of the tubes is made. If strain had caused a band gap to appear in a metallic nanotube, applying voltage would allow electrons to overcome the band gap, and the tube would become more conductive again. This voltage-induced conductance is observed nanotubes ability to sense tiny changes in pressure and relay the pressure change as an electrical signal allows to build nanotube switches sensitive to very small pressure changes.

Effect of Twisting on Nanotube

Effect of twisting on the properties of a metallic armchair (6, 6) nanotube. Three twisted configurations are shown (Fig. 8.13). Twisting is found to transform the metallic nanotube to a semiconducting one with a band-gap that varies with the twist angle as shown in Fig. 8.14.

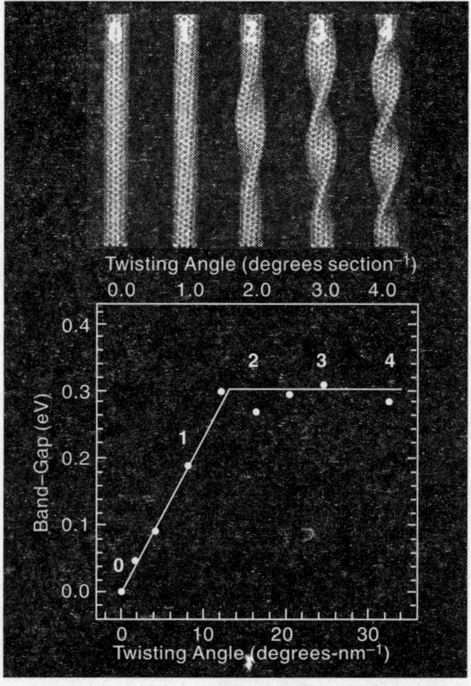

Fig. 8.13

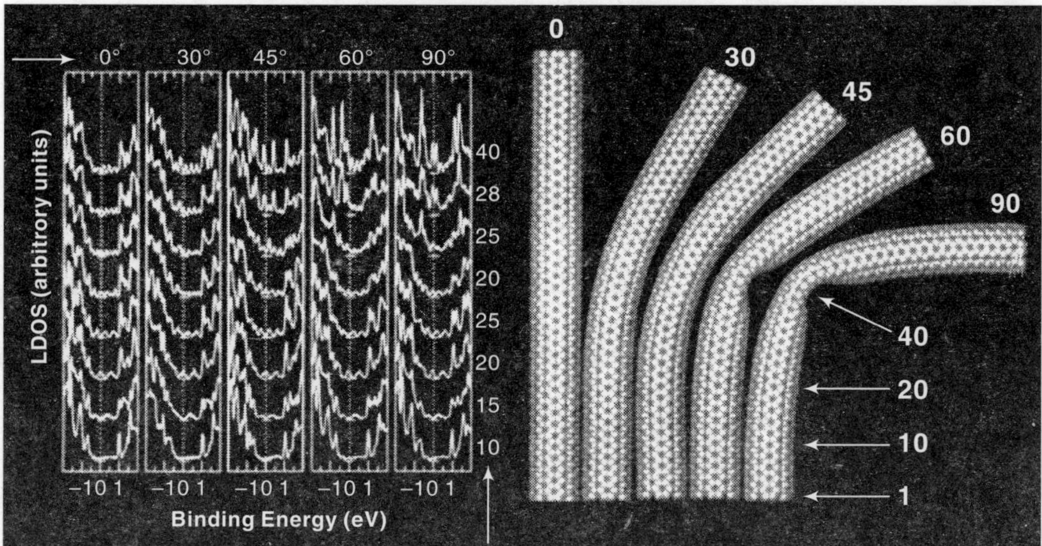

Fig. 8.14

Effect of Bending on Nanotube

Variation of the local density of states (LDOS) along the length of a (6, 6) nanotube as a function of the bending angle.

THERMAL CONDUCTIVITY

The thermal conductivity of carbon nanotubes is dependent on the temperature and the large phonon mean free paths. Phonons also propagate easily along the nanotube. Thermal conductivity seems to change; depending on temperature, current and vacancy concentration. The measured room temperature thermal conductivity for an individual MWNT (> 3000 W/m·K) is greater than that of natural diamond and the basal plane of graphite (both 2000 W/m·K). The graph of thermal conductivity vs temperature, the slope of the line at low temperature can be modeled using the heat capacity, sound velocity, and relaxation time of the tube.

There is a linear relationship.

$$K_{zz} = \Sigma CVZt$$

Where k_{zz} is the slope of the line on the graph (Fig. 8.15), C is the heat capacity, V is the sound velocity and t is the relaxation time, which is approximately 10^{-11} s. The thermal conductivity for a single rope at room temperature varies between 1800–6000 W/m-K. The conductivity is linear in temperature from 7 K to 25 K and from 25 K to 40 K, the line increases in slope. It arises monotonically with temperature to above room temperature.

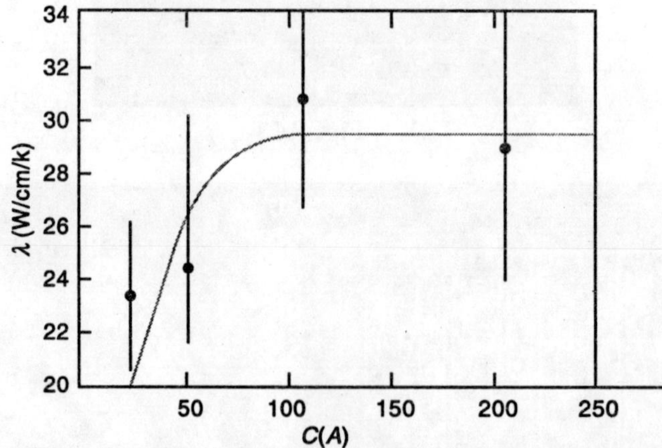

Fig. 8.15 *Thermal conductivity as a function of current passed through the nanotube*

Thermal conductivity of a (10, 10) nanotube is calculated to be 2980 W/m-K as the current applied to it is increased. The thermal conductivity of carbon nanotubes is 6,600 W/m-K at room temperature, is high valued due to the large phonon mean free paths. The graphs of the temperature dependence of thermal conductivity looks much less linear with a positive slope from low temperatures up to 100 K where it peaks around 37,000 W/m-K (Fig 8.16). Then, the thermal conductivity drops dramatically down to around 3000 W/m-k when the temperature approaches 400 K.

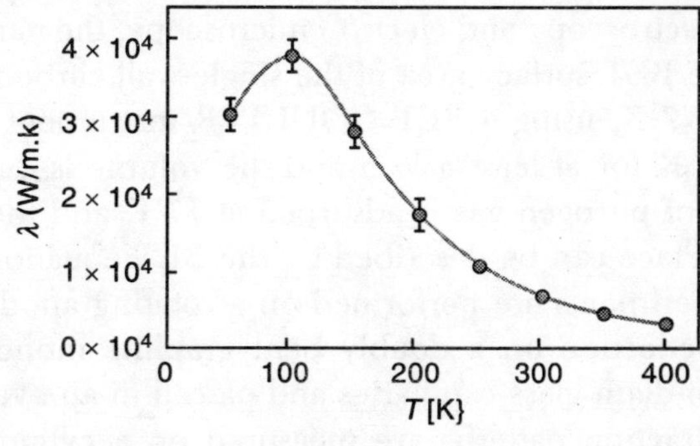

Fig. 8.16

Superconductivity has been observed, but only at low temperatures, with transition temperatures of ~0.55 K for 1.4-nm-diameter SWNTs and ~5 K for 0.5-nm-diameter SWNTs grown in zeolites.

EFFECT OF TEMPERATURE

I. Carbon Nanotube Diameter

The diameter of single-wall carbon nanotubes (SWCNTs) is an important property and is of interest for new applications that rely on diameter uniformity or the unique chemical properties of certain types. Understanding the factors that affect the diameter may lead to a better understanding of the growth mechanism of the nanotubes.

For diameter control of SWCNTs during the electric arc discharge process, the arc parameters must /be controlled carefully, like using different gas mixtures in the arc reactor. Also, use of microscopic HRTEM (high-resolution transmission electron microscopy and Raman (spectroscopy)] and macroscopic BET (Brunauer Emmett Teller), x-ray diffraction) analysis of the nanotubes is made.

To characterize the synthesized nanotubes, some resonant Raman studies are done using several excitation wavelengths to excite various diameter and chirality tubes following laser lines: 514.5 nm (argon laser) 632.8 nm (He-Ne laser), 676.4 nm (krypton laser) and 1064 nm [Nd: yttrium-aluminum-garnet (YAG) laser] are used. Raman spectra are recorded at room temperature and different spectrophotometers are used. For all spectra, the spectral resolution iss about 1 cm^{-1}, and spatial resolution iss close to 1 μm.

Transmission electron microscopy (TEM) samples are prepared by placing. tiny pieces of dry nanotubes on the dry TEM grid, followed by a droplet of methanol. The nanotube become flattened by a droplet of methanol. The nanotubes stick to the grid as the methanol dries; and in many places nanoropes from loops. When such loops are parallel to the electron beam of TEM, one can see little circles, from which the diameters are measured. Such measurements are more precise than measurements on the in-plane nanotubes. Images are scanned at 1200 dpi: TEM magnification factor is 23 nm/cm at 500 kX magnification.

In addition to Raman spectroscopy and electron microscopy, the nanotubes are characterized at a macroscopic level. The BET surface area of the single-wall carbon nanotube is determined by nitrogen adsorption at 77 K, using a BET COULTER instrument. Prior to absorption, the CNTs are degassed at 393 K for at least a ½ h and the volume is determined by introducing helium gas. Constant flow of nitrogen gas is adsorbed at 77 K and the pressure recorded. The adsorption process on a surface can be described by the BET equation.

The x-ray scattering experiments are performed on a rotating anode. Cu $K\alpha$ radiation (λ = 1.5417 A) is selected by reflection on a doubly bent graphite monochromator. The powder samples are scaled in 1-mm-diam glass capillaries and placed in an evacuated x-ray chamber to remove air scattering. Diffraction patterns are measured on a cylindrical imaging plate. The profiles are recorded along the equatorial line of the imaging plate. Intensities are corrected for polarization effects, and a broad diffuse scattering signal around 1.8 A^{-1}, coming from the glass capillary is subtracted from the x-ray scattering profiles.

Results

A. Raman spectroscopy

Raman spectroscopy turns out to be a powerful and non-destructive tool to characterize single-wall carbon nanotubes (SWNTs), since some vibrational modes are related to some of their structural features. The diameter (*d*), for example, can be assessed by measuring the radial breathing mode (RBM) frequency and applying the formula

$$\upsilon_{\text{RBM}}(\text{cm}^{-1}) - \frac{223.75}{d\,(\text{nm})}$$

The use of above formula slightly shifts the diameter distribution, but does not modify the observed trend. Therefore, the formula is used to estimate values of the nanotube diameter distribution.

In resonant Raman experiments, using different excitation laser lines distinguish between lines of SWNTs, the electronic density of states of SWNTs exhibit singularities, while energy positions are strongly dependent on the nanotube diameter (inversely proportional) and on the metallic (M) or semiconductor (SC) nature. Some electronic transitions are easily induced between singularities, allowing a selective characterization of SWNTs. These nanotubes are listed in Table 8.5.

Table 8.5 Resonant diameter and electronic character (M: metallic, SC: semiconducting) for each wavelength used in Raman experiments.

Wavelength	*514.5*	*632.8*	*676.4*	*1064*
Resonant diameter (nm)	1.32	1.15	1.45	1.35
Electronic character	SC	M	M	SC

For each wavelength, the diameter is calculated from the diameter distribution of the Raman spectra for different samples (see Fig. 8.17)

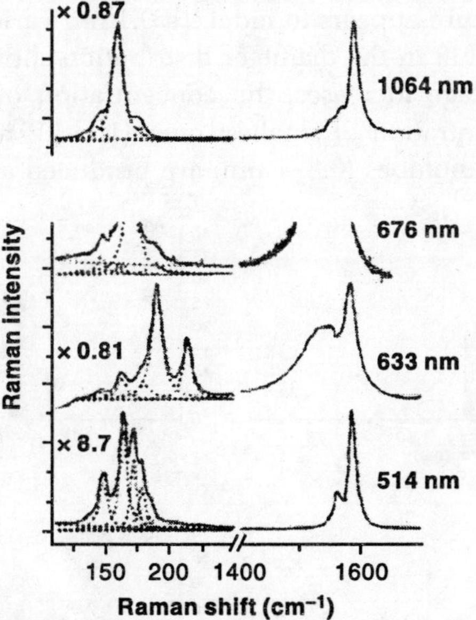

Fig. 8.17 *Relevant Raman spectra recorded on the 0% of Ar sample for four laser excitation wavelengths.*

In Fig. 8.18 the average diameter is shown as a function of argon percentage in the inert atmosphere used during the SWNT production. For the three wavelengths: 514.5, 632.8, and 676.4 nm, a continuous decrease of the diameter with increasing Ar percentage is noticed. The biggest diameter 1.4 nm is obtained with 20% of Ar, wereas 80% and 100% of Ar yield the smallest diameter, ≈1.3 nm.

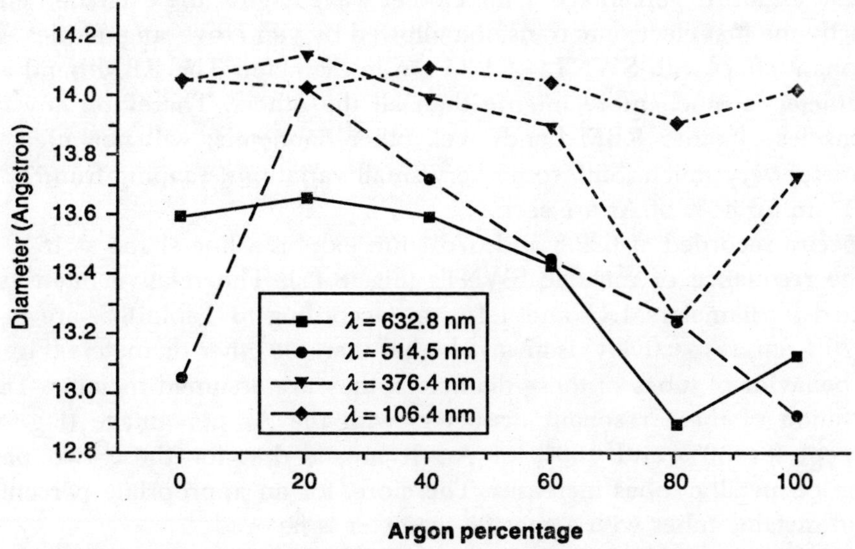

Fig. 8.18 *Variation of the average tube diameter with the argon percentage*

Thus, changing the gas mixture appears to induce a 0.1 nm variation in the SWNT diameter. This trend is confirmed by a shift in the diameter distribution shown in Fig. 8.19 for 514.5 and 676.4 nm. As the Ar percentage increases, the concentration of largest diameters (1.35–1.5 nm)decreases while the concentration of smallest ones (1.1–1.2 nm) increases. One can even notice that some very small nanotubes (0.8–1 nm) are produced at a high Ar percentage.

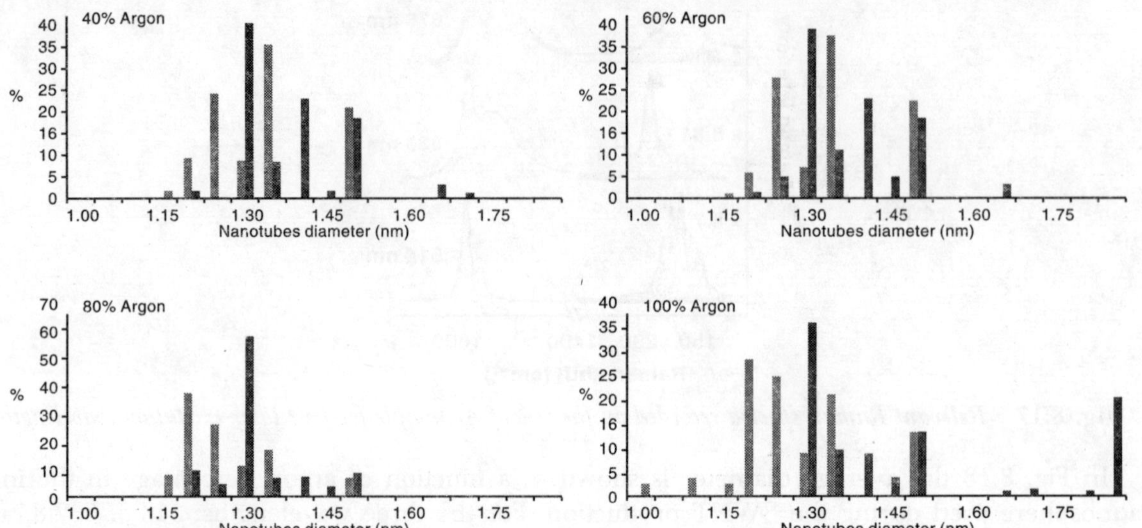

Fig. 8.19 *Nanotube diameter distribution from Raman spectra at 514.5 and 676.4 nm wavelength excitation*

For the 1064 nm excitation wavelength (i.e. 1.16 eV), the SWNT diameter appears to be nearly constant for all Ar percentage. Unlike other wavelengths, the excitation energy (1.16 eV) matches exactly the first electronic transition allowed by van Hove singularities. Hence, there is a strong resonant effect with SWNTs of 1.35 nm in diameter. The RBM band assigned to this resonant diameter is much more intense than all the others. Therefore, any changes in the relative intensities of other RBM bands (i.e., other diameters) will not affect the computed average diameter very much. Still, some very small variations, ranging from 1.39 nm for 80% of Ar to 1.41 nm for 40% of Ar are seen.

Raman spectra recorded at 632.8 and 676.4 nm exhibit a line shape at 1540 cm^{-1} which is related to the resonance of metallic SWNTs (Fig. 8.17). The relative intensity of the RBM bands assigned tc diameters 1.15 and 1.45 nm (according to Table 8.5, are in resonance for 632.8 and 676.4 nm, respectively) is always higher than for other diameters (Fig. 8.19). Because of resonant behavior of tubes of these diameters, they are assumed metallic. The evolution of the concentration of these resonant diameters with the Ar percentage (Fig. 8.20) is nearly constant except for 80% and 100% of Ar. It means that for these two percentages, the concentration of metallic tubes increases. Therefore, for an appropriate percentage of Ar, the production of metallic tubes with a specific diameter is favored.

Figure 8.19 also shows a specific behavior for 20% and 80% of Ar since these percentages, respectively, exhibit the biggest and the smallest diameters. Adding a small amount of another gas to a pure atmosphere seems to induce a step in the diameter distribution.

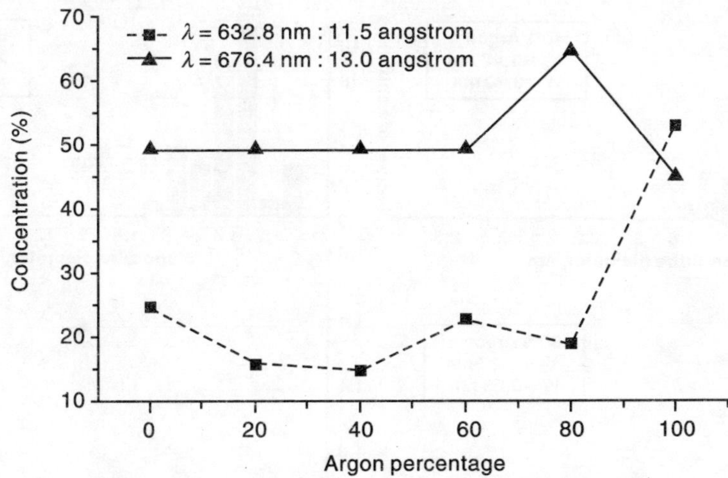

Fig. 8.20 *Variation of metallic tube concentration vs argon percentage obtained from Raman spectra excited at 632.8 and 676.4 nm wavelength*

B. TEM observations

Figure 8.21 shows HRTEM for nanotubes synthesized under different atmospheres. Nanotube yield is good for 0% Ar and 20% Ar, a little less in 40% Ar, and then decreases significantly in 60% Ar and 80% Ar. For Ar concentrations from 0% to 80%, the tube diameters are measured.

The diameter distributions are reported in Fig. 8.22, where they are fitted with Gaussian curves. For each sample, X_c is the center of the distribution (which is the most accurate measure of the mean diameter) and W is the distribution full width (not half width). This indicates a decrease of the tube diameter with argon percentage. Moreover, the diameter distribution measured is very close to the one observed by Raman spectroscopy at 514.5 nm Fig. 8.19).

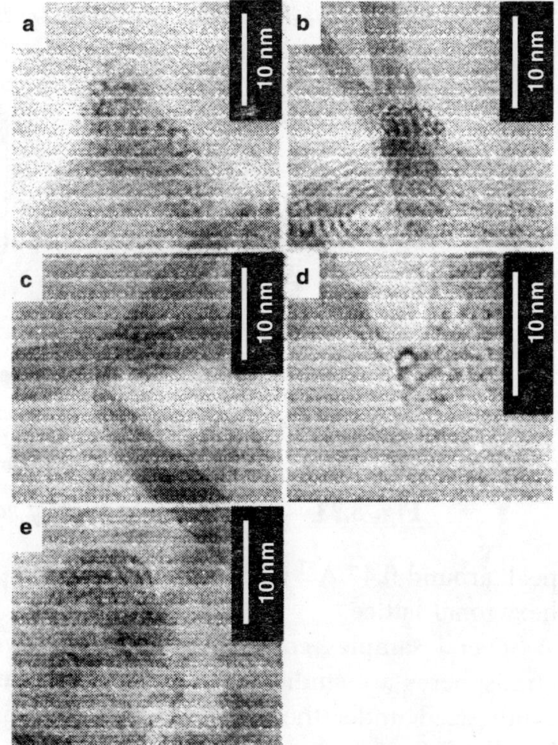

Fig. 8.21 *HRTEM images of the samples: (a) pure helium, (b) 20% argon, (c) 40% argon, (d) 60% argon, and (e) 80% argon*

C. X-ray scattering

To get further information on the tube diameter and bundle arrangement, some x-ray scattering experiments for all argon percentages are performed. Figure 8.23 shows an x-ray scattering profile obtained for nanotubes synthesized under a pure He atmosphere. The

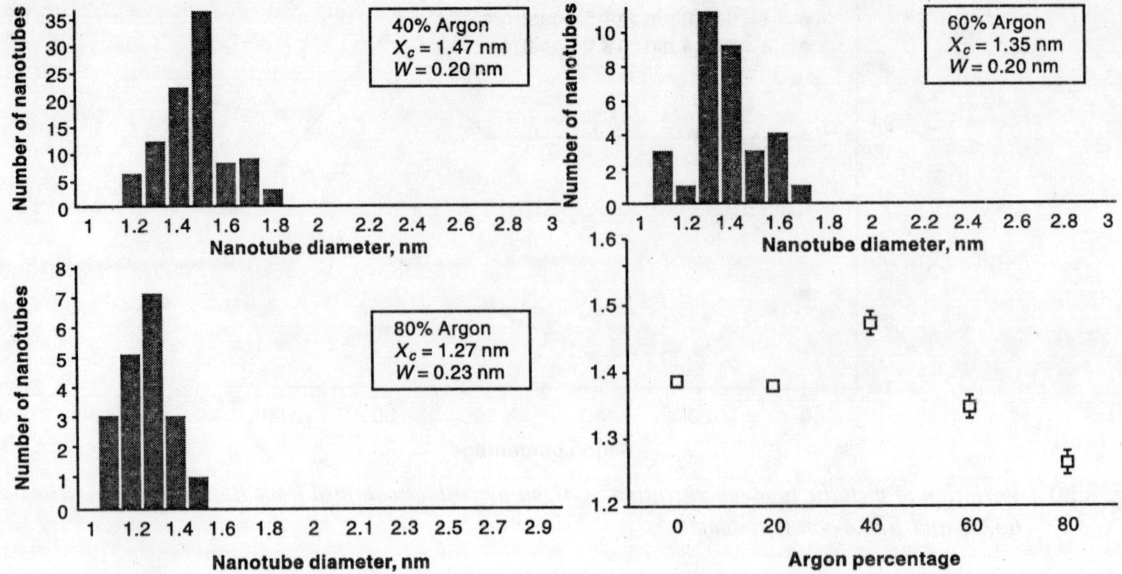

Fig. 8.22 *Diameter distribution of SWNT measured from TEM pictures. X_c is the center of the distribution and w is the distribution full width. On the botton roght, X_c is plotted vs the argon concentration and W is plotted as an error bar*

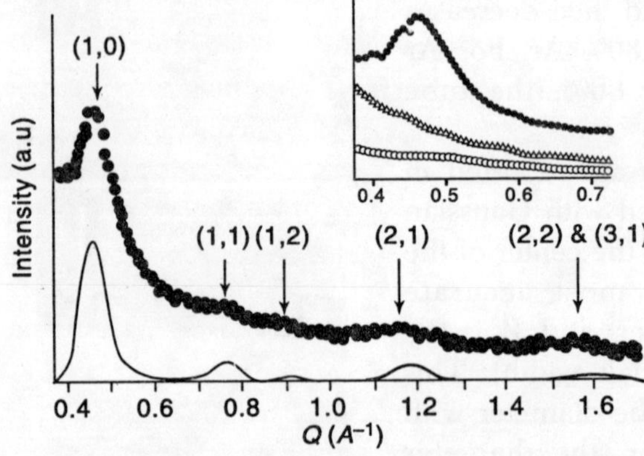

Fig. 8.23 *X-ray scattering study of nanotubes synthesized under 0% Ar atmosphere*

peak around 0.47 A^{-1} is the (1, 0) peak of nanotubes organized in bundles on a two-dimensional hexagonal lattice.

Several samples corresponding, respectively, to 0%, 20%, 40%, 60%, 80% and 100% Ar atmospheres are studied. Different x-ray scattering profiles are obtained for different samples synthesized under the same conditions. This is due to differences in nanotube diameters or bundle sizes.

It is deduced from the experiments, that when the amount of argon is increased, the (1, 0) peak becomes much weaker or disappears. This peak corresponds to relatively large bundles of about 10 nm in diameter for 0% Ar atmosphere with increasing Ar gas concentration, an increase of the (1, 0) peak position, it moves from 0.47 $Å^{-1}$ to 0.49 $Å^{-1}$ for 0% and 100% Ar atmosphere. These positions correspond to mean tube diameters of about 1.3 and 1.2 nm, respectively. The average tube diameter decreases by 0.1 nm as the inert gas changes from 0% Ar to 100% Ar, as also observed with spectroscopy.

II. On Growth Characteristics of Carbon Nanotubes (By SEM Images)

In order to study temperature effects on growth of carbon nanotubes on a Fe-deposited silicon oxidized film substrate, carbon nanotubes are synthesized 1023, 1123, and 1223 K temperature range. Before carbon nanotubes are synthesized, Fe is first deposited on a SiO_2 substrate at about 100 nm thickness. Then the substrate is dipped in distilled water diluted HF solution and nano-size fine catalytic particles are formed using NH_3. Fig. 8.24 shows SEM images of carbon nanotubes grown on SiO_2 substrate at 1223 K. It is found from Fig. 8.24(a) that when synthesis conditions are not optimized at the early stage growth density is poor, not vertically aligned, carbon nanofibers are threre along with carbon nanotubes, and a great amount of nanoparticles on the substrate surface. Fig. 8.24(b) shows carbon nanotubes are vertically aligned uniformly and carbon particles get thickly deposited on the surfaces of carbon nanotubes. Fig. 8.25 shows optimization growth of SEM images of carbon nanotubes on SiO_2 substrate at 1223 K. They are synthesized on nano-size fine catalytic fine particles with 100 sccm of NH_3 for 20 sccm of C_2H_2 for 10 minutes, Fig. 8.25(a) shows a 750× magnified SEM image taken at an angle of 45°. It is found that high quality carbon nanotubes are vertically aligned on SiO_2 substrate. It clearly shows that highly dense carbon nanotubes with about 55~60 μm in length were uniformly grown on the whole substrate, Fig. 8.25(b) shows a 5000× magnified image of carbon nanotubes. It is found from this image that there are no carbon particles on the surface at 1223 K and tips of carbon nanotubes are not vertically aligned, but grown upward and each tip is separated. As steric hindrances and Van der Waals forces are exerted between highly dense nanotubes per unit area, carbon nanotubes are vertically aligned. Fig. 8.26 shows a magnified SEM image of carbon nanotubes grown on SiO_2 substrate at 1223 K.

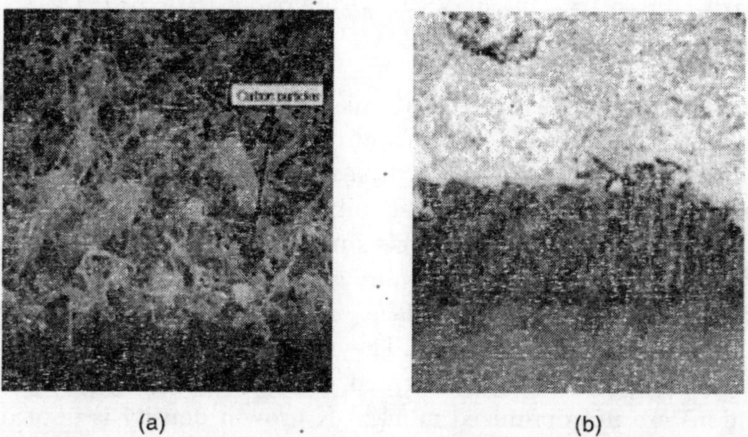

Fig. 8.24 *SEM images of carbon nanotubes grown on SiO_2 substrate at 1223 K*

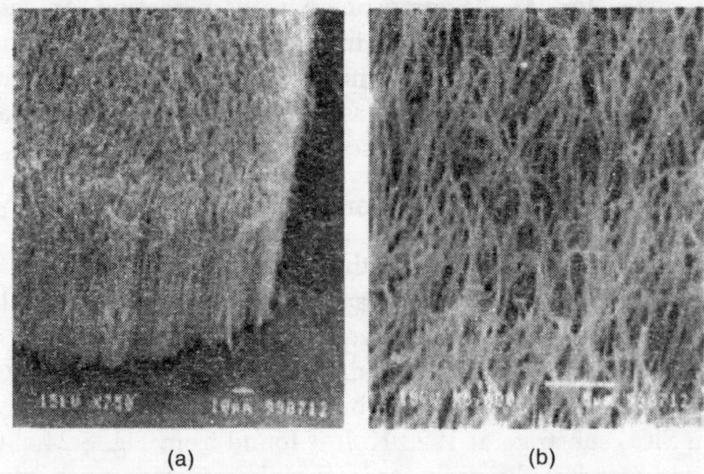

Fig. 8.25 *SEM images of carbon nanocubes grown on SiO_2* substrate at 1223 K. (a) Mag. 10,000, (b) Mag, 5,000

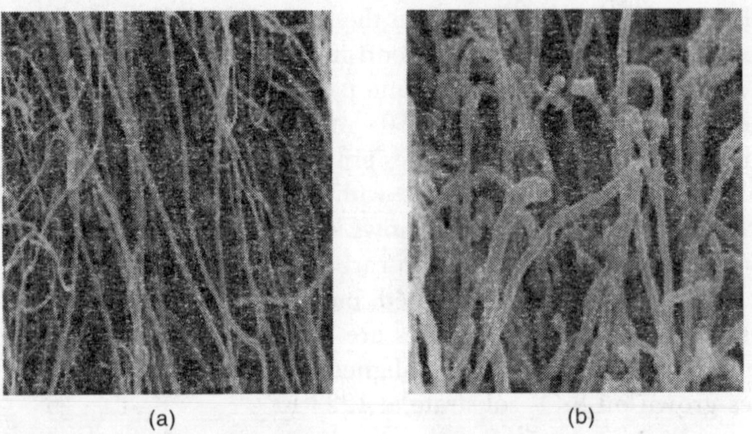

Fig. 8.26 *HR-SEM images of carbon nanotubes grown on SiO_2 substrate at 1123 K (a) Mag. 10,000, (b) Mag. 20,000*

Fig. 8.26(a) is a 10,000× magnified HR-SEM image and shows stem parts of carbon nanotubes. It is found that carbon nanotubes are grown with 150 nm in diameter and no carbon particles and deposits in external walls at all. Also, these images shows vertical alignment of carbon nanotubes by steric hindrance of carbon nanotubes. Fig. 8.26(b) shows a 20,000× magnified image of tips of carbon nanotubes. From this image tips are capped and are a little round. Although they are not randomly curved, they are good enough for electron emission. There are no carbon particles or deposits on the surface or tip at all. Fig. 8.27 show SEM images of carbon nanotubes grown on SiO_2 at 1123 K. They are synthesized on the above catalytic metal particles after catalytic metal particles are formed, using NH_3 From Fig. 8.27(a), it is found that if synthesis conditions are not optimized at 1123 K growth density is poor, carbon nanotubes

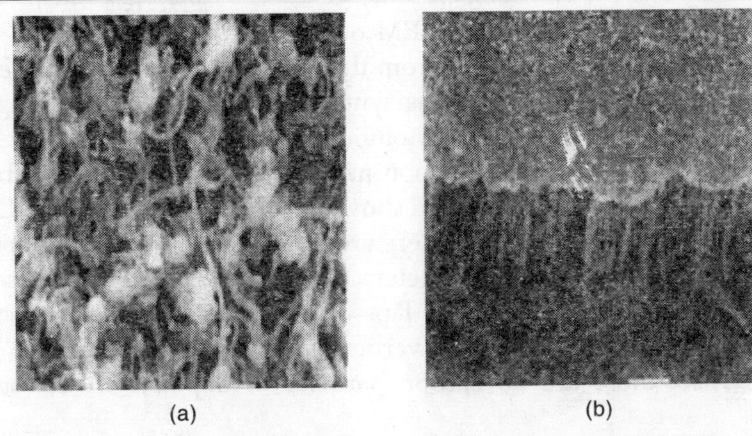

Fig. 8.27 *SEM images of carbon nanotubes grown on SiO_2 substrate at 1123 K. (a) Mag. 10,000, (b) Mag. 1,500*

are not vertically aligned, and a great amount of carbon particles exist. Although uniformly long carbon nanotubes are vertically aligned, thick carbon particles are covered on carbon nanotubes. However, in optimized synthesis conditions, high-quality carbon nanotubes are vertically aligned as in Fig. 8.28.

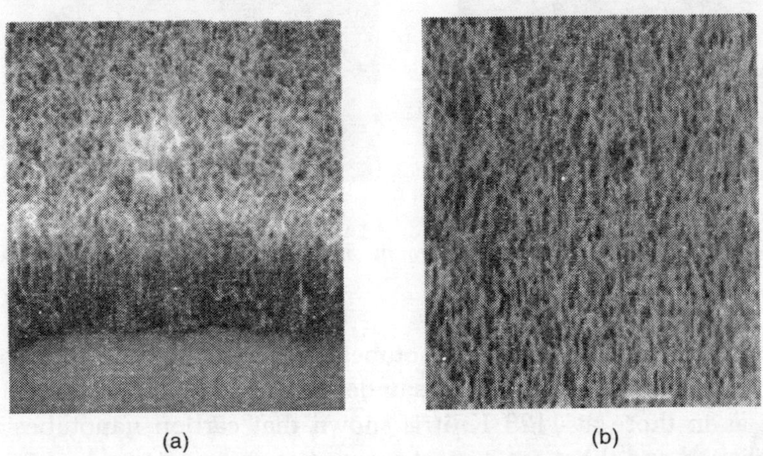

Fig. 8.28 *SEM images of carbon nanotubes grown on SiO_2 substrate at 1123 K (a) Mag. 1,000, (b) Mag. 3,500*

Fig. 8.28 shows SEM images of carbon nanotubes grown on SiO_2 substrate at 1123 K. The synthesis is carried out under 20 sccm C_2H_2 gas for 10 minutes after forming nano-size fine particles uncler 100 sccm NH_3 gas flow for 20 minutes. Fig. 8.28(a) shows a 1,000× magnified SEM image at 45°. High Quality carbon nanotubes are vertically aligned on SiO_2 substrate. The grown carbon nanotubes are highly dense, and about 20-25 μm in length. Carbon nanotubes grown at 1123K are shorter than those with 55-60 μm at 1223 K. The growth rate of carbon nanotubes shows a linear relationship with temperature. Carbon deposits on the tubes are fallen during scrubbing of carbon nanotubes for SEM observation. Fig. 8.27(b) is a 3500× magnified image of the surface of carbon nanotubes. From this image, it is found that there are no carbon particles on the surface of carbon nanotubes synthesized at 1123 K. Although tips of carbon nanotubes are not perfectly vertical: each nanotube is separated and vertically aligned. Fig. 8.29 shows highly magnified SEM images of carbon nanotubes grown on SiO_2 substrate at 1123 K. Fig. 8.29(a) and (b) are 30,000× magnified and show tips of carbon nanotubes. Carbon nanotubes are fairly uniform, 75 nm in diameter and there are no carbon particles or deposits on the walls or surfaces of carbon nanotubes. The diameters of carbon nanotubes are much shorter than those at 1223 K with 150 nm in diameter. Tips of carbon nanotubes are capped and a little round. Although they are curved, they are vertically aligned in general and good enough for electron emission. Also, there are no carbon particles or deposits on the surfaces or tips of carbon nanotubes.

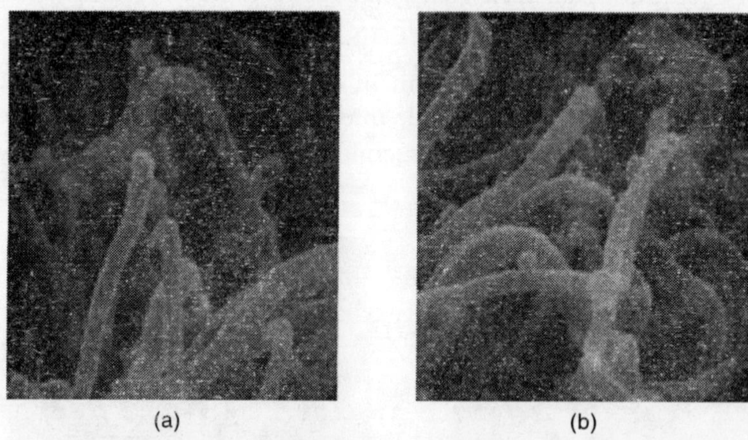

Fig. 8.29 *HR-SEM images of carbon nanotubes grown on SiO_2 substrate at 1123 K (a) Mag. 30,000, (b) Mag. 30,000*

Fig. 8.30 shows SEM images of carbon nanotubes grown on catalytic metal particles on SiO_2 substrate after forming catalytic metal particles under NH_3 gas at 1023 K. If synthesis conditions are not optimized as in those at 1123 K, it is shown that carbon nanotubes are not densely grown, vertically aligned and there are a great amount of carbon nanofibers and nanoparticles. In many cases. carbon nanoparticles were thickly covered on the surface of carbon nanotubes even though growth conditions were more or less optimized and vertical growth of highly dense carbon nanotubes possible. However. in optimum conditions. uniform high quality carbon nanotubes can be vertically aligned as in Fig. 8.31.

Fig. 8.31 shows SEM images of carbon nanotubes grown on catalytic metal particles on SiO_2 substrate after forming catalytic metal particles under NH_3 gas at 1023 K. The synthesis was carried out under 20 sccm C_2H_2 gas for 10 minutes after forming nano-size fine particles under 100 secm NH_3 gas flow for 20 minutes. Fig. 8.31(a) shows a 1,000 magnified SEM image at 15°. High quality carbon nanotubes were vertically aligned on SiO_2 substrate as those at 1123 K. The grown carbon nanotubes are highly dense. and about 15 μm in length. Carbon nanotubes

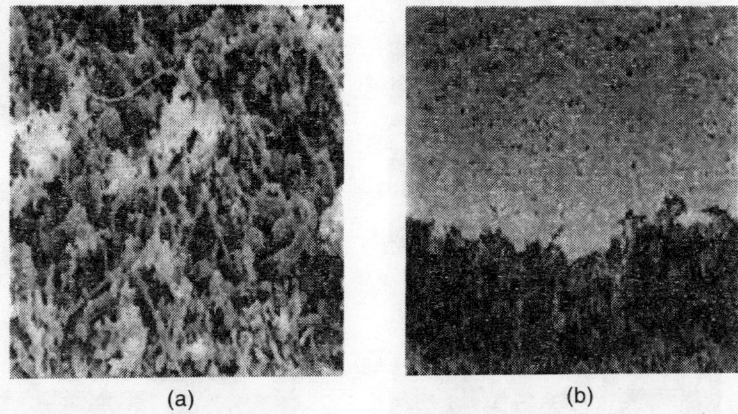

Fig. 8.30 *SEM images of carbon nanotubes grown on SiO_2 substrate at 1023 K. (a) Mag. 2,000, (b) Mag. 5,000*

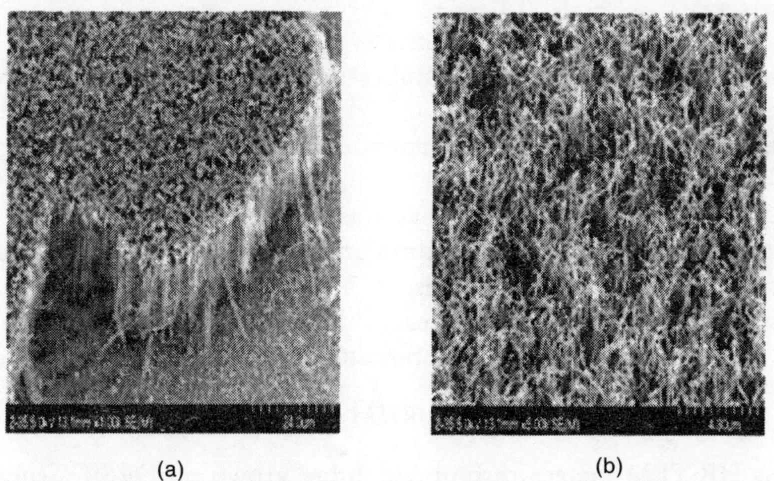

Fig. 8.31 *SEM images of carbon nanotubes grown on SiO_2 substrate at 1023 K (a) Mag. 1,000, (b) Mag. 5,000*

grown at 1023 K are much shorter than those at 1123 K or 1223 K. This means that the growth rate of carbon nanotuoes is highly affected by growth temperature of tubes. Fig. 8.31(b) shows a 5,000× magnified image of the surfaces of carbon nanotubes.

This image showstat, carbon nanotubes synthesized at 1023 K have no carbon particles on the surface. Although tips of carbon nanotubes are not perfectly vertical, each nanotube is separated and vertically aligned. The density of carbon nanotubes per unit area is lower than those synthesized at 1123 K or 1223 K.

Fig. 8.31. shows magnified SEM images of carbon nanotubes grown on SiO_2 at 1023 K. Fig. 8.32(a) and (b) are 40,000× and 60,000× magnified, showing the surfaces and tips of carbon nanotubes. It is found that there are no carbon particles or deposits on the external walls. Tips of carbon nanotubes are capped and curved. They are rather randomly and horizontally

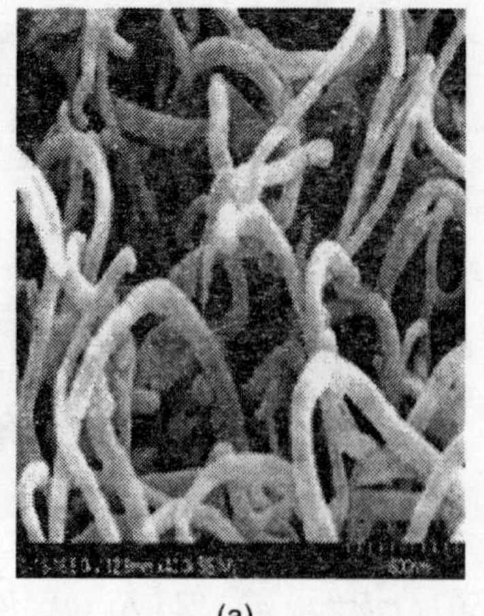

(a)

(b)

Fig. 8.32 *HR-SEM images of carbon nanotubes grown on SiO_2 substrate at 1023 K (a) Mag. 40,000, (b) Mag. 60,000*

placed, not vertically. Even in carbon nanotubes synthesized at 1023 K there are no carbon particles or deposits on the surfaces or tips.

Based on these results, temperature dependence of carbon nanotube growth can be summarized as follows.

(1) The density of carbon nanotubes increases as temperature increases. (2) The growth rate of carbon nanotubes increases as temperature increases. (3) Vertical alignment of carbon nanntubes enhances as temperature increases.

However, tips of carbon nanotubes synthesized by thermal CVD are capped and randomly curved. Vertical alignment of carbon nanotubes enhances as temperature increases.

III On Structural Characteristics of CNTs (By HRTEM Images)

Fig. 8.33 shows HR-TEM images carbon nanotubes grown on Fe/SiO_2 substrate at 1223 K. Fig. 8.33(a) is 70,000× magnified, and (b) 200,000× magnified. Fig. 8.33(a) shows that tips of carbon nanotubes are capped and curved. All carbon nanotubes are hollow in the middle with a jointed bamboo structures. There are various forms of bamboo structures. From Fig. 8.33(b), the spacings of joints in bamboo structure have a certain regularity and the internal diameter can be varied as the bamboo structure forms. but the external diameter is never changed.

Fig. 8.34 shows HR-TEM images of stems of carbon nanotubes grown on Fe/SiO_2 substrate at 1223 K. From these images, carbon nanotubes have a hollowed multi-walled structure and consist of about graphene sheets. In MWNT, the spacings between graphene sheets are 0.34 nm. same as those of typical graphene sheets. The crystal structure is very stable. However, graphene sheet between internal and external walls has a little defective crystalline structure. Fig. 8.35 shows HR-TEM images of a tip of carbon nanotubes grown on Fe/SiO_2 substrate at 1223 K. The tip is round. Each graphene sheet has a stable crystalline structure with multi-walls around the tip and the spacing between graphene sheets is also 0.34 nm.

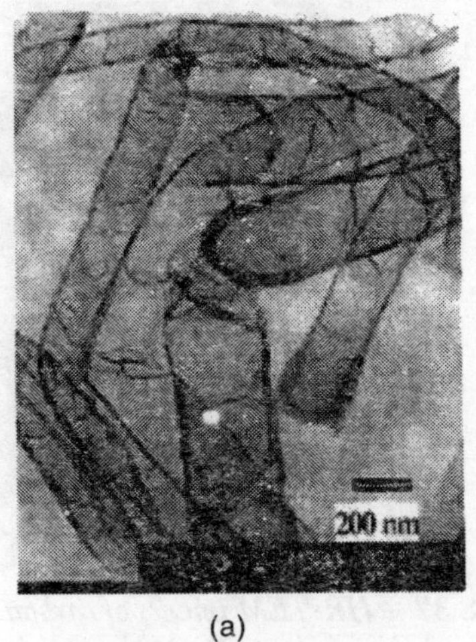

(a) (b)

Fig. 8.33 *HR-TEM images of carbon nanotubes grown on Fe/SiO$_2$ substrate at 1223 K (a) Mag. 70,000, (b) Mag. 200,000*

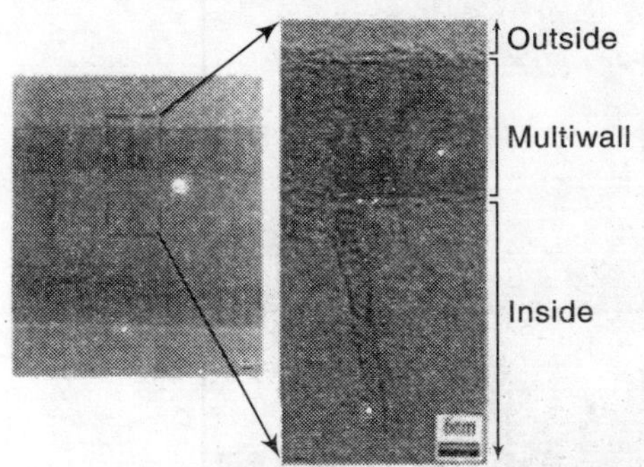

Fig. 8.34 *HR-TEM images of carbon nanotube stems grown on Fe/SiO$_2$ substrate at 1223 K*

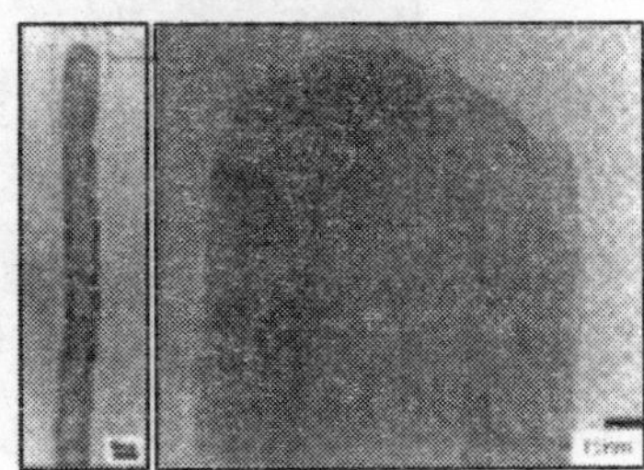

Fig. 8.35 *HR-TEM images of carbon nanotube tip grown on Fe/SiO$_2$ substrate at 1223 K*

Fig. 8.36 shows HR-TEM images of the root of carbon nanotube grown on Fe/SiO$_2$ substrate at 1223 K. It is found that the root is the place to which catalytic metal particles stuck and similar to catalytic metal particle in shape. In MWNT, the root of carbon nanotubes also has a stable crystalline structure in each graphene sheet and its spacing is 0.34 nm, too.

Fig. 8.37 shows HR-TEM images of the shunt of carbon nanotubes grown on Fe/SiO$_2$ substrate at 1223 K. Vertically grown graphene sheets and jointed bamboo-type graphene sheets were put together. In MWNT, each graphene sheet also has a stable crystalline structure and its spacing is also 0.34 nm. The inside graphene sheet shows a stable structure, but the one between internal and external walls is little defective crystalline structure.

Fig. 8.38 shows TEM images of carbon nanotubes grown on Fe/SiO$_2$ substrate at 1123 K. Fig. (a) is 40,000× magnified and (b) 50,000× magnified. Carbon nanotubes at 1123 K are a hollow MWNT structure. same as those at 1223 K. They have jointed bamboo structures and tips are capped and curved. There are various forms of bamboo structures. The spacings of

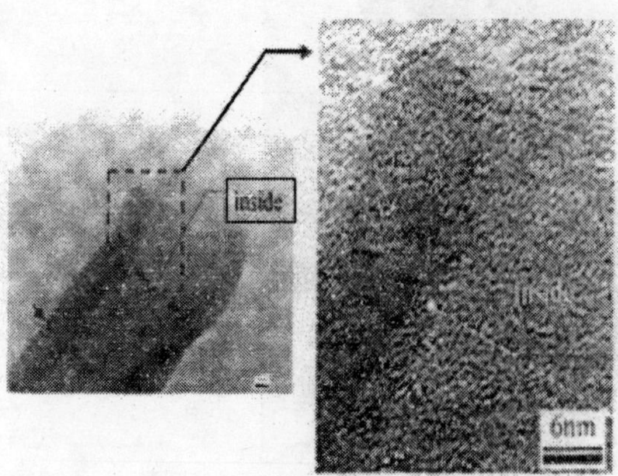

Fig. 8.36 *HR-TEM images of carbon nanotube roots grown on Fe/SiO_2 substrate at 1223 K*

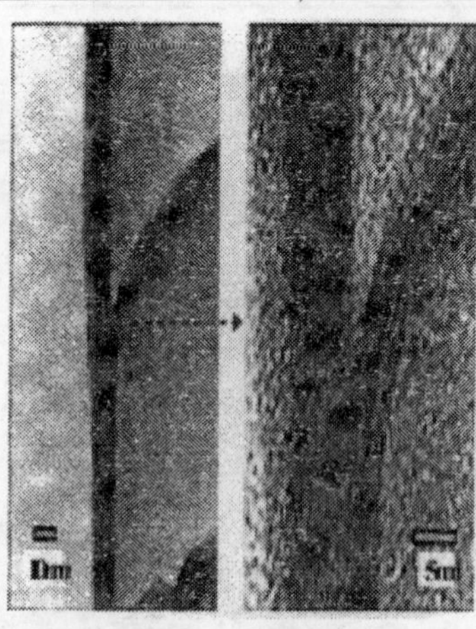

Fig. 8.37 *HR-TEM images of carbon nanotube shunt grown on Fe/SiO_2 substrate at 1123 K*

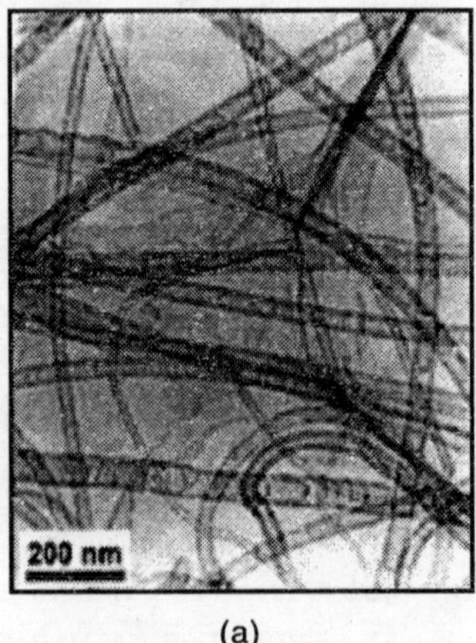

(a)

(b)

Fig. 8.38 *TEM images of carbon nanotubes grown on Fe/SiO_2 substrate at 1123 K (a) Mag. 40,000, (b) Mag. 50,000*

joints in bamboo structure have a certain regularity and the internal diameter can be varied as the bamboo structure forms, but the external diameter is hardly changed. From Fig. 8.38(b), catalytic metal chunk is got molten into the root of carbon nanotube. This happens because surface tension and capillary phenomena lead to catalytic metal chunks getting stuck to the hollow space of carbon nanotube with a small diameter.

Fig. 8.39 shows HR-TEM images of the tip of carbon nanotube grown on Fe/SiO_2 substrate at 1123 K. Carbon nanotubes at 1223 K have about 40 graphene sheets. but those at 1123 K about 20–25. The tip of carbon nanotube is a little round. each graphene sheet around the tip is stable crystalline. and the spacing is also 0.34 nm. However, graphene sheets between internal and externql sheets show a little defective crystalline structure.

Fig. 8.40 shows a 50.000 magnified TEM image of carbon nanotubes grown on Fe/SiO_2 substrate at 1023 K. As carbon nanotubes at 1123 K or 1223 K, those at 1023 K are a hollow

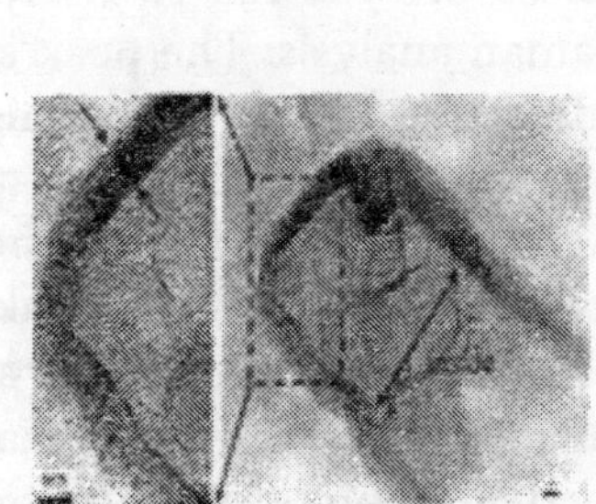

Fig. 8.39 *HR-TEM images of carbon nanotube grown on Fe/SiO_2 substrate at 1123 K*

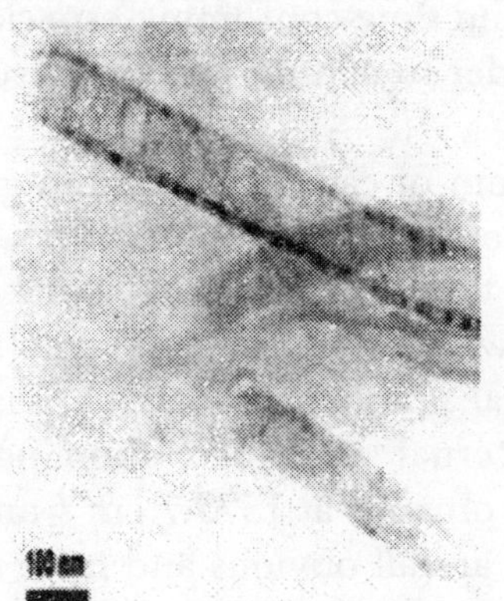

Fig. 8.40 *TEM image of carbon nano:ubes grown on Fe/SiO_2 substrate at 1023 K*

Fig. 8.41 *TEM image of carbon nanotube grown on Fe/ SiO_2 substrate at 1023 K*

multi-walled, jointed bamboo, structure, and tips are capped. There are various forms of bamboo structures. The spacings of joints in bamboo structure have a certain regularity and the internal diameter can be varied as the bamboo structure forms, but the external diameter is hardly changed.

Fig. 8.41 shows a TEM image of the tip of carbon nanotube grown on Fe/SiO_2 substrate at 1023 K. From this image, carbon nanotubes are a multi-walled hollow bamboo structure and the spacing between graphene sheets is 0.34 nm, same as that of a typical graphene sheet. The crystalline structure of the graphene sheet is very stable.

Characteristics of carbon nanotubes at different temperatures are summarized as follows based on the above results.

(1) Carbon nanotubes by thermal CVD has a hollow bamboo structure. regardless of synthesis temperature.
(2) As synthesis temperature increases, the number of graphene sheets increases and the structure is stable.
(3) The spacing between multi-walled graphene sheets is 0.34 nm, each graphene sheet has a stable crystalline structure, but that between internal and external walls is a little defective.
(4) Tips of carbon nanotubes by thermal CVD are capped and round-shaped.
(5) Fifth, in carbon nanotubes by thermal CVD, there are different forms of bamboo structures, but an individual nanotube has a certain regularity in form and joint spacing. The internal diameter can be changed for the formation of bamboo structure, but external diameter is almost invariant.

(C) Analysis by Raman

TEM and Raman analyses are typical among structural analyses of carbon nanotubes. A TEM analysis is very useful in crystalline structural analysis of bulk tubes. However, it has shortcomings–difficult in analysis and time-consuming-especially limitation in structural analysis of surface morphology. Thus in order to control and study surface reactions, Raman analysis is an important method.

Fig. 8.42 a Micro-Raman spectrum of carbon nanotubes grown on Fe/SiO_2 at 123 K. A He-Ne laser at 632.8 nm was used. There are two important peaks in Raman analysis. The peak at 1572.7 cm^{-1} is a typical G-line and that at around 1320 cm^{-1} a little defective D-line. According to Fig. 8.42, no typical peak is shown around 400-1100 cm^{-1}. This happens because an overlap effect is obvious, caused by several graphene sheets. A peak at 1324.8 cm^{-1} by defects and carbonaceous particles on the external walls of carbon nanotubes appears. A G-line peak, showing a multi-walled structure is obvious at 1572.7 cm^{-1} and a side peak around the G-line at 1598 cm^{-1}. Both D-line and G-line are all obvious and have a similar intensity. This shows that carbon nanotubes at 1223 K are structurally stable and have a defective graphite structure on surfaces. This result agrees well with the TEM analysis.

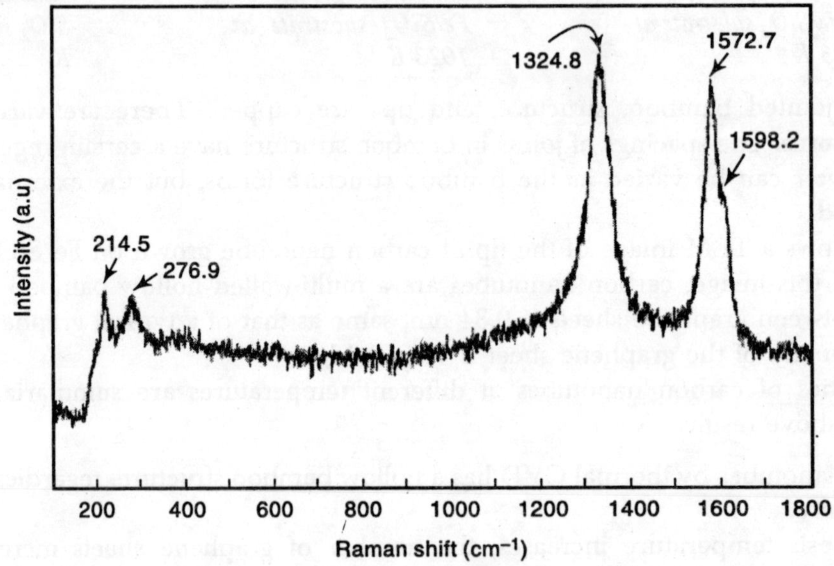

Fig. 8.42 *Raman spectrum of carbon nanotubes grown on* Fe/SiO_2 *substrate at 1223 K*

Fig. 8.43 shows a Micro-Raman spectrum of carbon nanotubes grown on Fe/SiO_2 substrate at 1123 K. On the whole, the aspect of peaks is similar to that of carbon nanotubes at 1223 K. However, the difference is that at 1123 K. D-line is relatively sharper and broader than G-line. This means that the graphene structure of carbon nanotubes are less stable and the number of structurally defective graphene sheets on the external walls increases as temperature decreases.

Fig. 8.44 shows a Micro-Raman spectrum of carbon nanotubes grown on Fe/SiO_2 substrate at 1023 K. The typical aspect of peaks is similar to that at 1123 K. However, it shows unlike at 1123 K or 1123 K. D-line is much sharper and broader than G-line. As mentioned before, this means that the graphite structure of synthesized carbon nanotubes gets more defects and the number of structurally defective graphene sheets on the external walls increases as synthesis temperature decreases. *Hence, as synthesis temperature increases. the crystallinity of graphene sheets of carbon nanotubes enhances and the number of defective grapheme sheets on MWNTs decreases.*

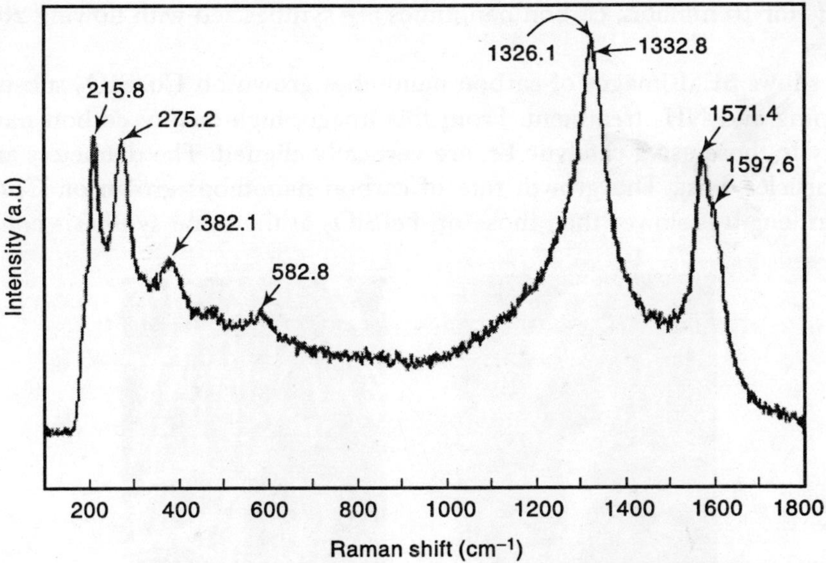

Fig. 8.43 *Raman spectrum of carbon nanotubes grown on Fe/SiO$_2$ substrate at 1123 K*

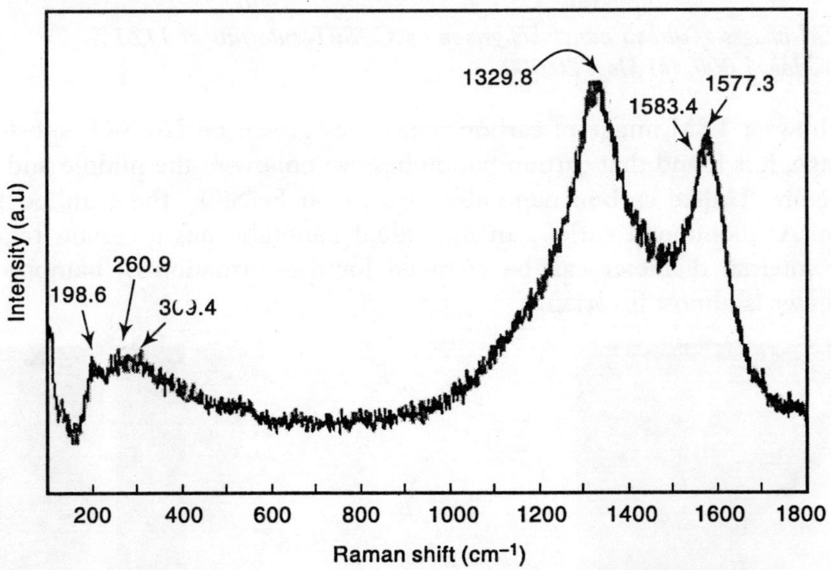

Fig. 8.44 *Raman spectrum of carbon nanotubes grown on Fe/SiO$_2$ substrate at 750°C*

On Different Catalytic Metals by Thermal CVD

Catalytic metals are needed to synthesize carbon nanotubes by thermal CVD. Prior to synthesis of carbon nanotubes, approximately 100 nm thick catalytic Co and Ni respectively were deposited on SiO_2 substrate. After dipping the catalytic metal deposited substrate into diluted HF solution for 140 seconds and forming catalytic metal particles in the atmosphere of

100 scem NH_3 for 10 minutes, carbon nanotubes are synthesized with flowing 20 scem of C_2H_2 for 10 minutes.

(A) Fig. 8.45 shows SEM images of carbon nanotubes grown on **Co/SiO_2** substrate at 1123 K, after HF dipping and NH_3 treatment. From this image, high-quality carbon nanotubes with a similar density to those using catalytic Fe, are vertically aligned. The diameters are uniform and no carbon particles exist. The growth rate of carbon nanotubes grown on Co/SiO_2 substrate with 20 μm in length is slower than those on Fe/SiO_2 at the same synthesis conditions.

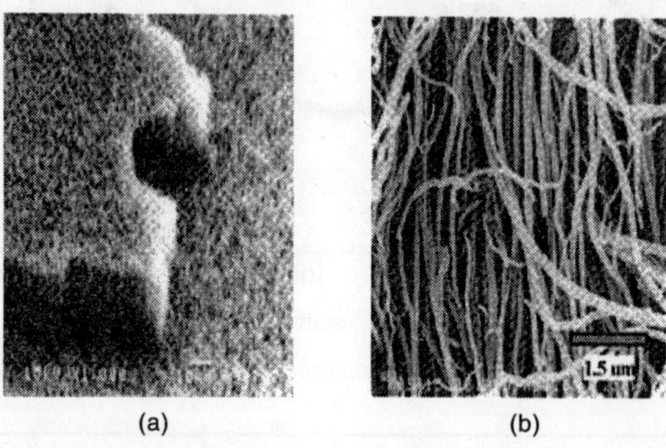

Fig. 8.45 *SEM images of carbon nanotubes grown on Co/SiO_2 substrate at 1123 K (a) Mag. 1,000, (b) Mag. 20,000*

Fig. 8.46 shows a TEM image of carbon nanotubes grown on Co/SiO_2 substrate at 1123 K. From the image, it is found that carbon nanotubes are hollow in the middle and have a jointed bamboo structure. Unlike carbon nanotubes grown on Fe/SiO_2, the bamboo form is a little conical shape. As mentioned earlier, an individual nanotube has a certain regularity in joint spacing. The internal diameter can be changed for the formation of bamboo structure, but external diameter is almost invariant.

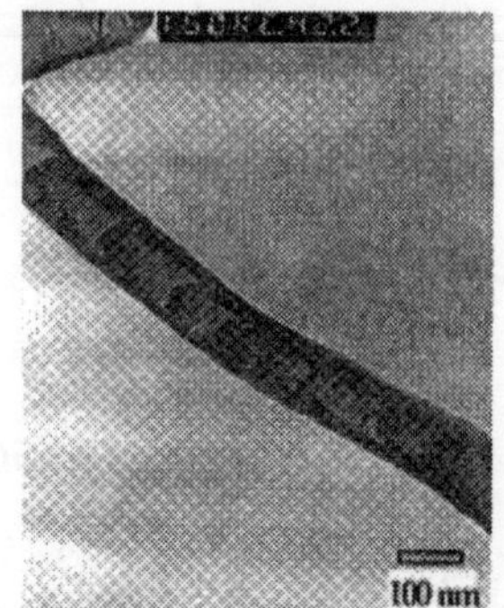

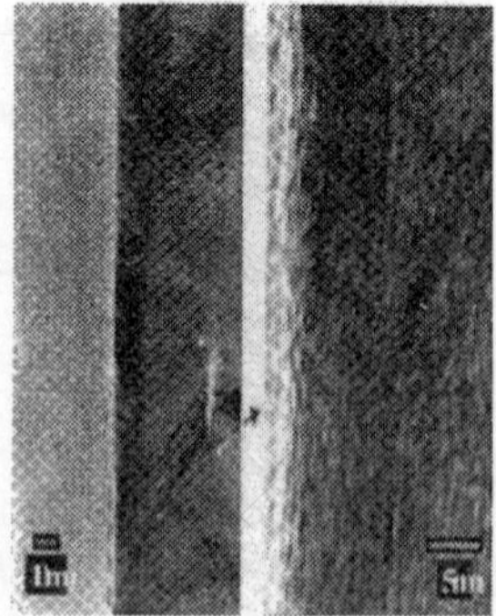

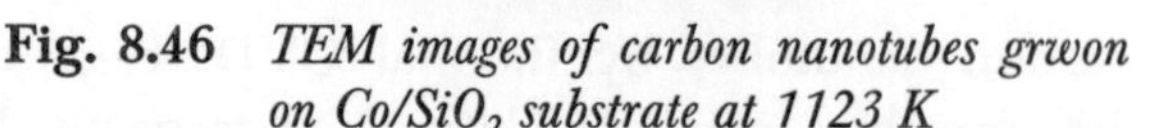

Fig. 8.46 *TEM images of carbon nanotubes grwon on Co/SiO_2 substrate at 1123 K*

Fig. 8.47 *HR-TEM images of carbon nanotubes grown on Co/SiO_2 substrate at 1123 K*

Fig. 8.47 shows a 4.000,000× magnified HR-TEM image of the bamboo structural shunts of carbon nanotubes grown on Co/SiO_2 substrate at 1123 K. Carbon nanotubes are a multi-walled hollow bamboo structure and the spacing between graphene sheets is 0.34 nm, same as that of a typical graphene sheet. The crystalline structure of graphene sheets is very stable, but a little defective in graphene sheets of internal and external walls of carbon nanotubes.

Fig. 8.48 shows a HR-TEM image of the tips of carbon nanotubes grown on Co/SiO_2 substrate at 1123 K. The tips have a hemi-spherical shape with a bigger curvature than those grown on Fe/SiO_2. Each graphene sheet in the multi-wall around the tip is stable and the spacing between sheets 0.34 nm.

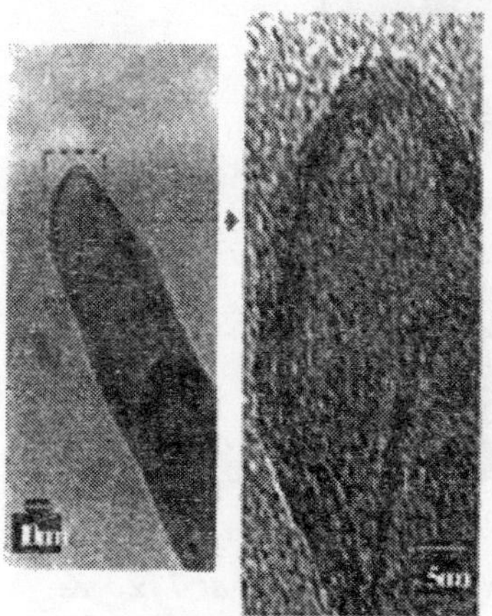

Fig. 8.48 *HR-TEM images of carbon nanotubes grown on Co/SiO_2 substrate at 1123 K*

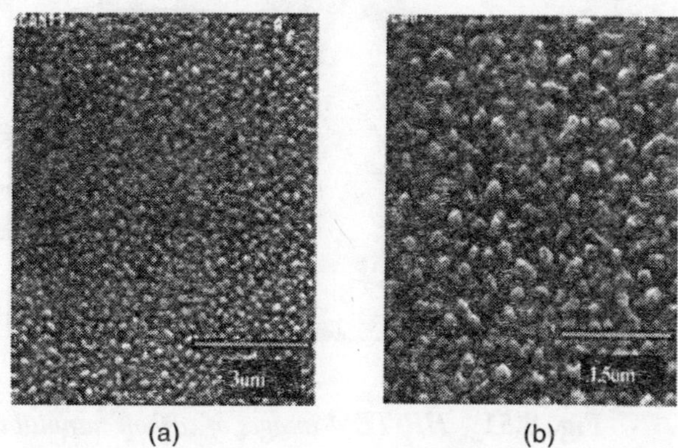

Fig. 8.49 *Surface morphology of Co/SiO_2 substrate. (a) Mag. 10,000, (b) Mag. 20,000*

Fig. 8.49 shows surface morphology of Co/SiO_2 substrate. Fig. 8.49(a) is 10,000× magnified and (b) 20,000× magnified. From the images, it is shown that uniform shape and size catalytic metal particles are densely formed on SiO_2 substrate. Surface morphology got rougher and no catalytic metal particles were formed when HF dipping without NH_3 treatment was carried out. Catalytic metals show more evident in curvature than those on Fe/SiO_2. That also affects the shapes of tips of carbon nanotubes.

(B) Fig. 8.50 SEM images of carbon nanotubes grown on **Ni/SiO_2** substrate at 1223 K after HF dipping and NH_3 treatment. From the images, it is found that carbon nanotubes on Ni have similar density. high-quality, and vertical alignment to those on Fe or Co. In this case, the growth rate with about 30 μm in length is slower than that on Fe, but a little faster than that on Co.

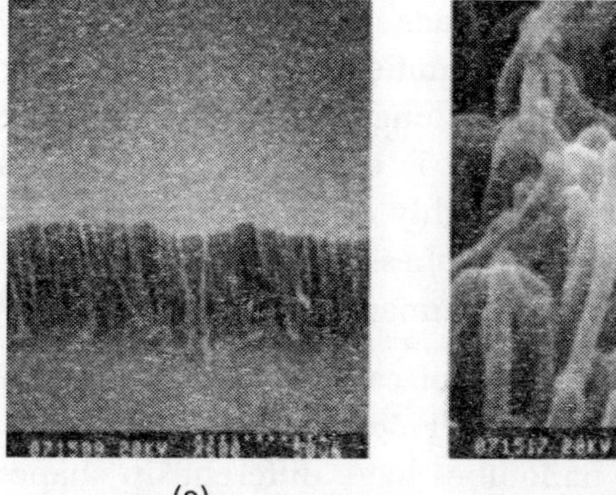

Fig. 8.50 *SEM images of carbon nanotubes grown on Ni/SiO_2 substrate at 1223 K (a) Mag. 600, (b) Mag. 80,000*

Fig. 8.51 shows HR-TEM images of carbon nanotubes grown on Ni/SiO_2 substrate at 1123 K. From this image, carbon nanotubes are hollow in the middle with a jointed bamboo structure inside. Unlike those on Fe or Co, the tips are a duck's beak shape. As mentioned before in growth on Fe or Co. an individual nanotube has some regularity in joint spacing. The number of graphene sheets. forming the bamboo structure is lower than that on Fe or Co. The internal diameter can be changed for the formation of bamboo structure, but external diameter is almost invariant.

(a)

(b)

Fig. 8.51 *HR-TEM images of carbon nanotubes grown on Ni/SiO_2 substrate at 1123 K*

Fig. 8.52 shows 4,000,000× magnified HR-TEM images of the tips of carbon nanotubes grown on Ni/SiO_2 substrate at 1223 K. The tips of carbon nanotubes on Ni/SiO_2 are much sharper than those on Fe or Co/SiO_2 and a duck's beak shape. The spacing of the carbon nanotube bamboo structure has some regularity as that on Fe or Co. Each graphene sheet in the multi-wall around the tip is stable and the spacing between sheets 0.34 nm.

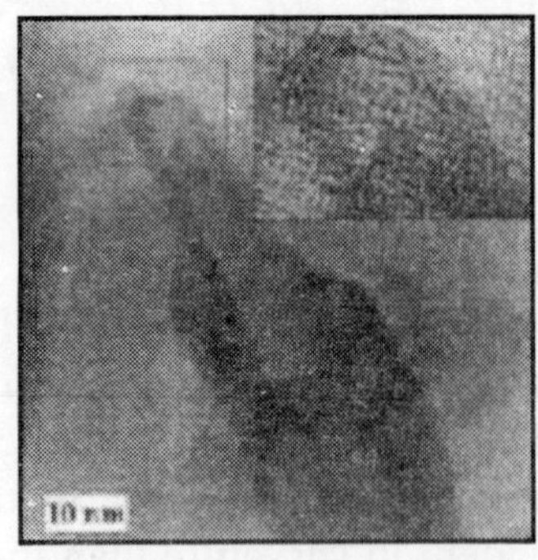

Fig. 8.52 *HR-TEM image of carbon nanotubes grown on Ni/SiO_2 substrate at 1223 K*

Fig. 8.53. shows surface morphology of Ni/SiO_2 substrate. Fig. 8.53(a) is 10,000× magnified and (b) 20,000× magnified. From the images. sharp and long oval-shaped catalytic metal particles are densely formed. It is explained that these shaped catalytic metal particles can highly affects the tip and bamboo structure forms of carbon nanotubes.

Followings are a summary of **(A)** and **(B)** given above.

(1) The growth rate of carbon nanotubes changes on different catalytic metals. The growth rate is as follows: Fe> Ni> Co.

(2) Carbon nanotubes have different tip shapes for catalytic metals. On Fe, smoothly round tips were formed. On Co, hemispherical tips with a big curvature were formed, and on Ni, duck's beak shaped sharp tips.

(3) The bamboo structural forms change for different catalytic metals. The forms tend to have a similar shape to the tip. However, the number of graphene sheets on Ni is much fewer than that on other catalytic metals.

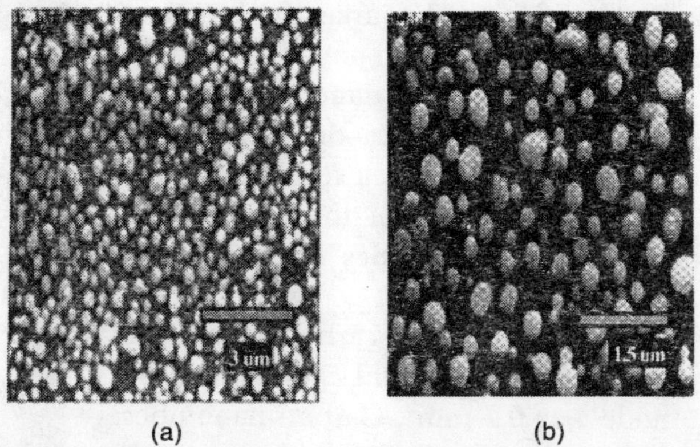

Fig. 8.53 *Surface morphology of Ni/SiO_2 substrate. (a) Mag. 10,000, (b) Mag. 20,000*

(4) If synthesis conditions of carbon nanotubes are optimized, it is possible that high-quality carbon nanotubes are vertically aligned, regardless of catalytic metals. However, no carbon nanotubes are grown on catalytic metals, but Fe, Ni, and Co.

ELECTRIC FIELD EMISSION CHARACTERISTICS OF CARBON NANOTUBES

Fig. 8.54 shows a standardized apparatus for measuring electron emission characteristics. High vacuum up to 10-9 mm Hg and authorized voltage up to 500 V are possible. The measurement of several specimens at a time can be carried out with position manipulation of

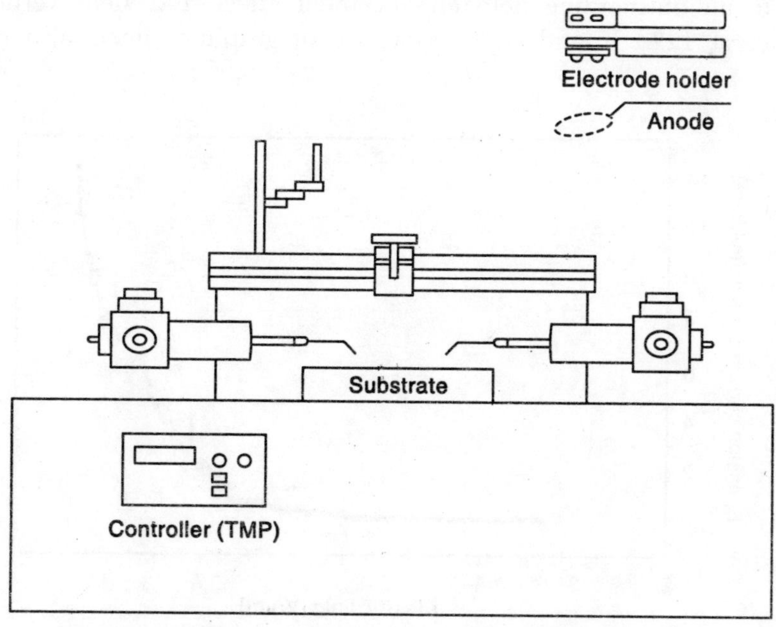

Fig. 8.54 *Schematic diagram of emission measurement system*

both electrodes from an external power source. In practice, electron emission characteristics were measured under 10^6 mmHg or less. A deposited Ti film below carbon nanotubes was as cathode and a round ring type electrode as anode. In addition, after a proper size anode plate covered with a fluorescent was fixed on carbon nanotubes on a receptor in the chamber to measure substance fluorescence characteristics from electron emission.

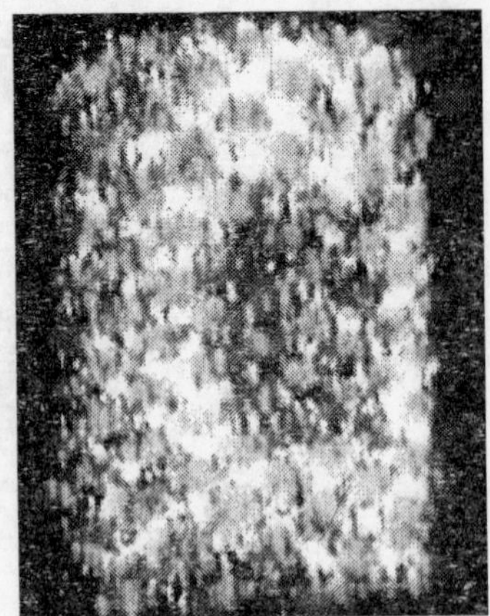

Fig. 8.55 *Emission test image of carbon nanotubes growth at 1223 K*

Fig. 8.55 shows a emission test image of carbon nanotubes grown at 1223 K. The gap between anode and cathode was 60 μm and the area of the anode was 0.2 mm². Carbon nanotubes for emission test were synthesized on Fe/Ti/SiO_2 substrate with 200 nm thick Ti film. Ti film was used as cathode and ring type metal electrode as anode. It is shown from the image that electron emission was fairly good on 20 mm × 30 mm.

Fig. 8.56 shows an emission current density of carbon nanotubes grown on Fe/Ti/SiO_2 at 1223 K. The threshold turn-on voltage for electron emission was 279 V. When electrons emitted from vertically aligned carbon nanotubes. the maximum current density was approximately 14.5 mA/cm², where the electric field was about 8.2 V/μm. It shows an excellent current density and good electric field emission characteristics. Electron emission using vertically aligned carbon nanotubes shows approximately 10-100 times better than conventional silicon tips, DLC thin films, or carbon thin films. These characteristics are attributed to the fact that carbon nanotubes are a good conductor and several nm size in diameter leads to an outstanding field enhancement effect. Excellent vertical alignment of carbon nanotubes at 1223 K and stable structure of graphite sheets also contribute to the characteristics.

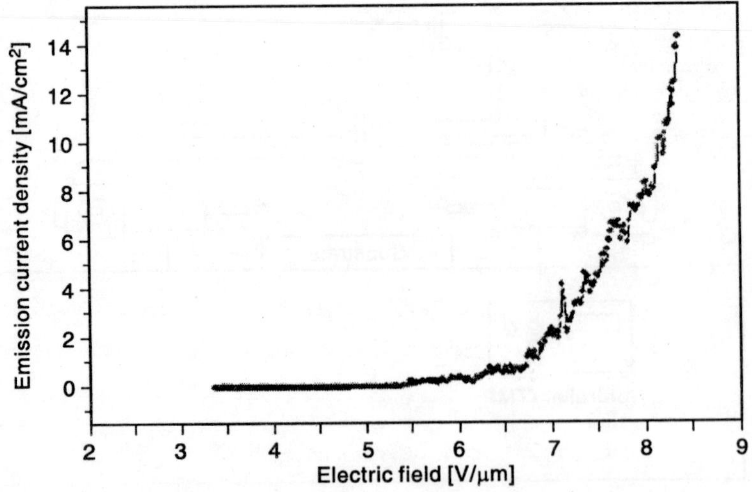

Fig. 8.56 *Emission current density of carbon nanotubes grown on Fe/SiO_2 substrate at 1223 K*

Fig. 8.57 shows an emission current density curve of carbon nanotubes grown on Fe/SiO_2 substrate at 1223 K. The turn-on voltage was 308 V, a little higher than that of carbon nanotubes at 1223 K. The maximum current density was approximately 3 mA/cm^2 where the applied electric field was 8.25 V/μm. Although the value was lower than that of carbon nanotubes at 1223 K, it is quite better than the values of conventional electron emission sources. Although a little higher turn-on voltage and much sparser current density at 850°C are observed than at 1223 K for electron emission. the values are good enough for application to various display and electron emission systems.

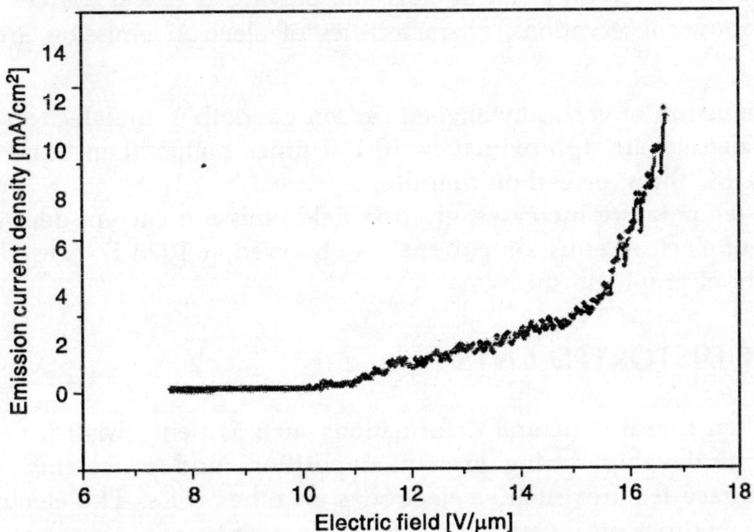

Fig. 8.57 *Emission current density of carbon nanotubes grown on Fe/SiO_2 substrate at 1223 K*

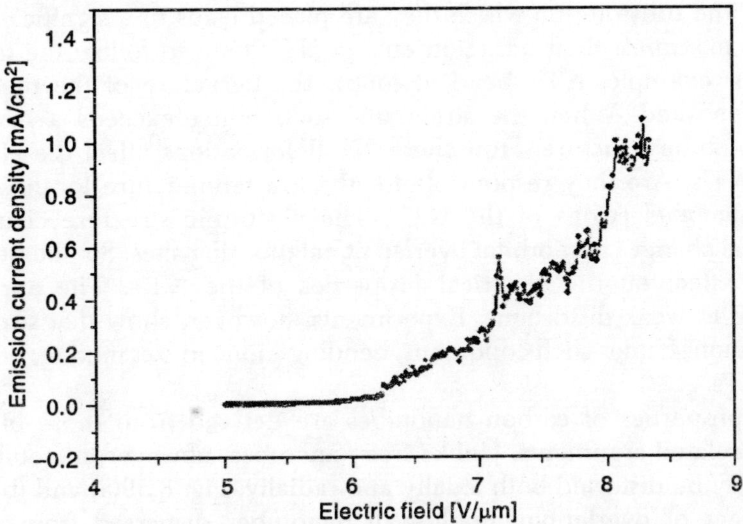

Fig. 8.58 *Emission current density of carbon nanotubes grown on Fe/SiO_2 substrate at 1023 K*

Fig. 8.58 shows an emission current density curve of carbon nanotubes grown on Fe/SiO_2 substrate at 1023K. The turn-on voltage was 350 V and the maximum current density approximately 1.2 mA/cm^2, where the applied electric field was 8.25 V/μm. Although higher turn-on voltage and lower current density were reported than those at 1223 K to 1123 K, electron emission voltage and density are tens of times higher than those reported using conventional silicon tips, DLC films, or carbon thin films, as mentioned earlier at 1123 K. As lower crystallinity of graphene sheets and incresed number of defective graphene sheets are observed at 1023 K, some fluctuations appeared in electron emission current density. These defective graphene sheets inhibit electron emission, playing as defect sources of emission.

Based on the above observations, characteristics of electron emission are summarized as follows.

(1) In electron emission of vertically aligned carbon nanotubes, the electron emission voltage and current density are approximately 10-100 times higher than those on conventional silicon tips. DLC films, or carbon thin films.
(2) As synthesis temperature increases, electron field emission current density increases.
(3) Fluctuations of electron emission current are observed at 1023 K, caused by some defects in crystallinity of graphene sheets.

PROPERTIES OF DISTORTED CNTS

Images of NT's often reveal structural deformations such as bent, twisted, or collapsed tubes. These deformations develop during growth, deposition, and processing; or following an interaction with surface features such as electrodes, or other NT's. The electronic structure of weakly distorted nanotubes on investigation using a low energy field has point defects such as vacancies or disorder which leads to significant modification of their electrical properties. AFM experiments and molecular mechanics simulations have shown that the van der Waals forces between NT's and the substrate on which they are placed leads to a significant deformation of their structure. To maximize their adhesion energy NT's tend to follow the topography of the substrate. Thus, for example, NT's bend to follow the curvature of the metal electrodes on which they are deposited. When the strain due to bending exceeds a certain limit, kinks develop in the nanotube structure. How these NT deformations affect the electrical transport properties of the NT's. Are they responsible for the low-temperature localization? Consider a model involving the π electrons of the NT's, The electronic structure changes induced by bending through the changes in π-orbital overlap at neighboring sites. So, the bending distortions have a negligible effect on the electrical properties of the NT's. The applicability of this approach is limited to weak distortions. Experiments, however, show that strong deformations and kinks are common. Under such conditions, bending-induced σ-π mixing, becomes important in strongly bent NTs.

The electronic properties of carbon nanotubes are derived from those of graphite on the basis of symmetry related arguments. Unlike free nanotubes, however, nanotubes supported on a solid substrate may be distorted both axially and radially. Fig. 8.59(a) and (b) shows two non-contact AFM images of overlapping multi-wall nanotubes dispersed from a dichloroethane solution on a H-passivated silicon surface. In this case, the nanotubes are expected to interact with the inert substrate by van der Waals forces.

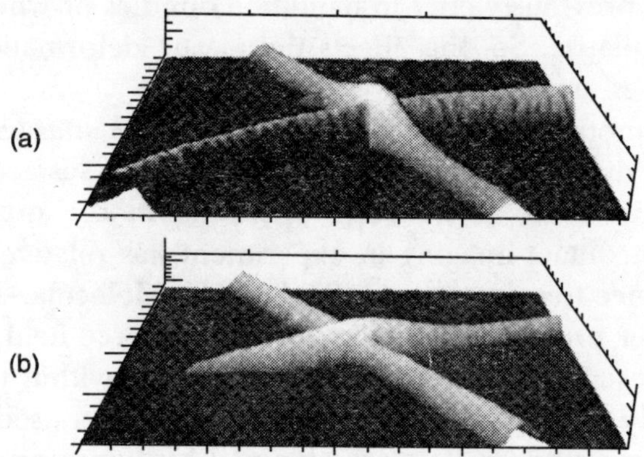

Fig. 8.59 *Atomic force microscope (AFM) non-contact mode images of two overlapping multi-wall nanotubes. The upper tubes are seen to wrap around the lower ones which are slightly compressed. The size of image (a) is 330 nm and that of (b) is 500 nm*

Deformation of CNTs by Surface Van der Walls Forces

The strength and effect of surface van der Waals forces on the shape of multiwalled and single-walled carbon nanotubes is investigated using atomic-force microscopy, continuum mechanics, and molecular mechanics simulations. Depending on the tube diameter and number of shells, the van der Waal interaction between nanotubes and a substrate results in high binding energies, Nanotubes on a substrate may consequently experience radial and axial deformations, which significantly modify the idealized geometry of free nanotubes, thus they have implication for electronic transport and the tribological properties of adsorbed nanotubes.

Carbon nanotubes have attracted growing interest due to their exceptional mechanical and electrical properties. Their high-tensile strength, flexural rigidity, 'and high-aspect ratio make them potentially interesting materials for strong fibers and tips in scanning-probe microscopy. The electrical properties which, depending on the diameter and helicity of the nanotubes, can be either metallic or semiconducting, also make them candidates for nanowires and electronic devices. However, the fabrication of carbon-nanotube/metal connections so as to estab.lish electrical contact to the outside world and to perfonn electrical measurements remains a major challenge. The character (Ohmic or Schottky type) and "quality" of nanotube substrate junctions depends upon the geometry of the contact region, which depends on the nanotube substrate interaction. Electronic transport through carbon nanotubes is generally discussed.m telms of the idealized geometry of free nanotubes, unperturbed by interaction with the environment. Experimental studies, however, are mostly performed using nanotubes that are dispersed on a substrate with which they therefore interact. Here we investigate the consequences of this interaction can have on the geometric structure of carbon nanotubes using, atomic-force microscopy and molecular-mechanics simulations is discussed below: Even though nanotubes only interact through van der Waals forces with the substrated the resulting elastic deformations are substantial. It is found that the extent of the observed axial and radial elastic deformations depends on the nanotube diameter and number of tube shells. Radial deformations of carbon nanotubes have been studied previously only in nanotube bundles or when free tubes with thin shells and large diameters collapse. So, the effects the elastic deformations may have on the properties of carbon nanotubes are discussed.

The multiwalled carbon nanotubes are dissolved in *methylene chloride* using an ultrasonic bath to agitate the solution. Once dispersed on a H passivated Si(100) surface, nanotubes could be imaged in air with an atomic force microscope operating in the noncontact mode. The commercial tips used for noncontact imaging in experiment has relatively low-force constants of 0.16–0.24 N/m and resonance frequencies near 25–30 kHz. Molecular-mechanics calculations are performed with the Tinker program employing the MM3 force field. Parameters based on alkene species are used to describe the interatomic interactions within the carbon nanotubes. These are found to accurately reproduce the C_{11} and C_{33} elastic constants of nanotubes. The surface van der Waals interaction parameters are obtained by summing atomic van der Waals interactions for an sp^2 hybridized carbon atom interacting with a graphite slab.

Tube-surface binding energies and geometries are round to converge rapidly with tube length. The atomic-force microscopy (AFM) image of a surface region densely covered with nanotubes, most of which are straight and undeformed, Fig. 8.60(a). A closer look at such images reveals that nanotubes that cross surface features such as other tubes are not entirely straight, but bend and deform elastically [see Fig. 8.60(b)]. The strain energy that thereby builds up in the tubes is compensated by a gain in binding energy as the tubes maximize their contact area with the substrate. Profiles like the one in Fig. 8.60(c) can actually be used as unique probes of the nanotube-surface binding energy, provided the Young's modulus of the tubes is known. The total energy of the system can be written as an integral of the strain energy $u(c)$ and the adhesion energy over the entire tube profile:

$$E = \int [u(c) + V(z)]\, dx, \tag{1}$$

with $c(x)$ being the local-tube curvature and $V(z(x))$ the nanotube-substrate interaction potential at a distance z above the surface (in eV/Å). The interaction potential V may depend on short-ranged chemical and long-ranged van der Waals interactions. However, the shape of profiles calculated from Eq. (1) is found to depend only on the depth and not the particular form of the interaction potential.

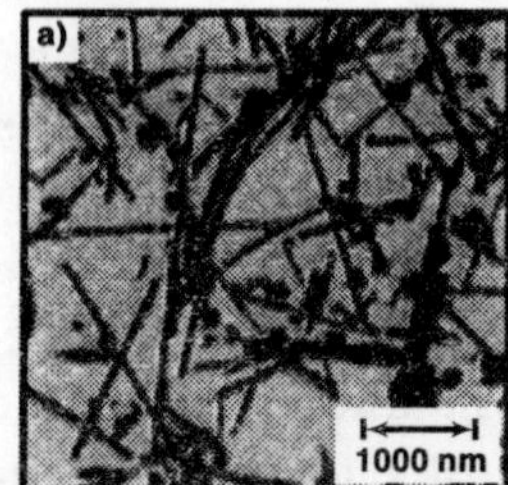

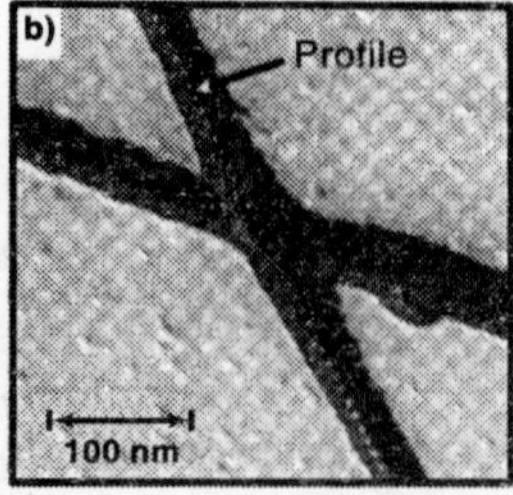

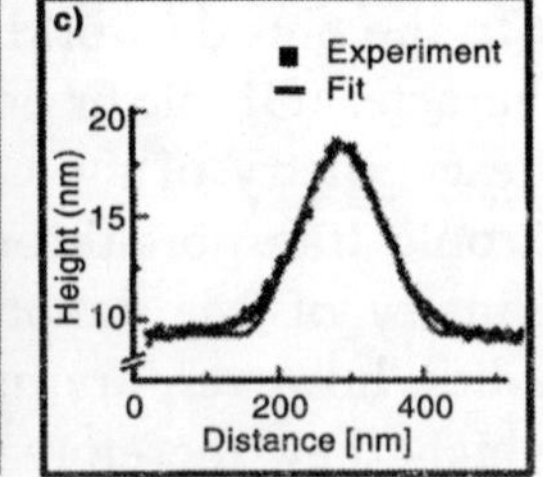

Fig. 8.60 *Multiwalled carbon nanotubes on an H-passivated Si(100) substrate imaged with the AFM. When nanotubes cross obstacles such as other tubes on the surface they bend and deform elastically, as shown in panel (b). Profiles like the one in panel (c) for a nanotube with 90 A diameter can be used to determine the binding energy of nanotubes to the substrate*

Calculated profiles to the experimental profile of a 90 A-diameter tubes as in Fig. 8.60(c), for example,is binding energy of ≈1.0 eV/Å. The average value for the binding energy of tubes with approximately 95 Å diameter is obtained from about a dozen such profiles and is 0.8±0.3 eV/A. Compared to smaller molecules interacting with surfaces, these binding energies are high but can be attributed entirely to van der Waals interactions. As a consequence of these large binding energies, the forces between nanotubes and perturbing features (such as defects, steps, or other nanotubes)on the surface can be extremely high. Using Eq. (1) the force which a 100 Å-diameter tube is pressed against an obstacle of similar height, a value of 35 nN is calculated a direct evidence for such high forces can be found in profiles along crossing tubes, clearly shows a compression of the lower tube in the contact region also that the total height at the top of such profiles as in Fig. 8.60(c) is occasionally up to 30% smaller than the sum of the two individual tube heights, suggests that one or both of the tubes may be compressed at the crossing point. However, unlike the obvious axial deformations discussed above, the AFM is not ideally suited to investigate cross-sectional deformations of nanotubes. Therefore, only in extreme cases–when tubes are collapsed entirely-may the AFM, be used to study their structure.The effect of such a compression on the measured nanotube profiles along the tubes axis can safely be neglected. In principle, van der Waals interaction potentials, and thus the force gradients measured in noncontact AFM do depend on the shape of the interacting objects, e.g., the shape of the tip and tube. But even for fundamentally different objects interacting with a surface such as a sphere or cylinder, for example, the observed change in force gradients should be small with a pronounced distance dependence of z^{-3} and $z^{-3.5}$, respectively. Since height differences along the axis of crossing nanotubes are of the order of 100 Å. Hence small variations of the tube diameter along such profiles will not have a significant effect on the measured topography, i.e., force gradient images should match the tube topography.

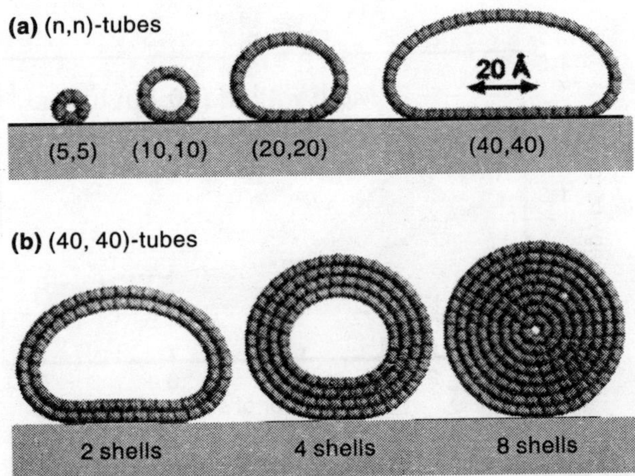

Fig. 8.61 *Radial deformations of adsorbed carbon nanotues caculated using molecular mechanics. The deformation shown are true representations of the results. (a) The radial compressions of adsorbed single-walled nanotubes with respect to the undistorted free tubes are: 0%, 2%, 13%, and 42%, for .7-, 13.5-, 27.1-Å, and 54.2-Å tubes, respectively. (b) When the number of inner shells is increased the compressions are reduced frmo 42% to 25%. 5% and to less than 1% for (40, 40) tubes with 1, 2, 4, and 8 shells, respectively*

To *explore radial and axial deformations* molecular -mechanics calculations on single and multiwalled carbon nanotubes are performed. The results of calculations on isolated nanotubes shown in Fig. 8.61(a) indicate that the extent of the cross-sectional deformation increases dramatically with increasing tube diameter. This trend can be reversed, however, if inner shells are added to the nanotube [see Fig. 8.61(b)]. The driving force behind these elastic deformations is provided by the gain in binding energy as the contact area with the substrate increases. The addition of more shells inside a tube increases its rigidity, makes these elastic deformations energetically more costly and, therefore, reduces the observed flattening of the tube wall in contact with the substrate. Such a decrease or increase of the tube's contact area with the substrate should have consequences for the tribological properties of adsorbed nanotubes. The computed binding energies for both, optimized single and multiwalled nanotube geometries reproduce these trends and are summarized in Fig. 8.62. The open squares give the binding

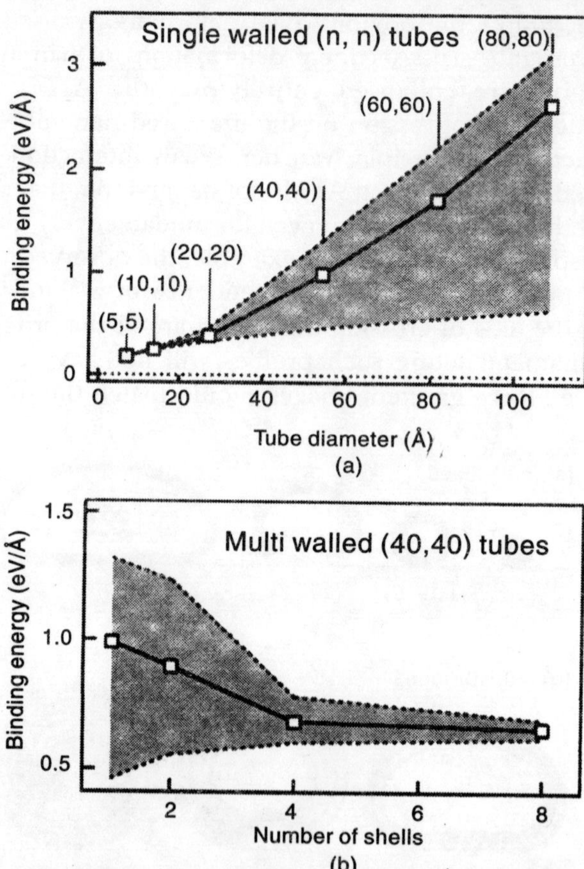

Fig. 8.62 *Binding energies of adsorbed nanotubes as a function of their diameter [panel (a)] or number of inner shells [panel (b)]. Open squares in the middle of the shaded areas give the binding energies as obtained from the molecular-mechanics calculation. The lower side of the shaded areas gives the van der Waals binding energies when no relaxation of the nanotubes is allowed. The upper boundary of the shaded areas gives the van der Waals contribution to the binding energy of the deformed tubes. The difference between this and the results from the MM3 calculation (open squares) thus reflects the cost for elastic deformations*

energies obtained for the optimized geometries based on the MM3 force field. The lower boundary of the gray area gives the van der Waals binding energy of a perfectly circular tube, i.e., when no elastic deformations are allowed. The results in Fig. 8.62 clearly indicate that the net gain in adhesion energy as a consequence of elastic deformations can easily reach 100%. The upper boundary of the gray area gives the van der Waals contribution to the binding energy. The difference between the latter and the binding energies for the optimized geometries thus reflects the strain energy that develops within the tube due to bond bending and bond stretching. When the number of inner shells increases, as illustrated for the (40, 40) tube, the cost for elastic deformation cannot be compensated as much by the gain in adhesion energy that results in a smaller flattening of the tube and a lower binding energy. Eventually the binding energy for the solid tube with eight shells is very close to that of a solid tube with no elastic deformation at all. The binding energy for a solid tube thus gives a *lower bound* to the expected adhesion energy. This dependence can be approximated by the empirical fit to our calculated binding energies $E_B = (-0.053 + 0.086d)$ eV/Å, where the diameter of the tubes d is expressed in Å. For solid tubes with around 100 Å diameter we calculate a van der Waals binding energy of 0.8 eV/Å, which agrees very well with the experimental value for the passivated silicon surface. This agreement of the calculated van der Waals binding energy to a graphite surface and the measured binding energy to a passivated silicon surface js not all that surprising due to the similarities of Hamaker constants observed for dissimilar media.

The radial distortions may also have an effect on the tubes' bending properties due to the coupling of the cross section to the flexural rigidity $B = YI$ of nanotubes, where Y is Young's modulus and I the geometrical moment of inertia. We may easily estimate the effect of radial distortions on the nanotubes' flexural rigidity if we assume that the compression leads to an elliptical deformation. The change in flexural rigidity for deformed tubes is then calculated from the geometrical moment of inertia $I = \iint y^2\, dx\, dy$ over the cross section of the tube. This estimate reveals that the flexural rigidity of tubes is reduced almost linearly with compression according to $B \cong 1.9B_0 a/_0$ where a_0 and a are the outer tube radii for undistorted and elliptically compressed tubes, respectively. Thus, tubes that have undergone significant radial compression can be bent much more easily than perfectly circular tubes.

Adsorbed on a substrate, where nanotubes are likely to encounter obstacles or defects; they will be subject to both radial and axial deformations. Therefore, simulating such a situation using molecular mechanics where one tube crosses another one lying on a surface, eg optimization for two crossing (10,10) nanotubes with altogether 6800 atoms are reproduced in Fig. 8.63. These tubes have a diameter of 13 Å and the longer one has a length of about 350 Å. The axial profile of the upper tube seen in Fig. 8.63(a) is very similar to the profiles of multi walled tubes imaged by the AFM. The molecular-mechanics calculations, however, give a more detailed account of the forces and resulting elastic deformations, in particular at the point of contact between the two tubes. It is found that the force of 5.5 nN with which the upper tube is pressed against the lower one leads to a cross-sectional compression of both tubes by almost 20% near

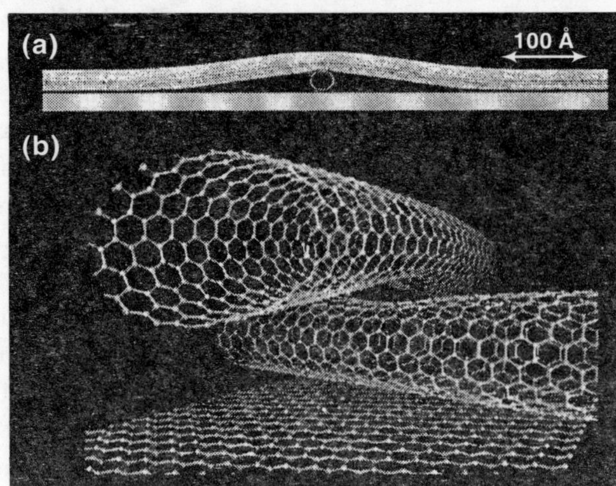

Fig. 8.63 *Results from the molecular-mechanics calculation of two crossing (10,10) nanotubes with axial and radial deformations. Panel (b) gives a perspective closeup of the crossing point. It shows that both tubes are deformed elastically near the contact region. The force acting on the lower tube is about 5 nN*

the point of contact [see Figs. 8.63(b) and (c)]. Further analysis of the molecular mechanics simulations shows that the'local strain associated with these deformations can be as high as 5-10 meV per atom. Such strains will lead to a modification of the electronic structure of adsorbed nanotubes and–at small enough temperatures–should have implications for electronic transport Rroperties through nanotubes. Experimental- and theoretical- studies also indicate a coupling between tube distortions and their electronic transport properties.

Thus, perturbing features on the surface–such as defects or steps–can induce substantial distortions in adsorbed nanotubes due to their strong adhesion to the substrate. The strain associated with such deformations is very likely the cause for the observed perturbations of the electronic transport properties. This poses a serious challenge for the control required for processing of nanotube devices if they are not to be affected by surface roughness. On the other hand, there is an opportunity to utilize the surface topography as a tool for tailoring the nanotube's transport properties. In this context we have estimated the critical surface roughness-expressed in terms of a critical surface curvature cc-up to which the tube will smoothly follow the substrate topography. This can be done by calculating the smallest curvature at which the axial strain energy, which develops as the nanotube bends, can still be compensated by the adhesion energy. Since the strain energy increases more rapidly with tube diameter d than the adhesion energy ($\propto d^4$ and $\propto d$, respectively) this critical curvature is only about $(12d)^{-1}$ for a tube with 13 Å diameter and decreases to $\approx (50d)^{-1}$ for tube diameters around 100 Å. Thus, a substrate with surface features having a curvature higher than c_c provides perturbing obstacles that lead to the kind of elastic defonnations seen in Fig. 8.63.

AFM and molecular mechanics calculations show that carbon nanotubes tend to adjust their structure to follow the surface morphology of the substrate. Let R_C be a critical radius of surface curvature, above which the nanotube can follow the surface structure or roughness. As the strain energy varies more strongly with tube diameter than the adhesion energy, the critical radius is a function of the tube diameter. Fig. 8.64 shows the approximate variation of R_c with d for (5,5) and (10,10) single-wall and larger multi-wall tubes based on binding and strain energies computed with molecular mechanics.

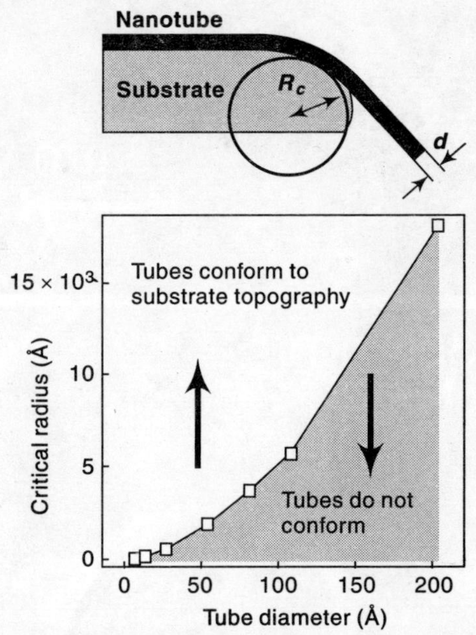

Fig. 8.64 *Criteria determining whether a nanotube with a particular diameter d conforms to the local topography of the substrate characterized by a radius of curvature* R_c

The van der Waals forces play an important role not only in the interaction of the nanotubes with the substrate but also in their mutual interaction. The different shells of a multi-walled tube interact by van der Waals forces; single-walled tubes form ropes for the same reason. Short ropes of single-wall nanotubes are converted into nanotube rings. (Fig. 8.65) Some of the loops appear to be single circular tubes. But they can be unwound by AFM manipulation which, suggests that there are no covalent bonds holding the ropes together. The loops are fairly large (about 400–800 nm radius). Therefore, the strain induced by the increased curvature is compensated by the gain in adhesion between nanotubes. This coiling behavior is similar to that observed in proteins and other biomolecules, where hydrogen bonding is thought to provide the main force for coiling. In the case of carbon nanotubes, however, only van der Waals forces are present.

In summary, the van der Waals interaction between nanotubes and a substrate can lead to substantial axial and radial defonnations of adsorbed nanotubes destroying the idealized shape of free tubes. This has important implications for tribological and electronic transport properties of adsorbed nanotubes. Experimentally detennined adhesion energies for nanotubes bound to a passivated silicon surface could be attributed entirely to van der Waals interactions. The observed flattening of adsorbed nanotubes on the side facing the substrate can lead to a substantial

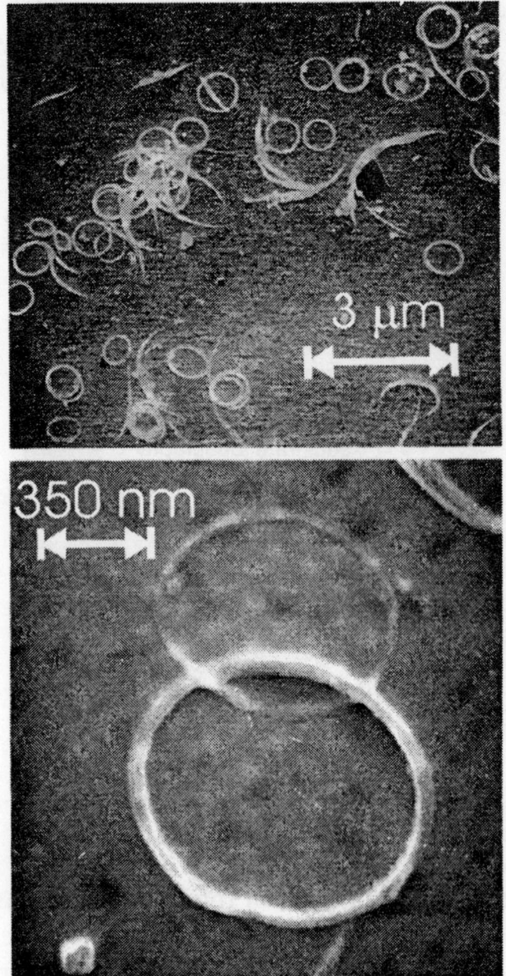

Fig. 8.65 *Top: Scanning electron microscope images of nanotube rings produced from ropes of single-wall nanotubes. Bottom: Images of two inter-connected nanotube rings*

increase in the contact area and binding energy with respect to undefonned tubes. Owing to the linear relationship between contact area and shear forces the larger contact area of defonned tubes allows to stabilize -highly strained tube configurations like those observed recently with the AFM. In fact, we found that collapsed tubes with the highest possible contact area could not be pushed around on the surface by the tip of an AFM-unlike their inflated "cousins." The elastic deformations and the resulting strains that we observe when nanotubes interact with surface features are likely responsible for irregularities in the electronic transport properties of adsorbed nanotubes.

II. Distortion by Bending and Twisting of CNT

Suppose a nanotube model is used in the computations that contains 948 carbon atoms arranged in an armchair (6,6) structure. The energetics of the deformations are determined with molecular mechanics with a modified force field. Calculations are also performed using a tight-binding density-functional theory (TBDFT) method. In both types of calculations the dangling bonds at the ends of the tubes are saturated by hydrogen atoms. The bond distance between carbon atoms in nondeformed regions of the NT is fixed to that in graphite 1.42 Å, leading to a tube length of 96 Å. The structures of the bent NT's are shown in Fig. 8.66(a). The twisted structures are generated by two different schemes. In the first, continuous twisting, the circular plane sections of carbon atoms along the tube axis were rotated sequentially by an (additive) specified angle. In the second, alternate twisting, a (nonadditive) rotation was performed on every second section. The molecular mechanics calculations showed that it is less energetically demanding to continuously twist the tube than to twist it alternatively.

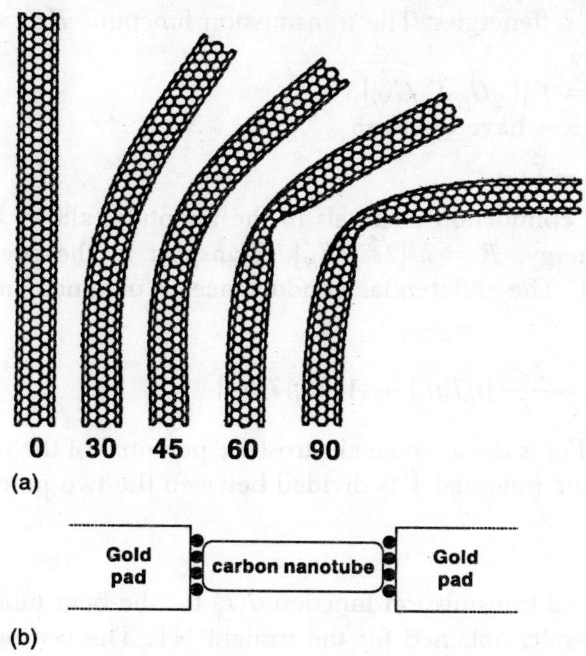

Fig. 8.66 *(a) Structures of the isolated bent nanotubes used in the computation. (b) Model used in computation involving a finite length carbon nanotube contained between infinite gold pads*

The electronic structures of distorted NT's are determined using the Extended Huckel Method (EHM).

The energy band gap is obtained as the difference between the highest occupied molecular orbital (HOMO) and lowest unoccupied molecular orbital (LUMO) energies. The metallic contacts consist each of 22 gold atoms in a (111) crystalline arrangement. The height of the NT over the gold layer is 1.0 Å, where the Au-C bond distances vary from 1.1 to 1.6 Å.

The conductance through a molecule or an NT cannot be easily computed; even if the electronic structure of the free molecule or NT is known, the effect of the contacts on it is substantial and needs to be taken into account. 'Typically, there will be two (or more) leads connected to the NT. The measurement are taken by a system as shown in model Fig. 8.66(b). The leads are macroscopic gold pads that are coupled to the ends of the NT through matrix elements between the Au surface atoms and the end carbon atoms of the NT. The NT's are laid on top of the metal pads. Then they are bend around the pads. Such bending deformations are first accounted calculations by introducing a single bend placed at the center of the tube.

The electrical transport properties are described in terms of the retarded Green's function (G_{NT}), which after some simplification is given by

$$G_{NT} = [ES_{NT} - H_{NT} - \Sigma_1 - \Sigma_2]^{-1},$$

where S_{NT} and H_{NT} are the overlap and the Hamiltonian matrices, respectively, and $\Sigma_{1,2}$ are self-energy terms that describes, the effect of the electronic structure of the two leads. The conductance of the NT to compute the transmission function, $T(E)$, from one contact to the other is evaluated from the selfenergies. The transmission function, $T(E)$, that is obtained from Green's function is given by

$$T(E) = T_{21} = Tr[\Gamma_2 G_{NT} \Gamma_1 G_{NT}]$$

In this formula, the matrices have the form

$$\Gamma_{1,2} = i(\Sigma_{1,2} - \Sigma_{NT})$$

The summation over all conduction channels in the nanotube allows the evaluation of the resistance at the Fermi energy, $R = h/[2e^2 T(E_F)]$. Transport in the presence of an applied potential is also computed. The differential conductance is obtained in this case using the approximation

$$\kappa(V) = \frac{\partial I}{\partial V} \approx \frac{2e^2}{h}[\eta T(\mu_1) + (1 - \eta)T(\mu_2)]$$

Here, $\eta = V_{NT}/V$ where V_{NT} is the average electrostatic potential of the nanotube. η is used to describe how the electrostatic potential V is divided between the two junctions.

(a) Nanotube bending

Fig. 8.67 shows the computed transmission function $T(E)$ for the bent tubes. In the upper-right are the raw transmission results obtained for the straight NT. The fast oscillations of $T(E)$ are due to the discrete energy levels of the finite segment of the carbon nanotubes used. At E_F $T(E)$ is about 1.2, leading to a resistance ($\approx$ 11 kΩ) higher than expected for ballistic transport [$\approx$ 6 kΩ for $T(E) = 2.0$]. This reduction in transmission is due to the contribution from the contact resistance. The increasing $T(E)$ at higher binding energies is due to the opening of new conduction channels. The asymmetry in the transmission $T(E)$ is a function of the NT-pad coupling (C-Au distance). A longer NT-Au distance increases the $T(E)$ above E_F while it decreases it below E_F and vice versa. Since the NT-pad geometry is kept fixed in all computations, this behavior does not obscure the effects induced by NT bending.

The contact resistance at E_F is only about 5 kΩ, much smaller than the $\approx$ 1 MΩ resistance typically observed in experiments on single-wall NT's. The dependence of $T(E)$ and contact resistance at E_F on the Au-NT distance is shown in the upper-left part of Fig. 8.67. $T(E_F)$ remains nearly constant between 1-2 Å and then decreases exponentially. For distances appropriate for van der Waals bonding ($\geq$ 3 Å) the contact resistance is in the MΩ range. The

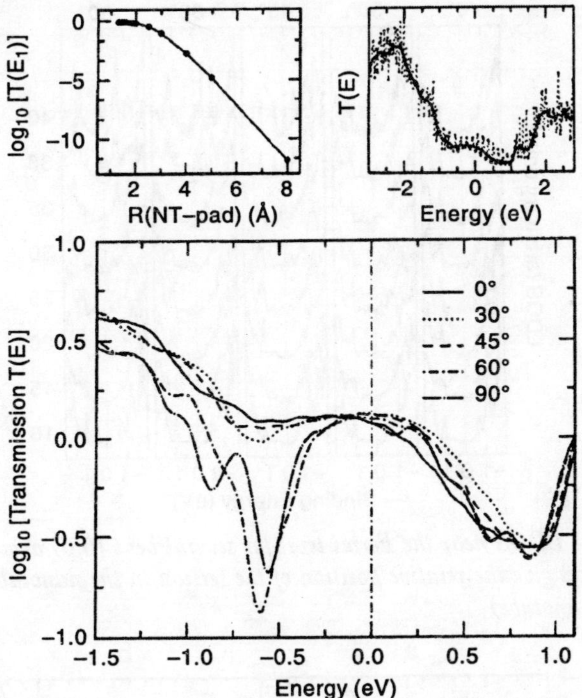

Fig. 8.67 *Transmission function T(E) for the different bent nanotubes. The upper-right inset gives the raw data obtained from the computation for the straight NT along with the smoothed curve. The upper-left inset shows the variation of the transmission function $T(E_F)$ with the nanotube-pad distance [R(NT-pad)]. In this last figure, one gold-NT distance is fixed at 1.0 Å while the other is varied*

above findings suggest that the high NT contact resistance observed experimentally is due to the topology of the contact. In most experiments, the NT is laid on top of the metal pad. The NT is at nearly the van der Waals distance away from the metal surface, and given that transport in the NT involves high k states, which decay rapidly perpendicular to the tube axis, the coupling between NT and metal is expected to be weak. Direct chemical bonding between metal and NT leads to a much stronger coupling. It is found that high-energy electron irradiation of the contacts leads to a drastic reduction of the resistance. Since the irradiation is capable of breaking NT C-C bonds so the resulting dangling bonds lead to a stronger metal-NT coupling.

The strongest $T(E)$ due to bending is observed at around $E = -0.6$ eV where a transmission dip appears. This dip is strongest in the 60° bent NT. Furthermore, its transmission function at higher binding energies is lower than those of the 0°–45° bent NT's, indicating that the transmission of higher conduction channels is also decreased. The nature of the dip at about −0.6 eV can be understood by examining the local density of states (LDOS) of bent tubes shown in Fig. 8.68. A change (increase) in the LDOS is seen in the same energy region (0.5–0.8 eV below E_F) as the transmission dip. This change is essentially localized in the vicinity of the deformed region. The new states result from the mixing of σ and π levels, have a more localized character than pure π states leading to a reduction of $T(E)$. As Fig. 8.68 shows, the

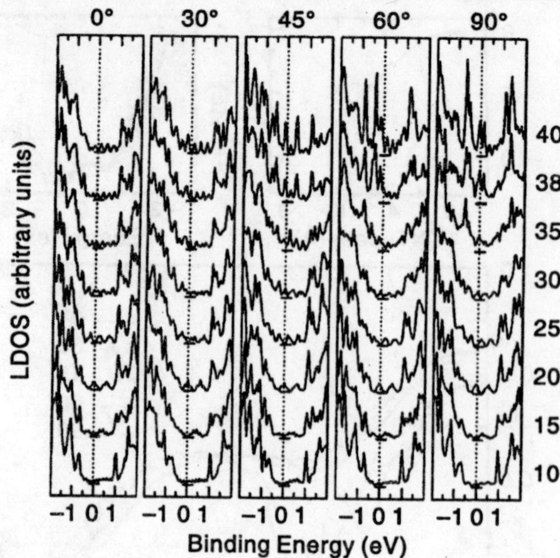

Fig. 8.68 *Variation of the LDOS near the Fermi level for several bent (6,6) armchair nanotubes (resolution = 0.05 eV). Indices give the relative position of the section in the nanotube structure (1: boundary; 40: middle of the nanotube)*

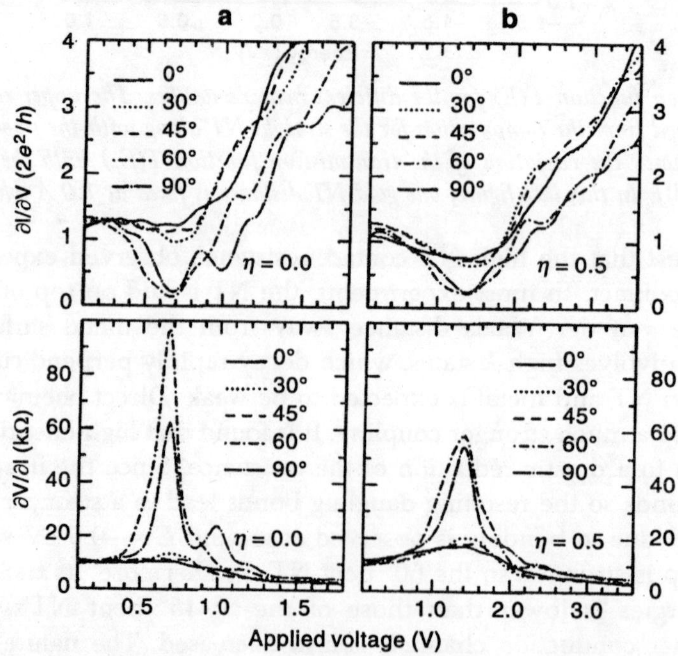

Fig. 8.69 *Differential conductance (top) and resistance (bottom) of bent tubes for two extreme cases where (a) $\eta = 0.0$ and (b) $\eta = 0.5$*

change in transmission with bending angle is not gradual; the transmission of the 30° and 45° is only slightly different from that of the straight tube. Apparently, large changes in DOS and $T(E)$ require the formation of a kink in the NT structure, as is the case in the 60° and 90° bent NT's.

Once the transmission function is computed, the differential conductance or resistance can be easily determined for a given electrostatic potential profile. Figure 8.73 shows the results for two extreme cases of equilibration of the Fermi levels. The first is when $\eta = 0$ [Fig. 8.69(a)], and the symmetric case $\eta = 0.5$ [Fig. 8.69(b)]. When $\eta = 0.0$, the electrochemical potential of the NT matches that of one of the electrodes and the conductance spectrum is directly proportional to $T(E)$. This potential profile is appropriate in the case where one of the electrodes is weakly coupled to the NT, as when the NT is probed with an STM tip. There is no large difference between the 0°, 30°, and 45° tubes, while the 60° and 90° tubes show the dip structure at around 0.6 V. The nonlinear resistance (NLR) spectra show clearly a sharp increase by almost an order of magnitude at 0.6 V. These features are also observed when $\eta = 0.5$, where the Fermi level of the NT is floating at half the voltage applied between the two gold pads. This value of η is more appropriate for the strong and symmetric electrode-NT-electrode coupling illustrated in Fig. 5.66(a). A broader conductance dip appears at about 1.2 V, and the NLR of the 60° bent tube increases by about a factor of 4 from the computed resistance of the straight tube. These results indicate that there exists a critical bending angle (between 45° and 60°) above which the conduction in armchair carbon nanotubes is drastically altered.

(b) Nanotube twisting

Now consider the modifications of the atomic and electronic structure brought about by twisting an NT. Fig. 8.70 is a graph between the potential energy for a continuous twisting deformation against the twist angle for a constrained (6,6) structure, and for a relaxed one where only the first two carbon sections at each end of the tube are held fixed. In the inset, the energies obtained with molecular mechanics for small angle twisting of the relaxed structures to the energies is determined with the TB-DFT method are compared. A number of observations are made. First, the energy required for even a moderate twisting of the tube is very large. At low twisting angles (inset) a quadratic law is obeyed. The twisting energies are calculated using the MM3 potentials and are lower than those computed using TB-DFT (inset). Therefore, in the following consider the MM3 energies as lower limits to the energetics of twisting deformations. Figure 8.70 shows that structural relaxation has a very strong effect on the potential energy of deformation, especially for large twisting angles where the relaxation energy is computed to be of about the same order of magnitude as the final energy. Also, the energy profile of the relaxed structure deviates from the quadratic law (dot-dashed line for $\Theta_T > 14°\ \text{nm}^{-1}$). This deviation corresponds to the onset of a structural transformation where the tube flattens and takes on a helical shape [such as in Figs. 8.71(c) and (d)]. A similar transition is observed in molecular dynamics simulations. In order to determine the effect of the diameter on the collapsing of the nanotubular structures,

MM3 energies of twisted (10, 10) models are computed. Taking into account the differences in the number of carbon atoms involved (10, 10)/(6, 6) = 1.67), and the NT diameters ($\phi_{(10,10)}/\phi_{1.6} = 1.67$), we find that the energy needed to twist a (10, 10) NT is similar to that needed to twist a (6,6) NT. The collapse of the (10, 10) tubular structure into the helical shape occurs at a lower twisting angle than the (6,6) structure, i.e. when $\Theta_T > 10°\ \text{nm}^{-1}$. However, due to its

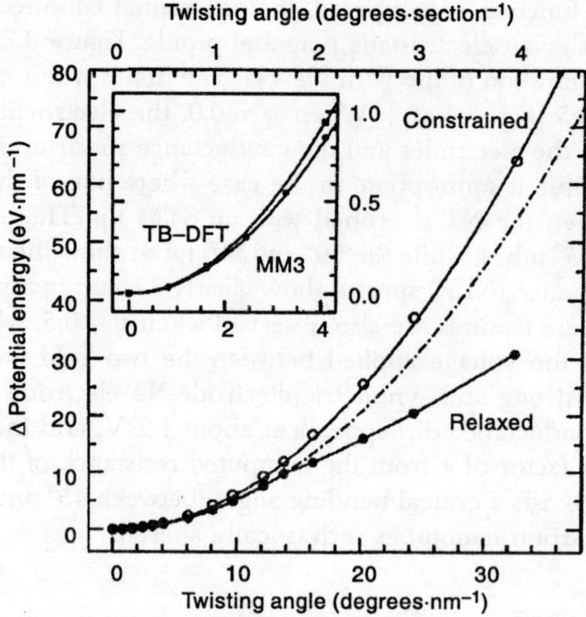

Fig. 8.70 *Molecular mechanics calculation of the energy required to twist a (6,6) armchair NT as a function of twist angle, for a constrained (open circles) and a relaxed geometry (closed circles). The dot-dashed line corresponds to the quadratic law determined from the low twisting angle region of the relaxed NT. The inset shows a comparison of the MM3 and TB-DFT energies for small twisting angles*

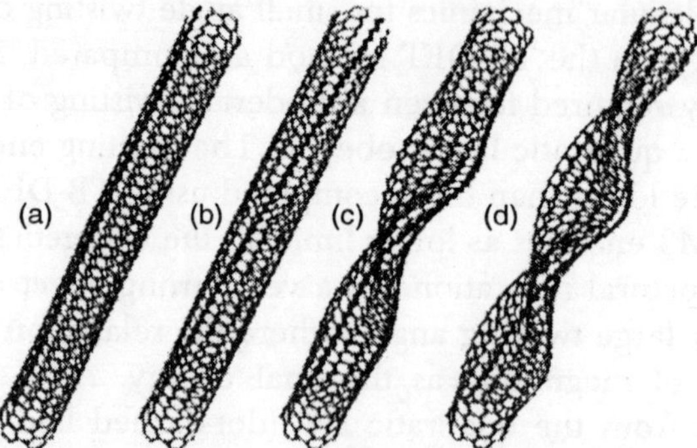

Fig. 8.71 *Relaxed NT structures obtained with force-field MM3 energy minimization on 0 (a), 1 (b), 2 (c), and 3° section^{-1} (d) twisted models*

larger diameter, the relative displacement of the carbon atoms during the tube to helix transition for a (10, 10) NT is larger than in the case of a (6, 6) NT.

As calculations show that, the twisting energies scale nearly linearly with the length of the nanotube, one can extrapolate and obtain the twisting energies ot longer nanotube segments, and use these energies to evaluate their Boltzmann distributions as a function of temperature. The relative population of twisted (6, 6) structures can be described by

$$F(\Theta_T) = \exp\left[-\alpha \frac{L\Theta_T^2}{T}\right]$$

where, α is a constant equal to 580 nm · deg−2 · K, L is the nanotube length in nm, Θ_T is the twisting angle (in deg · nm^{-1}), and *Tis* the temperature. From this equation, it is clear that large twists cannot be generated by thermal excitation. For example, to obtain by thermal excitation a relatively small population (1%) of a μm long (6, 6) nanotube twisted by 9.8° nm^{-1}, it requires a temperature larger than 10^6 K! Thus, large scale nanotube twists must be the result or either mechanical interactions taking place during the deposition or manipulation of the nanotube sample, or they are introduced during the high-energy growth processes and frozen in place by shear forces due to the interaction with the substrate and other tubes. Local twisting associated with tube collapse is, however, less energetically demanding and is encountered quite often in AFM Images.

The most important modification of the electronic structure induced by twisting a metallic armchair nanotube is the opening of a band gap. Figure 8.72 shows the variation of the band-gap value of a twisted (6,6) nanotube as a function of the twisting angle, for both constrained and relaxed NT structures. So for the constrained structure, the band gap increases linearly with twisting angle up to 16° nm^{-1}, then gradually decreases to zero at higher angles. Allowing relaxation in the twisted nanotube structure has a strong influence on the value of the band gap; the gap increases linearly (but with a different slope) with twisting angle up to 14° nm^{-1}, then reaches a stable value (≈0.3 eV) at higher angles. The first, linear, region corresponds to the opening of a gap produced by the straining, i.e., compression and dilatation of C-C bonds along the axis of the tube, brought about by twisting. The initial linear increase obtained in constrained structures is in general agreement with the predictions of low-energy theory which, however, does not take relaxation into account. For higher twisting angles, the band-gap

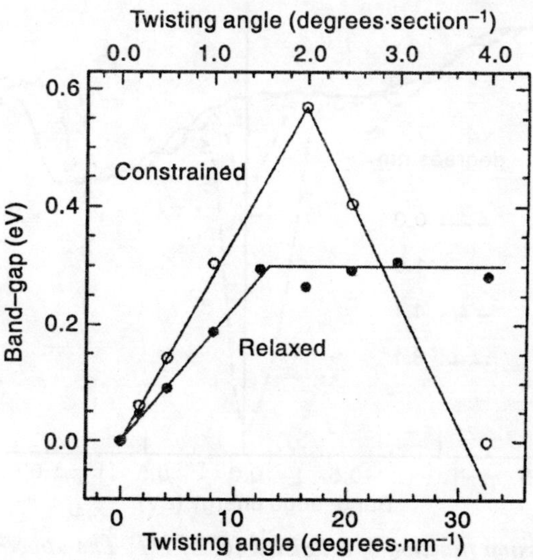

Fig. 8.72 *Band-gap opening in a (6, 6) nanotube as a function of the twist angle for constrained geometry (open circles), and a relaxed geometry (filled circles)*

decrease observed for the constrained geometry is mainly due to the presence of strong interaction between destabilized σ and π orbitals near the Fermi level. The plateau observed in relaxed structures is the result of the collapse of the tubular structure into a flattened helix shape which decreases the large repulsive overlap of strained NT's [Figs. 8.71(c) and (d)].

The opening of a band gap in originally metallic nanotubes leads to a drastic modification of the electrical properties ofnanotubes upon twisting by even low angles. In order to investigate the conduction of electrons in twisted (and relaxed) in (6,6) nanotubes, are connected both the tube ends (with dangling bonds on the end carbons) to gold electrodes. Figure 8.73 shows the transmission *T(E)* spectra of nanotubes twisted by 0, 1.6, 4.1, and 8.1° nm^{-1} angles (main panel). All these deformed tubes lie within the linear regime of the band-gap variation with twist angle. The variation of *T(E)* at E_F with twisting angle is shown at the upper-left panel, while the corresponding change in resistance at zero bias [= 12.9 $K\Omega/T(E_F)$] is given at the upper-right panel. Twisting induces important changes in the $T(E_F)$, a property that determines the linear conductance of the system. Already a twist of 4° nm^{-1} (or 0.50/ $section^{-1}$]), reduces the transmission by a factor of 14. It is interesting, however, that essentially no change is seen in $T(E_F)$ at low twisting angles ($< 2°\ nm^{-1}$) despite the fact that a finite band gap ($\approx$ 0.04 eV) is

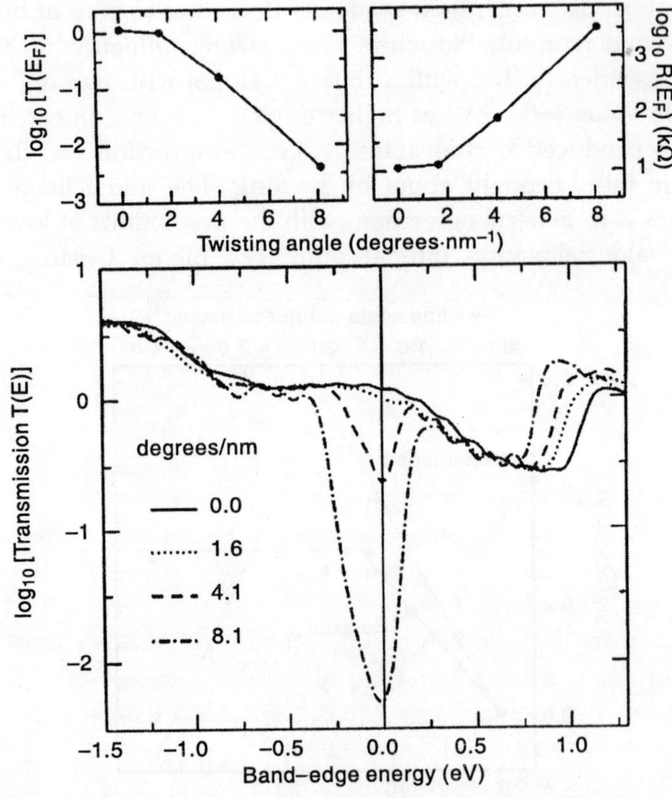

Fig. 8.73 *Computated transmission function of a twisted (6, 6) NT. The upper-left inset shows the logarithmic decay of T(EF) with twisting angle, and the upper-right inset gives the variation of he resistance at zero bias for the same twisted NT*

calculated for the free NT. This behavior due to the metal-induced gap states (MIGS) is produced by the interaction of the short (96 Å) NT segment with the gold metal electrodes and the presence of charge-transfer doping. The MIGS tend to bridge the distortion-induced gap when the latter is small. For twisting angles larger than 2° nm^{-1}, the transmission decreases exponentially (see the upper-left in Fig. 8.73). At 8° nm^{-1}, $T(E_F)$ is decreased by more than two orders of magnitude, and if transport is considered into a strongly twisted, helical shape nanotube [see Figs. 8.71(c) and (d)]; the decrease in $T(E)$ is about five orders of magnitude. A twisting-induced band gap is, most likely, responsible for the recently documented fieldeffect transistor action of a twisted, normally metallic multiwall nanotube.

Hence in the effects of structural distortions of armchair carbon nanotubes on their electrical transport properties, it is found that bending of the nanotubes decreases their transmission function and leads to an increased electrical resistance. The effect is particularly strong at bending angles higher than 45° when the strain is strong enough to lead to kinks in the nanotube structure. The electronic structure calculations show that the reduction in $T(E)$ is correlated with the presence at the same energy of localized states with significant (σ-π) hybridization due to the increased curvature produced by bending. Resistance peaks at or near E_F caused by bending can lead to localization at low temperatures or anomalous electrical transport properties.

Calculations of the resistance (including the contact resistance) of a perfect NT end bonded to gold electrodes give a value close to $h/2.4e^2$ instead of $h/4e^2$. This increase in resistance is due to the finite transmission of the contacts. The much larger contact resistances observed in many experiments on SWNT's are likely due to the weaker coupling of the NT to the metal when the NT is simply placed on top of the metal electrodes (this resistance, however, depends on the length of the contact). Therefore, NT's end bonded to metal pads will have contact resistances of only a few kΩ. Such low contact resistances will greatly improve the performance of NT-based devices and unmask the Q1D transport properties of NT's. Twisting a normally metallic armchair NT opens up a band gap which scales linearly with the twist angle up to a critical angle. Twisting above this angle leads to the collapse of the tubular structure and the formation of a flattened helix. The observed band gap is due to the asymmetric compression and dilatation of the C-C bonds along the NT.

Characteristics of Surface Treatment of Fe Catalytic Metal Film

The length, density, and structure of carbon nanotubes are highly affected by forms of catalytic metal films in synthesis of carbon nanotubes. Especially, as carbon nanotubes using chemical vapor deposition are grown on catalytic metal particles, the size, form. and density of carbon nanotubes need to be reviewed in detail. Fig. 8.74. shows surface morphology of Fe/SiO_2 substrates after Fe deposition and water diluted HF and NH_3 solutions of the substrate. Fig. 8.74(a) is surface morphology of Fe catalytic metal film by thermal chemical vapor deposition. Fe thin film is deposited on SiO_2 substrate. Although the surface morphology of the thin film is rough, there are no nano-size fine particles. Instead of nanotubes, nanoparticles are synthesized on this rough surface. Fig. (b) shows SEM surface morphology of the catalytic metal after dipping the Fe catalytic metal film on SiO_2 substrate into diluted HF solution for 60 seconds.

Although this image shows that catalytic metal particles began to form during etching of catalytic metal film, their sizes and forms are not uniform. Fig. 8.74(c) shows SEM surface morphology of catalytic metal film after dipping it into diluted HF solution for 140 seconds. As etching time increases, further etching of catalytic metal film leads to formation of nano-size fine catalytic metal particles. In this image, the sizes of catalytic metal particles are uniform, but very large in the range of 700–800 nm. Fig. 8.74(d) shows SEM surface morphology of the catalytic metal after dipping the Fe catalytic metal film on SiO_2 substrate into diluted HF solution for 140 seconds and flowing 300 sccm of NH_3 for 10 minutes at 1123 K. From the image, fine particles are uniformly grown and the catalytic metal particles are much smaller and better in uniformity than those in the image 8.74(c) simply under distilled HF treatment. Carbon nanotubes can be grown vertically on nano-size catalytic metal particles of these forms.

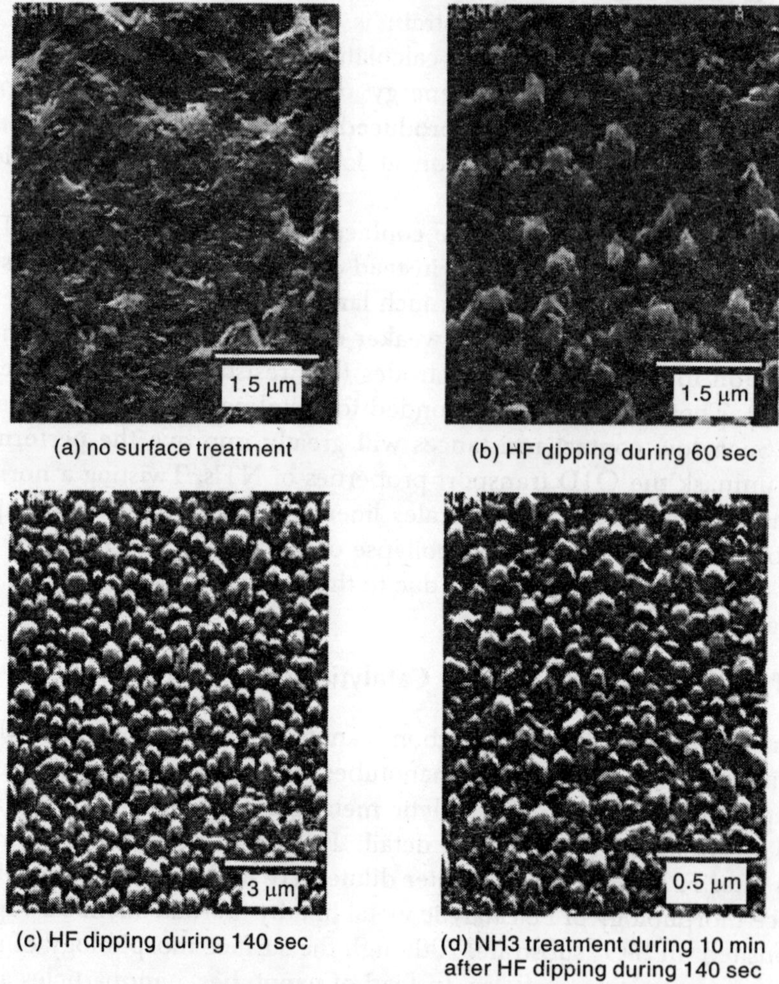

Fig. 8.74 *Surface morphology of Fe/SiO_2 substrates using various pretreatment conditions*

This explains that highly dense fine catalytic metal particles work as positions for nucleus formation of carbon nanotubes. After all, it is found that HF dipping and NH_3 treatment of catalytic metal film are very important in vertical growth of carbon nanotubes on silicon substrate and the sizes of fine catalytic metal particles have a great effect on diameters of carbon nanotubes vertically aligned. Thus the vertical alignment and adjustment of diameter and density of carbon nanotubes is done by properly etching the surface of catalytic metal and forming it into nano-size fine nanoparticles. The total effects of pretreatment of catalytic metal films is summarized on the above results as follows.

(1) HF dipping simply etches the surface of catalytic metal film. As time of HF dipping increases. surface morphology gets rougher.

(2) If HF dipping time is longer than 140 seconds. uniform particles with 700–800 nm in diameter are formed.

(3) When catalytic metal particles are formed by simple HF dipping without NH_3 treatment, carbon nanotubes grown on the particles are highly dense. but not vertically aligned.

(4) Synthesis of vertically aligned carbon nanotubes is possible in the conditions of HF dipping for 140 seconds and 300 sccm of NH_3 at 1123 K.

CHEMICAL PROPERTIES OF CNTs

(A) Solution Properties

SWNTs exist in the form of quasi-crystalline bundles containing tens to hundreds of individual nanotubes, with binding energy of the order of 950 meV/nm induced by van der Waals force between the tubes. In addition, SWNT ropes/bundles have highly entangled netwark structure. MWNTs and DWNTs exist as individual tubes; however, they are also highly entangled (Fig. 8.75). Therefore, to achieve full potential CNTs must be well dispersed in a solvent.

Individual SWNTs can be isolated in micelles using surfactants (catianic, anianic, palymeric). Di-and tri-black copolymers can also be used as a non-ionic surfactant such as a diblock copolymer of poly propylene oxide-co-polyethylene axide (PPO-co-PEO). The surfactants with a hydrophobic group equal to and longer than decyl (10 methyl carbon group) contribute to

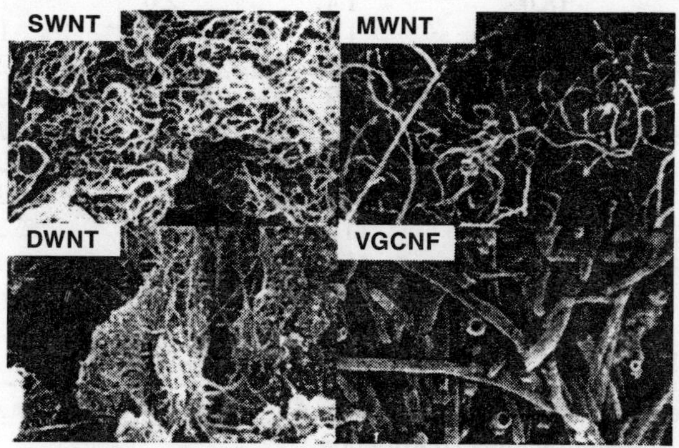

Fig. 8.75 *SEM images for SWNTs, MWNTs, DWNTs and VGCNFs*

the dispersion of nanotubes in an aqueous system. A molecular dynamic simulation shows SWNT embedded within sodium dodecyl sulfate (SDS) micelles, indicating that several tens of nanometer thick layer of surfactant surrounding individual SWNT.

Pristine SWNT dispersion in various organic solvents have been characterized by high values for electron pair donicity (β), negligible values for hydrogen bond donating parameter (α) and high values for solvatochromic parameter (π^*), eg: 1, 2-dichlorobenzene, N-methylpyrrolidone (NMP), and N, N-dimethylformamide (DMF).

These solvents can dissolve pristine SWNTs. On the other hand, solvents such as xylene, toulene, and alcohols exhibit poor solubility for pristime SWNTs (Table 8.6)

TABLE 8.6 Solvent parameters and solubility of pristine, as-purified SWNTs in various solvents.

Solvents	α (↓)	β (↑)	π*(↑)	*solubility* (mg^{-1})
N-methyl-pyrrolidone	–	0.754	0.921	10
N,*N*-dimethyl-formamide	–	0.710	0.875	7.2
1-butanol	0.732	0.850	0.503	<1
Ethanol	0.827	0.77	0.540	<1
Toulene	–	–	0.535	<1
p-Xylene	–	–	0.426	<1

Chemical oxidation of nanotubes with nitric acid or a mixture of sulfuric and nitric acids can provide functional groups such as carbaxylic acid en the nano.tube surface.

Nitric acid functionalized SWNTs with carboxylic groups exhibit improved solubility in hydrogen bondable solvents as well as in poor solvent mixtures such as butanal/toluene and ethanol/xylene mixture (Table 8.7). On the other hand, for pristine SWNTs, the polar companent of the solubility parameter is critical in governing SWNT dispersion in PMMA.

TABLE 8.7 Solubility parameter and measured solubility of nitric acid treated SWNTs in various solvents.

Solvents	δ_d	δ_p	δ_h	δ_t^a	*Solubility* (mgL^{-1})
Toulene	18.0	1.4	2.0	18.2	<1
Xylene	17.8	1	3.1	18	<1
Nitromethane	15.8	18.8	5.1	25.1	7
Acrylonitrile	16.5	17.4	6.8	24.8	13
N, N-dimethylformamide	17.4	13.7	11.3	24.8	18
1-butanol	16	5.7	15.8	23.1	25
Ethanol	15.8	8.8	19.4	26.6	27
Ethyleneglycol	I7	11	26	32.9	32
H_2O	15.5	16	42.4	47.9	35
ethanol/xylene[b]					17
butanoVtolune[b]					19
butanoVxylene[b]					20

[a]Total solubility parameter ($MPa^{0.5}$), which is defined as $\delta_t^2 = \delta_d^2 + \delta_p^2 + \delta_h^2$, δ_d is dispersive contribution, δ_p is polar contribution, δ_h is hydrogen bonding contribution.
[b]50/50 volume fraction.

The solubility of acid treated SWNTs in solvent mixtures such as ethanol/xylene provides a route for SWNT dispersion/exfoliation in polymers such as polypropylene.

SWNT solubility can be changed by sidewall or end-cap functionaization. CNTs are also being dispersed and exfoliated using various salts and ionic liquids. CNT/polymer dispersions have been achieved using polymerization in a number of polymer systems, uniform polymer coating on individual nanotubes has been observed, indicating nanotube wetting by the continuous polymer phase. For example, SWNT/polyaniline composite films are dispersed, prepared by electrochemical polymerization and a homogeneous SWNT/polyimide composite is obtained.

Also, polymerization of caprolactam/SWNT allows continuous spinning of SWNT/nylon 6 fibers.

In the absence of a dispersing agent, debundled SWNTs tend to reaggregate to form bundles. Figure 8.76 shows the current approaches (or dispersing or exfoliating bundled SWNTs into individual tubes and the concept for composite processing using the solution property of SWNTs.

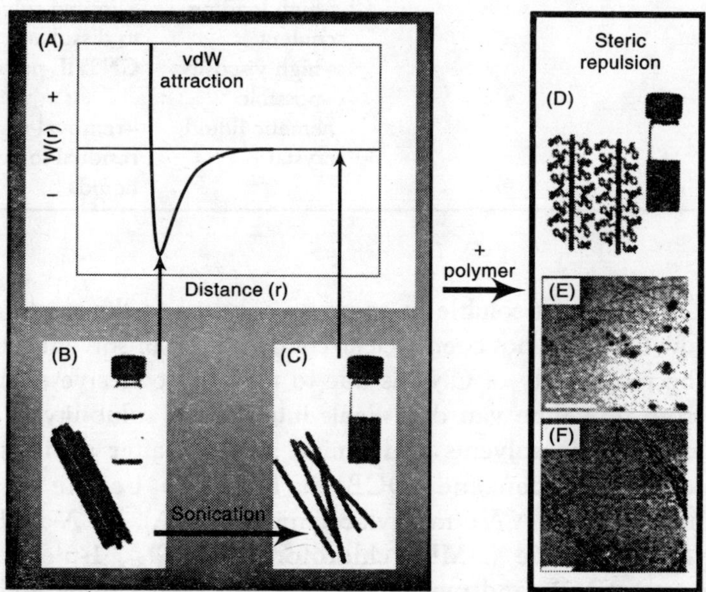

Fig. 8.76 *Schematic representation of the concept for composite processing using the solution property of SWNTS. (A) Total interaction energy W (r) versus the separation distance r, for two nanotubes. The attractive part of the potential leads to SWNT agglomeration in bundles, rendering the materials indispersible (A, B). While sonication leads to temporary exfoliation (C), adsorption of block copolymers in a selective solvent stabilizes the exfoliated tubes (D) and prevents re-aggregation. (E) cryo-TEM image (scale bar = 100 nm) and (F) HRTEM image (scale bar = 10 nm) show the existence of individual tubes*

SWNTs dispersion methods and applicable polymer systems are summarized in Table 8.8 below.

Table 8.8 Summary of SWNT dispersion in various systems.

Case	*Solution stability*	*Solubility*	*Advantages*	*Disadvantages*	*Applications*
Nitric acid treated SWNTs/ Organic solvent	poor to moderate (days ~ weeks)	10 ~ 30 mg L^{-1}	–broader solvent selection –nanotubes are functionalized	–electronic structure is compromised –removal of residual acid	PP, PE, Phenolic resin Epoxy, Nylon 6
Pristine SWNTs/ Surfactant	most stable (~ months)	100 ~ 1000 mg L^{-1}	–highest solubility –most stable	–removal of surfactant –applies only to aqueous system	PVA, bio polymer
Pristine SWNTs/ Organic solvent	moderate (~weeks)	5 ~ 20 mg L^{-1}	–simple preparation (sonication only) –no reaction	–low solubility –limitation of solvents	PAN PMMA, PS
Pristine SWNTs/ Ionic liquids	stable (high viscosity)	~10 mg^{-1} mL^{-1} (gel)	–stable bucky gels –high loading content –high viscosity –possible nematic liquid crystal	–high viscosity –limited solvent to dissolve CNT/IL paste –removal of residual ionic liquids	electrospinning, bucky films and fibers with polymer

(B) Solubility

Unlike fullerene, CNTs are not soluble in common solvents in the pristine state under normal conditions. Until now there has not been a clear-cut definition of solvent selectivity for fullerenes let alone CNTs. The insolubility of CNT is due to the high cohesive energy of more than 0.5 eV^{-1} nm, arising from tube-tube van der Waals interactions solubility of the pristine CNT in some aprotic polar aromatic solvents and amide solvents after sonication is there. Among all the solvents, *o*-dichlorobenzene (DCB) is found to be the best. In addition, *N*, *N*-dimethylformamide (DMF), *N*,*N*-dimethylacetamide (DMA), *N*, *N*-diethylacetamide (DEA), and N,N-dimethylpropanamide (DMP), chloroform ($CHCl_3$), 1-methylnapthalene (MNP), *N*-methylpyrrolidinone (NMP), and pyridine are found to have some ability to dissolve CNT. Table 8.8 compiles a list of compatible organic solvents and their dissolution limit (D_L).

Solubility is also found to be dependent on the source of CNTs; For example, as-produced *HiPCO-SWNTs* are observed to be stable in DCB to a greater extent than arc-and laser-produced SWNTs their differences in diameter distribution, purity, and related solvent effects influence the solubility. It is still difficult to understand the main driving force for the solubility of CNTs. Recently, two research groups thoroughly studied the interaction of SWNTs with DCB. One group confirmed that during ultrasonication DCB undergoes decomposition and polymerization, which made SWNTs stable in the solution.

TABLE 8.8 Solubility of SWNTs in different solvents at room temperature

Name	*Structural formula*	*Molecular mass* (gmol^{-1})	*Dielectric constant*	*Solubility* (mgL^{-1})
1, 2-Dichlorobenzene	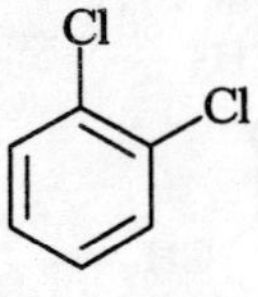	147	10.36	95
Chloroform	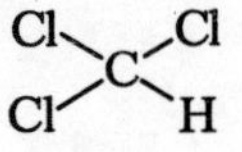 Cl_3C–H	119.38	4.89	31
1-Methylnaphthalene	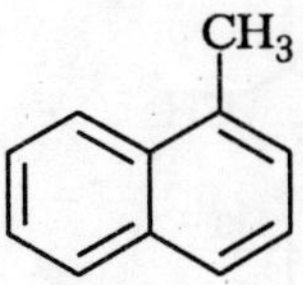 CH_3	142.20	2.92	25
1-Bromo-2-methylnaphthalene	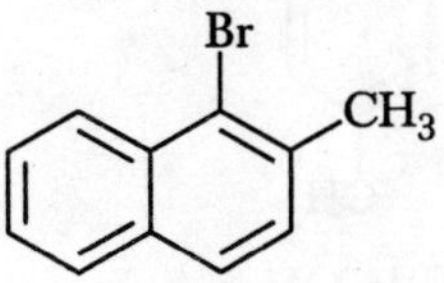 Br, CH_3	221.09		23
N-Methylpyrrolidione (NMP)	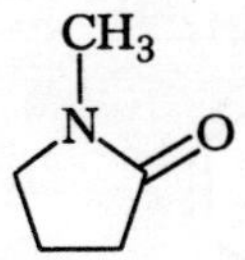 CH_3, N, O	99.13	32.58	10
Dimethylformamide (DMF)	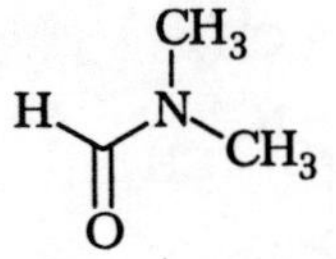 H, CH_3, N, CH_3, O	73.09	37.06	7.2
				1.56
Tetrahydrofuran (THF)	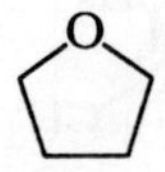 O	72.11	7.47	4.9
1, 2-Dimethylbenzene	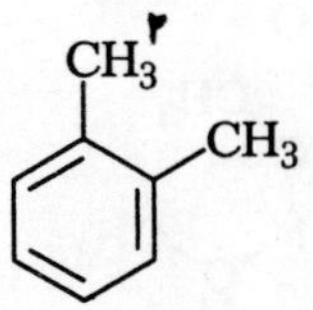 CH_3, CH_3	106.17		4.7
Pyridine	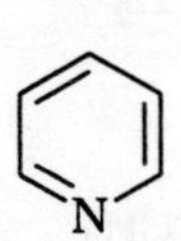 N	9.10	13.22	4.3
Carbon disulfide	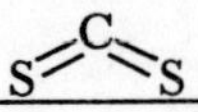 S=C=S	76.14	2.64	2.6

(*Contd.*)

(*Contd.*)

1, 3, 5-Trimethylbenzene	(structure: H_3C, CH_3, CH_3 on benzene ring)	120.19	2.40	2.3
Acetone	$H_3C-C(=O)-CH_3$	58.08	21.36	*a*
1, 3-Dimethylbenzene	(structure: CH_3, CH_3 on benzene ring)	106.17		*a*
1, 4-Dimethylbenzene	(structure: CH_3, CH_3 on benzene ring)	106.17	2.27	*a*
Ethanol	C_2H_5OH	46.07	24.55	*a*
Toulene	(structure: CH_3 on benzene ring)	92.14	2.43	*a*
N, N-dimethylacetamide (DMA),	$H_3C-C(=O)-N(CH_3)-CH_3$	87.12	38.30	6.25
N, N-dimethypropanamide (DMP),	$H_3C-CH_2-C(=O)-N(CH_3)-CH_3$	101.148	34.	3.13
N, N-diethylacetamide (DEA),	$H_3C-C(=O)-N(CH_2CH_3)-CH_2-CH_3$	115.71	31.3	3.13
Dimethyl sulfoxide	$H_3C-S(=O)-CH_3$	78.13	46.71	*a*
Acetonitrile	$H_3C-C\equiv N$	41.05	36.00	*a*

[a]The solubility in these solvents was <1 mg^{-1} L.

In addition, oxidized SWNTs were commonly found to be more soluble than pristine SWNTs. Theoretically showed in a theoretically study it is found that the interaction of DCB with SWNTs increases with an increase in defect sites, becaue the π-π interaction DCB has a relatively high dipole moment (2.14 D) and it interacts with the oxygen moieties present on the walls of SWNTs (induced during the oxidative treatment). This might be a favorable condition for stabilizing SWNTs in a solvent.

The effect of amide solvents on the dispersion of SWNTs and shows that it is a function of the polarity of solvents (as measured by their dielectric constants) and is likely to be an important feature because the resuting elctrostatic interactions can influence the van der Walls forces of the SWNTs with the solvents molecules. However, the electron donor character toward stabilizing the SWNTs in solution also plays a significant role. In comparison to DMF, the other alkyl amide solvents (DMA, DEA and DMP) have even a greater electron-donating ability, leading to a higher dispersion capability than that observed in DMF. However, disperison of SWNTs using acetonitrile and DMSO are found to be unsuccessful, indicating that the dispersion capability is not merely a function of polarity. Thus in addition to the dipole-dipole effects, the interaction between the alkyl amide solvents and the SWNTs may result from an increased π-orbital overlapping or "stacking", which is greatest for optimized solvent geometries. It is observed that the solvents DCB and DMA have a stmilar dispersion limit.

This difference in two separate studies, above is due to the difference in purity, source of SWNTs, or sonication time. Apart from that, it is likely that the equivalent dispersion limit is highly influenced by both polarity and π-orbital interactions. In the case of DCB, π-orbital interactions with SWNTs are higher compared to DMA. However, the polarity of DMA is higher than that of DCB (dielectric constant: 10.36) which possibly resulted in a similar D_L for these solvents. Hence, in the case of alkyl amide solvents, the presence of a highly polar π-system in conjuction with appropriate bond lengths and bond angles results in the highest SWNT dispersion limit.

REACTIONS OF CARBON NANOTUBES

A single SWNT is another carbon allotrope and is considered to be analogus with fullerene due to the chemical inertnes of its quasi-1D cylindrical aromatic system. However, it is well known that the curvature-induced pyramidalization and misalignment of the π-orbitals of the carbon atoms induce a local strain, and, hence CNTs are expected to be more reactive than a flat grephene sheet (Fig. 8.77). CNTs can have two types of reaction sites: end-caps and sidewalls. The end-caps resemble a hemispherical fullerene and will always be quite reactive, irrespective of the diameter of the CNT. Until now, there is no selectivity for performing reaction only at the end-caps or only on the side walls.

Nevertheless, significant progress has been made in the area of derivation of functiolaization of SWNTs using various chemistries: Despite having the same chemical composition as that of fullerenes, they have very poor solubility the CNT end-caps, defect sites, and the interior part are the major sites for chemical derivation, are accomplished by different methods such as direct covalent functionalization, indirect covalent functionalization, and non-covalent functionalization.

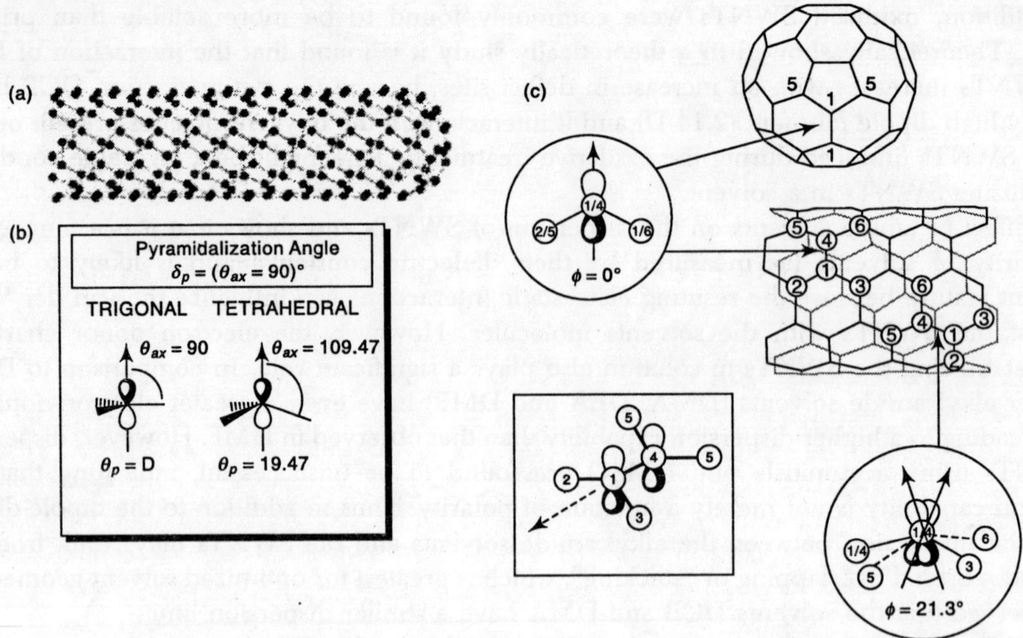

Fig. 8.77 *Diagrams of (a) metallic (5, 5) SWNT, (b) pyramidalization angle (Θ_P), and (c) the π-orbital misalignment angles (ϕ) along the C1-C4 in the (5, 5) SWNT and its capping fullerne, C_{60}*

Covalent functionalization reactions can be categorized into two subgroups: reactive *via* the direct attachment of functinal groups to the surface of CNT (direct method) and reactions *via* oxidized CNT products (indirect method).

(A) Direct Methods

Halogenation

Among the direct methods, halogeneration is the first type of reaction, example is BIRCH REDUCTION using dichloro carbene (Fig. 8.78) This carbene is a strong electrophile and can easily add to deactivated double bonds.

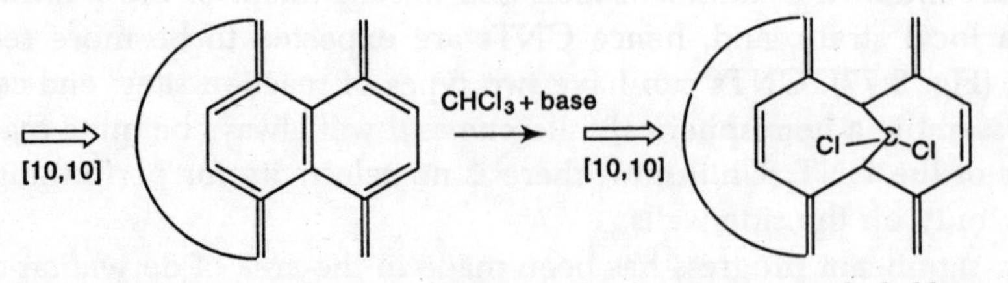

Fig. 8.78 *Halogenation of 10, 10 SWNT by Birch reduciton with dichlorocarbene*

By another method fluorine-functionalized SWNTs are obtained by heating SWNTs in the presence of elemental fluorine (Fig. 8.79). Due to the solubility of these functionalized SWNTs in variouis alcohols, it is possible to carry out further solution chemistry. However, this product

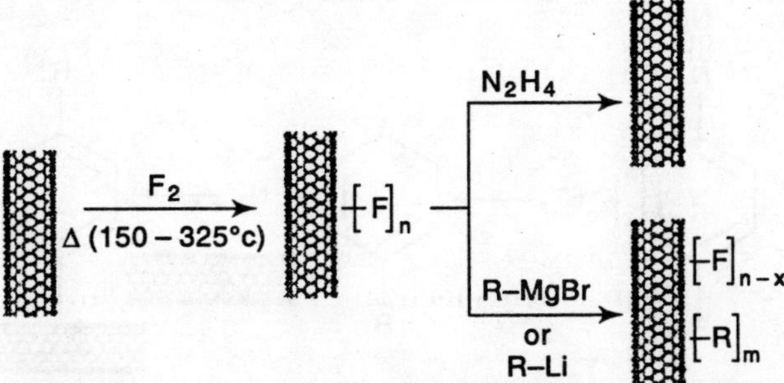

Fig. 8.79 *Halogenation of SWNT by fluorine and further defluorination by treatment with hydrazine or reaction wih strong nucleophiles, such as alkyllithium reagents and metal alkoxides*

became an insulator after the reaction. These functionalization nanotubes were further defluorinated by treatment with hydrazine or allowed to react with strong nucleophiles such as alkyllithium reagents and metal alkoxides.

An electrophilic addition to SWNTs yielding a chlorine-functionalized product in aluminum trichloride ($AlCl_3$), a molecule of chloroform is found to be added to sidewalls of CNT. (Fig. 8.80)

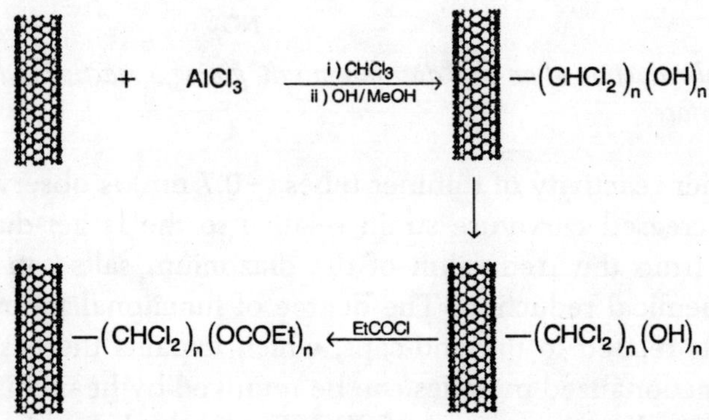

Fig. 8.80 *Halogenation by an electrophilic addition to SWNT*

Hydrogenation

Hydrogenation of CNTs is done by reduction in ammonia. The reactionis performed using the metal reduction method with Li and methano in liquid ammonia. The evidence of the C–H bond formation is confirmed by observing the formation of methane duringn dehydrogenation. The result is further supported by the observation of corrugated and disordered structures on the functionaized nanotube product from the TEM image. Due to th eformation of C–H bonds, the neighbuoring C–C bods became elongated due to the sp^3 character of the reacted C atoms, which resulted in a disorder of the surface of nanotubes.

Arylation

The arylation of SWNTs is done (Fig. 8.81) by electrochemical reduction using a variety of

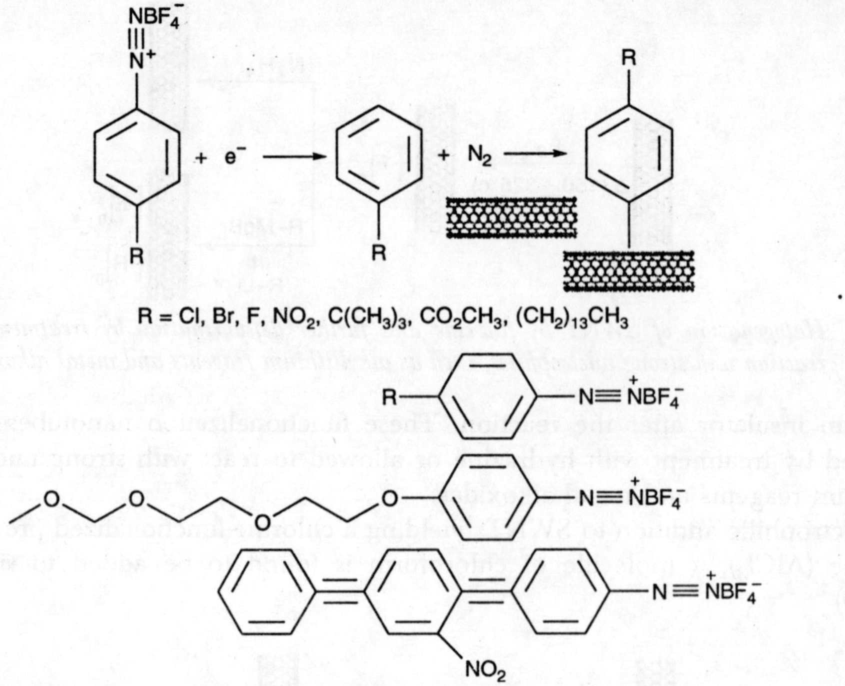

R = Cl, Br, F, NO_2, $C(CH_3)_3$, CO_2CH_3, $(CH_2)_{13}CH_3$

Fig. 8.81 *Electrochemical reduction of an aryl diazonium salt, giving a reactive radical that covalently attaches to a carbon surface*

diazonium salts. A higher reactivity of thininer tubes (~0.7 nm) is observed during the reaction, which is due to the increased curvature strain relative to the larger-diameter nanotubes. The aryl radical generated frmo the reduction of the diazonium salts can react successfully with nanotubes by electrochemical reduction. The degree of funcitonalization is found to be higher and exceeded the limits related to the end-caps, which expains the easy reactivity of the salts with the CNTs. The functionalized moieties can be removed by heaing in an organ atmosphere.

Thermal technique is a direct treatment of SWNTs with aryl diazoium salts. This reaction is initiaed via the injection of an electron from a SWNT into the aryldiazonium salt, formed a reactive aryl radical that then reacted with the SWNT to form a new SWNT-arene bond Fig. 8.82. It is observed that both the direct treatment of SWNTs with aryl diazonium terafluoroborate salts in solution and the *in situ* generation of the diazonium ion with an alkyl nitrite resulted in the desired product. This *in situ* generation of the diazonium species could easily avoid teh necessity of a multi-step procedure, including isolating and storing light-sensitive aryl diazonium salts.

The most important part of the arylation *via* diazonium chemistry is that it reacts selectivity with *met*-SWNTs, leaving behind the *sem*-SWNTs.

Thus, there is the possibility of the separation of CNTs with respect to their chirality. For covalent functionalization of SWNTs in fuming sulfuric acid (Fig. 8.83), the product is debundled as water-soluble SWNT-arylsulfonic acid. In super acids, SWNTs get surrounded by a double layer of protons and counterions, thus the tube is unlikely to donate an electron to an acceptor

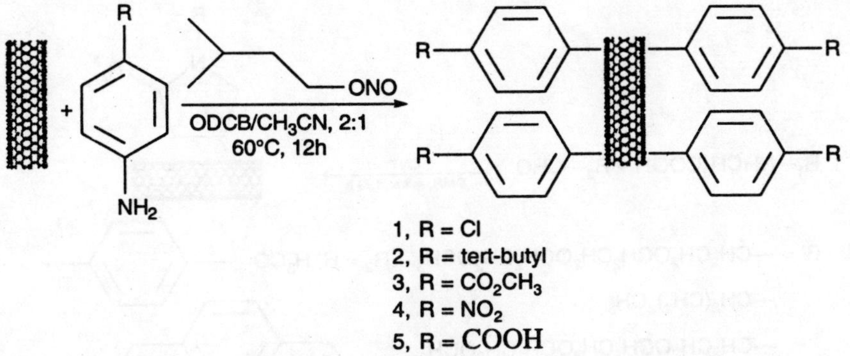

1, R = Cl
2, R = tert-butyl
3, R = CO_2CH_3
4, R = NO_2
5, R = COOH

Fig. 8.82 *Reaction sequence for derivatization of SWNTs by an in-situ generation of the diazonium species and the functionalized phenyl moieties employed*

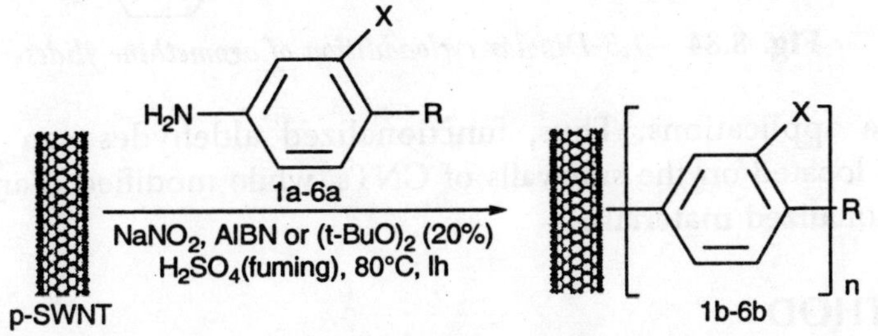

Fig. 8.83 *Functionalization of SWNTs dispersed in oleum by arylation*

such as a diazonium salt. This is overcome by adding a radical initiator, e.g., azobisisobutyronitrile (AIBN). This leads to the dispersion of individual tubes in a varierty of solvents including water, DMF, and ethanol. This process does not require any conventional techniques for individualization such as surfactant or polymer pre-wrapping, centrifugation, or sonication. Hence, this method is very helpful for the use of SWNTs as promising building blocks for high performance composite materials.

Cycloaddition

SWNTs are functionalized by the 1.3-bipolar cycloaddition of azomethine ylides (Fig. 8.84). The decarboxylation of immonium salts, derived from the condensation of an α-amino acid and an aldehyde, results in remarkably bundled CNTs that are soluble in organic solvents ($CHCl_3$, CH_3Cl_2, acetone, methanol, and ethaol). Furthermore, the technique is reported to be successfully with seveal types of CNT such as oxidised, non-osidized, and from different sources.

A similar technique is also applied to obtain water-soluble CNT derivatives by reacting N-substituteed α-amino acids with a tert-butoxycarbonyl (Boc) protected amino gruop and paraformaldehyde. Ammonium-modified CNTs, can be derivatized with N-protected amino acids, which is the first step toward the synthesis of peptide-basd CNT materials.

The main adantage of this reaciton is the easy attachment of pyrrolidine rings subsituted with chemical function to be sidewalls of the NT which allows the construction of novel

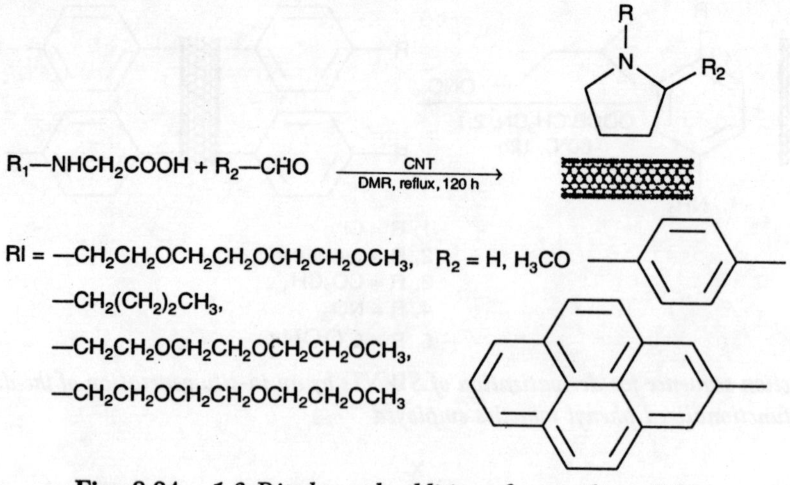

R_1—$NHCH_2COOH$ + R_2—CHO $\xrightarrow[\text{DMR, reflux, 120 h}]{\text{CNT}}$

RI = —$CH_2CH_2OCH_2CH_2OCH_2CH_2OCH_3$, R_2 = H, H_3CO
—$CH_2(CH_2)_2CH_3$,
—$CH_2CH_2OCH_2CH_2OCH_2CH_2OCH_3$,
—$CH_2CH_2OCH_2CH_2OCH_2CH_2OCH_3$

Fig. 8.84 *1,3-Dipolar cycloaddition of azomethine ylides*

materials for divere applications. Thus, functionalized aldehydes can yield 2-substituted pyrroliding moieties located on the sidewalls of CNTs, while modified α-amino acids can also lead to divere functionalized materials.

B. INDIRECT METHOD

Amidation and Acylation

The shortening of the tubes with different functionally at their ends is of great interest due to the easy manipulation in a variety of applications. The introduction of carboxylic acid functionalities at the tube ends during the oxidative purification process provides a good basis for amidation and acylation of these shortented tubes. In order to design CNT-based devices, the shortened amidated or acyltated tubes are assembled using the self-assembled monolayer (SAM) technique on different surfaces, including gold, silver, highly-oriented pyrolitic-graphic (HOPG), and silicon. The development of the surface chemistry has stimulated many studies on the further derivatization of oxidized CNTs with different moieties. They can be roughly divided into two groups.

The carbodimide coupling route (Fig. 8.85). This involves two stages: the activation by the carbodimide reagent to form an o-acyl isourea intermediate and the nucleophilic displacement of the intermediate by the amine to form the final amide bond. By using this coupling route, one can obtain SAMs of shortened CNTs on a cysteamine-assembled surface of gold. The other way is *via* the acid chloride route, where carboxylic groups are converted to acid chloride by thionyl chloride (or phophorus pentachloride, oxalyl chloride) in an aprotic solvent. This can be easily reacted with an amine to yield the amide. The attachement of gold nanoparticles onto oxidized shortened SWNTs through 11-aminoundecane-1-thiol and the use of the acid groups for attaching long alkyl chains to SWNTs *via* an amide linkage has been demonstrated. Also, the. Functionalized SWNTs can be purified using high performance liquid chromatography

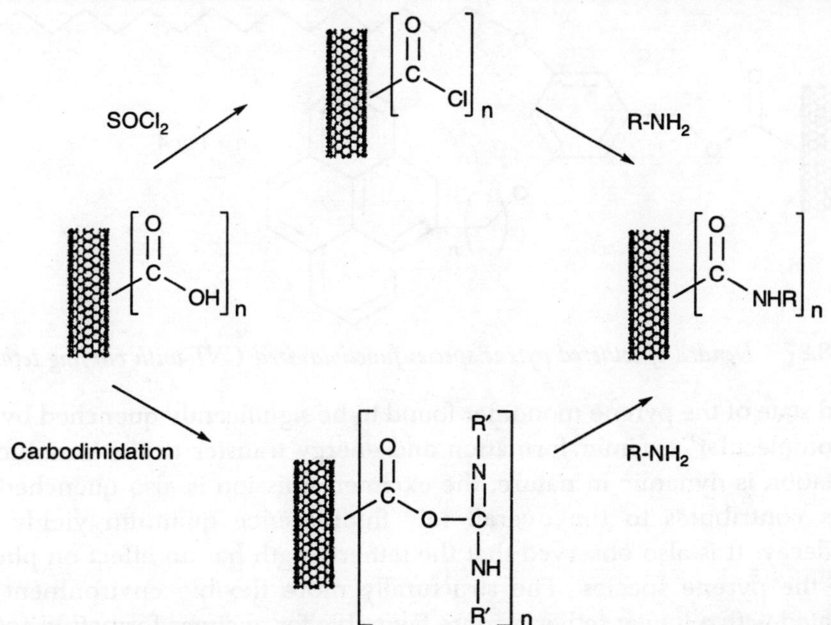

Fig. 8.85 *Amidation of carboxylic acid-functionalized CNT b ytwo different routes*

(HPLC). Similarly, the esterification of carboxylic acid with alcohol containing polymers (Fig. 8.86), has resulted in soluble products with CNTs of any length. An additional advantage of these ester linkages is that they can be easily defunctionalized *via* acid- or base-catalyzed hydrolysis, allowing the recovery of the CNTs.

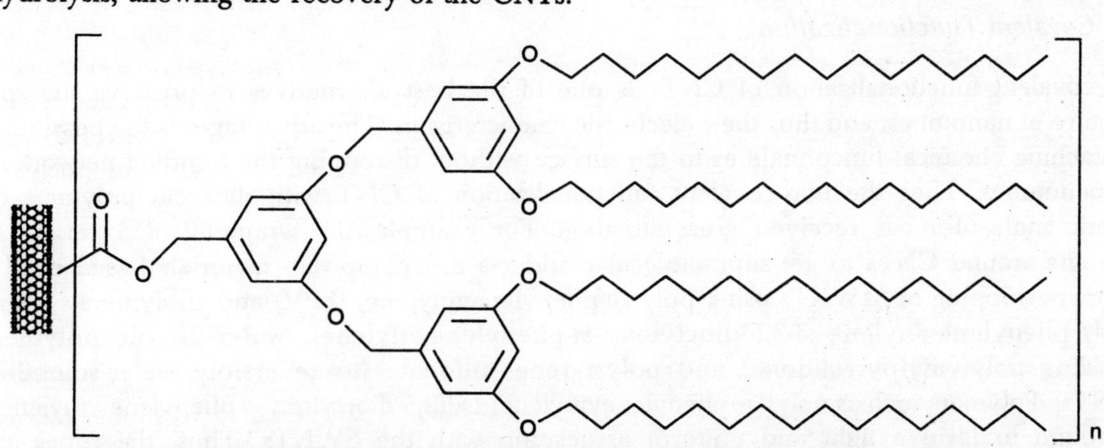

Fig. 8.86 *Esterification of the carboxylic acid groups with hydroxyl-containing polymers*

For the development of nanoscale photo-active materials, functionalized CNT with dendra of tethered pyrene species of different tether lengths have been reported Fig. 8.87.

These functionalized products are found to be soluble in common organic solvents, which allowed solution-based characterization and the investigationof the photophysical properties of the tethered pyrene moieties in solution.

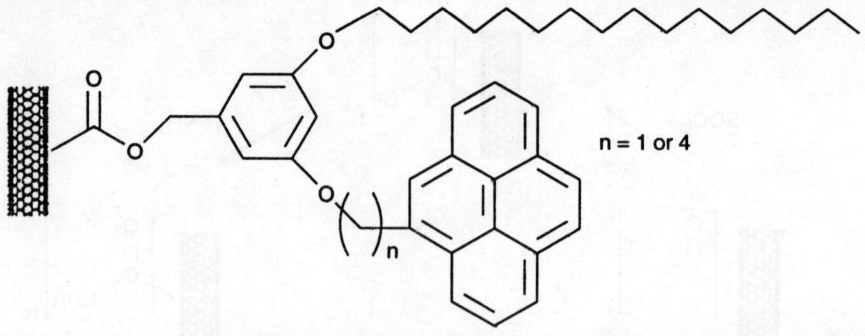

Fig. 8.87 *Dendra of tethered pyrene species functiolanized CNT with varying tether length*

The excited state of the pyrene monomer found to be significantly quenched by two competing process: 'intromolecular' excimer formation and energy transfer to the nanotube. Because the excimer formation is dynamic in nature, the excimer emission is also quenched by the energy transfer. This contributes to the overall low fluorescence quantum yields and the rapid fluorescence decay. It is also observed that the tether length has an effect on photoexcited-state properties of the pyrene species. The structurally more flexible environment of the pyrene moiety associated with a longer tether is more favorable for excimer formation and less favorable for excited-state energy transfer.

In soluble SWNTs functionalized with the highly fluorescent chromophore 1,8-naphthalimide by covalent modification (Fig. 8.88) significant fluorescence quenching is observed for the naphthalimide chromophore bonded to the SWNT backbone, which indicates the existence of electronic interactions between the two chromophores in their ground and/or excited states.

Non-Covalent Functionalization

Non-covalent functionalization of CNTs is one of the best alternatives to preserve the sp^2-structure of nanotubes, and thus their electronic characteristics. The advantages is the possibility of attaching chemical functionalities to the surface without disrouping the bonding network of the nanotubes. Thus the non-covalent functionalization of CNT with different polymers or organic molecules has received great attention. For example, the wrapping of a variety of polymers around CNTs to get supramolecular adducts e.g. composite materials based on the polymer wrapping of MWNTs using poly (*p* phenylenevinylene) (PPV) and cpolymers such as poly(*p*-phenylenevinylene-*co*-2,5-dioctyloxy-*m*-phenylenevinylene), water-soluble polymers including poly(vinylpyrrolidone) and polystyrene sulfonate (to reversibly water-solubilize SWNTs). Polymers such as poly{(*m*-phenylenevinylene)-*co*-[(2,5-dioctyloxy-*p*-phenylene)vinylene]} are found to form a tight and uniform association with the SWNTs. Thus, the ropes are effectively broken up into single stands resulting in their dissolution in chloroform. Optical imaging and opto-electronic measurement on single, isolated PmPV-wrapped SWNT ropes revelated that polymer-wrapped SWNTs do not further aggregate into thicker ropes, and the polymer and nanotube components of a wrapped structure are in intimate electrical contact with each other.

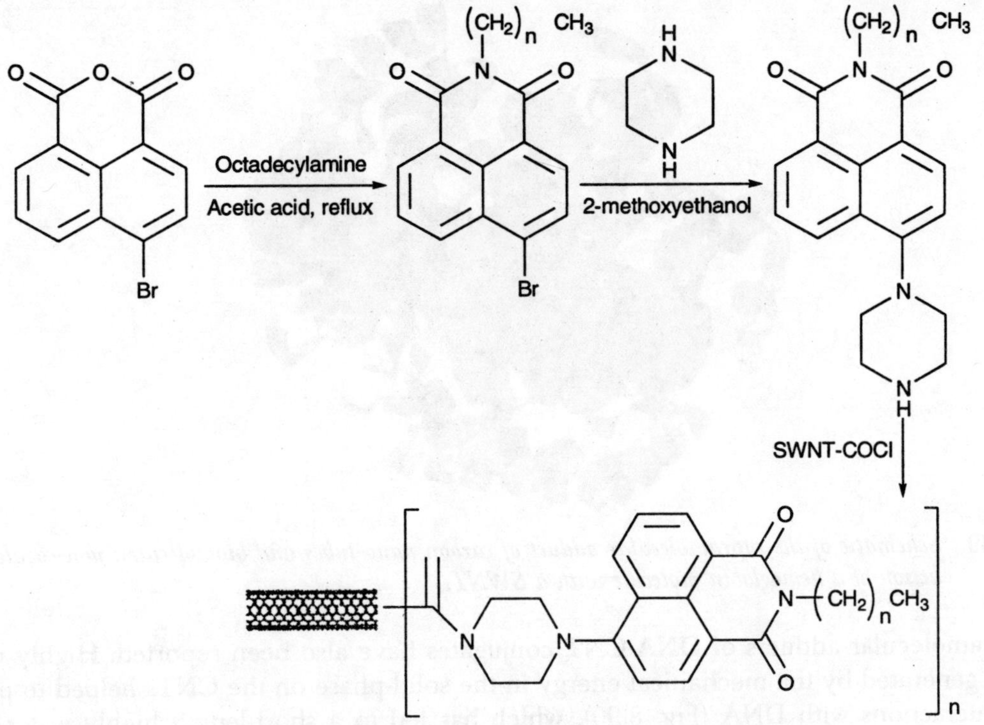

Fig. 8.88 *Amidation of oxidately etched SWNTs wiht napthalimide cromophores*

CNTs wrapped with a polymer, which are specifically linked to the accessible C-5 position on the *meta*-disubstituted phenylene ring, such as poly (5-alkoxy-*m*-phenylenevinylene)-*co*-[(2,5-dioctyloxy-*p*-phenylene)vinylene]} (PA*m*PV) derivatives are prepared. These polymers change the physical properties of nanotubes such as their solubility and their ability to be further functionalized.

Recently by observing a selectivity toward *sem*-SWNTs, by taking advantages of the non-convalent interactions (π-π interactions) of porphyrin derivatives with SWNTs has resulted in significantly enriched *sem*-SWNTs in the solubilized sample and in predominantly met-SWNTs in the residual solid sample when the reaction is done in a heterogeneous mixture of reactants in THF. This selectivity is believed to be based on the difference in the surface properties of *sem*-SWNTs and *mel*-SWNTs. It is presumed that *sem*-SWNTs are more like conjugated macromolecules, whereas the surface properties are conceptually similar to those found in redical ion pairs and thus amenable to interactions with the free-base porphyrin molecules.

The potential of CNTs for the development of novel bioelectronic devices is achieved and biomolecules immobilized which make them processable, and to simulataneously utilize their unique properties to design novel artificial systems. In this respect, a variety of biomolecules hold great promise for such kind of development. A schematic of the non-covalent functionalization of SWNT with hemoglobin is shown in Figure 8.89. DNA, (which possess unique properties for both biological and non-biological applications), have been, is covalently bonded with SWNT; sonication-assisted DNA-SWNT conjugate, field effect transistor, and DNA-assisted purification have been reported.

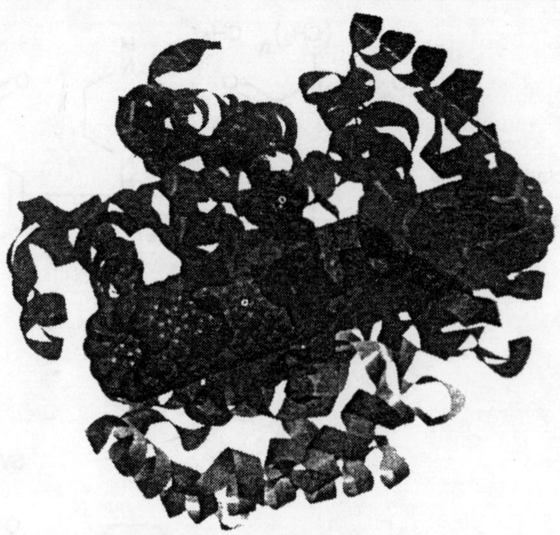

Fig. 8.89 *Schematic of the supramolecular adduct of carbon nano-tubes and biomolecules: non-covalent interaction of a hemoglobin molecule with a SWNT.*

Supramolecular adducrs of DNA-CNT conjugates have also been reported. Highly reactive centers generated by the mechanical energy in the solid-phase on the CNTs helped to promote good interactions with DNA (Fig. 8.90), which has led to a short-length highly water-soluble CNT product in a single step. This technique has also been successfully used to make fullerene derivatives as well as alcohol-functionalized CNTs. Interestingly, this soluble CNT product looks fully wrapped with DNA.

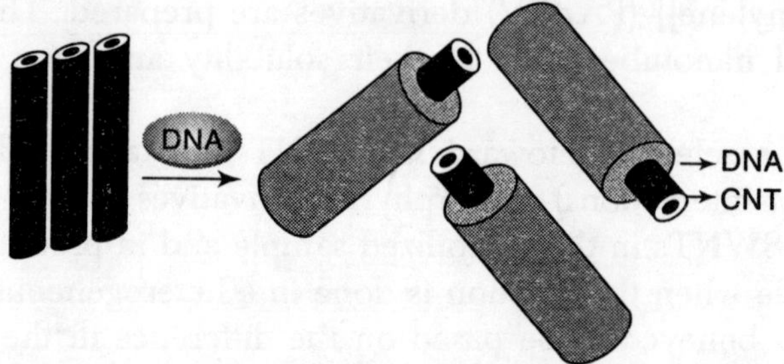

Fig. 8.90 *Schematic of DNA-wrapping of carbon nanotubes.*

Recently, a water-soluble product is utilized to design an electrode for bioelectrochemical sensors by taking advantage of the non-covalent interactions of chitosan with CNTs. Similarly, cyclodextrim, among several different macrocycles, has attracted significant interest both due to its supramolecular complex formation as well as to its ability to enhance the solubility of CNTs. Furthermore, η-cyclodextrin is found to be capable of forming pseudo-rotaxane complexes with SWNTs.

Biopolymers such as Schizophyllan (s-SPG) and curdian (β-1,3-glucan) are capable of wrapping SWNTs. A "periodical" helical structure was created, reflecting the helical nature of the SPG main-chain on the SWNT surface. Also, a simple and efficient process for the solubilization of SWNTs with amylose in aqueous DMSO using sonication is made. The former step separates the SWNT bundles and the latter step provides a maximum interaction of SWNTs with amylose, leading to an immediate and complete solubilization.

APPLICATIONS OF CNTs

Carbon nanotubes have many potential applications, depending on the following properties:

1. Structural
2. Electromagnetic
3. Chemical
4. Mechanical
5. In electrical circuits e.g:
   5.1 Metallic and semiconducting nanotubes
   5.2 Carbon Nanotube Interconnects
   5.3 Carbon Nanotube Transistors
   5.4 Challenges in Electronic Design and Design Automation and
6. Fiber and films

Applications of Carbon Nanotubes are in almost all the scientific areas, such as aerospace science, bio-engineering, environmental energy, materials industry, medical and medicine science, electronic computer, security and safety and science education.

Being superior to any other existing material in mechanical, electrical, and hydrogen storage characteristics, carbon nanotubes thus find main applications in.

1. Molecular Electronics e.g:
   (a) Semi conductors
   (b) Transistors
   (c) Logic Gates
   (d) Nano electronics, displays, devices.

2. Fuel Energy
   (a) Hydrogen storage
   (b) Lithium battery
   (c) Super capacitors

Application is also found in super computers, space elevators, launch vehicles etc. Other applications include, quantum, wires, nanofibres/ropes/ test tubes; high functional composites, nanowire, nanocapsule, nanotweezer, AFM/STM tip, single electronic device, gas sensor, probes (STM and AFM) medical/engineering microscopic parts, etc.

Single-walled carbon nanotubes (SWNTs) show promise for applications ranging from ultra-small electronic and sensing devices to multifunctional materials. To date, a number of SWNT-based proof-of-concept devices such as field effect transistors field emission displays and chemical sensors have been fabricated, and the integration of the nanotube materials in all cases relies on one's ability to control the placement, orientation, and shape of the nanotube components within the context of the device on the micrometer-to nanometer-length scale. Such positional control over large areas is extremely challenging and currently quite limited. Depending on the intended application, one wants to be able to pattern SWNTs as individual tubes, small bundles or thin films.

Following are some examples of future devices and their salient attributes.

Structural

- **Clothes:** waterproof tear-resistant cloth fibers
- **Combat jackets:** combat jackets are made by using carbon nanotubes (as ultrastrong fibers) monitors the condition of the wearer.
- **Concrete:** In concrete, they increase the tensile strength, and halt crack propagation.
- **Polyethylene:** Adding CNTs to polyethylene increases the polymer's elastic modulus by 30%.
- **Sports equipment:** Adding CNTs gives stronger and lighter tennis rackets, bike parts, golf balls, golf clubs, golf shaft and baseball bats.
- **Space elevator:** This will be possible only if tensile strengths of more than about 70 GPa can be achieved. Monoatomic oxygen in the Earth's upper atmosphere would erode carbon nanotubes at some altitudes, so a space elevator constructed of nanotubes would need to be protected (by some kind of coating). Carbon nanotubes in other applications would generally not need such surface protection.
- **Ultrahigh-speed flywheels:** The high strength/weight ratio enables very high speeds to be achieved.
- **Bridges:** In suspension bridges (CNTs will be able to replace steel), or bridges built as a "horizontal space elevator".

Electromagnetic

- **Artificial muscles**
- **Buckypaper:** A thin sheet made from nanotubes that are 250 times stronger than steel and 10 times lighter that could be used a heat sink for chipboards, a backlight for LCD screens or as a faraday cage to protect electrical devices/aeroplanes.

- **Chemical nanowires:** Carbon nanotubes additionally can also be used to produce nanowires of other chemicals, such as gold or zinc oxide. These nanowires in turn can be used to cast nanotubes of other chemicals, such as gallium nitride. These can have very different properties from CNTs for example, gallium nitride nanotubes are hydrophilic, while CNTs are hydrophobic, giving them possible uses in organic chemistry that CNTs could not be used for.
- **Computer circuits:** A nanotube formed by joining nanotubes of two different diameters end to end can act as a diode, suggesting the possibility of constructing electronic computer circuits entirely out of nanotubes. Because of their good thermal properties, CNTs can also be used to dissipate heat from tiny computer chips. The longest electricity conducting circuit is a fraction of an inch long.
- **Conductive films:** Drawing transparent high strength swathes of SWNT is a functional production technique. Transparent, electrically conductive films of carbon nanotubes are divided which replace indium tin oxide (ITO) in LCDs, touch screens, and photovoltaic devices. Carbon nanotube films are substantially more mechanically robust than ITO films, making them ideal for high reliability touch screens and flexible displays. Nanotube films show promise for use in displays for computers, cell phones, PDAs, and ATMs.
- **Electric motor brushes:** Conductive carbon nanotubes have been used for several years in brushes for commercial electric motors. They replace traditional carbon black, which is mostly impure spherical carbon fullerenes. The nanotubes improve electrical and thermal conductivity because they stretch through the plastic matrix of the brush. This permits the carbon filler to be reduced from 30% down to 3.6%, so that more matrix is present in the brush. Nanotube composite motor brushes are better-lubricated (from the matrix), cooler-running (both from better lubrication and superior thermal conductivity), less brittle (more matrix, and fiber reinforcement), stronger and more accurately moldable (more matrix). Since brushes are a critical failure point in electric motors, and also don't need much material, they became economical before almost any other application.
- **Light bulb filament:** alternative to tungsten filaments in incandescent lamps.
- **Magnets:** MWNTs coated with magnetite
- **Optical ignition:** A layer of 29% iron enriched SWNT is placed on top of a layer of explosive material such as PETN, and can be ignited with a regular camera flash.
- **Solar cells:** Carbon nanotube diodes have a photovoltaic effect. They replace ITO in some solar cells to act as a transparent conductive film in solar cells to allow light to pass to the active layers and generate photocurrent.
- **Superconductor:** Nanotubes have been shown to be superconducting at low temperatures.
- **Ultracapacitors:** Use of nanotubes to charge plates of capacitors in order to dramatically increase the surface area and therefore energy storage ability.
- **Displays:** One use for nanotubes that has already been developed is as extremely fine electron guns, which could be used as miniature cathode ray tubes in thin high-brightness low-energy low-weight displays. This type of display would consist of a group of many tiny CRTs, each providing the electrons to hit the phosphor of one pixel, instead of having one giant CRT whose electrons are aimed using electric and magnetic fields. These displays are known as field emission displays (FEDs).

Chemical

- Air pollution filter: *Future* applications of nanotube membranes include filtering carbon dioxide from power plant emissions.
- **Biotech container:** Nanotubes can be opened and filled with materials such as biological molecules, raising the possibility of applications in biotechnology.
- **Hydrogen storage:** Research is currently being undertaken into the potential use of carbon nanotubes for hydrogen storage. They have the potential to store between 4.2 and 65% hydrogen by weight. This is an important area of research, since if they can be mass produced economically there is potential to contain the same quantity of energy as a 50l gasoline tank in 13.2l of nanotubes.
- **Water filter:** Recently nanotube membranes have been developed for use in filtration. This technique can purportedly reduce desalination costs by 75%. The tubes are so thin that small particles (like water molecules) can pass through them, while larger particles (such as the chloride ions in salt) are blocked.

Mechanical

- **Oscillator:** Fastest known oscillators (> 50 GHz).
- **Liquid flow array:** Liquid flows up to five orders of magnitude faster than predicted through array.
- **Slick surface:** Slicker than Teflon and waterproof.

ELECTROSTATICS OF CARBON NANOTUBE DEVICES

With the scaling limit of conventional silicon transistor in sight, there is rapidly growing interest in nanowire transistors with one-dimensional channels, such as carbon nanotube transistors and silicon nanowire transistors. Due to the one-dimensional channel geometry, the electrostatics of nanowire devices can be quite different from bulk silicon devices. Previous studies of carbon nanotube p/n junctions and metal/semiconductor junctions demonstrated unique properties of nanotube junctions.

For example, the charge transfer into the nanowire channel from the metal contacts (or heavily doped semiconductor contacts) can be significant. By looking the dependence of the charge transfer on the metal/semiconductor Schottky barrier height, the insulator dielectric constant, and the metal contact geometry it is shown that if an intrinsic nanowire is attached to bulk metal contacts to two ends, large charge transfer is achieved (if the Schottky barrier is low and the insulator dielectric constant is high). If, however, the intrinsic nanowire is attached to one-dimensional metal contacts, the charge density on the nanowire depends critically on the electrostatic environment rather than the properties of the metal contacts. Reducing the gate oxide thickness and the contact size decreases the distance over which the source/drain field penetrates into the nanowire channel and can, therefore, help to suppress the short channel effects and improve the transistor performance.

The modeled coaxially gated carbon nanotube transistor is shown in Fig. 9.1(a). Although the calculations given below are for carbon nanotube transistors, the general conclusion should apply to other nanowire transistors with one-dimensional channels. The equilibrium band profile and charge density are obtained by solving the Poisson equation in cylindrical coordimnated self-consistently with the equilibrium carrier statistics of the carbon nanotube. The charge density per unit length on the nanotube, $Q_1(z)$, is calculated by integrating the "universal" nanotube density-of-states (DOS) $D(E)$, over all energies,

$$Q_L(z) = (-e).\int_{-\infty}^{+\infty} dE \cdot \ \text{sgn}(E)\ D(E) f(\text{sgn}(E))[E - \tilde{E}_F(z)]. \quad (1)$$

Where e is the electron charge, sgn(E) si the sign function, and $\tilde{E}_F(z) = E_F - E_m(z)$ is the Fermi energy level minus the middle gap energy of the nanotube, Em(z). Since the source/ drain

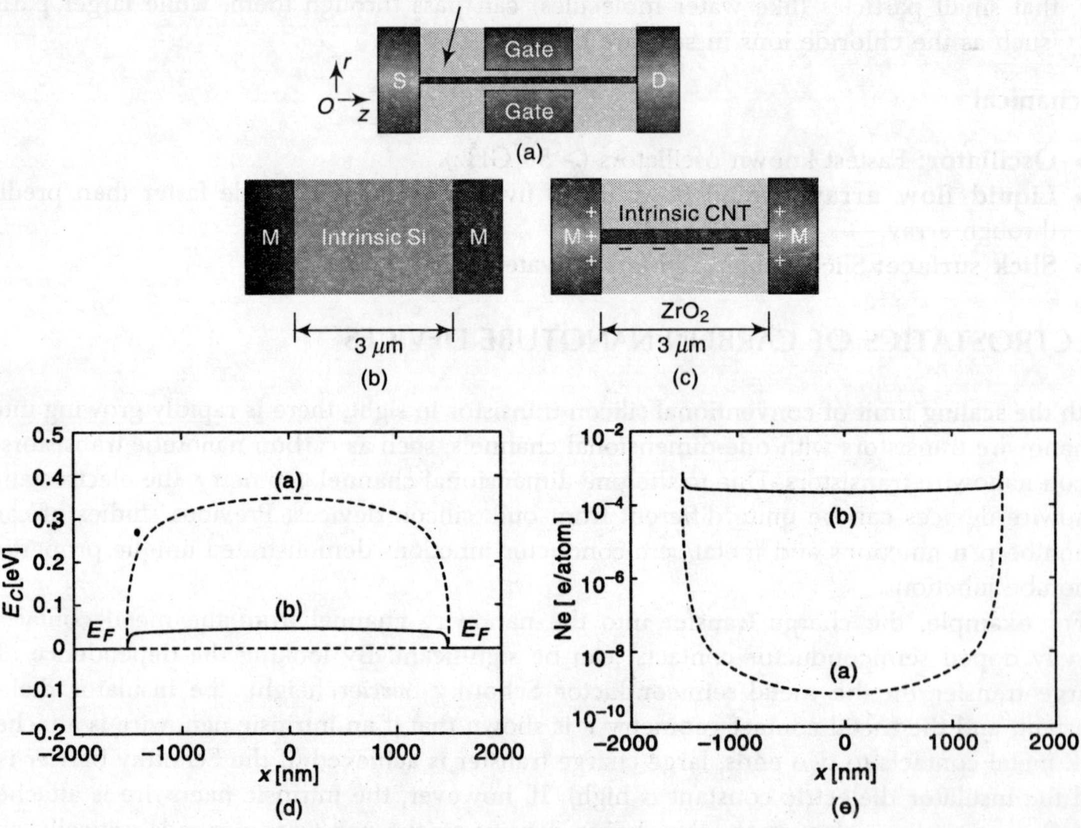

Fig. 9.1 *(a) The modeled, coaxially gated carbon nanpotube transistor. The intrinsic nanotube channel has a diameter of 1.4 nm and the gate work function of zero. The cylindrical coordinates of solving the Poisson equation is also shown. (b) a bulk Si structure where the cross-sectional area is assumed to be large a carbon nanotube channel between bulk metal electrodes. (c) The Schottky barrier heights for elctrons are zero. (d) The conduction band edge and (e) the electron density in the units of doping fraction. Results for the bulk Si structure are shown as dashed lines and for nanotube as solid lines.*

electrodes are grounded, the Fermi level is set to zero, $E_F = 0$. The nanotube middle gap energy is computed from the electrostatic potential at the nanotube shell, $E_m(z) = -eV(z, r = r_{cnt})$, where r_{cnt} is the nanotube radius. The electrostatic potential, V, satisfies the Poisson equation,

$$\nabla^2 V(z, r) = \frac{\rho}{\varepsilon} \tag{2}$$

where ρ is the charge density, ε the dielectric constant. The following boundary conditions are used,

$V = (E_g/2 - \phi_{bn})/e$ at the left metal contact,
$V = (E_g/2 - \phi_{bn})/e$ at the left metal contact, and
$V = V_G$ at the gate cylinder (the flat band voltage is assumed to be zero),

where E_g is the nanotube band gap, ϕ_{bn} is the Schottky barrier height for electrons between the source/drain and the nanotube, and V_G is the gate voltage.

Poisson equation is solved numerically by two methods, the finite difference method and the method of moments. In order to improve the convergence when interatively solving eqns. (1) and (2), the Netwton-Ralphson method is used. The results obtained by the finite difference method and by the method of moments agree.

By comparing the charge transfer from bulk contacts to the one-dimensional carbon nanotube to the charge transfer to a bulk silicon channel (1) an intrinbsic bulk Si channel sandwiched between two metal contacts as shown in Fig. 9.1(b) and intrinsic carbon nanotube channel between metal contacts as shown in Fig. 9.1(c). In both cases, the Schottky barrier heights between the metal contacts and the semiconductor channel are zero, which aligns the metal Fermi level of to the conduction band edge of the semiconductor. Electrons are transferred from metal contacts into the intrinsic channel due to the work function difference between the metal and the semiconductor. Fig. 9.1(d) plots the conduction bands, and Fig. 9.1(e) plots the charge densities in the unit of electron per atom for the bulk SI and nanotube channel. Compared to the bulk Si channel, the barrier in the nanotube is much lower, and the charge density is much higher. Although the nanotube is 3 μm long, the charge density at the center of the tube is still as high as 10^{-4} e/atom, about 5 orders of magnitude higher than that of the bulk Si in terms of electron fraction. As the result, the carbon nanotube channel is more conductive.

The charge transfer to the tube is significant because the charge on the doesn't effectively screen the potential produced by the bulk contacts. Compared to the bulk channel, the charge element on the nanotube only changes potential locally. For example, in the bulk channel, the charge element is a two-dimensional sheet charge, which produces a constant field. The charge dipole formed by charge sheet in bulk Si and metal contacts shifts the potential far away. In contrast, for the nanowire channel, the charge element is a point charge, which produces a potential decaying with distance ~1/r and has little effect far away (the potential of a point charge dipole decays even faster as $\sim/r^2$). As the result, for the one-dimensional channel, the potential produced by the bulk contacts is not screened by the charge on the nanotube near the metal/semiconductor interface. The bulk contacts tend to put the conduction band edge near the Fermi level over the whole 3 μm-long tube if the metal/CNT barrier height is zero.

Now, estimate the charge density in the channel because it provides a simple way to understand how the charge density of the tube varies with the contact and insulator properties. For the

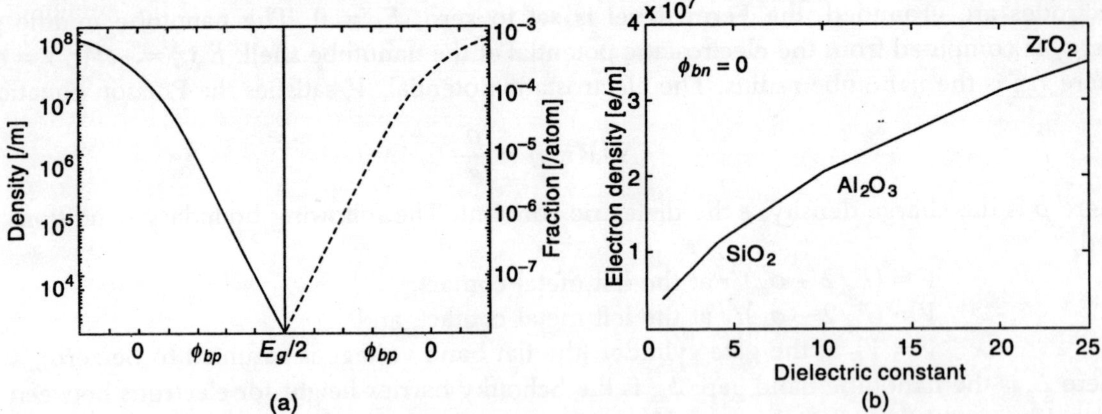

Fig. 9.2 *(a) The electron density (the dashed line) and hole density (the solid line) at the center of the 3 μm-long CNT (in Fig. 9.2(b) vs. the Schottky barrier height for electrons, θ_{bn}, and the for holes, θ_{bp}. The left axis shows the charge density ion the unit of number of electrons (holes) per unit length and the right axis shows the same quantity in the unit of charge fraction. (b) The electron density at the center of 3 μm-long tube (in Fig. 9.1(c) vs. the insulator dielectric constant. The Schottky barrier height for electrons, θ_{bn}, is zero.*

device structure shown in Fig. 9.1(c), if the metal contacts are grounded, and the metal/semiconductor work function differene is $U_0 = \phi_{CNT} - \phi_M$, where ϕ_{CNT} (ϕ_M) is the nanotube (metal) work function, the electron density is

$$n(z) = \bar{D}(U_0 - U(z)), \tag{3}$$

where $U(z)$ si the electron potential energy produced by charge in the channel, and $\bar{D}$ is the average density-of-states for the energy between the nanotube middle gap energy and the Fermi level. The charge element in the one-dimensional channel only shifts the potential locally. The potential, $U(x)$, to the electron density at the same position, $n(x)$ is related as:

$$U(x) = e^2 n(z)/C_{ins} \tag{4}$$

where C_{ins} is the electrostatic capacitance per unit length between the nanotube and the bulk contacts. The electron density due to the charge transfer from the bulk contacts are obtained from eqns. (3) and (4) as

$$n(x) = \frac{U_0/e^2}{1/C_{ins} + 1/\bar{C}_Q} \tag{5}$$

where the quantum, capacitance is defined as, $\bar{C}_Q = e^2\bar{D}$, which is proportional to the average DOS of the nanotubes. Equation (5) is also interpreted in a simple way. The bulk electrodes modulate the charge density of the nanotube through an insulator capacitor, C_{ins}, which is in series with the quantum capacitance of the nanotube.

Now examine how the charge transfer varies with the Schottky barrier height and the insulator dielectric constant. Fig. 9.2(a) which plots the charge density at the center of the tube as shown in Fig. 9.1(c) vs. the barrier height, shows that when the barrier height decreases, the charge density first increases. Fig. 9.2(b), which plots the charge density at the center of the tube vs. the insulator dielectric constant, shows the charge density increases as the dielectric constant increases. The dependence of the charge density on the barrier height and the dielectric constant can be easily understood based on eqn. (5). Lowering the barrier height increases the metal/CNT work function difference, U_0, and increasing the insulator dielectric constant increases C_{ins}, both of which increase the electron density, $n(x)$ (or hole density if the metal/semiconductor barrier height is lower for holes).

One consequence of the significant charge trandfer is that nanowire transistors with large gate underlap can still operate. Fig. 9.3(a) shows a coaxially gated CNTFET with a 500 nm gate underlap and the bulk electrodes. Fig. 9.3(b) plots the conduction band profile at $V_G = 0$ and 0.3 V. At the off state ($V_G = 0V$), a large barrier is created in the channel and the transistor is turned off. AT the on-state, (VG = 0.3V), the barrier under the gate is pushed down. Because the low dimensional charge on the ungated nanotube doesn't effectively screen the potential produced by the gate and S/D electrodes, the potential at the ungated region is close to the Laplace potential produced by the source and gate electrodes. The conduction band edge is approximately linear in the ungated region. If the Schottky barrier height between S/D and the channel is ~50meV, the barrier height at the ungated region at the on-state is low enough to deliver an on-current of ~1 μA. This mechanism provides a possible explanation for the operation of the *n*-type CNTFET with large, intrinsic gate underlaps with had a good on-off ratio.

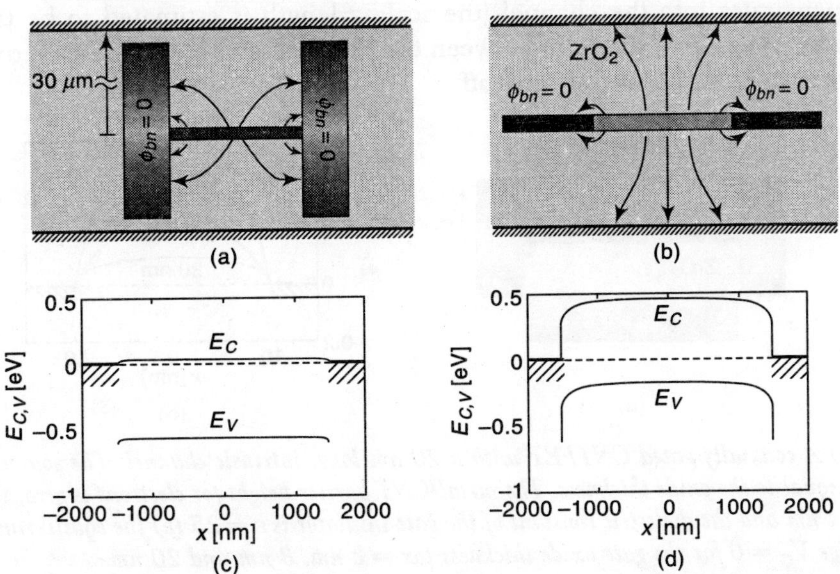

Fig. 9.3 *Contact geometry. A 3 μm-long CNT between (a) the bulk contacts and (b) the one-dimensional wire contacts. The tube diameter is 1.4 nm. and Schottky barrier heights for electrons are zero. A coaxial gate for away with a 30 μm radius is grounded. The workfunction of the gate metal equal to the semiconductor affinity plus the band gap, so that the gate tends to dope the CNT to p-type. (c) The band profile (a). (d) The band profile for (b).*

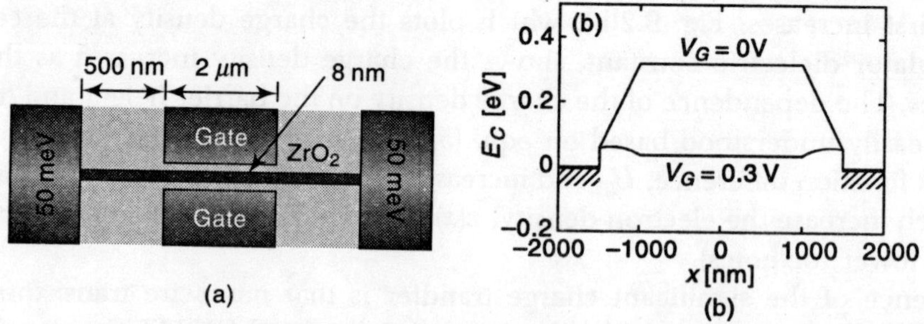

Fig. 9.4 *(a) A coaxially gates CNTFET with bulk electrodes (with a radius of 500 nm) and a large gate underlap. (b) The conduction band profile at $V_G = 0$ V and 0.3V. The metal/CNT barrier height for electrons is 50 meV, the ZrO_2 gate oxide thickness 8 nm, the tube diameter is 1.4 nm, the gate length is 2 μm, and the gate underlap is 500 nm.*

One concern about the nanowire transistors with low meta/CNT Schottky barriers is that due to the significant charge transfer, it might be difficult to turn off the rtransistor. To examine the concern, we simulated the coaxially gated CNTFET as shown in Fig. 9.5(a) with different gate oxide thickness. Fig. 9.5(b) which plots the equilibrium band profile, show that when the gate oxide thickness is the same as the channel length, the source/drain field penetrates into the channel the channel and the transistor cannot be turned off. When the gate oxide is thin, however, the gate still has very good control over the channel and the transistor is well tuned off. By solving the Poisson equation for the CNTFET in Fig. 9.5(a), the length by which the drain field penetrates into the channel (the scaling length is estimated to be the radius of the cylindrical gate, $\Lambda \sim R_G$. If the ratio between the channel length and the gate oxide thickness is large, the transistor can be well turned off.

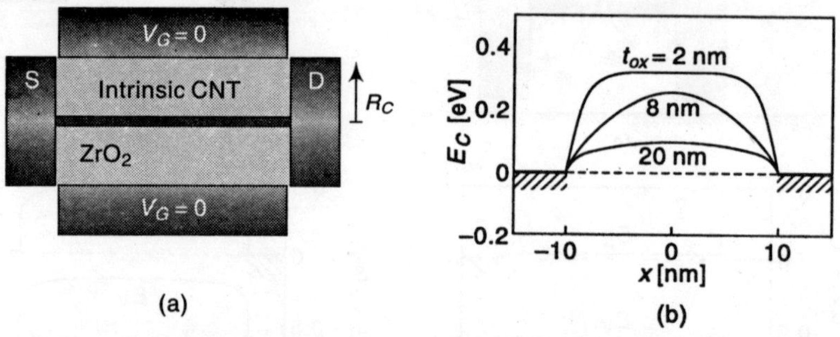

Fig. 9.5 *(a) A coaxially gated CNTFET with a 20 nm-long, intrinsic channel. The source/drain radius, R_C, is equal to the oxide thickness. The metal/CNT barrier height for electrons is zero, the tube diameter is 1.4 nm and the dielectric constant of the gate insulator is $\varepsilon = 25$ (b) the equilibrium conduction band edge $V_G = 0$ for the gate oxide thickness tox = 2 nm, 8 nm and 20 nm.*

Fig. 9.3 illustrates the important role of the contact geometry. (1) a CNT between grounded bulk contact is shown in Fig. 9.3(a). A CNT between grounded wire contacts is shown in Fig. 9.3(b). In both cases, the tube length is 3 μm and a grounded, coaxial gate cylinder is far away with a radius of 30 μm. The S/D contacts have zero Schottky barrier heights for electrons thus tend to dope the tube *n*-type, while the gate has a high work function and zero barrier height for holes thus tends to modulate the tube to *p*-type. For the bulk contact case, the whole tube is doped to *n*-type by bulk contacts and the charge density on the tube is independent of the voltage on the gate cylinder. In contrast, for the contracts, the tube is lightly modulated to *p*-type and the charge density on the tube is very sensitive to the potential on the gate, although it is far away. The results are shown in Fig. 9.3. For the bulk contacts, because the gate cylinder is far away, contacts at the ends collect all field lines and image all charge on the tube, Fig. 9.3(a). For the wire contacts, however, the potential produced by the charge in the one-dimensional wire decays rapidly with distance, thus several nanometer away from the metal/semiconductor interface, the wire contacts have little effects. The capacitance between the gate cylinder and the tube decays slowly (logarithmically) with the tube radius, thus several nanometer away from the metal/semiconductor interface, the charge on the tube images on the gate rather then the wire contacts nearby. As a result, the charge density is determined by the potential on the gate. The charge density on the nanotube channel is essentially determined by the electrostatic environment.

(a) Metallic and Semiconducting Nanotubes

Most experimentally observed CNTs are multi-walled structures with outer most shell diameters exceeding 10 nm. Since current conduction in a MWCNT is known to be mostly confined to the outermost single-walled nanotube and since band gap of a SWCNT varies inversely with its diameter, MWCNTs are metallic in nature. SWCNTs can be either metallic or semiconducting depending on the way the roll-up of the graphene sheet occurs–an aspect termed as Chirality, and if all the roll-up types are realized with equal probability, 1/3 of the SWCNTs end up being metallic and 2/3 semiconducting. Thus, when CNTs are fabricated either by arc growth, laser ablation or chemical vapor deposition (CVD), a mixture of metallic and semiconducting nanotubes is formed.

(b) Carbon Nanotube Interconnects

Metallic CNTs have a lot of interest in their applicability as Very-large-scale integration (VLSI) interconnects future because of their desirable properties of high thermal stability, high thermal conductivity and large current carrying capacity. An isolated CNT can carry current densities in excess of 1000 MA/sq-cm without any signs of damage even at an elevated temperature of 523 K, thereby eliminating electromigration reliability concerns that plague Cu interconnects. Recent modeling work comparing the performance, power dissipation and thermal/reliability aspects of CNT interconnect to scaled copper interconnects have shown that CNT bundle interconnects can potentially offer advantages over copper Additionally, the concept of hybrid CNT/Cu interconnects-employing CNT vias in tandem with copper interconnects has been shown to offer advantages from a reliability/thermal-management perspective.

(c) Carbon Nanotube Transistors

Semiconducting CNTs have been used to fabricate field effect transistors (CNTFETs), which show promise due to their superior electrical characteristics over silicon based MOSFETs.

Since the electron mean free path in SWCNTs can exceed 1 micrometer, long channel CNTFETs exhibit near-ballistic transport characteristics, resulting in high speed devices. In fact, CNT devices are projected to be operational in the frequency range of hundreds of GHz. Recent work detailing the advantages and disadvantages of various forms of CNTFETs have also shown that the tunneling based CNTFET offers better characteristics compared to other CNTFET structures. This device has been found to be superior in terms of subthreshold slope–a very important property for low power applications.

(d) Challenges in Electronic Design and Design Automation

Although CNT devices and interconnects have been separately shown to be promising in their own respects, there have been few efforts to successfully combine them in a realistic circuit. Most CNTFET structures employ the silicon substrate as a back gate. Applying different back gate voltages might become a concern when designing large circuits out of these devices. Several top-gated structures have also been demonstrated, which can alleviate this concern. A fully integrated logic circuit built on a single nanotube has been made. However, this circuit also employs a back-gate. Additionally, there are still several process related challenges that need to be addressed before CNT-based devices and interconnects can enter mainstream VLSI process. This makes it an exciting and open field for research. Problems like purification, separation of carbon nanotubes, control over nanotube length, chirality and desired alignment, low thermal budget as well as high contact. (See table on next page) some examples of proposed device and their salient atuibutes is given ht Table 9.1.

MOLECULAR ELECTRONICS WITH CNTS

1. Field Emitting Devices

If a solid is subjected to a sufficiently high electric field, electrons near the Fermi level can be extracted from the solid by tunnelling through the surface potential barrier. This emission current depends on the strength of the local electric field at the emission surface and its work function (which denotes the energy necessary to extract an electron from its highest bounded state into the vacuum level). The applied electric field must be very high in order to extract an electron. This condition is fulfilled for carbon nanotubes, because their elongated shape ensure a very large field amplification.

CNTs have good field emitting properties. They are such good field emitters because of their length to diameter ratio and low voltage emission. Due to this property one finds good and cheap techniques for the production of CNT field emitting guns and even entire displays.

Also, for technological applications, the emissive material should have a low threshold emission field and large stability at high current density. Furthermore, an ideal emitter is required to have a nanometre size diameter, a structural integrity, a high electrical conductivity, a small energy spread and a large chemical stability. Carbon nanotubes possess all these properties. However, a bottleneck in the use of nanotubes for applications is the dependence of the conductivity and emission stability of the nanotubes on the fabrication process and synthesis conditions.

TABLE 9.1 Examples of proposed devices and their salient attributes

Device	*Possible applications*	*Advantages*	*Disadvantages*	*Remarks*
Single-electron transistors (SET)	Logic element	1. Small size 2. Low power	1. Sensitive to background charge instability. 2. High resistance and low drive current. 3. Cannot drive large capacitive (wiring) loads. 4. Requires geometries $<<$ 10 nm for room-temperature operation.	Use of Coulomb blockade in nanocrystal "floating-gate"-type nonvolatile memory demonstrated. May improve retention time.
Quantum dot (quantum cellular automata)	Logic element	Small size	1. Multiple levels of interconnection across long distance difficult. 2. Room-temperature operation difficult. 3. New computation algorithms required. 4. Method of stting the initial styate of the system not available. 5. Single defect in line of dots will stop propagation.	Devices demonstra ted at low temperatures. QCA architectures extensively investigated.
Resonant tunneling diode (RTD)	1. Logic element 2. Dynamic memory	1. Small size	1. Tunneling process sensitive to small film thickness (tunneling distance) variation, leading to process control difficulties. 2. Requires dc bias, large standby power consumption. 3. Multivalue logic sensitive to noise margin 4. Speed of RTD circuits likely to be determined by the conventional devices required in the circuit.	Small- to medium-scale circuits demonstrated. Most demonstrations on III-V compound semiconductors.
Rapid single-flux quantum (RSFQ) device	Logic element	Very high speed possible	1. Requires liquid helium temperature. 2. Lacks a high-density random-access memory. 3. Requires tight process tolerance.	Very-high-speed (THz) circuits demonstrated.
Two-terminal molecular devices	1. Logic element 2. Memory	1. Small size	1. No inherent device gain. 2. Scaling to large memory size may be difficult without gain. 3. Placement of molecules in a circuit difficult and not yet demonstrated. 4. Temperature stability of organic molecules may be problematic.	Sixteen-bit cross-point memory demonstrated.
Carbon nanotube FET	Logic element	1. Ballistic transport (high speed) 2. Small size	1. Placement of nanotubes in a circuit difficult and not yet demonstrated. 2. Control of electrical properties of carbon nanotube (size, chirality) difficult and not yet achieved.	Device scaling properties not yet explored. Inverter circuit demonstrated.
DNA computing	Logic element	1. High parallelism	1. Imperfect yield. 2. General-purpose computing not possible.	

Examples of potential applications for nanotubes as field emitting devices are flat panel displays, gas-discharge tubes in telecom networks, electron guns for electron microscopes, AFM tips and microwave amplifier.

2. CNT Electron Field Emitters

Two different kinds of electron sources can be made from CNTs, single and multiple electron beam devices. For example single beam as a electron bundle in an electron microscope and multiple beam in a flat panel display.

(a) Single Nanotube Emitters

- Single MWNT emission source:
  Individual MWNTs can be mounted on an etched wire under an optical microscope with two, independently moving, 3-axis micromanipulators. The individual nanotubes stick to the tip only by van der Waals forces or by applying a bit of conductive adhesive to the tip. Because of the too low resolution of an optical microscope SEM is necessary to observe the single MWNT.
- SWNT emission from raw material:
  It is more difficult to obtain a single SWNT from ropes or unpurified material, so a macroscopic fragment of raw material is picked up with a pair of tweezers and attached to a tungsten tip so you actually have many emitters but for measurements one can use a probe hole in the counter electrode to study the behaviour of only a single SWNT as field emitter. Use of SEM is necessary mandatory.
- SWNT emission from rope of SWNTs:
  A rope is picked out of a mat of SWNT material and detached from it by making electrical contact and applying a voltage difference (~ 10V) over the SWNT and the mat. Use of SEM is mandatory.
- SWNT emission from purified material:
  Commercially available material with well separated tubes in a solution allows mounting of tubes with very little contaminating material. The disadvantage are the unknown effects of the purification process on the morphology of the CNTs.
- A SWNT attached to an AFM tip:
  By using the electron bundle of a SEM for electron beam irradiation one can weld the SWNT to the AFM tip. The welds are mechanically strong and have a good electronic contact.
- Single SWNT growth on AFM tip:
  Direct growth of a SWNT by CVD on an AFM tip can be done and looks promising.

(b) Continuous Film Emitters

Planar films of emitting nanotubes can be made rather easily:

- Film from MWNTs in a colloidal suspension:
  A film is made by drawing a colloidal suspension of MWNTs through a nanopore alumina filter. Next, the film is transferred by pressing the filter face down on a Teflon or Teflon-

coated metal surface. This technique is not restricted to MWNTs but can be used for all types of nanotubes. The same result is also obtained by spraying the suspension on a heated substrate.

- Purified MWNTs dispersed in a matrix:
  Nanotubes are dispersed in an epoxy matrix and emitters are prepared (of $(50 \times 50)\ \mu^2$) under pressure and polishing is done to obtain a reproducible and macroscopically flat surface.
- Crushed and mixed MWNTs:
  Crushed MWNTs are mixed with products to form a paste which can be applied to a metallic cathode and cured.

(c) Patterned Films using Post Deposition Techniques

For many applications it is necessary for the emitting film to have a certain pattern. Different post deposition techniques to obtain patterned films are presented here:

- A nanotube/epoxy paste is pressed into channels etched in a glass substrate. Cathode surface is polished to expose the micro channels.
- Nanotubes mixed in an organic mixture of nitrocellulose, the resulting paste is squeezed onto a metal-patterned sodalime glass through a metal mesh. Subsequently the emitter was heated to remove the organic binder.
- A resist layer is patterned bye-beam lithography on a substrate. After this a nanotube dispersion is applied to the substrate. Nanotubes precipitate in the resist openings and after lift-off a patterned film is obtained.
- It is also possible to position individual nanotubes at specific locations and with specific orientations by applying a nanotube suspension on self-assembled monolayer templates defined by lithography

(d) Patterned Films using Direct Growth on Catalyst Support

Predefined catalyst patterns can be used to grow CNTs on specific locations. The patterns are made from i.e. cobalt films that are deposited from laser ablation. The obtained film is etched with single laser pulses using a cylindrical lens, while being exposed to air. The MWNTs only grow on the coparticles along the edges of the eroded tracks.

A second technique is based on standard photolithography or e-beam lithography. Basically a resist is coated on the substrate, exposed and developed. The resist can be used in two ways:

- Resist as mask over the catalyst and CNT growth over resist after which the resist is removed and a pattern is left over
- Resist as protection, unprotected catalyst is evaporated and after removal of resist the CNTs are grown.

The techniques to grow the CNTs after making a catalyst pattern also differ and are much alike described in the continuous film production.

One method that is different from the others is the use of a stamp with ink, called microcontact printing. Liquid phase catalyst precursor is dropped on a patterned elastomer stamp. The stamp is then pressed against a substrate and afterwards MWNTs can grown by CVD on the pattern of catalyst. Important is the ability to vary the catalyst precursor concentration, hydrocarbon gas and the temperature. All these parameters can be optimised for the desired MWNT density.

(e) Electronic Behaviours of CNTs

Carbon nanotubes have a interesting electronic behaviour. The nanotube emits electron from its tip at a prodigious rate. Because they are so sharp, the nanotubes emit electrons at lower voltages than electrodes made form most other materials, and their strong carbon bonds allow nanotubes to operate for longer periods without damage.

It is easy to make a high-current field emitter from nanotubes: just mix them into a composite paste with plastics, smear them onto an electrode, and apply voltage. In variably some of the nanotubes in the layer will point towards the opposite electrode and will emit electrons. Clusters of upright nanotubes in neat little grids are grown. At optimum density, such clusters can emit more than one amp per square centimeter, which is more than sufficient to light up the phosphors on a screen and is even powerful enough to drive microwave relays and high-frequency switches in cellular base stations.

Using nanotube composites, prototype vaccumtube lamps in six colors that are twice as bright as conventional lightbulbs, longer-lived and at least 10 times more energy-efficient have been made.

(g) Field Emission from CNT Films

Field emission is the emission of electrons from a solid by tunnelling through the surface potential barrier. The emission current depends on the electric field, E, and the workfunction, ϕ. As the dependence of the emitted current on the local electric field and the workfunction is exponential. So, a small variations in the shape or surrounding of the emitter have strong impact on the emitted current.

All types of nanotubes have excellent field emission properties with threshold fields, that produce 10 $\mu A/cm^2$, if only 1 $V/\mu m$ and turn on fields, that produce $10 mA/cm^2$, around $5V/\mu m$. Nanotube film are also able to emit a current density up to a few A/cm^2 at fields below 10V/m.

Differences occur due to structural differences and differences in treatment, diameter size, open/closed cap, surface treatment etc.

Density and orientation of the tubes on the film also influences the emission.

When comparing samples with different densities (Figure 9.6), the influence can be seen in the I-V curves (Figure 9.7).

Analysis of eleven samples shows that films of medium densities with nanotubes protruding over the film surface show emission at the lowest fields. Which also can be seen in the scanning field emission images, Figure 9.8.

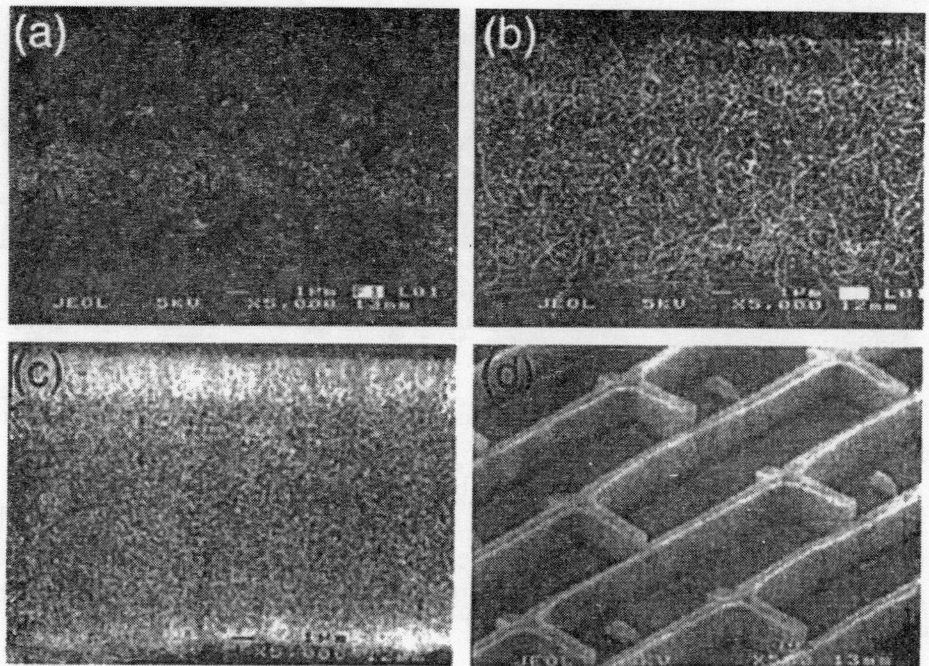

Fig. 9.6 *SEM micrographs of patterned films of MWNTs prepared by μCP of a catalyst followed by CVD growth. The concentration of catalyst in the ink solution was (a) 1 mM, (b) 5 mM, (c) 40 mM, (d) 50 mM.*

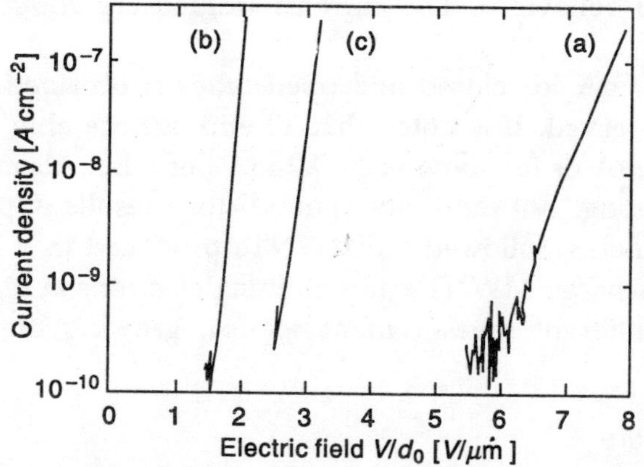

Fig. 9.7 *I-V curves for different MWNT densities as seen in Figure 9.6.*

Here it is also clear that the medium density film has the highest density.

(h) Degradation of CNT Films

Several different processes can lead to degradation of the CNT film, resistive heating, bombardment from gas molecules, arcing electrostatic deflection and mechanical stresses etc.

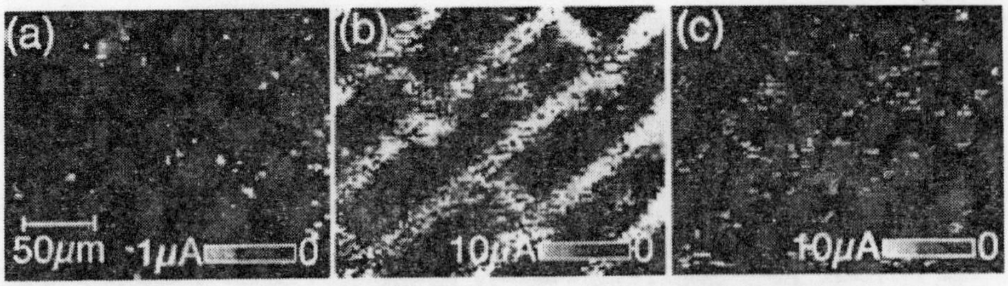

Fig. 9.8 *Scanning field emission images of MWNT films of different densities acquired at constant voltage and tip height. The greyscale represents current intensity. Images were taken from samples similar to those from Figure 9.1.*

Also chemical degradation phenomena such as adsorption and desorption of molecules on the emitter surface can occur.

Stable emissions over more than 20 h are observed and an emission of 8000 h of 10mA/cm^2 with an 11% field increase has been possible. The origin of degradation because the residual gasses which have a significant influence as well as the emitted current density. In addition the intrinsic properties of the CNTs have an importance. Tests have shown that degradation of SWNTs is much faster than of MWNTs.

(h) Field Emission from Single CNTs

Experiments on single nanotube emitters give a reliable way to investigate the emission mechanism and workfunction. On an arc produced MWNT mounted on a carbon fibre, the emission follow a rough behaviour. The currents emitted are about 100nA at 0.12 Vμm field strength.

Behaviour up to 5-20 nA for closed and open tubes is obtained. As the current increases, saturation effects are observed. It is noted that all emitters are able to emit over an incredible current range up to 0.1 mA or for some even 0.2 mA per tube. Also that open tubes need twice the voltage to start emitting. But there are contradictory results. Open MWNTs start emitting electrons at the lowest fields, followed by MWNTs produced in a hydrogen atmosphere and finally the closed arc discharge MWNTs start emitting electrons at higher fields. The differences observed are due to different measurement set-ups, growing methods, purity and emitter geometry.

Emitter Stability and Failure

FEM measurements on single SWNTs show two regions in the I-V curve. Step like fluctuations at low emitted currents with a switching frequency increases with the current and becomes maximal at the onset of saturation, followed at higher currents by stable emission with flicker noise.

Failure of single nanotubes is in almost all cases abrupt. For a single closed MWNT stable emission is observed at 2 mA for more than 90 h at 10-7 mbar. Failure occurres on most emitters in a catastrophic irreversible way in less than 10 ms.

However measuring one MWNT at 0.4 μA at 10^{-9} mbar without any degradation leads to catastrophic failure as it appears that the failure occurs at a very short time scale (<1 ms) at currents above 0.1 mA and that it involves an irreversible damage to the tube, in all cases the failure occurs on the defects of the nanotube.

FUEL ENERGY FROM CNTs

I. ENERGY STORAGE

Graphite, carbonaceous materials and carbon fibre electrodes are commonly used in fuel cells batteries and other electrochemical applications. Advantages *of* considering nanotubes for energy, storage are their small dimensions, smooth surface topology and perfect surface specificity. The efficiency of fuel cells is determined by the electron transfer rate at the carbon electrodes, which is the fastest on nanotubes following ideal Nemstian behaviour.

Two elements that can be electrochemically stored in CNTs are hydrogen and lithium. Hydrogen can also be stored in CNTs by gas phase intercalation. Three units are commonly used to describe the hydrogen and lithium contents of storage materials with:

1. [H/C] ([Li/C]) as the ratio of hydrogen (lithium) atoms per atom of storage material, in this case carbon;
2. [wt%] as the ratio of the mass of hydrogen (lithium) to the mass of storage material (the gravimetric density);
3. [kgH_2m^{-3}] as the ratio of the mass of molecular hydrogen to the volume of storage material (volumetric density).

Hydrogen Storage

The advantage of hydrogen as energy source is that its combustion product is water. In addition, hydrogen can be easily regenerated. For this reason, a suitable hydrogen storage system is necessary, satisfying a combination of both volume and weight limitations. The two commonly used means to store hydrogen are gas phase and electrochemical adsorption.

Because of their cylindrical and hollow geometry, and nanometre-scale diameters, it has been predicted that carbon nanotubes can store a liquid or a gas in the inner cores through a capillary effect. For economical storage, the storage requirements of 6.5 % by weight is the minimum level for hydrogen fuel cells. SWNTs are able to meet and sometimes exceed this level by using gas phase adsorption (physisorption). But, high storage capacities are rather to difficult achieve.

(a) Electrochemical Storage of Hydrogen

Chemisorption

Another possibility for hydrogen storage is electrochemical storage. In this case not a hydrogen molecule but an hydrogeń atom is adsorbed. This is called chemisorption.

There are two methods to store hydrogen atoms reversibly in CNTs. One method is called gas phase intercalation and The second method is based on a electrochemical charge-discharge process, in which the hydrogen absorption is controlled by the potential.

The hydrogen storage capacity of the CNT samples is analysed by means of electrochemical galvanostatic measurement in a 6 M KOH electrolyte. There are commonly three electrodes in the set-up: a work electrode (negative), often made of a mixture of gold or nickel with the nanotube material pressed into a pellet, a reference electrode (Hg/HgO/OH^-) and a counter electrode, usually made *of* nickel. In Figure 10.1, the reference electrode is left out. Instead, a polymer separator separates the working and the counter electrode.

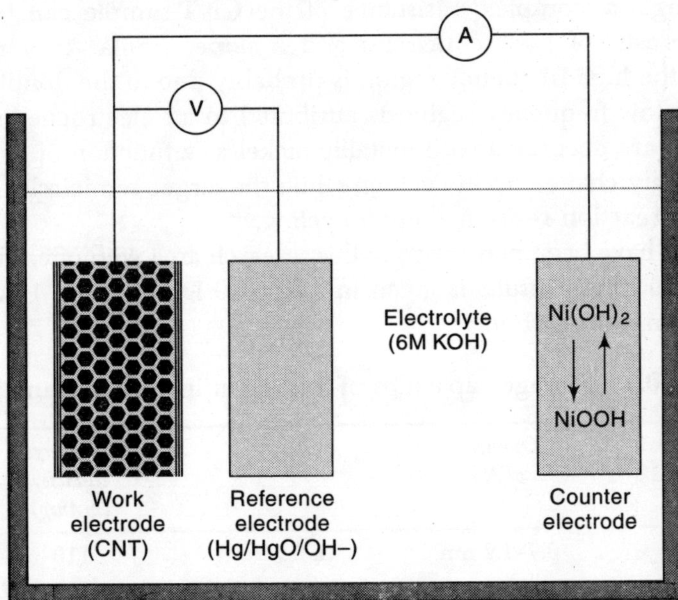

Fig. 10.1 *A schematic diagram of a charge-discharge cycling apparatus.*

During the charging process, the water in the electrolyte dissociates at the work electrode. The adsorbed atomic hydrogen may be absorbed or intercalate in the electrode *or* recombine at the surface to molecular hydrogen and diffuse into the electrode *or* form gas bubbles at the surface of the electrode. During the discharge process, the hydrogen in the electrode recombines with the OR ions in_the electrolyte to form water molecules. The amount of hydrogen desorbed from the electrode can therefore be measured by measuring the electric charge, which is equal to the product of current and time in a galvanostatic setup.

The following equation gives the reaction in the working electrode:

$$\text{CNT} + xH_2O + xe^- \underset{\text{oxidation}}{\overset{\text{reduction}}{\rightleftharpoons}} \text{CNT} + xH^+ + xOH^-$$

The reaction at the counter electrode, in the case of a Ni-electrode is given by:

$$Ni\,(OH)_2 \underset{\text{oxidation}}{\overset{\text{reduction}}{\rightleftharpoons}} NiOOH + H^+ + e^-$$

When measuring the electrochemical hydrogen storage capacity of a CNT sample, a double-layer capacitance is measured too. This effect has to be taken into account when looking at the experimental results of the capacity measurements.

Measurement methods are charge-discharge measurements, cyclic voltammetry, constant current measurements, electrochemical impedance measurements and chrono amperometry. The charge/discharge experiments are commonly performed at a constant current. During the discharge, the cell resistance leads to an overpotential. With cyclic voltarnmetry, different voltages are applied at a certain sweep rate and the current is measured. From the thus obtained oxidation and reduction peaks it can be speculated if the hydrogen adsorption is due to the CNTs or not. With chrono amperometry the current behaviour is studied after applying a voltage step. From impedance spectra, performed by applying sinusoidal voltage over a certain frequency range, a 'complex' resistance of the CNT-sample can be calculated. These spectra commonly consist of two semicircles and a slope, similar to a metal hydride. The smaller semicircle in the high frequency region is probably due to the double-layer capacitance. The semicircle in the low frequency region is attributed to an electrochemical reaction. When these spectra of CNTs are decorated with metallic nickel as a function of cycling found that the smaller semicircle hardly change upon cycling, while the larger semicircles increase, indicating an increase in surface reaction resistance upon cycling.

Several experiments have been performed in this research area with often differing conclusions. An overview of most of these results is given in Table 10.1. Note that 1 Ah/g corresponds to 3.54 wt% hydrogen stored in carbon.

TABLE 10.1 Storage capacities of hydrogen in carbon nanotubes.

Material	*Processes*	*Diameter of NT*	*Purity*	*Discharge capacity (mAh/g)*	*Stored hydrogen (wt%)*
SWNT	Arc	0.7–1.2 nm	a few%	110	0.39
MWNT	CVD		?	200	0.8
SWNT	Arc		80%	800	2.9
SWNT	LA		90%	440	1.616
SWNT (Li)	LA		90%	640	2.333
MWNT	CVD		?	297	1.051
SWNT	?		?	141	< 0.5
SWNT-ropes	Arc	50 nm (100 mm length)	60–70%	503	1.84
SWNT	Arc		50%	552	2
SWNT	Arc		70%	550	2

The electrochemical hydrogen storage behaviour of ropes of aligned SWNTs (up to 100 mm in length and 50 μm in diameter) with a large sample quantity of 200 mg (with a purity in the range 60–70%) at 298 K under normal atmosphere, is found to have higher capacity which is due to the purity and larger mean diameter of the SWNTs (1.72 nm, resulting in larger microscopes.)

(b) Gas Phase Intercalation of Hydrogen

Gas phase intercalation of hydrogen in CNTs concerns the adsorption of H_2, called *physisorption* instead of chemisorption (involving H^+ and chemical bonds). This adsorption of H_2 (other gases are possible too) on the surface of CNTs is a consequence of the field force at the surface of the solid, called the adsorbent, which attracts the molecules of the gas or vapour, called adsorbate. The forces of attraction emanating from a solid can be either physical (Van der Waals) or chemical (thus chemisorption, involving the electrochemical storage of hydrogen).

Carbon nanotubes have attracted considerable interest due to several reports of high hydrogen storage capacities at room temperatures, even higher than the goals set for vehicular storage (being an H_2-storage capacity of 6.5 wt% and 62 kg H_2/m^3). but room temperature storage capacities do not exceed ~0 wt%. A few important results found on hydrogen storage are:

1. Hydrogen storage with a H/C ratio larger than 2, thought to be incorrect (> 14 wt%). For example, a storage capacity of 67 wt% in GNFs (see Table 10.1) is reported, but this seems highly irrational because this corresponds to an average of about 24 hydrogen atoms for each carbon atom;
2. Hydrogen storage consistent with expectations based on findings for activated and other, conventional high-surface area carbons (0–2 wt%);
3. Hydrogen storage results are in the intermediate range and are not obviously incorrect (2–14 wt%).

There are several possibilities for the hydrogen to be stored in the CNT samples. For hydrogen storage in closed CNTs, the structure has two possible sites: inside the tubes and in the interstitial sites between the tube array (Figure 10.2). In case of a long SWNT closed with fullerene-like end caps, the hydrogen can only get access to the tube interior via the hexagons of the graphite-like tube wall. An opened tube with removed caps gives an easier access for the hydrogen molecules to the tube. Typically, the tubes are very long and therefore a good diffusivity of the hydrogen inside the tube will be required in order to fill the whole tube volume. Cutting the rope in shorter pieces may therefore help to improve hydrogen storage and its kinetics.

Mainly, three different techniques' are applied to study the hydrogen storage in solids: (1) the volumetric method, (2) the gravimetric method and (3) thermal desorption spectroscopy.

(1) The volumetric method measures the pressure drop owing to hydrogen adsorption after applying a hydrogen pressure to the specimen, contained in a constant volume. Similarly the pressure increase due to desorption can be measured. For good accuracy and reliable results, this method requires typically specimen masses of 500 mg or higher. Leakage or temperature instability of the apparatus may give rise to large experimental errors. The advantage of this technique is that both adsorption and desorption can be measured.

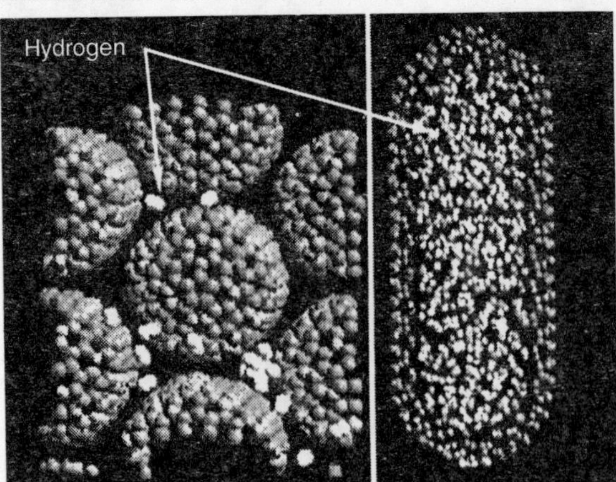

Fig. 10.2 *Schematic representation showing potential sites for hydrogen adsorption within a nanotubes -bundle: (left) hydrogen atoms occupying the interstitial spaces between the tubes, and (right) hydrogen atoms inside the tube interior.*

(2) The gravimetric method measures the weight change of the specimen due to absorption or desorption of hydrogen. A high accuracy can be achieved even at sample masses of 10 mg. However, this technique is sensitive to all gasses adsorbed or desorbed since it is only based on weighing.

(3) Thermal desorption spectroscopy (TDS) measures only the hydrogen desorption in high vacuum utili sing mass spectrometry. This method is selective and highly sensitive, which can be improved by using deuterium-loaded specimens. In this case no disturbing background from water or other hydrogen containing adsorbents occurs.

SWNTs are highly promising for H_2 adsorption, even at room temperature. H_2 storage capacity of SWNTs is synthesised by a hydrogen arc-discharge method, with a relatively large sample quantity (about 500 mg) at ambient temperature under a modestly high pressure, which is soaked in hydrochloric acid and then heat-treated in vacuum. A H_2 uptake of 4.2 wt%, which corresponds to a H/C atom ratio of 0.52, is obtained by these SWNTs with an estimated purity of 50 to 60 wt%. Also, 78.3 % of the adsorbed hydrogen (3.3 wt%) released under ambient pressure at room temperature, even if the release of the residual stored hydrogen (0.9 wt%) requires; some heating of the sample. Furthermore, after four cycles of adsorption/desorption, the H_2 uptake capacity of the SWNTs is found to remain unchanged and that less than 1 wt% is stored at pressures lower than 5 MPa. This relatively high H_2 adsorption capacity of SWNTs is related to their larger mean diameter of 1.85 ± 0.05 nm, while typical SWNT diameters are in the range of 1.2 to 1.4 nm.

A maximum capacity for adsorption of hydrogen on SWNTs is ~8 wt%. This is achieved by preparing purified SWNTs (sonicated) in a dilute nitric acid solution with a high-energy probe. Here, SWNTs are cut into shorter segments and this introduces a Ti_6AI_4V alloy due to the disintegration of the ultrasonic probe. The hydrogen adsorption on the incorporated alloy. Assuming that the alloy in the SWNT sample behaved like the pure probe-generated alloy sample of Ti-6AI-4V, the hydrogen uptake on the SWNT fraction still reached ~8 wt%. Cross-calibration of the experimental apparatuses with three different standards establish the validity of the measurements, and repeated measurements on a given sample yield reproducible results, but nominally similar sample preparation procedures do not repeatedly produce samples that exhibit the same hydrogen storage capacities. A possible explanation is that the degree of tube cutting, and how, where and in what form the metal particles are incorporated can vary dramatically even if identical sonication parameters are employed. It is also found that the desorption of hydrogen fitted first order kinetics and the activation energy for desorption measured is 19.6 kJ/mol. Also, adsorption measurements on samples having differing distributions indicate a link between SWNT size and type and hydrogen capacity. Recently, a controlled dry cutting technique is developed, that does not employ ultrasonication. Transmission electron microscopy analyses and Raman electroscopy showed that significant cutting occurred without extensive damage to the SWNTs.

(c) Lithium Intercalation

The basic principle of rechargeable lithium batteries is electrochemical intercalation and de-intercalation of lithium in both electrodes. An ideal battery has a high-energy capacity, fast charging time and a long cycle time. The capacity is determined by the lithium saturation concentration of the electrode materials. For Li, this is the highest in nanotubes if all the interstitial sites (inter-shell van der Waals spaces, inter-tube channels and inner cores) are accessible for Li intercalation. SWNTs have shown to possess both highly reversible and irreversible capacities. Because of the large observed voltage hysteresis, Li-intercalation in nanotubes is still unsuitable for battery application. This feature can potentially be reduced or eliminated by processing, i.e. cutting, the nanotubes to short segments.

(d) Electrochemical Storage of Lithium

Lithium intercalated graphite and other carbonaceous materials are commercially used in Li-ion batteries. In this case, the specific energy capacity is partially limited by the thermodynamically determined equilibrium saturation composition of LiC_6. During intercalation (discharging) the following adsorption takes place:

$$-CC- + Li^+ + e^- \rightleftharpoons -C-Li_{ad}$$

Or written in a different way:

$$xLi + C_6 \rightleftharpoons Li_xC_6$$

During de-intercalation (charging) the process is reversed. In graphite this adsorption could be visualised as in Figure 10.3.

Carbon nanotubes are interesting intercalation hosts because of their structure and chemical bonding. Nanotubes might have a higher saturation composition than graphite as guest species can intercalate in the interstitial sites and between the nanotubes. Therefore, carbon nanotubes are expected to be suitable high energy density anode materials for rechargeable Li-ion batteries.

SWNTs spontaneously form bundles that are called nanoropes. These bundles are kept together by Van der Waals forces. Reversible, electrochemical intercalation of SWNT bundles

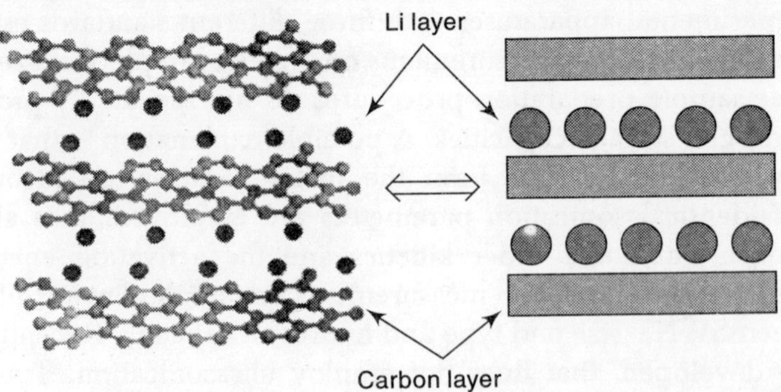

Fig. 10.3 *Structure of graphite intercalated with lithium.*

with lithium has been demonstrated in the past few years. If SWNTs have been purified, they show a reversible saturation composition of $Li_{1.7}C_6$ (632 mAh/g). In any case, this is higher than LiC_6, which is the ideal value for graphite this corresponds to a capacity of 372 mAh/g. This LiC_6 value also holds for MWNTs. However, ball-milling can further increase the reversible saturation Li composition of SWNTs. This process induces disorder within the SWNT bundles and fractures the individual nanotubes. After ball-milling the saturation composition can be as high as $Li_{2.7}C_6$ which corresponds to a capacity of 1000 mAh/g.

Experiment

SWNT bundles synthesized by laser ablation are used. The crude materials are purified by filtering off the impurities over a micro-pore membrane while keeping the nanotubes in suspension. The purified material is for 80% of SWNT bundles with a bundle diameter varying between 10 and 40 nm. The individual nanotube diameter in the bundles is between 1.3 and 1.6 nm. The purified SWNTs from the same batch are divided into several groups. Every group is treated differently by impact ball-milling in air for up to 20 minutes. The charge-discharge data of each group is observed.

A composition of $Li_{5.4}C_6$ is obtained after the first discharge with the untreated purified SWNTs. The reversible capacity, C_{rev}, is $Li_{1.6}C_6$. A large voltage hysteresis between discharge and charge cycle is observed in all samples measured. This hysteresis is related to the kinetics of the occurring electrochemical reaction. Preliminary experiments show that cutting nanotubes to smaller segments can reduce hysteresis.

The ball-milled samples show a better performance. C_{rev} increases with increasing processing time showing a maximum after 10 minutes of ball-milling. In this case, C_{rev} equals $Li_{2.7}C_6$. Further processing of the nanotubes lead to a lower C_{rev}.

It is found that SWNTs are fractured and shortened by ball-milling. With Raman spectroscopy, different samples (before and after ball-milling) are analysed (Figure 10.4). The peak around 1600 cm^{-1} is associated with some of the SWNT vibrational modes. The peak at 1340 cm^{-1} is associated with disordered carbon. One can see that the major phase in the sample ball-milled for 10 minutes still consists of SWNTs. This is in contrast with the 20 minutes ball-milled sample. It seems that ball-milling increases the C_{rev} of SWNTs, but ball-milling too long gives the opposite effect. Ball-milling the SWNTs for 10 minutes gives the best results for Li intercalation.

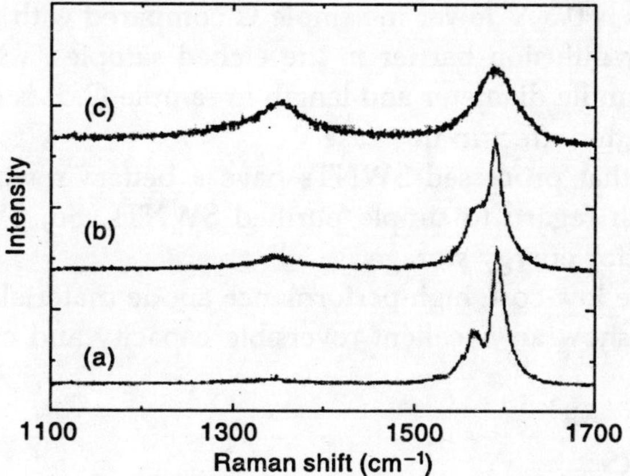

Fig. 10.4 *Raman spectra of purified [a], 10 minute [b] 20 minute [c] ball-milled SWNT materials.*

The purified samples contain over 90% SWNT bundles with a length of 10 μm and a bundle diameter of 30–50 nm, while the average diameter of an individual nanotube is at 1.4 nm. The purified bundles are processed to shorter ones by sonication in a solution of H_2SO_4 and HNO_3 for 10–24 hours. The average lengths of the bundles are reduced to 4 μm after 10 hours and to 0.5 μm after 24 hours of processing. After chemical etching, the samples are electrochemically reacted with lithium. Then, the reversible capacity of the samples is analysed. This capacity turns out to be Li_2C_6 (744 mAh/g) for the sample with a bundle length of 4 μm (sample B) and $Li_{2.1}C_6$ (781 mAh/g) for the sample with a bundle length of 0.5 μm (sample C). This is an increase of a factor 2 with regard to sample C (Figure 10.5). This effect is attributed to Li filling of the interior space of the chemically etched SWNTs. However, the difference in capacity between samples Band C is not very big despite their difference in length. So, it seems that etching for more than 10 hours has no additional value.

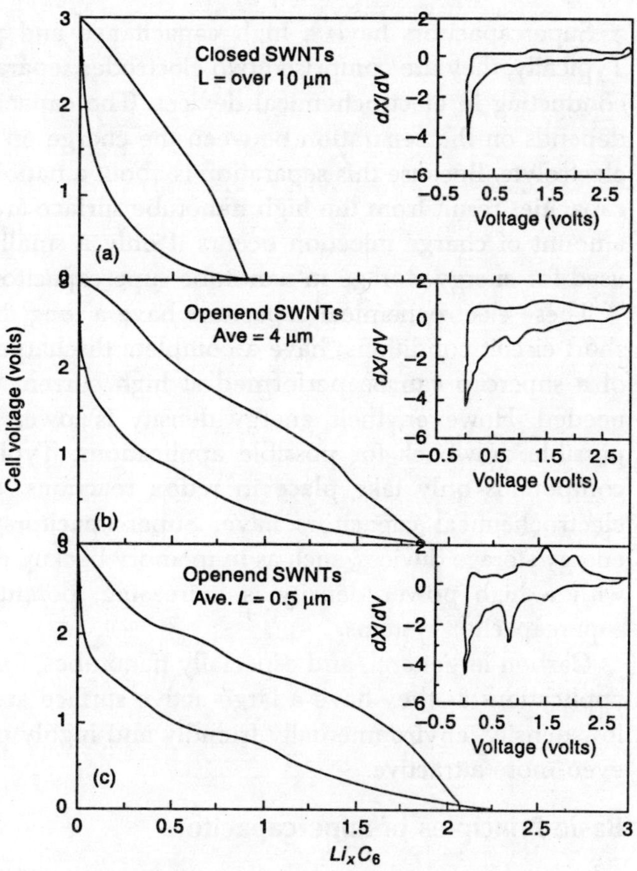

Fig. 10.5 *Second cycle intercalation (discharge) and deintercalation (charge) data collected frmo the purified an detched SWNTs.*

The voltage hysteresis is 0.5 V lower in sample C compared with sample A. This indicates a shorter diffusion path or diffusion barrier in the etched samples, which is consistent with the observation of smaller bundle diameter and length in sample C. It is clear that etching for more than 10 hours gives a better effect in this case.

It can be concluded that processed SWNTs have a better reversible capacity concerning lithium intercalation with regard to simple purified SWNTs. So, SWNT bundles seem to be attractive host materials for energy storage.

MWNTs could also be low-cost, high-performance anode materials for rechargeable lithium-ion batteries since they show an excellent reversible capacity and cycle ability during lithium insertion and extraction.

2. SUPERCAPACITORS

Electrochemical Supercapacitors

Supercapacitors have a high capacitance and potentially applicable in electronic devices. Typically, they are comprised two electrodes separated by an insulating material that is ionically conducting in electrochemical devices. The capacity of an electrochemical supercap inversely depends on the separation between the charge on the electrode and the counter charge in the electrolyte. Because this separation is about a nanometre for nanotubes in electrodes, very large capacities result from the high nanotube surface area accessible to the electrolyte. Thus, a large amount of charge injection occurs if only a small voltage is applied. This charge injection is used for energy storage in nanotube supercapacitors.

These electrochemical capacitors have a long durability (over 10^6 cycles); don't suffer from short circuit conditions; have a complete discharge and possess a high power density. Loading of a supercap can be performed at high current densities, which decreases the loading time needed. However, their energy density is lower than for conventional batteries, which is a possible drawback for possible applications. Typical electrochemical accumulators, in which compounds only take place in redox reactions, cannot fulfil these good characteristics that electrochemical capacitors have. Supercapacitors have already been applied in small-scale energy storage devices, such as in memory backup devices. Now the capability of supercapacitors with a high power density is increasing, potential applications extend to hybrid battery/ supercapacitor systems.

Carbon in general, and especially nanotubes, form an attractive material for electrochemical applications as they have a large active surface area. In addition, carbon is a relatively cheap, low density, environmentally friendly and highly polarisable material which makes application even more attractive.

Basic Principles of Supercapacitors

The basic principle of energy storage in a supercapacitor is based on creating a charge-separated state in an electrochemical double layer. In this case, energy storage is based on the separation of charges in the double layer across the electrode/electrolyte interface. The positive electrode is electron deficient whereas the negative electrode contains a surplus of electrons (Fig. 10.6). The energy (W) stored in a capacitor as a function of the voltage applied (U) and the capacity (C) is equal to: $W = \frac{CU^2}{2}$.

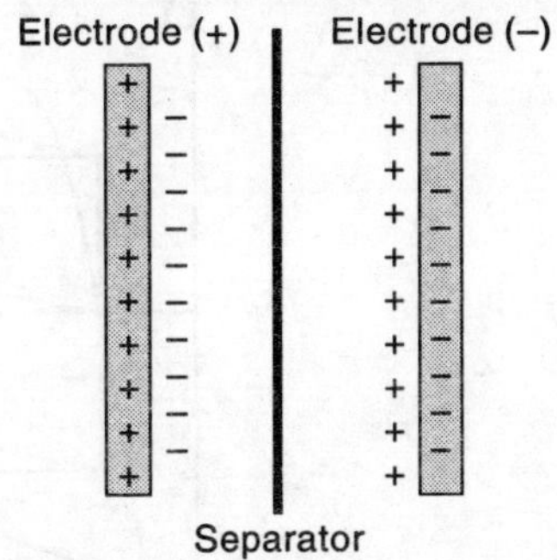

Fig. 10.6 *Scheme of an electrochemical capacitor*

The electrodes of a supercapacitor must be electrochemically stable, which is the case for chemically unmodified carbon. The decomposition voltage of the electrolyte determines the maximum operating voltage of a supercapacitor. For the generation of high voltages, aprotic electrolytes with a decomposition range between 3 and 5 V should be used. However, these liquids only have a fraction of the conductivity that water has. In addition, the use of an aprotic electrolyte has technological, economical and safety barriers. The final electrolyte choice depends on the demanded specific power and energy values.

Electrochemical capacitors based on carbon are of two different types depending on the type of energy storage. The first type is the electrical double layer capacitor (EDLC) where only a pure electrostatic attraction between ions and the charged surface of an electrode takes place. The second type is a supercapacitor (SC), which is additionally based on faradaic supercapacitance reactions. The total capacitance C is determined by the series capacitances of the anode (C_A) and cathode (C_C) according to the equation: $\frac{1}{C} = \frac{1}{C_A} + \frac{1}{C_C}$.

In the EDLC, the contact between the electrode surface and the electrolyte plays an important role and determines the amount of charge stored. The capacitance C is proportional to the surface area S and the permitivity ε of the electrolyte and reciprocally dependent on the distance of charge separation: $C = \frac{S\varepsilon}{d}$. The surface area determined by gas adsorption differs from the active surface area available for charged species. When ions are solvated by water molecules, their mean diameter is approximately 15A. This explains the need for a relatively large pore size of the electrode material for good interaction of ions with the electrode. Thermal treatment of the electrodes results in significant alteration of the pore size distribution and thus to enhanced interaction. Increasing the capacitance values 10 to 100 times is possible by using pseudocapacitance effects. These depend on the surface functionality of carbon and/or on the presence of electro-active species.

In supercapacitors faradaic reactions similar to processes in accumulators occur. but Pseudocapacitances arise when the charge q, is a continuously changing function of potential U. These pseudocapacitance effects, strongly depended on the chemical affinity of carbon materials to the ions sorbed on the electrode surface example electrosorption of H or metal ad-atoms and redox reactions of electroactive species, example electrosorption of H or metal ad-atoms and redox reactions of electroactive species.

An ideal double layer capacitance results in ideally rectangular shaped cyclo-voltammetry diagrams. This phenomenon is ideal if the current density is independent of the potential applied and if this effect is purely electrostatic of nature. Due to redox peaks, pseudocapacitances result in deviations from this ideal shape as is shown in Figure 10.7.

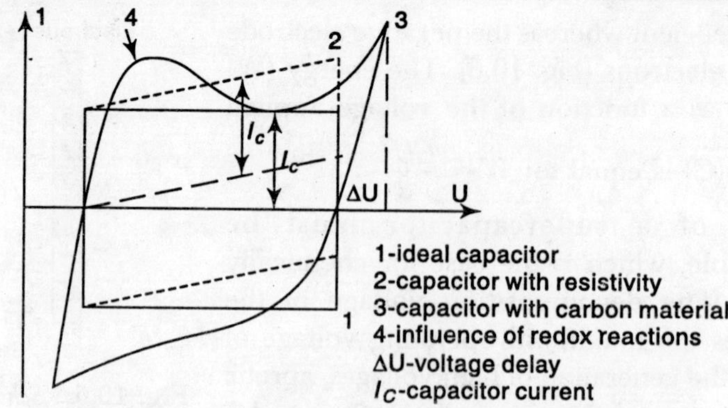

Fig. 10.7 *Typical charge/discharge voltammetry chractenstlcs of an electrocchemical capacitor*

ECDL capacitors can be represented in an equivalent electrical circuit as is shown in Figure 10.8. This circuit consists of a capacitance *C*, a parallel resistance R_F which is responsible for the self-discharge, an inductance *L* and an equivalent series resistance R_S that models the internal resistance. A maximum value of *C* and a minimum value of R_S result in a high value for the power density and energy value.

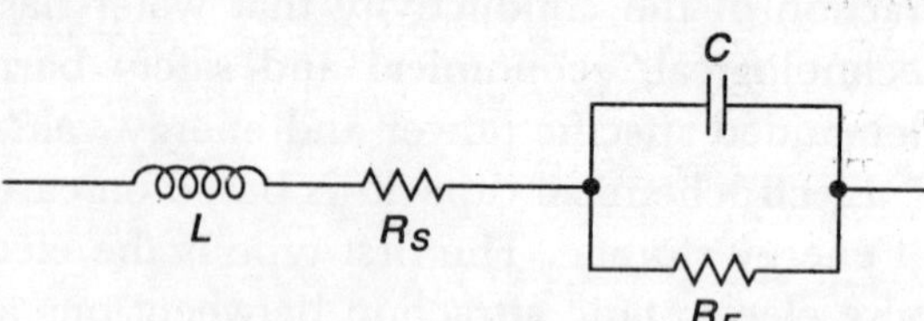

Fig. 10.8 *Equivalent circuit for a real ECDL capacitor*

The time constant for the charging and discharging cycles is equal to $R_S C$, while the self-discharge time constant is equal to $R_F C$. In order to minimise self-discharge the value for R_F should be as large as possible.

CNTs as Capacitors

Carbon nanotubes have been proposed as electrodes for supercapacitors. Different values of capacitance depend on the kind and purity of CNTs. For purified nanotubes specific capacitance varies from 5 to 80 F/g. Pure carbon nanotubes have a moderate surface area (120 to 400 m^2/g) because of their highly mesoporous character. The more graphitised nanotubes show smaller values of capacitance. However, presence of defects causes an increase of ability for accumulation of charges.

There is a great difference in capacitance between SWNTs and MWNTs. However, the many different ways of producing the nanotubes also create differences in capacitance values. This can be seen in Table 6.3. Surface groups noticeably enhance the measured capacitance value, as is the case with MWNTsCo/973 K modified by HNO_3.

TABLE 10.2 Specific capacitances of differently produced carbon nanotubes

	SWNTs Rice	*SWNTs 1923 K*	*MWNTs templ/1023 K*	*MWNTs Co/923 K*	*MWNTs Co/1123 K*	*MWNTs Co/923 K mod.* HNO_3
6 Mol L^{-1}KOH	40	18	36	80	62	137

The capacitance of nanotubes is increased by increasing the electrode surface area or by to increasing the pseudocapacitance effects obtained by addition of special oxides or Electrically Conducting polymers (ECP) like polypyrrole (PPy). The ECPs have the advantage of lower costs compared to oxides. Another advantage is that the pseudocapacitance effects of ECPs are quite stable.

Charging of the electrical double layer proceeds mainly in the micropores. These micropores are present in the pure CNTs. So, to get mesopores as well as micropores they are modified by activation of the nanotubes by KOH. Both activation and coating increases the capacitance of the nanotubes.

The mesoporous character of as-produced nanotubes determines their electrochemical properties. For charging of the double layer a developed surface area is needed, thus the presence of micropores is necessary. With chemical activation of pure MWNTs by KOH, microporosity is introduced (Figure 10.9).

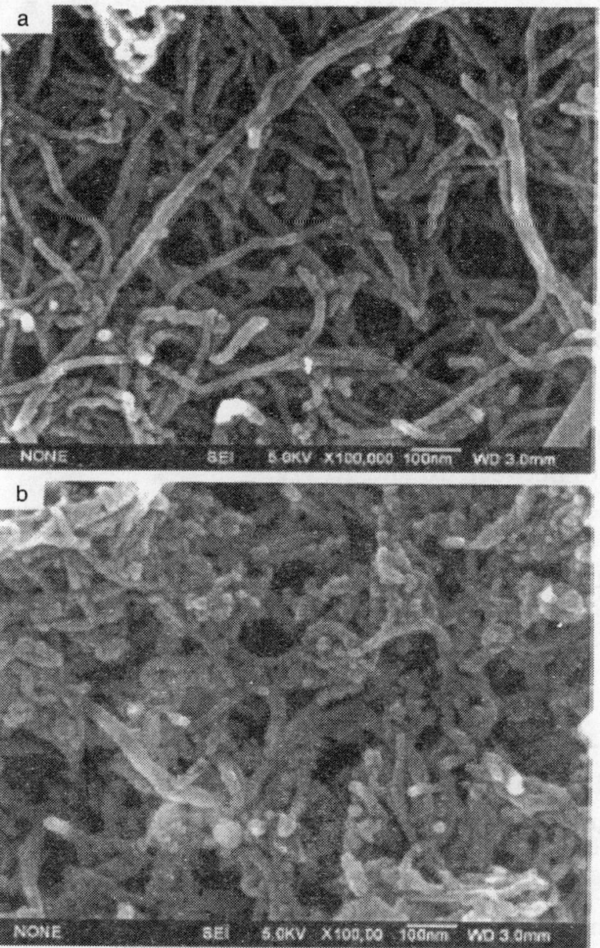

Fig. 10.9 *SEM images of (a) normal CNTs and (b) activated CNTs.*

This KOH activation is perfotmed at 1073 K under argon flow. The KOH:C weight ratio is 4:1 during the entire process. Afterwards, the samples are washed with demineralised water.

Specific surface area is measured by nitrogen adsorption at 77 K after the samples are outgassed at 623 K for 24 hours until pressure reaches 10.6 mbar.

The electrodes are prepared in the form of pellets with 85 wt% nanotubular material, 5 wt% acetylene black and 10 wt% polyvinylidene fluoride. The mass of the electrodes varies from 4 to 8 mg.

Three different types of electrolytic solutions are used: aqueous 6 M KOH, aqueous 1 M H_2SO_4 and organic 1.4 M $TEABF_4$ in acetonitrile.

Two different types of MWNTs are used. One type is obtained by decomposition of acetylene at 973 K on cobalt (12,5 %) supported on silica (ACoSi700). The other type is prepared by decomposition of acetylene at 873 k on Co particles from a solid solution of cobalt oxide and magnesium oxide (ACoxMg$_{(1-x)}$O).

The values of capacitance are estimated by voltammetry, galvanostatic charge/discharge cycling and impedance spectroscopy.

The specific surface of the A/CoSi700 nanotubes increases from 430 to 1035 m^2/g and the micropore volume changes from nearly zero to 0.47 cm^3/g after activation. For $ACo_xMg_{(1-x)}O$ nanotubes the specific surface area increases from 220 to 885 m^2/g and the micropore volume increased to 0.40 cm^3/g from almost zero. These are significant changes. For the latter type an additional effect is that the tips are opened by KOH activation. Microporosity is now enhanced by activation, but the mesoporous character is still present because of the entanglement and presence of a central canal.

The results of the capacitances of the different nanotubes in three different electrolytes are given in Table 10.3. The capacitances for the unactivated nanotubes are between 10 and 15 F/g.

Table 10.3 Capacitances of the activated MWNTs. The unactivated nanotubes show capacitances between 10 and 15 F/g.

Electrolyte	*A/CoSi700 activated*		*A/CoxMg (1-x) O activated*	
1 M H2SO4	95 F/g		85 F/g	
6 M KOH	—		90 F/g	
1.4 M TEABF4	65 F/g		65 F/g	
	not act.	act.	not act.	act.
surface area (m2/g)	430	1035	220	885

The values of capacitance are estimated by voltammetry with a scan rate of potential from 1 to 10 mV/s and galvanostatic charge/discharge cycling from 0 to 0.6 V or higher voltage limitation until 1.2 V.

The modification of carbon material by a specific additive providing quick pseudo-capacitance redox reactions is another way to enhance capacitance. This is possible with metal oxides, but in this case the addition of ECP is used. ECP itself has a capacitance of about 90 F/g. Pseudocapacitance effects of ECP are relatively stable. If we coat a nanotube with, for instance, polypyrrole we take the profit of the good electronic conducting properties and keep the advantage of ionic conductivity in the opened mesoporous network of the nanotube. These are perfect conditions for a supercapacitor.

Take five nanotube, all of them made in a different way and measured the capacitance of them with and without a layer of PPy coated on the surface. Now, measure the specific surface area of the nanotube materials by nitrogen adsorption at 77 K and perform chemical polymerisation of pyrrole on the nanotubes with ammonium persulfate as an oxidant. The thickness of the PPy layer is found to be about 5 nm.

Electrodes are either from bucky paper or pellets of a mixture of MWNTs (85 wt%), acetylene black (5 wt%) and polyvinylidene fluoride (10 wt%). The aqueous electrolyte is 1 M H_2SO_4.

The results of capacitance measurements on the different nanotubes are given in Table 10.4. It can be concluded that the nanotubes with electrochemically deposited polypyrrole give much higher values of capacitance than the untreated samples. Figure 10.10 the difference in voltammograms of Hyperion with and without coating.

TABLE 10.4 Capacitance values (F/g) of the nanotubular material in acidic medium (1 M H_2SO_4) with and without PPy.

Kind of nanotubes	C *(F/g) without PPy*	C *(Fig) with PP*
A/CoSi700	65	141
Hyperion™	78	146
P/800Al	5	123
Pyrograf III	16	130
A/CoNaY600	50	163

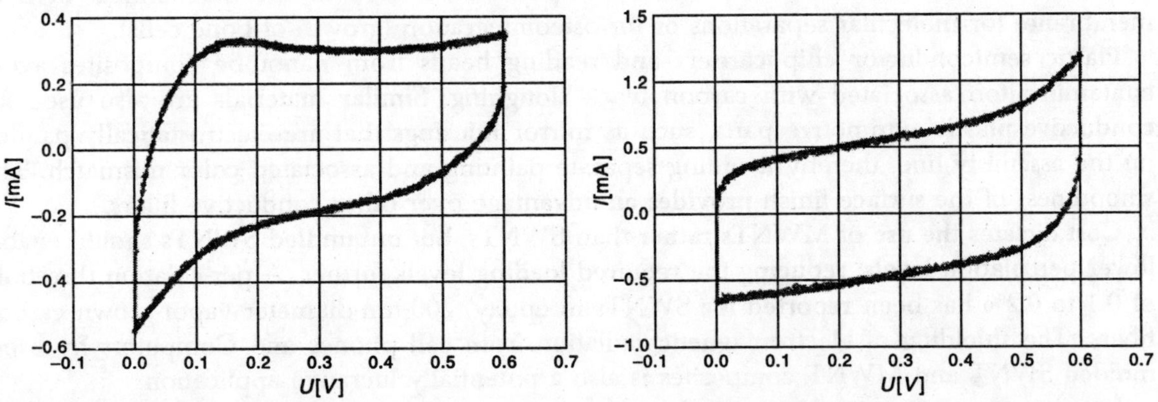

Fig. 10.10 *Potentiodynamics characteristics (2 mV/s) of a capacitor assembled in 1 M H_2SO_4 from Hyperion nanotubes without PPy (left) and with PPy (right). Mass of each electrode was 3.5 mg.*

A couple of the capacitors have been cycled over 2000 cycles and the charge loss never exceeded 20%. So, coating of the nanotubes with polypyrrole seems a nice way to enhance capacitance and efficiency for a long durability.

CARBON NANOTUBE AS COMPOSITE

Carbon nanotubes, are used in structural applications due to their stiffness; example, as reinforcements in high strength, low weight, and high performance composites.

SWNTs have a Young's Modulus of 1 TPa. MWNTs are less strong because the individual cylinders slide with respect to each other. Ropes of SWNTs are also less strong. The individual tubes pull out by shearing and at last the whole rope breaks. This happens at stresses far below the tensile strength of individual nanotubes. Nanotubes also sustain large strains in tension without showing signs of fracture. In other directions, nanotubes are highly flexible.

An advantage of using nanotubes for structural polymer composites is that nanotube reinforcements increase the toughness of the composites by absorbing energy (during their highly flexible elastic behaviour). If the density of the nanotubes is low, an increased electrical conduction and better performance during compressive loads is there. The first realized major commercial application of MWNTs is their use as electrically conducting components in polymer composites. Depending on the polymer matrix, conductivities of 0.01 to 0.1 S/cm can be obtained for 5% loading; much lower conductivity levels suffice for dissipating electrostatic charge. The low loading levels and the nanofiber morphology of the MWNTs allow electronic conductivity to be achieved; while avoiding or minimizing degradation of mechanical properties and the low melt flow viscosity needed for thin-wall molding applications. In automotive gas lines and filters, the nanotube filler dissipates charge buildup, that can lead to explosions and better maintains barrier properties against fuel diffusion.

Another possibility, which is an example of a non-structural application, is filling of photoactive polymers with nanotubes example. PPV (Poly-phenylenevinylene) filled with MW and SWNTs are a composite S. These composites show a large increase in conductivity with only a little loss in photoluminescence Nanotube-polymer composites are used in the biochemical field as membranes for molecular separations or for osteointegration (growth of bone cells).

Plastic semiconductor chip carriers and reading heads from nanotube composites avoid contamincaiton associated with carbon black sloughing. Similar materials are also used for conductive plastic automotive parts, such as mirror housings that are electrostatically painted on the assembly line, thereby avoiding separate painting and associated color mismatch. The smoothness of the surface finish provides an advantage over other conductive fillers.

Cost dictates the use of MWNTs rather than SWNTs, but unbundled SWNTs should enable lower percolation levels, reducing the required loading levels further. A per-colation threshold of 0.1 to 0.2% has been reported for SWNTs in epoxy, 200-nm-diameter vapor-grown carbon fibers. The shielding of electromagnetic radiation from cell phones and Computers by using molded SWNT and MWNT composites is also a potentially lucrative application.

Incorporation of nanotubes into plastics can potentially provide structural materials with dramatically increased modulus and strength. The critical challenges lie in uniformally dispersing the nanotubes, achieving nanotube-matrix adhesion that provides effective stress transfer, and avoiding intratube sliding between concentric tubes within MWNTs and intrabundle sliding within SWNT ropes a monotonic increase of resistance to indentation (Vickers hardness) is up by 3.5 times on loading up to 2% SWNTs and a doubling of thermal conductivity with 1%

SWNTs. Also, 1% MWNT loading in polystyrene increases the modulus and breaking stress by up to 42 and 25% respectively.

Clay-polymer Nanocomposites

Clay-polymer nanocomposite materials have enhanced mechanical and thermal properties as compared to conventional materials. They find application in the electronics, automobile, and furnishing industries.

One preparation method of clay-polymer nanocomposites is *in situ* intercalative polymerization. This process involves mechanical mixing of the clay mineral with the required monomer. The monomer then intercalates within the interlayer and promotes delamination. Polymerization follows, initiated by a number of ways, to yield linear or cross linked polymer matrices. Often the clay mineral needs to be dispersed by a preswelling step of long-chain alkylammonium ion intercalation to aid exfoliation.

Another method of preparation, called is self-catalyzed *in situ* intercalative polymerication, is similar to *in situ* intercalative polymerization, but differs in the fact that no pretreatment of the clay is required.

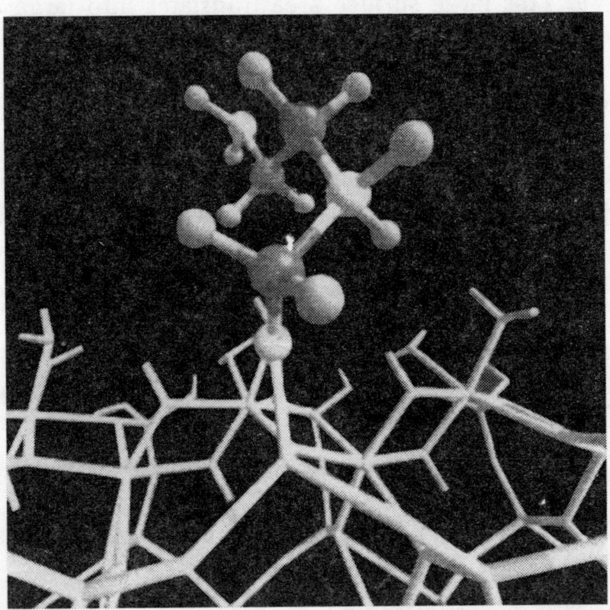

Fig. 10.11 *Methanal coordinated to aluminium in the tetrahedral layer of montmorillonite is activated towards nucleophilic attack by ethylenediamine*

Example in methanal with ethylenediamine within a sodium montmorillonite interlayer, the clay structure acts as catalysis and the potential sources of acidity in sodium montmorillonite. By using DFT this method is suitable for crystalline materials (e.g. clays and zeolites). Using DFT as the optimization method on periodic model systems the proximity to isomorphic substitution sites within the silicate layers has a marked effect on the acidity of the hydroxy groups. The hydroxy groups being further from the sites of the magnesium-for-aluminium substitution being more acidic, although it was shown that it was energetically favourable for the to be nearer the substitution sites. The simulations also revealed that the monomer catalysis occurs at the mineral lattice edge. Here exposed Al^{3+} ions and hydroxy groups act as Lewis and Bronsted acid sites, respectively.

APPLICATION OF CARBON NANOTUBES AS SENSORS

I: CNT as Chemical Sensors

There is a tremendous need for chemical sensors with high sensitivity in the parts per million (ppm) to parts per billion (ppb) level, like sensor for cosmochemistry (identification of atmospheric constituents of various planets), earth observation (greenhouse gases), and leak detection in the shuttle and future crew exploration vehicle (CEV) that involves sensing of hydrazine N_2O_4. Chemical sensors are needed for the detection of toxic industrial chemicals and nerve agents for homeland security applications.

Although, nanomaterials in general have a very large surface area, SWNTs possess a surface area of about 1600 m^2/g. This large surface area translates into large adsorption rate of gases and vapors. This adsorption can lead to a change in some specific properties of the SWNTs, example: conductance, capacitance etc. Indeed SWNT field effect transistors have been fabricated wherein a single SWNT or a film of multiple SWNTs forms the conducting channel. The conductivity of this device has been shown to change reproducibly upon adsorption of gases like NH_3 and NO_2 (which forms the basis for chemical sensing). However, three terminal transistors are expensive and complex to fabricate. An interdigitated electrode is shown in Fig. 10.12. This device is fabricated using single microfabrication steps and purified SWNTs are solution-cast across the electrode to form the conducting channel.

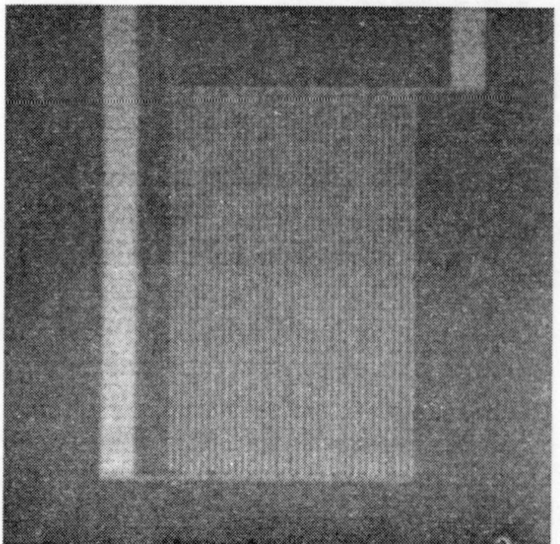

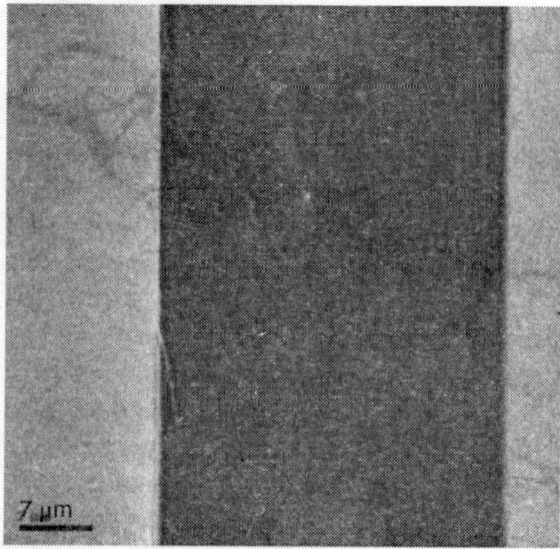

Fig. 10.12 *Interdigitated electrode structure for chemical sensors.*

Figure 10.13 shows the sensor signal for repeated doses of NO_2 wherein the response is fairly rapid. The projected senstivity is about ~40 ppb. The sensor recovery time (the time to recover to baseline after the source is removed) needs further improvement; shining a UV source accelerates the desorption of the vapor from the nanotubes thus accelerating sensor recovery.

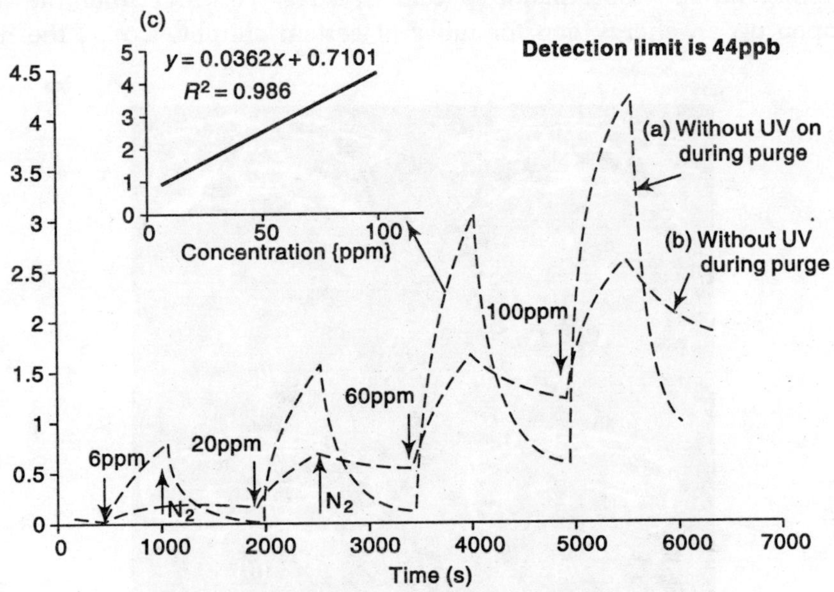

Fig. 10.13 *Sensor response to* NO_2

The sensors have been tested for NO_2, NH_3, CH_4, Cl_2, HCl, toulene, benzene, acetone, formaldehyde and nitrotoulene.

II. CNT as Electromechanical Sensors

In electromechanical sensors, bonding differences, between two types of nanotube deformation i.e. Bending and pushing is examined using D Mol^3 models. For this, ultrasensitive sensors are designed which detect the smallest mechanical purburtations and strain gauges, based on carbon nanotubes. Here, a metallic nanotube is suspended over a 600 nm long trench. When the middle part of such a suspended nanotube is pushed with the tip of an atomic force microscope (AFM), the conductivity is found to increase by almost two orders of magnitude. This drop in conductance is much higher than the predicted values for tubes bent under mechanical duress.

It is found that beyond a critical deformation several C-atoms close to the AFM tip become sp^3-coordinated. This leads to the tying up of n- electrons into localized σ-states, which explains the large drop in electrical conductance.

Also, the first-principles DFT code $DMol^3$ is used to examine bonding differences between two types of nanotube deformation: bending, and pushing with atomically sharp AFM tips. The smallest models of nanotubes necessary in such simulations typically involve a few thousand atoms, which make a pure DFT simulation unfeasible.

By combination of DFT and classical molecular mechanics, bond reconstruction, is likely to occur only in the highly deformed, middle part of the tube. For such atoms (~ 100–150 atoms including AFM-tip atoms) a DFT description is used, while the long and essentially straight part away from the middle is described accurately using the Universal Forcefield (UFF). Thus for AFM-pushed tubes, no sp^3 coordination occurs (see Fig. 10.14). Rather the tube stretches, which might open up an energy gap for tubes of certain chiralities, e.g., the metallic zigzag tubes.

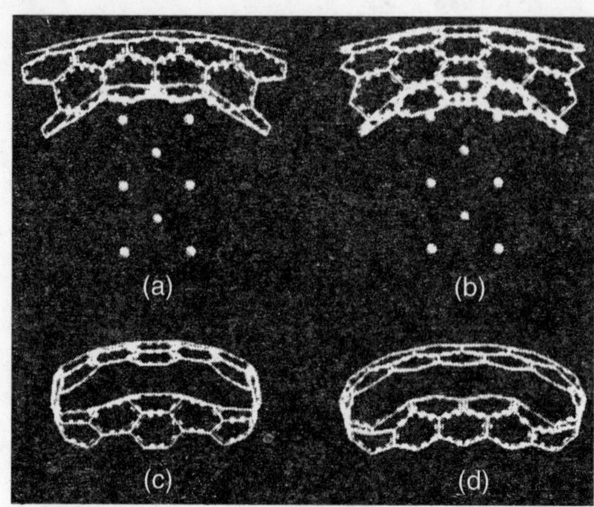

Fig. 10.14 *DMol³-relaxed Li-tip-deformed Om clusters for: (a) the (6, 6z0 amchair (132 C-atoms); and (b) the (12, 0) zigzag (144 C-atoms), in side views. The deformation angle is 25° for both tubes. Fig. (c) and (d) are respective views along the tube length, clearly displaying the absence of sp^3 coordination. The rest of the tube is described by a classical forcefield and is hidden for clarity.*

III. CNT as Biosensor

The MWNFs are ideal for developing electrodes. They range in 20–100 nm diameter. Their density and hence, the average spacing between individual MWNFs can be varied by varying the thickness of the catalyst films. Of course, if one wishes a precise control on the spacing, then catalyst patterning by e-beam lithography would be ideal. In applications as electrodes, particularly with electrochemical approaches involving electrolytes and indicators, it is important to maintain the rigidity of the nanostructures. For this reason, gap-filling techniques are developed to fill the spacing between MWNFs with a dielectric such as SiO_2 spin-on-glass. TEOS CVD is used to deposit SiO_2 followed by chemical mechanical polishing to planarize the top surface (see Fig. 10.15). Only the tip of the MWNFs are exposed while the rest are buried inside the SiO_2. Electrical characterization of these tips indicate that they function as electrodes as desired and hence, are suitable for attaching DNA or other chemical groups. These nanoelectrode arrays are useful for biosensor development as described below.

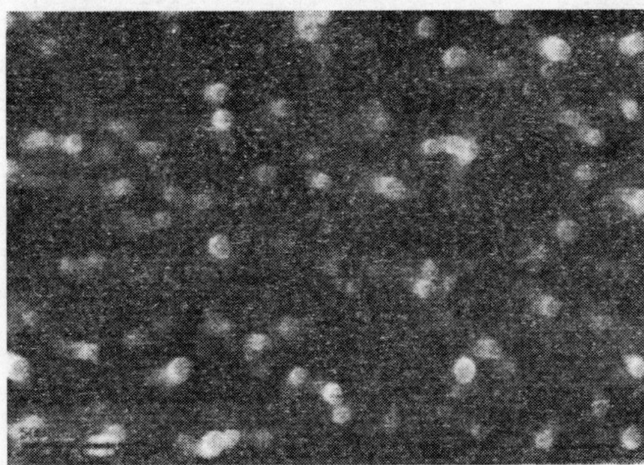

Fig. 10.15 *SEM image of MWNFs after encapsulated in SiO_2*

The above electrode is attached to DNA and then to the MWNF tips. The attachment is due to covalent bonding. By bringing in a complementary strand for the probe DNA molecule, hybridization is also demonstrated. In addition to the evidence from fluorescene microscopy for these events, by using an electrochemical indicator signals that are proportional to the amount of guanine in the target sequence are obtained. A major area of focus in CNT applications is development of biosensors for various applications using the above demonstration; example: a 30 sensor die on a 4" wafer has been fabricated. The application focus ranges from water quality and bacteria monitoring to biothreat monitoring for homeland security and cancer diagnostics, involves development of a prototype biosensor catheter that permits detection of specific oligonucleotide sequences which serve as molecular signatures of cancer cells. A critical element of this technology involves the ability to functionalize the tip of a nanotube array with (a monolayer of) probe molecules.

The nanoelectrode array is also being used for deep brain stimulation (DBS) to treat disorders such as Parkinson's disease and epilepsy by developing a 3-d interface between the neural tissues and solid state electronics using the nanoelectrode array.

DOPED CARBON NANOTUBE SENSORS

A new type of nanoscale molecular sensors using carbon nanotubes with modified electronical and chemical properties through doping impurity atoms into them is given below:

Using ab initio simulations this type of sensors can overcome the weaknesses of existing intrinsic and functionalized carbon nanotubes sensors. Different impurity atoms are doped into intrinsic carbon nanotubes because the impurity atoms will be able to detect different molecular species. Two representative chemical species–carbon monoxide and water molecules, which are not detectable to either intrinsic or functionalized CNTs–demonstrate the capability of this new type of sensors.

Figure 10.16 depicts a schematic structure of this new type of devices that can detect the presence of certain gases, such as carbon monoxide and water molecules. Boron- or nitrogen-doped carbon nanotubes experience a drastic change in electrical properties when exposed to carbonmonoxide or water molecules. When an intrinsic carbon nanotubes is doped with either boron atoms of nitrogen atoms through a replacement of carbon atoms, the physical properties around the impurity atoms (boron or nitrogen) undergo a significant change, resulting in the change of the chemical reactivity. This chemical reactivity change increases the binding energy and the possibility of gas molecules interacting with the doped nanotube. Gas molecules bound on the doped CNTs can be detected through a charge transfer between the molecules and CNT that induces a conductance change of the CNT

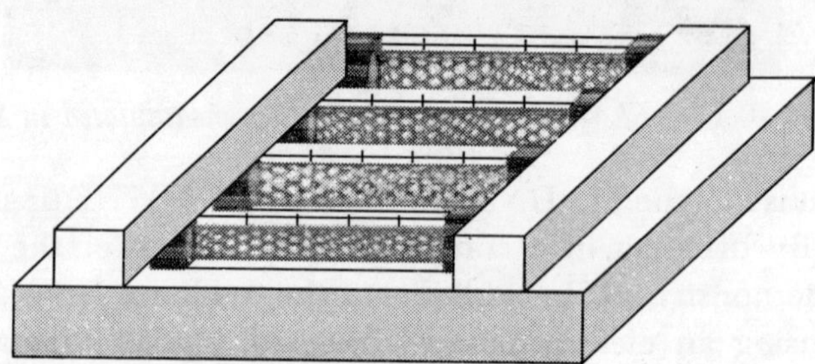

Fig. 10.16 *Schematic structure of doped nanotubes as a chemical sensor array to detect molecules such as carbon monoxide and water molecules*

The charge transfer is driven by the chemical potential realignment when the doped nanotubes and gas molecules are put into one system. Furthermore, the charge-transfer capacities are different for the nanotubes doped with different impurity atoms. Thus a wide range of nanotube sensor devices can be designed which are capable of detecting a variety of gas species by using different impurity atoms and their complexes as doping agents.

The sensitivity of these devices to the presence of carbon monoxide molecules, as compared to that of the devices using intrinsic carbon nanotubes, is given in Table 10.5. All the configurations for

Table 10.5 Calculated Data for Adsorption of CO on Doped Carbon Nanotubes or $B_xC_yN_z$ Nanotubes

	$E_g(eV)^c$	*configuration*	*Eb (eV)*	*D(Å)*	$E_T(e1)^d$
CNT	0.56	CO[e]		no binding	
		OC[f]		no binding	
B-CNT[a]	0.44	CO	–0.85	1.52	0.59
		OC	–0.27	2.83	0.02
N-CNT[b]	0.43	CO	–0.22	2.99	0.05
		OC	–0.18	2.82	0.50
BC_3	0.61	CO	–1.72	1.50	0.50
		OC	–0.28	2.53	0.04

[a]Intrinsic nanotube doped with born atom in 2-unit cell. [b] Instrinsic nanotubes doped with nitrogen atom in 2-unit cell. [c] HOMO-LUMO band gap.[d] Electron charge transfer from nanotube to molecules. [e] CO molecule binds to the nanotube with the carbon atom close to the tube.[f] CO molecule binds to the nanotube with the oxygen atom close to the tube.

carbon monoxide molecules are shown in Figure 10.17. To determine the conductance changes through the nanotubes, charge-transfer calculations are carried out. When the molecule is adsorbed on the nanotube surface, some charge is transferred between the molecule and the nanotube, resulting in local charge changes within the atomic spheres of the molecule. The resulting charge transfer is calculated through the charge difference between the molecule adsorbed on the nanotube and an isolate molecule itself. Table 10.5 shows that carbon nanotube with boron doping and intrinsic BC_3 nanotube serve as good sensor devices for carbon monoxide due to the strong binding between them and a large electron charge transfer from a nanotube to carbon monoxide molecule, which leads to a large change in nanotube conductance.

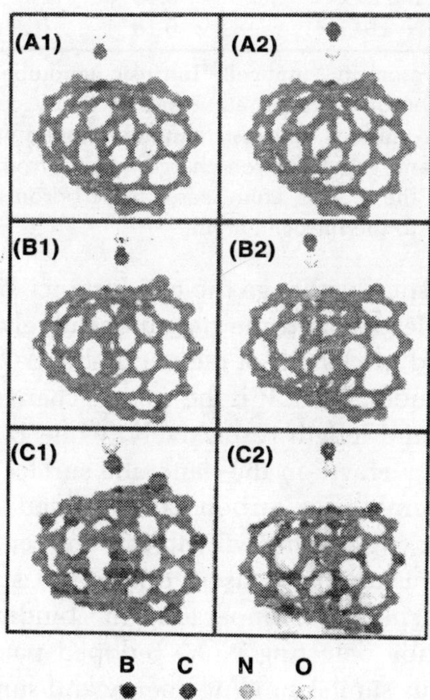

Fig. 10.17 *Semiconducting (8, 0) CNT with one boron atom substition that binds to CO molecules. (A1) Boron atom is close to the arbon atom of a CO molecule. (A2) Boron atom is close to the oxygen atom of a CO molecule; semiconducting (8, 0) CNT with one nitrogen and substitution that binds to CO molecule. (B1) Nitrogen atom is close to the carbon atoms of a CO molecule. (B2) Nitrogen atom is close to the oxygen atom of a CO molecule; semiconducting (8, 0) BC_3 nanotube that binds to CO molecule. (C1) Boron atom is close to the carbon atom of a CO molecule. (C2) Boron atom is close to the oxygen atom of a CO molecule*

The sensitivity of these devices to the presence of water molecules, as compared to that of the devices using intrinsic carbon nanotube see Table 10.6. All the configurations studied for water molecules are shown in Figure 10.18. Carbon nanotubes with boron doping and intrinsic BC_2N type 2 nanotube serve as good sensor devices for water molecules due to the strong interaction between the molecules and the nanotubes, and a large electron charge transfer from

Table 10.6 Calculated Date for Adsorption of H_2O on the Doped Carbon Nanotubes or $B_xC_3N_2$ Nanotubes.

	$E_g(eV)^f$	*Configuration*	$E_b(eV)$	$D(Å)$	$ET(e1)^B$
CNT	0.56	H_2O		no binding	
B–CNT[a]	0.44	B–H_2O[h]	–0.56	1.70	–0.12
N–CNT[b]	0.43	N–H_2O[£]	–0.23	3.12	–0.02
BN–VNTc	0.48	B–H_2O	–0.48	1.72	–0.20
		N–H_2O	–0.24	3.14	–0.02
B2N–CNT[d]	0.44	N–H_2O	–0.64	3.15	–0.28
BN2–CNT[e]	0.42	B–H_2O	–0.53	1.51	–0.08
BC_2N-type 2	0.93	B–H_2O	–0.48	1.72	–0.21
		N–H_2O	–0.18	3.14	–0.05

[a]Intrinsic nanotube doped with boron atom in 2-unit cell. [b]Intrinsic nanotube doped with nitrogen atom in 2unit cell. Intrinsic nanotube doped with boron–nitrogen atom pair in 2-unit cell. [d]Intrinsic nanotube doped with boron–nitrogen–boron atom pair in 2-unit cell. Intrinsic nanotube doped with nitrogen–boron–nitrogen atom pair in 2unit cell. [f]HOMO–LUMO band gap. [g]Electron charge transfer from the nanotube to molecules. [h]H_2O molecule binds to the nanotube with the oxygen atom close to the boron atom. IH_2O molecule binds to the nanotube with the oxygen atom close to the nitrogen atom.

molecules to nanotubes that drastically change the conductance of the nanotubes. The resulting charge transfer between the molecules and the nanotubes is related to the gate capacitance of the devices and the diameter and the length of the nanotube: $\Delta Q = (C_g \Delta V_g = \delta\theta(\pi/dl/\sigma)$ where, C_g is the capacitance of the nanotubes, and ΔV_g is the voltage change measured in the experiments, d and l are nanotube diamater and length respectively, while σ is molecule cross-section area and θ represents the molecule coverage on the nanotube surface. Additionally, a simultaneous doping of boron and nitrogen atoms into a carbon nanotube can lead to B_mN_n defect complexes. The possibility of defect complex engineering will increase the sensitivity of the doped nanotubes and thus make them very promising nanosensors due to the strong binding energy between doped nanotube and molecule species. Comparating the binding energy and charge-transfer resulted for the intrinsic nanotube detecting NO_2, B-doped nanotubes give under ppm level sensitivity CO and H_2O due to its similar binding energy and similar amount of charge-transfer per molecule.

The binding of CO and H_2O molecules with the boron-doped CNTs indicate that these molecules undergo chemical adsorption, while physical adsorption occurred when CO or H_2O molècules adsorbed on the nitrogen-doped CNTs. Reversing the adsorption of these molecules can be realized through applying UV light to the doped nanotube to desorb the molecules from the tube surface. Boron-doped CNTs show stronger binding energies and shorter binding distances for both CO and H_2O sensing cases. The stronger binding energies and shorter binding distances are consistent with the larger charge transfer between molecules and doped CNT systems. This indicates that the binding characteristic between molecules and doped CNT system is an ionic type. And the electron charge transfer is an important mechanism in changing the conductivity in the doped CNTs when adsorbed with CO or H_2O molecules.

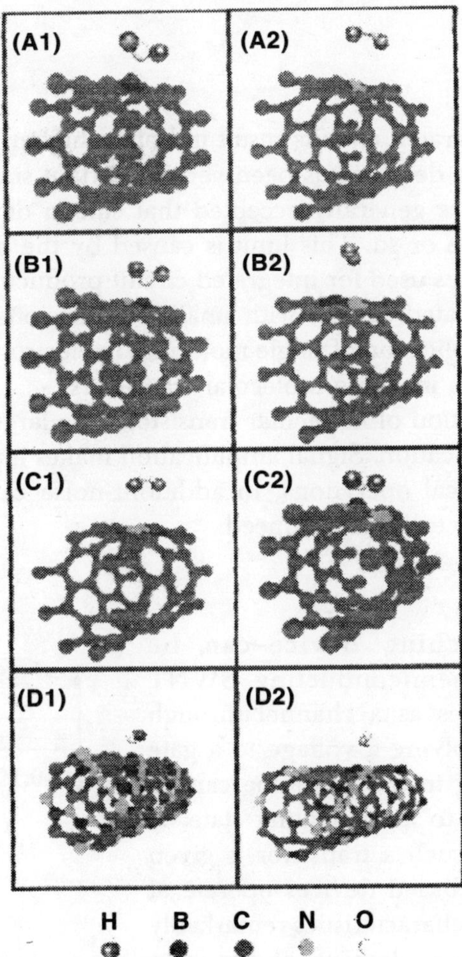

Fig. 10.18 *(A1) Semiconducting (8, 0) CNT with one carbon atom replaced by a boron atom that binds to an H_2O molecule, where the boron atom is close to the oxygen atom. (A2) Semiconducting (8, 0) CNT with one carbon atom replaced by one Nitrogen atom that binds to an H_2O molecule, where the nitrogen atom is close to the oxygen atom. Semiconducting (8, 0) CNT with one carbon atom replaced by a boron atom and another by a nitrogen atom that binds to an H_2O molecule, where the boron atom is close to the oxygen atom as in (B1), and the nitrogen atom is close to the oxygen atom as in (B2). (C1) Semiconducting (8, 0) CNT with two carbon atoms replaced by two boron atoms and one by a nitrogen atom that binds to an H_2O molecule, where the nitrogen atom is close to the oxygen atom. (C2) Semiconducting (8, 0) CNT with one carbon atom replaced by one boron atom and two others by two nitrogen atoms that bind to an H_2O molecules, where the boron atom is close to the oxygen atom. Semiconducting (8, 0) BC_2N type 2 nanotube that binds to an H_2O molecules, where (D1) BN_2 group binds to an H_2O molecule with the boron atom close to the oxygen atom, and (D2) B_2N group binds to an H_2O molecule with the nitrogen atom close to the oxygen atom*

Electronic Uses of CNTs

(A) Transistor

Smaller silicon based integrated circuits result in both a higher speed and device density. As a result, downscaling of these devices has been very important since their first implementation. However, at the moment it is generally accepted that silicon devices will reach fundamental scaling limits within a decade or so. This limit is caused by the minimum wavelength of light used in lithographic techniques used for integrated circuit production nowadays. For this reason a quest for alternative, integrated circuits with smaller dimensions has started. A major step in downscaling would be the application of single molecules in electronic devices. Carbon nanotubes have shown promising results in single molecular transistors.

For successful implementation of molecular transistors in a large and complex logic systems, they must show signal amplification. Signal amplification makes it possible to reference separate signals along a chain of logical operations. In addition, noise caused by thermal fluctuations and environmental disturbances is also reduced.

FET The Field-effect Transistor

A three-terminal switching device–can be constructed of only one semiconducting SWNT between two metal electrodes as a channel through which electrons flow. By applying a voltage to a gate electrode the current flowing in the nanotube can be switched from a conducting to an insulating state. A schematic representation of such a transistor is given Figure 10.19. The nanotube-based devices operate at room temperature. Electrical characteristics remarkably similar to off-the-shelf silicon devices of the gate electrode can change the conductivity of the nanotube channel in an FET by a factor of one million or more, comparable to silicon FEts. Because of its tiny size, however, the nanotube FET should switch reliably use much less power than a silicon-based device.

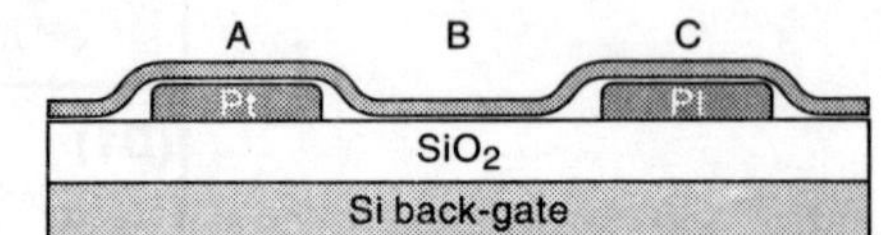

Fig. 10.19 *A single semi-conducting nanotube is contacted by two electrodes. The Si substrate, which is covered by a layer of SiO2 300nm thick, acts as a back-gate.*

Such carbon nanotube transistors can be coupled together, working as a logical switch, which is the basic component of computers.

Three terminal nanotransistors, in special, field-effect transistor show amplifying behaviour.

CNTs IN—MOSFET and CNTFET

Many applications for CNTs are electronic switch, field emitter for flat-panel displays electrochemical energy storage for batteries and fuel cells, interconnect for electronics, probe tips for scanning probe microscopy and lithography. Discussion on the use of a carbon nanotube as an electronic switching element which performs basic logic functions is given below.

Single-wall carbon nanotubes (SWNT) are two-dimensional graphene sheets rolled into nanometer-diameter cylinders that can either be 1D metals or semiconductors. With the appropriate chirality, SWNT can be semiconducting, with a bandgap inversely proportional to the diameter of the tube: $E_{gap} = 2\gamma_0 a_{C\text{-}C}/d$, where E_{gap} is the bandgap, γ_0 is the carbon-to-carbon tight-binding overlap energy, $a_{C\text{-}C}$ is the nearest-neighbor carbon-to-carbon distance (0.142 nm), and d is the diameter of the nanotube. A typical semiconducting SWNT has a diameter of 1.4 nm and a bandgap of about 0.5–0.65 eV. A straightforward application of this semiconducting property of the carbon nanotube is to form a field-effect transistor (FET) Fig. 6.29 which uses gold electrodes (as source and drain contacts) and a back-gated structure. Both p- and n-channel carbon nanotube transistors (CNFETs), as well as CNFETs with ambipolar behavior, are made simple circuits such as inverters, and ring oscillators have also been successfully fabricated.

CNT based transistors can be divided into two main categories: CNT field-effect-transistor (CNTFET) and CNT single-electron-transistor (CNTSET). CNTFET behaviour and performance is also compared with conventional the Metal-Oxide-Semiconductor. Field Effect Transistor (MOSFET).

A basic MOSFET Fig. 10.20 involves two main electrodes, designated as "source" and "drain", connected by a semi-conducting channel. In conventional devices this channel is made of silicon. In CNTFETs the channel is formed by a single, semi-conducting CNT. A third electrode, the "gate" is separated from the channel by a thin insulator film and can also be placed on top of the nanotube.

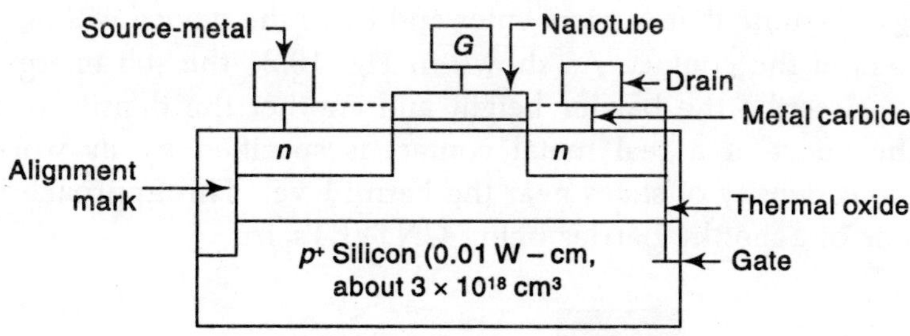

Fig. 10.20 *Set-up of a MOSFET*

If no charge is present on the gate, the channel is not conducting. The channel of a p-type (hole-conducting) MOSFET becomes conductive when a negative charge is placed on the gate and a certain threshold voltage is exceeded. This is the very basic principle of this type of transistor. Almost the same accounts for n-type MOSFETs in which a positive charge must be put on the gate. A CNTFET works in a slightly different way.

SIMULATING QUANTUM TRANSPORT IN BALLISIT CARBON NANOTUBES

Carbon nanotubes show promise for applications in future electronic systems, and the performance of carbon nanotube transistors, in particular, has been rapidly advancing. From a scientific perspective, carbon nanotube electronics offers a model system in which to explore and understand the effects of detailed microstructure of contacts, interfaces, and defects. Theory and computational techniques for the atomistic simulation of small electronic devices in general are also developed. A detailed treatment of carbon nanotube electronics an atomistic description of the nanotube along with a quantum mechanical treatment of electron transport, both ballistic and with the effects of dissipative scattering included.

A carbon nanotube can be viewed as a rolled-up sheet of graphene with a diameter typically between one and two nanometers. The nanotube can be either metallic or semiconducting, depending on how it is rolled up from the graphene sheet (i.e. depending on its chirality). *Semiconducting nanotubes are suitable for transistors.* In order to correctly treat carbon nanotube transistors, strong quantum confinement around the tube circumferential direction, quantum tunneling through Schottky barriers at the metal/nanotube contacts, and quantum tunneling and reflection at barriers in nanotube channel need to be considered. The non-equilibrium Green's function (NEGF) formalism, which solve Schrödinger equation under noon-equilibrium conditions and can treat coupling to contacts and dissipative scattering process, provides a sound basis for quantum device simulations. The NEGF simulation approach has demonstrated its usefulness for simulating nanoscale transistors from conventional is MOSFETs. MOSFETs with novel channel materials to CNTFETs and molecular transistors.

In carbon nanotube transistors, the metal source and drain are typically attached directly to the intrinsic nanotube channel, and the gate modulates the source-drain current by changing the transmission through the Schottky barrier at the source end of the channel. To properly simulate such devices, the metal/CNT junction must be treated quantum mechanically, by defining self-energy. Assume that carriers enter and leave the device without the need to tunnel through any barriers at the contact. As shown in Fig. 10.21 the self energy must contain two parameters, one to describe the barrier height and another the density of metal-induced gap states (MIGS). The effect of a real metal contact is specified by its work function and by injecting a continuous density of states near the Fermi level. This approach is in understanding transistor operations of Schottky barrier being CNTFETs.

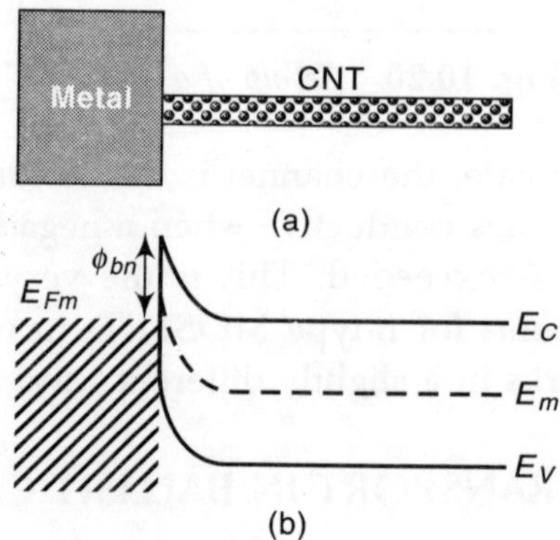

Fig. 10.21 *(a) The metal-carbon nanotube junction. (b) The band diagram of the junction. E_C, E_V and E_m, are the conduction band edge, the valence band edge, and the middle gap energy in the nanotube, respectively. E_{Fm} is the metal Fermi level, and ϕ_{bn} is the Schottky barrier height for electrons.*

The overall simulation must be done self-consistently, with Poisson's equation. Figure 10.22 shows the modeled, coaxial gate CNTFET, which provides the theoretically best gate control over the channel. The source and drain are heavily doped, semi-infinite carbon nanotubes, and the gate modulates the conductance of the channel, just like in a conventional Si MOSFET. By using a self-energy for metal/NT contacts the simulation scheme can be applied to Schottky barrier CNTFETs.

The transistor I-V characteristics strongly depend on the interplay of quantum transport and electrostatics. A self-consistent iteration between the NEGS transport equation and the Poisson equation is performed. For a given charge density, the Poisson equation is solved to obtain the electrostatic potential in the nanotube channel. Next, the computed potential profile is used as the input for the NEGF transport equation, and an improved estimate for the charge density is obtained. The iteration between the Poisson equation and the NEGF transport equation continues until self-consistence is achieved. Finally, the current for the self-consistent potential profile is computed.

For the coaxially gated carbon nanotube transistor, it is convenient to solve Poisosn's equation in cylindrical coordinates. Since the potential and charge density are invariant around the nanotube, the Poisson equation is essentially 1 2D problem along the tube (*x*-direction) and the radial direction (r-direcion) as shown in Fig. 10.22

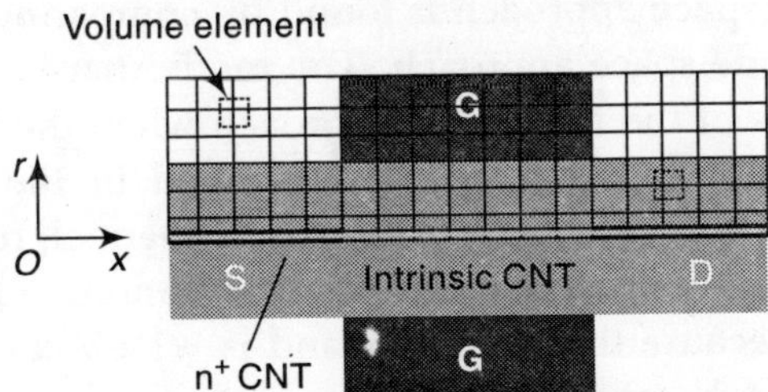

Fig. 10.22 *The modeled, coaxially gated carbon nanotube transistor with heavily-doped, semi-infinite nanotubes as the source/drain contacts. The channel is intrinsic and the gate length equals the channel length. Also shown are the simulated area, the simulation grid and the cylindrical coordinate system used for solving the Poisson equation. The dashed rectangular area shows the element used to discretize the Poisson equation at* (x_i, r_j).

Simulating a coaxially gated, CNTFET is shown in Fig. 10.22. The transistor channel is a (25, 0) intrinsic CNT, which results in a band gap of ~0.42 eV and a diameter of ~2 nm. The nanotube length is ~50 nm, consisting of ~1.2×10^4 carbon atoms. A self-consistent Poisson-NEGF simulation in the real space (using the recursive algorithm for computer the Green's function) is performed. Fig. 10.23(a) shows the energy-resolved local-density-of-states (LDOS), and the energy band profile. The band gap region with extremely low LDOS (darker in the grayscale plot) can be clearly identified. Due to the existence of the barriers, the source/drain incident wave is reflected and the quantum interference pattern between the incident and reflected waves is apparent. A quantum well is formed in the valence band of the channel, and the 1st and 2nd confined stated with one or two LDOS maxima, respectively, can be clearly seen. The band edge of the second subband is also observed. Figure 10.23(b) shows the energy

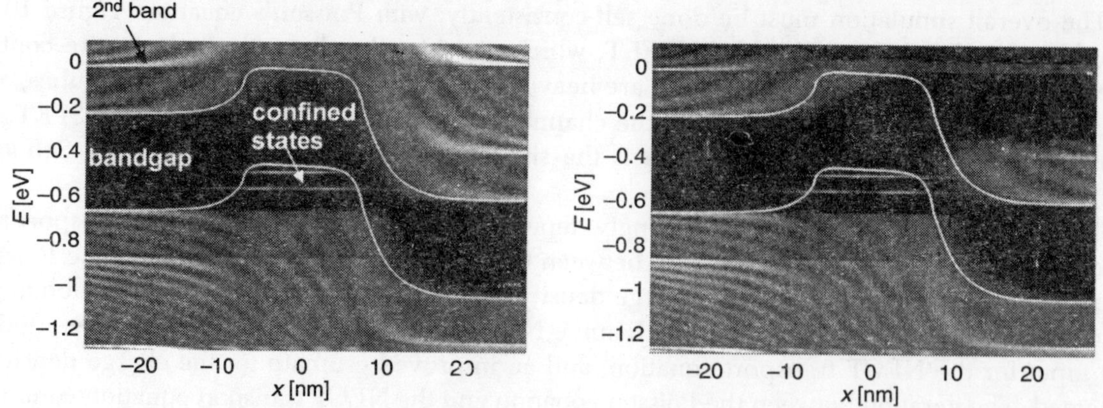

Fig. 10.23 *(a) The local-density-of-states (LDOS) and (b) the electron density spectrum computed by the real space approach at $V_G = 0.25$ V and $V_D = 0.4$ V. The modeled transistor is shown in Fig. 71. The nanometer is (25, 0) CNT with a diameter d~2 nm and bandgap Eg~0.4 eV.*

resolved electron density (electron density spectrum), which is obtained by filling the LDOS with the source or drain Fermi level. The bandgap, quantum interference, quantum confinement, and the second subband can still be clearly seen.

The validity of the mode space approach is found by comparing the results of the real space approach to those of the mode space approach. The mode space approach theoretically should exactly reproduce the results of the real space approach when the potential is invariant around the tube, and a sufficient number of modes is included in the mode space simulation. A CNTFET as shown in Fig. 10.22 with a (13, 0) nanotube channel, results in a band gap of ~0.83 eV and a diameter of ~1 nm, is simulated. The carbon nanotube length is ~50 nm, consisting ofg ~6000 carbon atoms. Because the third subband is ~1 eV away from the lowest subband and the applied bias is ≤ 0.4 V, only the lowest two subbands are treated in the mode space simulation. The Hamiltonian matrix for the lowest subband is small (~500 × 500), and computing the Green's function for a subband using the recursive algorithm is fast even on a single CPU PC. Fig. 10.24 compares the I-V characteristics of the real space and mode space approaches. It shows that the mode space simulation excellent reproduces the results of the real space approach. Figure 10.25(a) and (b) plots the band profile and the charge density at on-state, again shows that the mode space approach excellently reproduces the results from the real space approach results. The good agreement between the real and more space approach results from the equal potential and around the tube direction when it's coaxially gated. The mode space is highly advantageous in reducing the

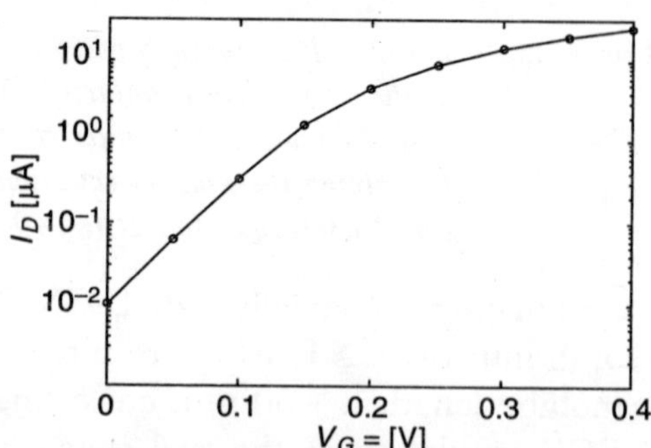

Fig. 10.24 *(a) The I-V characteristics computed by the real space approach (the solid line) and the mode space approach with 2 subbands (the circles) for a CNTFET. The (13, 0) nanotube channel length is 15 nm.*

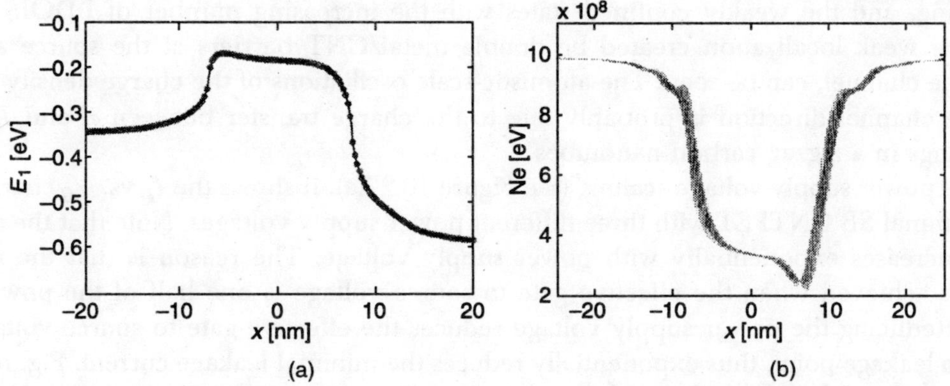

Fig. 10.25 *(a) The conduction band profile computed by the real space approach (the solid lines) and the mode space approach (the circles) at V_G -= V_D = 0.4 V. (b) The charge density computed by the real space approach (the solid line) and the mode space approach (the dashed line). The solid and dashed lines lie on top of each other.*

computational burden, and it is valid when the potential variation around the tube is much smaller than the spacing between the subbands.

For an SBFET-like CNTFET by self-consistent, quantum, simulation (see Fig. 10.26(a) the simulated transistor structure), the metal source/drain is directly attached to a (13, 0) intrinsic nanotube channel, so a Schottky barriers forms between the source/ drain and channel. A mid-gap Schottky barrier, with equal barrier height for electrons and holes, is simulated. Fig. 10.26(b) shows the local density of states at $V_D = V_G = 0.4$ V. The metal-induced gap states (MIGS) near the metal/CNT interfaces are apparent and decay rapidly with a tail of a few nanometers inside the channel. The tunneling states under Schottky barrier in the conduction band at the source end of the channel are clear. The metal-nanotube interface is not perfectly

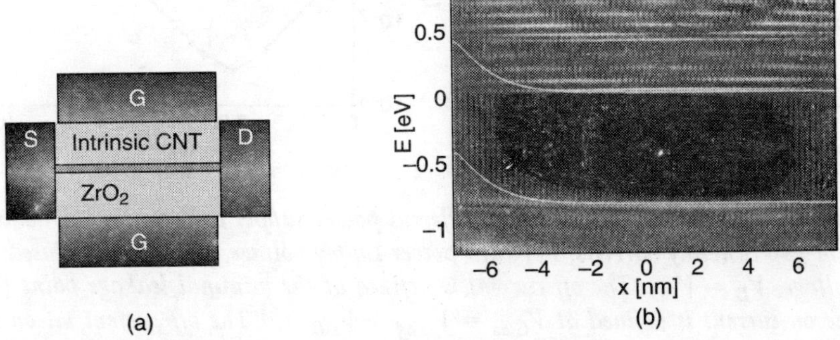

Fig. 10.26 *(a) The coaxially gated Schottky barrier carbon nanotube transistor with an intrinsic nanotube channel directly attached to metal source and drain contacts. The nanotube channel is a (3.0) zigzag CNT with a diameter d~1 nm and bandgap Eg~0.83 eV. The gate insulator is a 2 nm-thick ZrO_2. (b) The local-density-of-states (LDOS) at $V_D = V_G = 0.4$ V, which clearly shows tunneling through the Schottky barrier at the source end of the channel, and metal induced gap states (MIGS) at the metal/CNT interfaces.*

transmitting, and the weakly confined states with the increasing number of LDOIS maxima, due to the weak localization created be double metal/CNT barriers at the source and drain ends of the channel, can be seen. The atomistic-scale oscillations of the charge density spectrum along the channel direction is probably due to the charge transfer between *A* and *B* types of carbon rings in a zigzag carbon nanotubes.

For the power supply voltage scaling, (see Figure 10.27(a). It shows the I_D vs. V_G characteristics of the nominal SB CNTFET with three different power supply voltages. Note that the minimum current increases exponentially with power supply voltage. The reason is that the minimum leakage is achieved when the effective gate to source voltage is one half of the power supply voltage. Reducing the power supply voltage reduces the effective gate to source voltage at the minimum leakage point, thus exponentially reduces the minimal leakage current. Figure 10.27(a) also shows that the on-current increases with V_{DD}. The off-current vs. on-current for different power supply voltages is plotted in Fig. 10.27(b). The trade-off for reducing off-current by lowering the power supply voltage is the degradation of on-current. The choice of power supply voltage will depend on the type of circuit applications. For a low power design, the off-current must be small and the on-off current ratio needs to be large, which sets an upper limit for the power supply voltage. Although the low power supply voltage guarantees a small leakage current and large on-off ratio (because for small power supply voltage, the transistor operates in the exponential portion of the I_D vs. V_G characteristics), the on-current may still be too small for high-performance applications. To achieve a large on-current and a reasonable off-current simultaneously, which is required for high performance applications, the power supply voltage must be large.

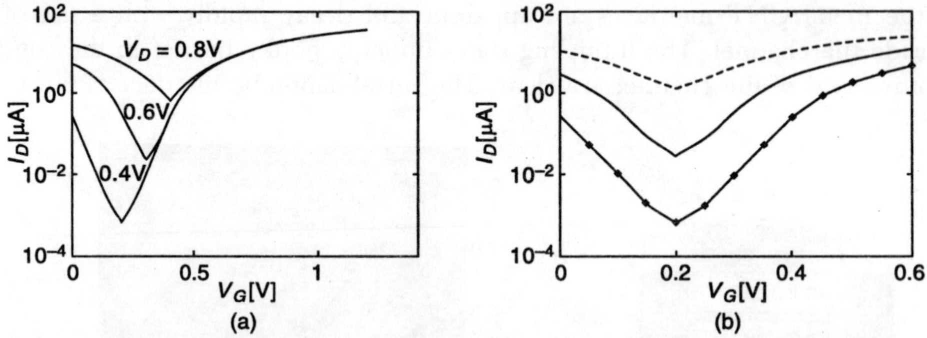

Fig. 10.27 *(a) I_D vs V_G characteristics under different power supply voltages for the nominal CNTFET with mid-gap Schottky barriers. For each power supply voltage, the drain is biased at the power supply voltage, $V_D = V_{DD}$. The off-current is defined at the minimal leakage point ($V_{G.off} = V_D/2$), and the on-current is defined at $V_{G,on} = V_{G,off} + V_{DD}$. (b) The off-current vs. on current for different power supply voltages.*

Figure 10.28 explores the issue of channel length scaling. In order to establish the ultimate scaling limit imposed by source-drain tunneling, very tin gate oxide (tox = 2 nm for the channel length of 10 nm and longer and tox = 1 nm for the channel length of 5 nm) is used to ensure excellent gate controlled electrostatics. Although we assume mid-gap Schottky barrier

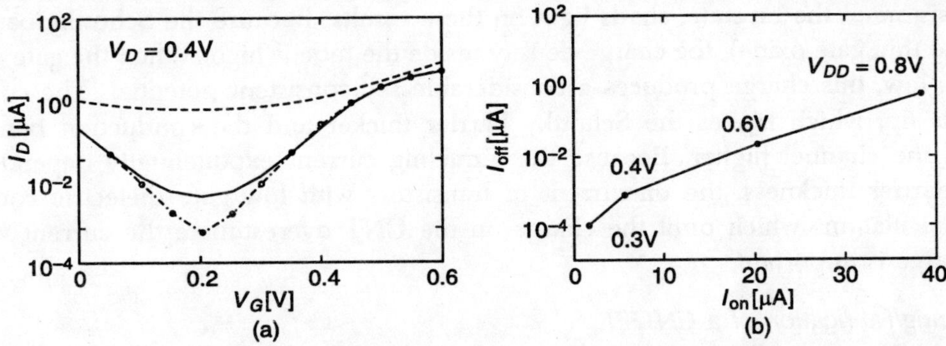

Fig. 10.28 *I_D vs. V_G characteristics of CNTFET with different channel length. The circles are for channel length (Lch) 30 nm and gate ZrO2 thickness (t_{ox}) = 2 nm, the dash-dot line for L_{ch} = 15 nm and t_{ox} = 2 nm, the solid line for L_{ch} = 10 nm and t_{ox} = 2 nm, and the dashed line for L_{ch} = 5 nm and t_{ox} = 1 nm. The flat band voltage is zero and mid-gap Schottky barriers are assumed for all transistors.*

height, similar observations apply top other barrier heights. When the channel length is larger than about 15 nm, the I-V characteristics are independent of the channel length because the channel is ballistic and the quantum tunneling resistance of the Schottky barrier at the source dominates the total channel resistance. Scaling the channel length down to 10 nm significantly increase the off-current, but the on-off current ratio still exceeds 100, which is probably acceptable for digital logic. If the channel length is aggressively scaled down to 5 nm, the on-off current ratio decreases to less than 10 due to significant source-drain quantum tunneling. Compared to Si MOSFETs with the similar channel lengths, the tunneling leakage of the CNTFET is more sever, partly due to typically smaller band gap and lighter effective mass for carriers in the CNT channel. (A parabolic E-k fit the very bottom of the conduction band of a 1 nm diameter CNT gives an effective mass of ~0.08.)

Fig. 10.29 explores the role of the gate dielectric constant. A previous study, which omitted the charge on the nanotube, concluded that simply increasing the gate dielectric constant everywhere did't change the band profile thus had no effect on the I-V characteristics of SB-CNTFETs. Those conclusions apply below and near threshold. Fig. 10.29, which shows the I_D vs. I_G characteristics for mid-gap SB-CNTFETs with a 2 nm thick oxode and three different dielectric constants, shows that a high-κ gate insulator does increase the on-current. The inset, which plots the conduction band profiles for

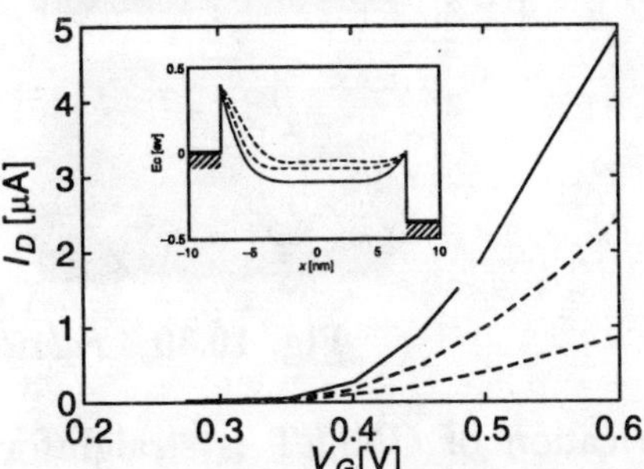

Fig. 10.29 *I_D vs. V_G characteristics at V_D = 0.4 V for the nominal CNTFET with different gate dielectric constant. The solid line is for ε_{ox} = 25, the dashed line for ε_{ox} = 4 and the dash-dot line for ε_{ox} = 1. The inset shows the corresponding conduction band profile at V_G = 0.6 V. The flat band voltage is zero and mid-gap Schottky barriers are assumed for all transistors.*

three transistors at the on-state, sheds light on these results. Because the Schottky barriers are thin (due to thin gate oxide), the charge density inside the tube is high. When the gate dielectric constant is low, this charge produces a considerable self-consistent potential. The conduction band floats up, which makes the Schottky barrier thicker and the conduction band in the interior of the channel higher. Because the tunneling current exponentially depends on the Schottky barrier thickness, the on-current of transistors with low gate dielectric constants is smaller. Calculations which omit the charge on the CNT overestimate the current when the mobile charge is important.

Manufacturing/Fabrication of a CNTFET

In 1998 Semi-conducting SWNTs and MWNTs were used in CNTFETs for the first time. Due to the basic principles of a CNTFET a metallic CNT cannot be used since it is always conducting. An example of the manufacturing of one of the first CNTFETs is given below.

As grown CNTs are purified by sonication in a H_2SO_4/H_2O_2 to remove catalyst particles. The purified nanotubes are dispersed in dichloroethane and spread over a silicon substrate, which is covered with siliconoxide (an insulating compound). On top of this layer, predefined electrode areas are created by lithographic techniques. The nanotubes are randomly spread over the surface, and in some cases a nanotube connects two electrodes. A different approach is to spincoat CNTs on a substrate first, after which the electrodes are deposited with electron beam lithography. The latter method is also used to create top gated CNTFETs, which is shown in Figure 10.30.

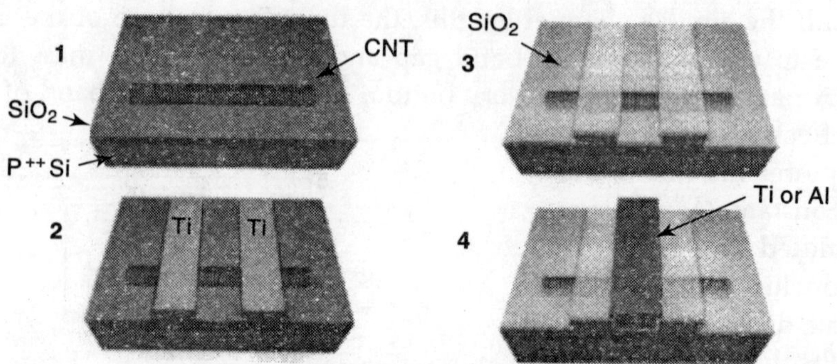

Fig. 10.30 *Fabrication of a top gate CNTFET*

The fabrication of CNFET is straightforward: First, the gold electrodes are formed on a silicon wafer with a thermally grown oxide (the gate insulator); then carbon nanotubes are dispersed on the wafer. The wafer substrate serves as the gate electrode of the CNFET Fig. 10.31. While this process is simple and suitable for a first demonstration of FET action, the source/drain contact resistance clearly limited the current drive of the CNFET. A recent embodiment of the CNFET is shown in the inset of Figure 10.32(a), where ideas borrowed from conventional microelectronics fabrication are employed to improve the contact resistance. In this process, carbon nanotubes are dispersed onto a heavily doped silicon wafer with a thermally grown oxide (serving as the gatę insulator). Then a transition metal (titanium or cobalt) is patterned on the wafer (by lift-off) as the source/drain contact metal. A subsequent

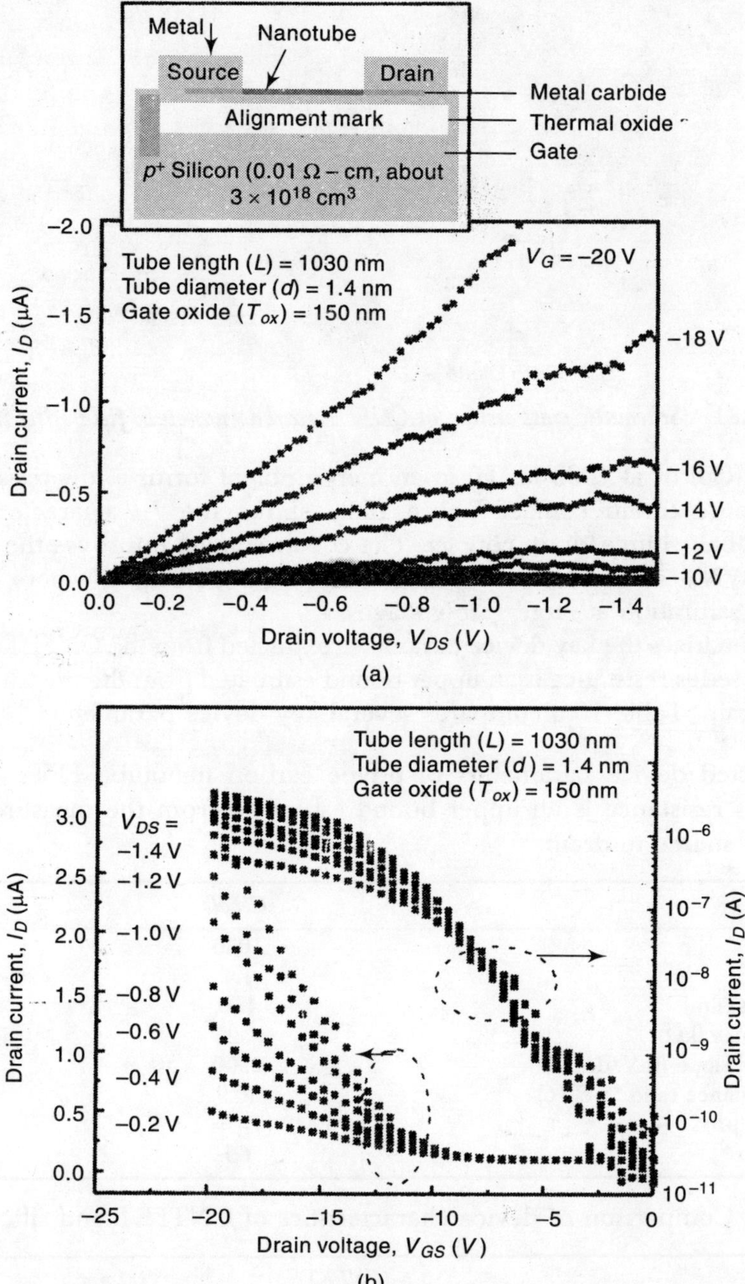

Fig. 10.32 *Room-temperature current–voltage characteristics of a carbon nanotube FET (CNFET) with a cobalt source/drain contact. (a) Drain current vs. drain voltage. The inset shows the schematic cross section of a carbon nanotube FET with a metal carbide source and drain contact and a back-gated structure. (b) Drain current vs. gate voltage. Carbon nanotube length = 1030 nm, tube diameter = 1.4 nm, gate-oxide thickness = 150 nm.*

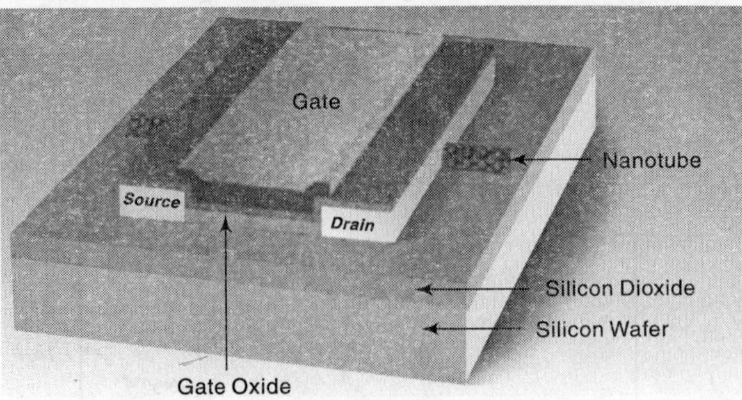

Fig. 10.31 *Schematic cross section of CNFET (carbon nanotube field effect transistor)*

anneal, at 673 K (Co) or at 1093 K (Ti) in an inert ambient forms a low-resistivity cobalt or Ti carbide at the source/drain contact Figure 10.32 shows the *I-V* characteristics of a p-type CNFET with a cobalt source/drain contact. The cobalt contact improves the contact resistance compared to SWNT material as evidenced from Figure 10.32(b), where the drain current shows no sign of saturation at large gate voltages.

Table 10.7 summarizes the key device parameter extracted from the CNFET *I-V* characteristics. The source/drain series resistance is an upper bound estimated from the measured total resistance from source to drain. Table 10.8 compares several key device parameters.

TABLE 10.7 Selected device parameters of p-type carbon nanotube FETs. The source/drain series resistance is an upper bound estimated from the measured total resistance from source to drain.

Contact scheme	*Cobalt*	*TiC*
Channel length (nm)	1030	800
Nanotube diameter	1.4	1.4
Gate-oxide thickness (nm)	150	150
Source/drain resistance (kΩ)	< 25	*
Inverse subthreshold slope (mV/decade)	2500	730
Interface trap capacitance ratio,** C_{it}/C_i	40.9	11.2
Transconductance# (μS)	0.342 (V_D = 1.4 V)	(V_D = 1.0 V)
"Mobility'" ($cm^2V^{-1}s^{-1}$)	60	68

Table 10.8 Comparison of device characteristics of CNTFET and silicon MOSFET.

	p-(CNFET 1030 nm	*100-nm MOSEFT*	*25-nm MOSEFT*
Transconductance (μS/μm)	122	100 (n-FET) 460 (p-FET)	1200 (n-FET) 640 (p-FET)
External resistance (Ω-cm per side)	< 70	~66 (n-FET) ~143 (p-FET)	~40 (n-FET) ~86 (p-FET)
Gate insulator (nm)	150	2.0	0.8

At high packing densities, the shielding from neighboring nanotubes renders the electric field pattern similar to the parallel-plate situation. It is therefore important to explore in future work the use of high-*k* gate dielectric materials and to study the electrostatics of alternative gate and channel geometries (such as the coaxial geometry) to maximize the gate-to-channel charge coupling.

In principle, the carbon nanotube has a perfect interface with the gate dielectric because all bonds of the carbon atoms are satisfied in a carbon nanotube. In other i.e., there are no dangling bonds which form interface states. However, the subthreshold turn-off characteristics shown in Figure 10.32(b) are very gradual (hundreds of mV/decade). This behavior is probably caused by traps in the gate dielectrics that arc close to the interface. Reducing the impurities in the gate dielectric will probably lead to improved subthreshold characteristics.

Characteristics of a CNTFET

CNT based transistors have excellent properties that are comparable with silicon based MOSFET transistors. The high stability and conductivity of single carbon nanotubes result in excellent transistor behaviour. High current densities for the 'on' state (2100 μA/μm) were achieved together with excellent on/off ratios of approximately 10^5. In a similar MOSFET the maximum current density is only 650 μA/μm under the same bias conditions.

Channel length of CNTFETs can be reduced to 5 nm, before tunnelling at room temperature results in unacceptable leakage currents. This means that very high device densities could be achieved.

Top Gate Devices even show better performance than transistors in which the substrate is the gate electrode. Switching in top gated CNTFET happens at much lower gate voltages. Another advantage of top gate transistors is the ability to switch different transistor individually. On/off ratios as high as 10^6 can be achieved with top gate CNTFETs.

CNT—Inter-and Intermolecular Logic Gates

Single wall carbon nanotubes (SWCNTs) are used as the active channels of field effect transistors (FET).

One of the most important findings is that they can be used as the channel of CNTFETs. Individual CNTFETs, and, arrays of CNTFETs with good switching characteristics have been produced. The suitability of these devices for computer electronics is in the integration of individual CNTFETs to form logic gates example: building logic gates out of a single nanotube or nanotube bundle. To build such logic circuits, one needs nanotube devices that use electrons (n-type CNTFETs) and holes (p-type) as the carriers of electricity. However, in all of the CNTFETs made of as-prepared invariability shows p-type characteristics; i.e., they are ON for negative gate bias.

n-type CNTFETs can be made by direct doping or the tube with an electropositive element such as potassium.

The *p*-type electronic character of CNTFETs is not an intrinsic property of the nanotubes annealing the CNTFETs in a vacuum, can change their electrical character and turn CNTFETs initially behaving as p-type into fully n-type ones, producing CNTFETs in a controlled manner any intermediate situation between p- and n-type will result in ambipolar (i.e., electron and hole) transport.

Using vacuum annealing or doping to make n-type CNTFETs, complementary devices, i.e., p- and n-CNTFETs on the same substrate can be fabricated. These complementary CNTFETs are then assembled to form intermolecular logic gates. The fabrication of a "NOT" gate or voltage inverter is demonstrated, using spatially selective doping. Complementary CNTFETs on a single nanotube bundle is fabricated; thus forming the first intramolecular logic gate.

The nanotubes are side-bonded to electrodes semiconducting SWNT (diameter 1.4, nm, band gap ~0.6 eV) dispersed from a 1,2-dichloroethane solution on an oxidized silicon substrate prepatterned with gold electrodes.

The conductive backside of the Si wafer is used as the gate. The data presented below, is obtained at room temperature and in a vacuum. Potassium deposition is also performed.

Figure 10.33(a) shows the result of a single flushing to 673 K in a vacuum of an SWNT - based p-CNTFET on its source/drain current (I_{DS}) vs gate voltage (V_G) characteristics. A striking transformation has taken place with the device now acting as a fully type transistor (ON for positive gate bias and OFF for negative bias). A positive gate bias allows electron conduction, while a negative gate bias turns the device OFF. For comparison, the inset in Figure 10.33(a) displays the data on a logarithmic scale taken before and after the annealing step. The ON-OFF current ratio of the n-CNTFET is more than 3 orders of magnitude, which is sufficient for use in logic circuits. Similar results have been obtained from more than 20 independent devices composed of either one single nanotube or a small bundle of nanotubes.

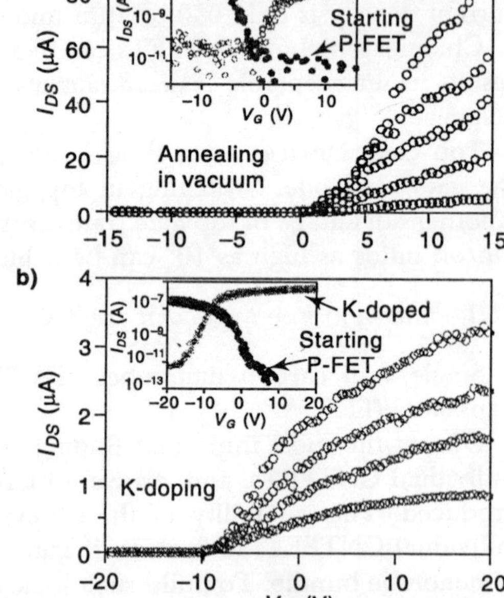

Fig. 10.33 *Electrical characteristics of n-type nanotube FETs made by (a) annealing the device in a vacuum at 700 K for 10 min and (b) doping with potassium. The curves are obtained at different values of V_{DS} starting at (a) 300 mV, step size 200 mV and (b) 200 mV, step size 200 mV.*

Figure 10.33(b) shows the effects of potassium (an electron donor) doping on the I-V characteristics of a CNTFET.

Doping produces n-type carbon nanotubes by shifting the Fermi level to the conduction band. As a result, the electron concentration in the conduction band increases and the conductance of the FET increases for a given positive gate voltage.

The charge transfer from oxygen to the nanotube leads to p-doping but the effects of O_2 on the electrical characteristics or SWCNTs are distinct from those caused by doping the nanotube, For example, while doping is known to shift the threshold voltage for turning ON an FET, oxygen treatment shows no such threshold shift. Instead, oxygen exposure simply decreases gradually the electron conduction. Although doping may take place, it is the contact barriers that control the carrier injection. The independence of these barriers on oxygen determines the electrical character of the CNTFETs.

The two different methods used here to produce n-CNTFETs, i.e., potassium doping and vacuum annealing, require vacuum and are reversed by exposure to air or oxygen. The stability of the devices in air is achieved capping the nanotube with an insulating film. In silicon technology, the need for capping ensures good stability over time and environment changes. In the case of nanotube devices, the requirements for the capping material are to be dense enough to prevent gas penetration at room temperature but permeable enough at high temperature to allow outgas sing of the device through the cap. PMMA is used as a capping material because it is a common resistant can be easily patterned. The found that p-CNTFET protected by PMMA can be converted into an n-FET by annealing in a vacuum at 473 K for 10 h. This n-FET is stable upon low oxygen exposure (10^{-2} Hgmm) but not under ambient air pressure. But by using a thin silicon oxide layer (10 nm) for capping, air stable n-type devices can be produced. PM.MA can also be used to provide an effective protection against potassium doping.

By bonding together a p- and an n-FET, the first logic gate based on carbon nanotube transistors is built a "NOT" gate or voltage inverter. The NOT gate is a fundamental one but along with the "AND" gate, it can be used to generate all other logic gates. This inverter works exactly the same way as the ordinary CMOS inverters. Figures 10.34(a)–(d) show the sequence

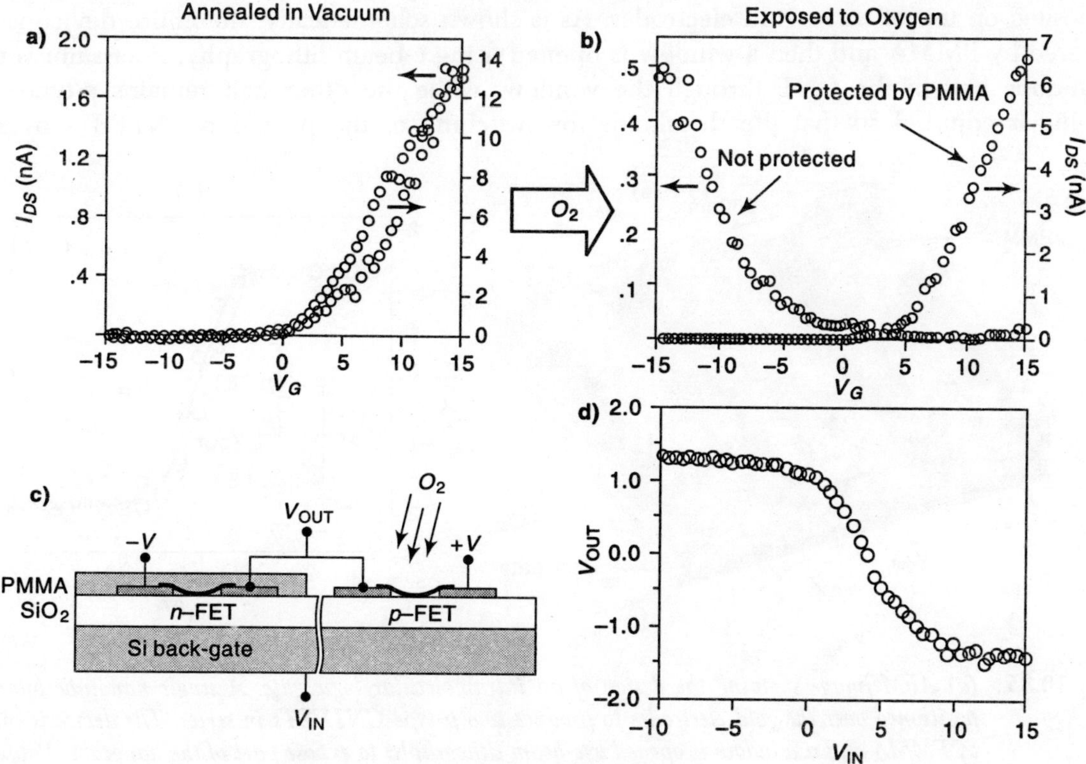

Fig. 10.34 *Fabrication of a voltage inverter ("NOT" logic gate) using two p-Typw CNTFETs (a) After vacuum annealing both CNTFETs are converted to n-type. (b) The two CNTFETs are exposed to oxygen (for 3 min). The unprotected n-CNTFET (black curve) converts back to the original p-type, while the protected CNTFET remains n-type. (c) The two complementary CNTFETs are wired as shown. (d) Characteristics of the resulting intermolecular inverter ($V = \pm 1.5$ V)*

of steps by which an intermolecular inverter (two independent CNTFETs are used) is fabricated using the vacuum annealing technique. First, one of the two CNTFETs is protected by PMMA, the other is not. After vacuum annealing, both CNTFETs are converted into n-CNTFFTs. Next the device is exposed to a pressure of 10^{-3} mm.Hg with oxygen for 3 min. The unprotected n-CNTFET converts back to the original p-type, while the protected CNTFET remains n-type. In the completed circuit, the input voltage is applied simultaneously to the gates of the two complementary CNTFETs. The p-CNTFET is polarized by a positive voltage, the n-FET by a negative voltage. A common contact is used as the intermolecular inverter's output. A positive input voltage turns the n-FET ON (the p-FET being OFF), resulting in the transmission of the negative polarization voltage to the output. A negative input. On the other hand, turns the p-CNTFET ON, inducing a positive output. In the intermediate input bias region, some current can flow through both FETs; that is, the device acts as a voltage divider. Thus, the transfer curve $V_{out}(V_{in})$ of this logic circuit is characteristic of a "NOT" logic gate, i.e., it changes an input of I or 0 to 0 and 1, respectively.

The ultimate integration of CNTFETs is to build logic gates using the same nanotube, i.e., produce an intramolecular logic circuit. Implement this integration is shown in Figure 10.35(a). This figure shows an atomic force microscope (AFM) image of a single-walled nanotube bundle deposited on top of three gold electrodes. As is shown schematically, the entire device is first covered by PMMA and then a window is opened using e-beam lithography. Potassium is used to n-dope half of the CNT through the window, while the other half remains p-type. The doping is adjusted so that the thresholds for switching of the p- and n-CNTFETs overlap.

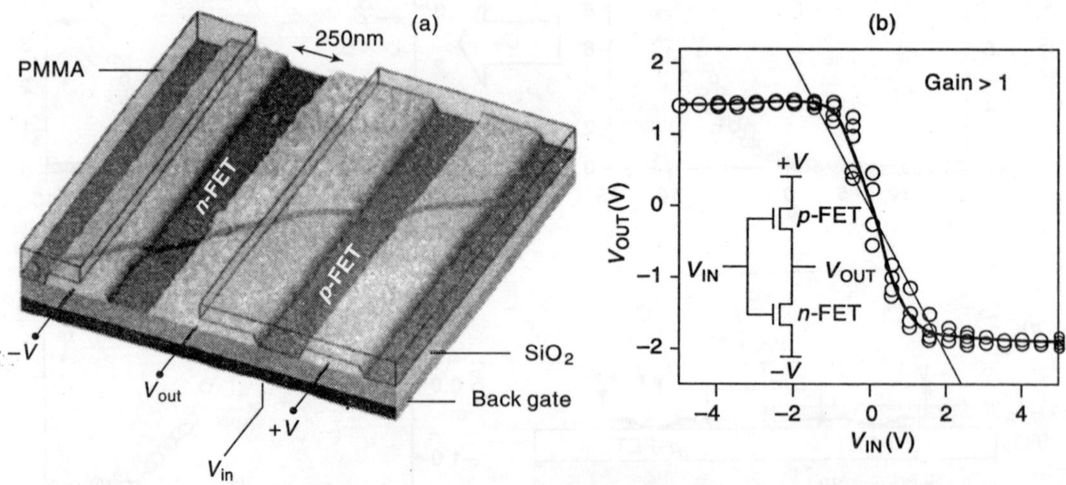

Fig. 10.35 *(a) AFM image showing the design of an intramolecular logic gate. A single nanotube bundle is positioned over the gold electrodes to produce two p-type CNTFETs in series. The device is covered by PMMA and a window is opened bye-beam lithography to expose part of the nanotube, Potassium is then evaporated through this window to produce an n-CNTFET, while the other CNTFET remains p-type. (b) Characteristics of the resulting intramolecular voltage inverter. Open red circles are raw data for five different measurements on the same device (V = ±2 V). The blue line is the average of these five measurements. The thin straight line corresponds to an output/input gain of one.*

Figure demonstrates that this design leads to an intramolecular NOT gate. The ability to adjust the relative position of the two thresholds by choosing the appropriate doping level allows a much sharper transition from the up to the down logic state.

To use a logical gate as part of a more complicated computing system, a gain (output/input ratio) of one is required. The intramolecular inverter presented here has a gain higher than one (1.6) and can thus be used to drive another gate or a more complicated logic circuit such as a ring oscillator.

These logic circuits have good output characteristics, and a gain larger than one is possible and the impact of oxygen on the electrical characteristics of tube devices is a contact effect rather than a bulk doping effect.

Carbon Nanotube Tips as Electrodes for Local Anodization of Surfaces

The unique structure, mechanical and electrical properties of carbon nanotubes make them promising materials for use as STM or AFM tips. Their shape allows them to probe crevices and image structures_with large curvature gradients. It is found that the electrical conductivity of such tips makes them useful not only for STM imaging but also in device fabrication. In particular, they can be used as the negative electrode in nanoscale tip-induced anodization. In Fig. 10.36 shows an example where a bundle of multi-walled nanotubes is used as a tip to oxidize a H-passivated silicon surface and generate the oxide pattern 'C- Tube'. For this purpose, the nanotube tip is biased at – 10 V while being scanned in contact with the surface under ambient conditions, (relative humidity of 20%). The oxidant is the atmospheric H_2O, which is condensed by capillary action near the apex of the tip.

Fig. 10.36 *Use of conducting carbon nanotube AFM tips as nanofabrication tools. In this case, a nanotube-bundle tip was used as the negative electrode to locally oxidize silicon and write the oxide pattern 'C-Tube'.*

The OH^- ions are driven by the strong field into the solid and induce the oxidation by reacting with Si holes in bulk Si. Hence, the resolution obtained is comparable to that achieved with conventional tips, but it can be enhanced significantly by using single-nanotube tips.

NANOBUD

In nanotechnology, NanoBuds are a newly discorerd material combining carbon nanotubes with fullerenes wherein the fullerenes are covalently bonded to the outer sidewalls of the underlying nanotubes. This new materials exhibits properties of both carbonnanotubes and fullerenes and is expected to replace both materials in may commercial application. For instance, Nanobud films have been shown to exhibit extremely high conductivity while being both optically transparent and flexible. Moreover, randomly oriented NanoBuds film have been demonstrated to have an extremely low work function for field emission of electrons thus making them ideal candidates for field emitters in, for example in, for example, field emission display screens.

DOPED CARBON NANOTUBES

The shells of MWCNT have different chiralities. The incommensurate character of the stacking of the different shells and comentum conservation requirements led to weak coupling between the shells of a MWCNT. As a results, the current at low bias and temperature in a bulk-contracted MWCNT flows only through the outermost shell. With respect to the transport mechanism, It is found that transport in MWCNT's is ballistic even at room temperature, while others concluded that the transport is Ohmic.

The transport characteristics of MWCNTs ie the role of defects on electrical transport, in particular, substitutional impurities may influence transport through scattering and active doping. So MWCNTs are doped with boron. The B-doped multiwalled carbon nanotubes (B-MCWNT's) are produced by are discharge between graphite electrodes containing boron. The content of boron in the MCWNT is about 1%. The nanotubes are purified using microfiltration and suspended in 1,2-dichloroethane. This suspension is used to disperse the nanotubes on SiO_2 substrates bearing the 30nm thick Ti/Au metal electrodes. The electrodes consist of 200 nm wide finger separated by gaps of 250nm. So as to perform both 2P and 4P resistance measurements on single B-MWCNT. The a. c. (23.3Hz) probe voltage is kept below k_BT/e over the whole temperature range (4.2 K to 300 K)

The strength of electron-phonon scattering and electron-electron scattering is examined and the role of electron coherence and may-body effects on the low-temperature electrical behavior of MWCNT's is evaluated here, use of four-probe (4P) and 2P resistance and magnetoresistance (MR) measurements covering the temperature range from 300 to 4.2 K is done and evidence of boron-induced multimode conduction in MWCNT's is observed. It is found to have linear temperature dependence upon cooling at lower temperature. Coherence lengths are extracted from the magnetoresistance (MR) measurements, and weakly inelastic electron-electron collision are found to dominate the decoherence mechanism. While a power law can fit the temperature dependence of the conductance.

Liquid formation, coherent backscattering effects account for the overall transport behavior.

Below is presented electrical data on two samples of B-MWCNT's. Their diameters are 27 nm and 17nm for samples *S*1 and *S*2, respectively important observations are (1) the low bias conductance of the sample are significantly higher than expected. At room temperature, sample *S*1 has $G_{2P} = 174$ μS and $G_{4P} = 443$ μS while *S*2 has $G_{2P} = 222$ μS and $G_{4P} = 500$ μS. (2) the maximum conductance per carbon shell is expected at low bias is $G = 2\ (2e^2/h) = 155$ μS since a metallic nanotube shell does not have more than two bands crossing at the Fermi energy (E_F). Thus more that two conduction channels contribute to the low bias conductance in the

B-MWCNT's (3) Scanning tunneling microscopy (STM) spectra of the B-MWCNT's on graphite show a lowering of the Fermi level (E_F) of the nanotube by ≤ 0.1 eV. The energy states of metallic nanotubes with diameters D are separated by $3a_0\gamma_0/D$, where a_0 is the C-C bond length (0.141 nm), and γ_o is the nearest-neighbor C-C interaction energy (2.9 eV). Thus, from sample *S1* and *S2* it is found; that upon a 0.1 eV E_F shift, at least 6 and 4 channels can participate, respectively, in condition this number of channels account for most of the resistance measured at 300 K. Moreover, as in graphite, substitutional boron may also enhance the coupling between the carbon shells of the MWCNT. It is, therefore, possible that more than one shell contributes to conduction even at low bias.

Upon cooling, the B-MWCNT samples in Fig. 10.37 show an initial linear decrease of both the 2P and 4P resistances down to a crossover temperature T^*. The linear decrease of the tube resistance with decreasing temperature is characteristic of metals, and has been ascribed to electron-phonon (*e*-ph) scattering which is observed in single-walled CNT (SWCNT) bundles and is due to the reduce influence of the contact resistance in 4P and in 2P measurements, along with the lowering of the Fermi level by the boron doping. The linear slope of the resistance vs temperature curves, *dR/dT*, is in the range of 0.3 Ω/K to 0.5 Ω/K in the different B-MWCNT's.

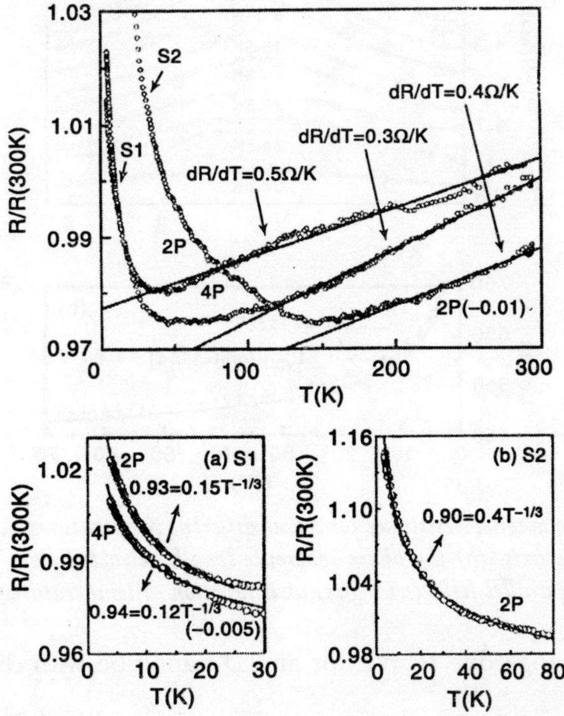

Fig. 10.37 *Top: Temperature dependence of the resistance R (T) normalized by R (300 K) for two multiwalled nanotubes: S1 (d = 27 nm) and S2 (d = 17 nm). For S1: R_{2P} (300 K) = 5.63 kΩ,and R_{4P} (300 K) = 2.23 kΩ, while for S2: R_{2P} (300 K) =4.5 kΩ and R_{4P} (300 K) = 2.0 kΩ. Bottom: the low-temperature reistance is fitted to one-dimensional weak-localization theory.*

The vibrational relaxation time $\tau_{e\text{-}ph}$ is estimated assuming the tube is a one-dimensional (1D) conductor with a conductance $G = (2e^2/h)(v_F\tau_{e\text{-}ph})N/L$, where v_F is the Fermi velocity, L is the electrode separation, and N is the number of the participating conduction channels. The elastic mean free path L_{el} = 220–250 nm i.e., similar to the spacing between the electrodes. Since the average separation between B atoms is estimated to be below 10 nm, the large L_{el} indicates that substitutional boron produces little scattering in MWCT's. The relaxation time $\tau_{e\text{-}ph}$ is about 0.4 ps at 300 K. This value is about 10 times longer than the relaxation time in copper, but about 10 times shorter than a similarly determined relaxation time in undoped SWCNT ropes.

At $T < T^*$ the resistance starts to increase with decreasing temperature. Figure 10.38(a) shows the MR (magneto resistance) of sample *S2* measured at different temperatures with the magnetic field perpendicular to the tube axis (circles). A negative MR, i.e., a decrease in R with increasing field, is observed. Negative MR is characteristic of a material in a WL state due to coherent backscattering processes.

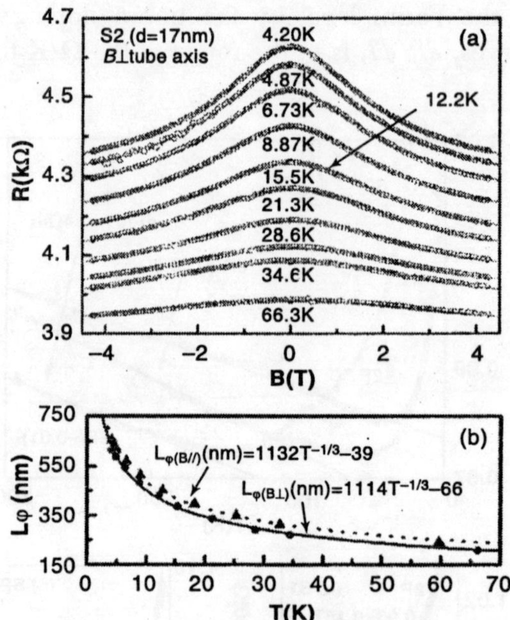

Fig. 10.38 *(a) Two-probe magnetoresistance for S1 at different temperatures in the magnetic field perpendicular to the tube axis, (b) the phase coherence length extracted from the MR in perpendicular field (circles), and parallel field (triangles) as a function of temperature.*

The conductances change due to WL for an 1D nanotube with diameter D and length L is given by

$$\Delta G(B) = \frac{-0.62\,Ne^2}{\hbar L}\left(\frac{1}{L_\varphi^2} + \frac{D^2}{12 l_B^2}\right)^{-1/2}, \tag{1}$$

where L_φ is the electron phase coherence length, and $l_B = (h/eB)^{1/2}$ is the magnetic length. L_φ is determined by fitting the MR results to Eq. (1).

The dominant dephasing mechanism can be determined by analyzing the temperature dependence of L_φ. The coherence lengths for *S*2 at different temperatures are plotted as circles in Fig. 10.38(b). The fit shows that the L_φ between 4 K and 70 K can be described very well by L_φ (*n m*) = 1114$T^{-1/3}$ – 66 (solid line). When the magnetic field is parallel to the tube axis, the conductance change is given by

$$\Delta G(B) = \frac{Ne^2}{\pi^2 \hbar L}\left(\frac{1}{L_\varphi^2} + \frac{D}{2l_B^2}\right)^{-1/2} \tag{2}$$

The coherence lengths obtained from the fits of MR in a parallel magnetic field with Eq (2) are plotted in Fig. 10.38(b) as triangles. They can be described by L_φ (*n m*) = 1132$T^{-1/3}$ – 39(dashed line). The two measurements of the coherence lengths agree remarkably well with each other over the entire temperature range studied. The temperature dependence of the phase coherence length, L_φ (*n m*) $\propto T^{-1/3}$, indicates that at low temperatures dephasing is dominated by weakly inelastic electron-electron interactions as found in many other 1D and 2D systems. Completely analogous behavior was observed in sample *S*1, where the L_φ in the range of *T*= 4.2-30 K can be expressed as L_φ (*n m*) =1861$T^{-1/3}$ –226 in the perpendicular field and as L_φ (*n m*) = 1434$T^{-1/3}$ – 82 in the parallel field.

Thus, this shows that the coherence lengths are quite long, and that there is an excellent agreement between the data and the predictions of dephasing over a wide temperature range up to about 70 K. The L_φ obtained here are larger than the tube circumference but smaller than the tube length, and also $L_\varphi >> L_{el}$ at low *T*. This indicates that the MWCNT's act as 1D conductors with respect to WL despite their large diameter. Under these conditions the MWCNT's are in the weak-localization regime where transport is diffusive. From the intercept of the L_φ vs T curve, the value of the diffusion coefficient D= 0.5 m^2 s - 1, and the dephasing time τ ~1 ps (at 4 K) are obtained.

Consider the low-temperature behavior of the resistance (Fig. 10.37). The theoretical expression for R (T) due to the first-order quantum correction at B = 0 for a 1D - WL system can be expressed as.

$$\frac{R(T) - R_0}{R_0} = \frac{e^2}{\pi^2 \hbar \sigma_u}[6.1L_\varphi(T) + 4.91\beta L_T(T)], \tag{3}$$

where σ_u is conductance of the 1D system per unit length, β is a screening factor, and L_T = $(\mathrm{hD}/2\pi k_B T)^{1/2}$ is the thermal length. The temperature dependence of the 2P and 4P resistances for both *S*1 and *S*2 tubes can be well described by Eq. (3), as shown by the solid lines in Fig. 1. In the fitting, $L_\varphi(T)$ obtained from MR data was used, and it was found that the contribution from the second term in Eq. (3), which describes large energy transfer in *e-e* collisions is negligible.

The low-temperature dependence of the conductance of SWCNT's has also been discussed in terms of Luttinger liquid (LL) theory where the presence of the LL state is the existence of a power-law dependence of the tunneling conductance on temperature: G$\propto T^\alpha$.

LL behavior in SWCNT bundles in the range of 8 to 300 K, where values of α are in the range of 0.33 to 0.58 LL state is also be present in MWNCT's.

α exponent of the outer shell reduces by the interaction with the other metallic shells in the MWNCT. To test the possibility of a LL-based interpretation of the low temperature behavior of B-MWCNT's, plot the conductance of *S*1 and *S*2 as a function of *T* on a log-log scale [Fig. 10.39(b)]. A good power-law fir is obtained the lowest temperatures, giving $G(2P)_{S1} \propto T^{0.02}$, and $G(4P)_{S1} \propto T^{0.03}$ (T below ~30 K) and $G(2P)_{S2} \propto T^{0.05}$ (*T* below ~50 K).

The α values obtained as very low. These imply screening by an unrealistically large number of metallic shells (N > 15) in he MWCNT.

dI/*dV* as *V* curves [Fig. 10.39(b)] show a zero-bias nanomaly. Such an anomaly is usually considered as a manifestation of the LL state. However, disorder-enhanced electron correlation can also give rise to a zero-bias anomaly. Finally, there is a weak peak near E_F in Fig. 10.39(b) (see arrow). A similar peak is observed in the STM spectra of B-doped MWCNT's. From the temperature dependence of this one concludes that it originates from the free acceptor BC_3 local structure.

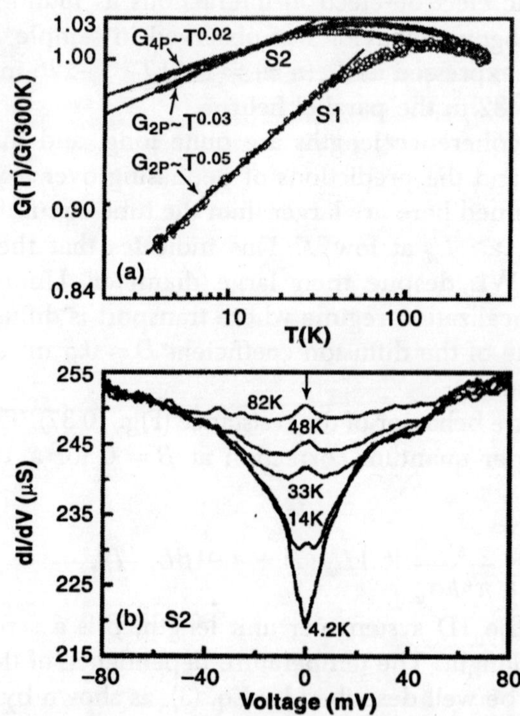

Fig. 10.39 *(a) Temperature dependence of the conduclance G(T) normalized by G (300 K) for S1 and S2 (b) Zero-bias anomaly (ZBA) seen in dI/ dV vs V plots as a function of temperature. The boron-induced peak is indicated by the arrow*

In conclusion, the importance of impurity, phonon, and electro-electron scattering on the transport in B-doped MWCNT's, B-doped affects transport by shifting the Fermi level and thus changing the number of participating conduction channels without introducing strong backscattering. Upon cooling, the resistance of the tubes initially shows a linear decrease with temperature characteristic of metal. The electron-phonon scattering is weak and leads to only 2–3% change in the resistance over the range of 300–4 K. At lower temperatures, the resistance starts to increase. A negative magnetoresistance characterizes this regime indicating the formation of a 1D weakly localized state. The extracted coherence lengths are long, and their temperature dependence indicates that the dominant dephasing mechanism involves electron-electron scattering. A comparison of our results with the expectations of the Luttinger liquid and of the coherent backscattering (weak localization) models shows that only the latter can account for all of the observed low-temperature transport characteristics. The strong influence of boron on the electrical characteristics of the tube, and the similarity of the effects (i.e., electron trapping) produced by boron and by structural defects in graphite, suggest that the puzzling *p*-type character of nanotubes may be due, at least in part, to defects and foreign acceptor atoms.

Electronic Structural Techniques with CNT

(A) COMPUTATIONAL TECHNIQUES

Electrons in carbon nanotubes are confined along their circumference. From the electronic structure of finite length arm-chair CNT it is found that the armchair tubes (that are metallic) when infinitely long, develop a band gap. The value of the band-gap decreases with increasing tube length, but the decrease shows oscillation in short tubes. These oscillation save due to the changes in the bonding characteristics of the HOMO and LUMO orbitals of the tubes. Finite-size graphene sheets are also found to have finite band-gaps, but no clear oscillation is observed. The DOS spectra of short nanotubes evolve with increasing length from that characteristic of a 0-D system to that characteristic of a delocalized 1-D system. This transformation is complete already for tubes about 10 nm long. The chemical stability of the nanotubes, expressed by the binding energy of a carbon atom, is low for short tubes but increases and saturates at tube lengths < 10 nm. In the carbon nanotube models, the C–C and C–H bond lengths are fixed at the values observed in bulk graphite, which are 1.42 and 1.09 A, respectively. All calculations are performed at fixed geometries without optimization of the atomic structures. The elementary unit used is defined as a single circular plane of carbon atoms which are packed along the length of the nanotube (note: the distance between two consecutive sections in the nanotube is 1.22 Å). The dangling bonds at the ends of the tube are saturated with hydrogen atoms. The energy band-gap is obtained as the difference between the LUMO and HOMO energies, while the Fermi energy ($E_F = 0$ eV) is taken as the mean energy between LUMO and HOMO.

Density of states (DOS) plots are generated by convoluting the computed electronic structure with a 50:50 combination of Gaussian and Lorentzian function. In order to analyze the nature of the energy bands, series of projections to obtain the local density of states (LDOS) is performed where each molecular orbital is weighted by the contribution obtained from analysis of specific carbon atoms. The general exp. ession used to generate the DOS plots is

$$\mathrm{DOS}(\epsilon) = \sum_{i=1}^{N}\left(\left[\frac{n_i}{\omega(\pi/2)^{1/2}}\exp\left(\frac{-2(\epsilon-\epsilon_i)^2}{\omega^2}\right)+\left(\frac{2n_i\omega}{\pi(\omega^2+4(\epsilon-\epsilon_i)^2}\right)\right]\right)$$

where ω is the resolution, n_i is the population of level i, and $(\epsilon - \epsilon_i)$ is the energy difference between the energy ϵ and the eigenvalue ϵ_i.

Electronic structure calculations are performed using density functional theory (DFT) and semiempirical model 3 (MNDO-PM3) and extended Huckel molecular orbitals (EHMO)] computational methods.

To evaluate the influence of finite length on the electronic properties of the nanotubes, the electronic structure of different-length segments of a (6, 6) nanotube is calculated. Figure 11.1 shows, that, unlike the infinite tube that is metallic, very short (<100 Å) nanotubes have an energy band-gap (E_g). Figure 11.1 also reveals that E_g shows a regular oscillatory dependence on tube length. The oscillation of the band-gap values is particularly strong in the DFT method where E_g varies from 2.8 to 9.0 eV for tubes shorter than 20 Å. The semiempirical MNDO-PM3 method gives results comparable to those obtained by DFT. On the other hand, EHMO and density functional theory–generalized gradient approximation (DFT-GGA) methods lead to quite similar values of the band-gap. The oscillation amplitude decreases, and the band-gap value converges slowly to zero for larger nanotubes. For a 96 Å long nanotube, EHMO predicts a metallic behavior while MNDO-PM3 shows a band-gap of 2.0 e V. The comparison of the different quantum-chemical methods used to determine E_g will be discussed in more detail below.

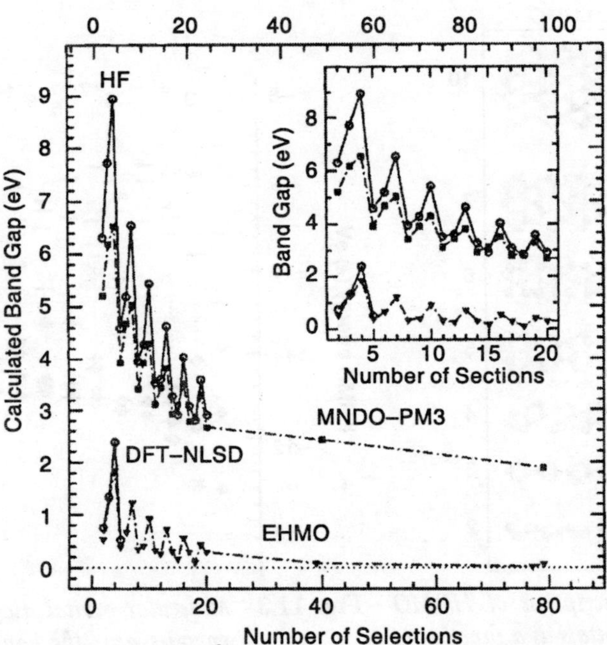

Fig. 11.1 *Variation of the band-gap of a (6, 6) nanotube as a function of its length determined using different computational techniques. The inset gives an expanded view of the band-gap behavior of very short tubes.*

The inset in Figure 11.1 gives an expanded view of the bandgap oscillations for nanotubes shorter than 25 Å. All computational methods predict a regular and converging oscillation of E_g. The oscillation is related to the structure of the nanotube; high band-gap values are invariably obtained for tubes having $(3n + 1)$ (where $n = 1, 2, \ldots$) sections (a section contains a single circular plane of carbon atoms). Furthermore, consecutive medium or low band-gap values are periodically observed for nanotubes having $(3n)$ and $(3n - 1)$ sections. The main difference between the oscillatory behavior of E_g obtained with the different computational methods lies in the different length needed for an inversion between medium and low bandgap behavior to occur. The different oscillations observed, as well as the low to medium band-gap inversion, for small nanotubes can be explained in terms of the changing bonding characteristics of the HOMO and LUMO orbitals.

Figure 11.2 shows the nodal structure of the frontier π orbitals for the nanotube models, while Figure 11.3 gives the energies of 10 molecular orbitals (obtained from EHMO theory) which lie near the band-gap region. The relative stability of frontier orbitals (HOMO and LUMO) for the first three models (S2, S3, and S4 with respectively two, three and four sections) is based on the nature of the bonding along the circumference of the tube, i.e. intra-section interactions, as well as the bonding along the tube axis, i.e. inter-section interactions. Furthermore, the S2, S3, and S4 models constitute the repeat units from which the HOMO/LUMO orbitals of the longer tubes may be constructed. Figure 11.2 shows that the HOMO for S2 (two-section model), the first member of the S$(3n - 1)$ group, has bonding character along the circumference, while the LUMO has bonding character between sections, i.e. along the tube axis.

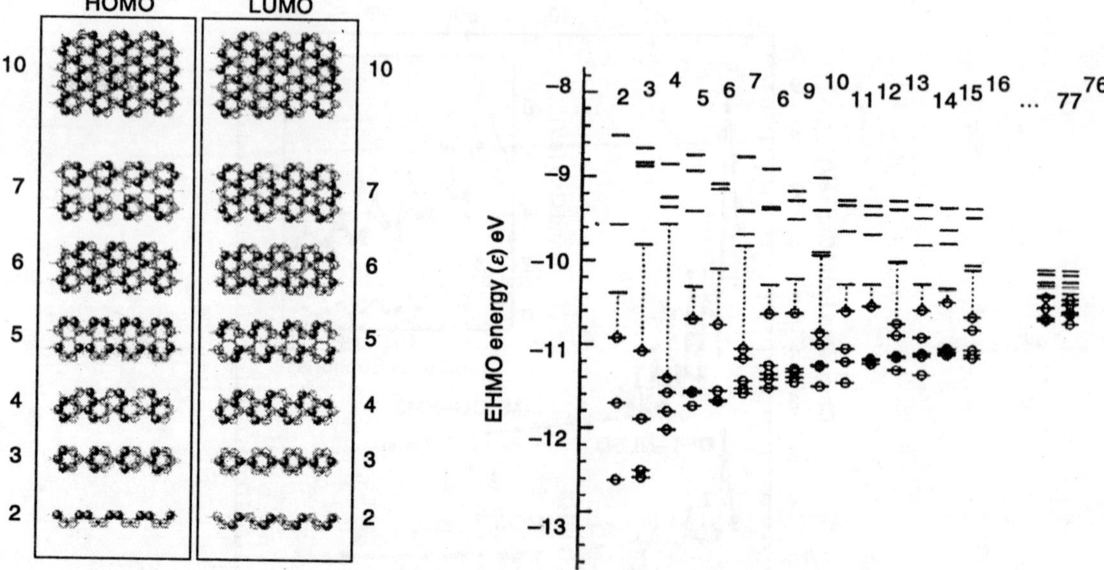

Fig. 11.2 *Qualitative description of HOMO and LUMO orbitals as a function of the nanotube length. The filled and empty circles indicate opposite phases of the molecular wave function.*

Fig. 11.3 *Molecular orbital diagram for ten orbitals with energies near the band-gap as determined using EHMO theory.*

So, the energy difference between the HOMO and LUMO of the S2 structure is small, leading to a low band-gap. On going from the S3 (S(3*n*) group) to the S4 (S(3*n* + 1) group) structures, Figure 11.2 shows that the HOMO acquires an additional bonding character along the tube axis and hence is stabilized more. But the LUMOs of both S3 and S4 type structures show antibonding inter-section interactions which lead to a larger band-gap. The gap is lower for S3 because of the net bonding character of the LUMO along the circumference. The structure and the stability of the frontier orbitals for longer nanotubes are directly related to the characteristics of those three first structures. The *S*(3*n* – 1) group (i.e. S2, S5, S8, ...) is characterized by non bonding inter-section interactions, as is evident by considering the orbital nodal properties in Figure 11.2. The stability of the frontier orbitals in the *S*(3*n* – 1) group is weakly influenced by inter-section interactions. The slight destabilization of the frontier orbitals is due to long-range anti bonding lateral interactions that reduce slowly the band-gap with increasing tube length. In the S(3*n*) group (53, S6, S9, ...), the decreasing band-gap is caused by a destabilizing lateral anti bonding interaction in the HOMO coupled with a stabilizing lateral interaction in the LUMO. The lateral interactions have a strong influence on the value of the band-gap of the S(3*n*) group so that it actually becomes lower than the band-gap value of the S(3*n* – 1) group at *n* = 3. This behavior occurs at different tube lengths depending on the computational method used to determine the energies of the molecular orbitals. Finally, the S(3*n* + 1) group that is characterized by high band-gaps follows a trend similar to that of the 5(3*n*) group with increasing nanotube length a destabilization of the HOMO through anti bonding interactions between sections and stabilization of the LUMO through partly bonding lateral interactions.

Band-gap oscillations for 2-D polyphenanthrenes on the basis of EHMO structure calculations are made. Here, instead of the carbon section, a phenanthrene-edge structure is used as the elementary unit, high band-gaps are found in systems with (6*n* – 2) sections rather than (3*n* + 1) sections. Nevertheless, although the compositions of the frontier orbitals of the (6, 6) armchair nanotube are to a certain extent different from those of the periodic 2-D polyphenanthrenes, both structures give rise to a similar pattern of band-gap oscillations. In Figure 11.4 shows the variation of the band-gap value for a perfect (6, 6) nanotube to that of an equivalent 2-D graphene fragment (whose broken C–C bonds were saturated with hydrogen atoms) and that of the periodic sheet of 2-D polyphenanthrenes. The growth of the finite 2-D graphene sheet does not lead to a clear band-gap oscillation pattern and the E_g rapidly converges to low values. Furthermore, the finite graphene sheet gives E_g values quite different from those of the infinite polyphenanthren sheet. Figure. 11.5 shows the influence of the nanotube diameter on the magnitude of the band-gap oscillations with respect to the fluctuations observed for a periodic graphene sheet increasing the nanotube diameter by going from the (6, 6) to the (10, 10) tube brings its band-gap value closer to that of the periodic 2-D graphene sheet. However, the 1/(diameter) scaling of E_g predicted for infinite tubes, is not observed in very short tube segments.

Crucial to the understanding of the nanotube growth process are changes in the binding energy of *C* atoms as a function of the increasing nanotube length. Figure 11.6 compares the binding energy (BE) of carbon atoms in a finite (6, 6) nanotube as determined by different computational techniques. A converged BE value indicates that all carbon atoms are nearly

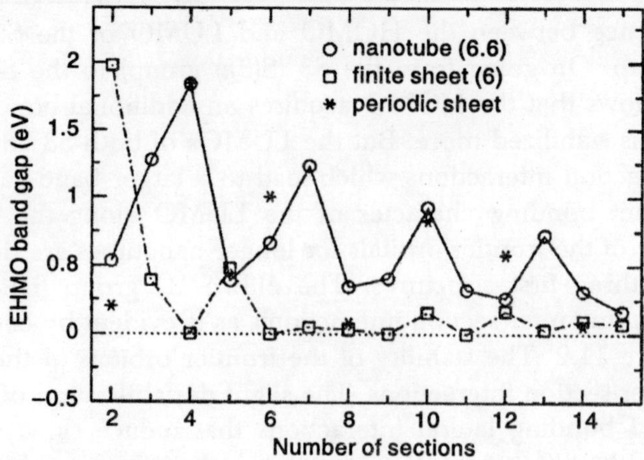

Fig. 11.4 *Comparison of the band-gap variation for finite and periodic graphene sheets with respect to the (6, 6) nanotube structure. The "finite" sheet is the unwrapped structure of the (6, 6) nanotube where broken C–C bonds were saturated with hydrogen.*

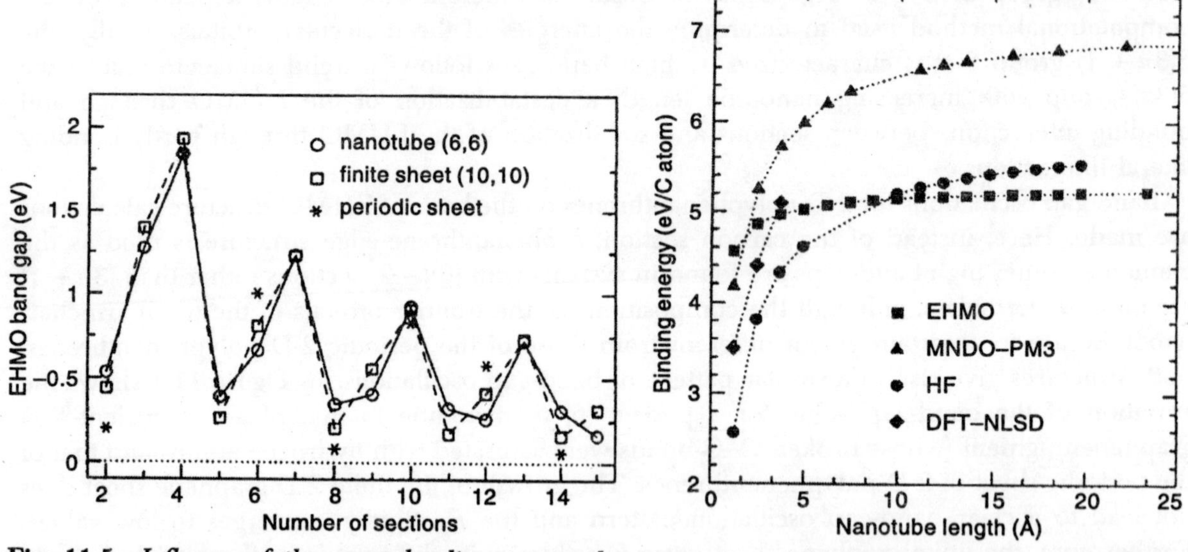

Fig. 11.5 *Influence of the nanotube diameter on the band-gap oscillations and comparison for an equivalent periodic 2-D graphene sheet.*

Fig. 11.6 *Variation of the carbon atom binding energy as a function of the length of a growing (6, 6) nanotube.*

equivalent to each other. Major changes are observed only for very short tubes where the BE is found to increase sharply up to about 15 Å. The smaller changes in BE observed with the EHMO method can be understood as due to the treatment of long-range interaction in EHMO theory; carbon atoms are weakly perturbed by second and higher neighbor carbon atoms. The highest estimate of BE is obtained with MNDO-PM3 (7.0 eV/C atom), and the lowest, with

EHMO (5.5 eV/C atom). The BE values lie between these two extremes. The low binding energy obtained for short tubes suggests that carbon in short nanotubes would be more reactive.

A sufficiently long nanotube reproduces the main characteristics of the electronic structure of a quasi-1-D periodic system. The influence of the length on the general electronic structure of a (6, 6) armchair nanotube in Figure 11.7 is obtained using the EHMO method. The nanotube electronic structure is characterized in terms of the position and the intensity of the main DOS peaks, the magnitude of DOS at the Fermi level, and the presence of fine structure. The spectrum of the very short 4 Å (four carbon sections) Fig. 11.7(a) is characteristic of a 0-D (quantum dot) system with discrete quantum levels. However, the band position and its overall DOS profile are not drastically different from those of more extended systems. The DOS profile for a 96 Å Fig. 11.7(d), clearly reveals the presence of characteristic fine structures at high binding energy (BE) between –22 and –12 eV. These structures are "van Hove singularities" characteristic of periodic one-dimensional (1-D) systems. Each of these peaks is characterized by a specific number of nodes in the wave function along the circumference of a single nanotube section, while the $1/(E - E_0)^{1/2}$ tail reflects the free-electron character along the tube axis. In fact, these singularities are becoming evident even in the 47 Å nanotube Fig. 8.7(c) that represents the minimum length nanotube that exemplifies such 1-D-like characteristics. For nanotubes larger than 47 Å, the main changes in electronic structure occur around the Fermi level, where the DOS is found to increase slowly with increasing nanotube length. Although it gives a different estimate for the band-gap, the semiempirical MNDO-PM3 method shows trends similar to those of EHMO concerning the electronic structure of the growing nanotubes.

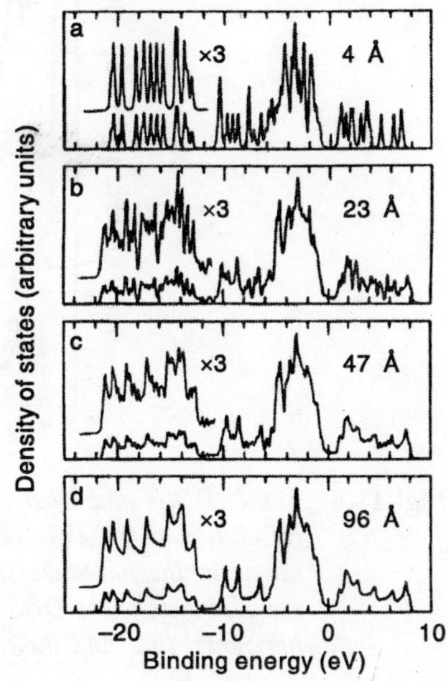

Fig. 11.7 *Electronic stmcture of a growing (6, 6) carbon nanotube obtained using the EHMO technique. Energy resolution $\omega = 0.2$ eV.*

Figure 11.8 gives a more detailed description of the electronic structure of the 96 Å nanotube. In the general DOS diagram, the van Hove singularities at high BE are σ states that originate from C(2s) atomic orbital combinations. The σ states formed from C(2p) orbitals lie at lower BE between –11 and –4 eV, while π states extend from –5 to 0 eV. These states are important because they are responsible for the electrical properties of nanotubes. The states above the Fermi level are mainly π^* states, while σ^* states lie above 10 eV. The inset gives an expanded view of the fine structure of the 96 Å nanotube near the Fermi energy. In addition, the values of the local density of states (LDOS) near the Fermi energy for sections ranging from the tube boundary (section 1) to the middle of the tube (section 40) are plotted. To construct these LDOS plots, we summed over the density of the 12 carbon atoms contained in that section of the nanotube. From Figure 11.8 it can be seen that the LDOS of the π states near E_F hardly changes along the length of the tube, indicating that these states are indeed well delocalized within this short tube.

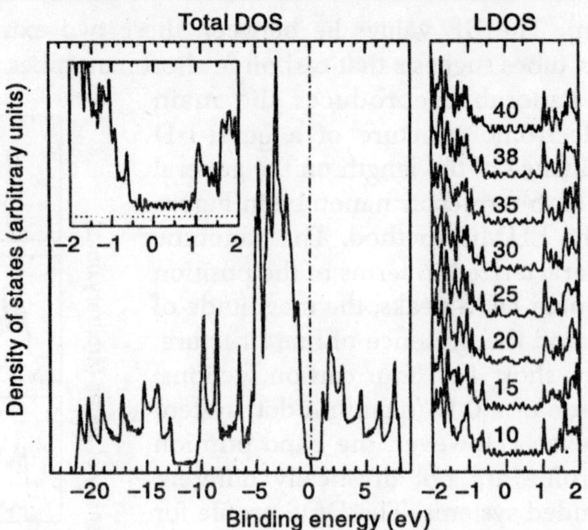

Fig. 11.8 *Total (DOS) and local density of states LDOS) diagrams of a 96 Å armchair (6, 6) nanotubes (resolution = 0.2 eV). The indicates in the LDOS diagram give the relative position of the carbon atoms in the nanotube structure (1, boundary; 40 middle of the nanotube). The insert gives an expaded view of the DOS near the Fermi level (E = 0 eV). The zero of the DOS scale is indicated by horizontal line, and the energy resolution is 0.05 eV.*

On comparing briefly the results obtained with the different computational techniques; Figure 11.1 shows, both HF and MNDO-PM3 methods give large band-gaps (9.0 and 6.5 eV, respectively) for a short 4 A nanotube, while DFT and EHMO give moderate band-gap values (2.4 and 1.8 eV, respectively). Whereas the MNDO-PM3 method, the parametrization includes a treatment, of correlation through the fitting with experimental data and MNDO band-structure calculations have predicted a metallic behavior ($E_g = 0$ eV) for periodic 2-D graphite.

This suggests that the computed band-gap should converge to zero for extended nanotubes. But Figure 11.1 shows that nanotubes larger than ~450 Å would be metallic. On the other hand DFT-based methods tend to systematically under-estimate the band-gap value of low-band-gap polymers. However, developments of DFT/hybrid functionals that include a weighted contribution of DFT exchange allow a better evaluation of the band-gap value. Given that the computed band-gaps with EHMO are similar to the DFT results, the band-gap values are underestimated with EHMO. On the basis of DFT and EHMO results, nanotubes longer than 100 A would be metallic. However, given the above discussion on the limitations of the computational techniques, the transition to the metallic state likely occurs at longer lengths, probably in the range of 10–20 nm.

(B) Single Nanotube field-effect Transistors, Electrical Transport, and Coulomb Blockade

Carbon nanotubes are thought to be examples of ideal one-dimensional quantized conductors. As, electrons in π-states are involved in the conduction, the low bias resistance of the nanotube for ballistic transport 6 kΩ. Electrical measurements on individual nanotubes became possible by preparing electrical contacts, dispersing nanotubes on them expecting that a nanotube will end up bridging the electrodes.

Fig. 11.9 shows an AFM image of two gold electrodes and several multi-wall nanotubes dispersed on them. By AFM tip manipulation one of the nanotubes is placed such that it bridges the two electrodes and its *I–V* characteristics are evaluated. An ohmic behavior is seen in the bias range of ±100 me V from which a resistance of 38 kΩ is obtained. This resistance is higher than expected for ballistic transport. The *I–Vs* in the cases of high resistance are non-ohmic and suggest that the measured two probe resistance is dominated by tunneling through a barrier at the electrode–nanotube contact. This contact resistance is very sensitive to contamination and roughness of the gold surface. A selective electron irradiation of the contact areas of such a device in the SEM is found to drastically improve the conductance. Fig. 11.10 illustrates this behavior. Initially, the low bias two-terminal resistance of this device is extremely high, ~ 1 TΩ, and the *I–V* curve is highly non-linear. After irradiation experiments with 25 keV, the resistance is reduced to ~ 100 kΩ and the *I–V* developes ohmic characteristics.

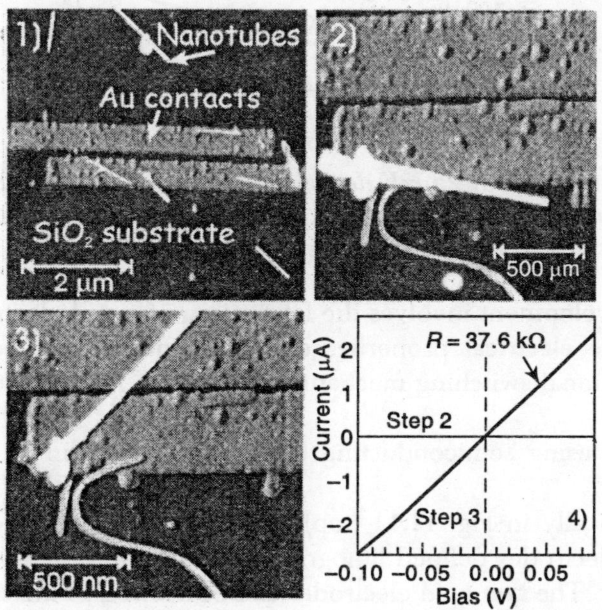

Fig. 11.9 *(1) A pair of gold electrodes and a number of multi-wall nanotubes.*
(2) Close up f the electrodes and of a tube partially overlapping one of the electrodes.
(3) The tube is manipulated so as to bridge the gap between the two electrodes.
(4) The current flowing between the electrodes at step 2 and after step 3.

For electrical applications, it is of great interest to know how much current can be passed safely through an individual nanotube. A coupling between the electrons and the vibrations of the nanotube would lead to energy dissipation, thus raising the temperature of the tube. For high enough currents the tube will be destroyed. Considering the MWNT in Fig. 11.9, it is found that it can carry a current of up to 60 μA or a current density of the order of 10^{11}–10^{12} A/m^2 (depending on the actual area the current flows through). The tubes tend to break in the section which is suspended above the substrate. between the electrodes, i.e., where heat dissipation is expected to be the lowest.

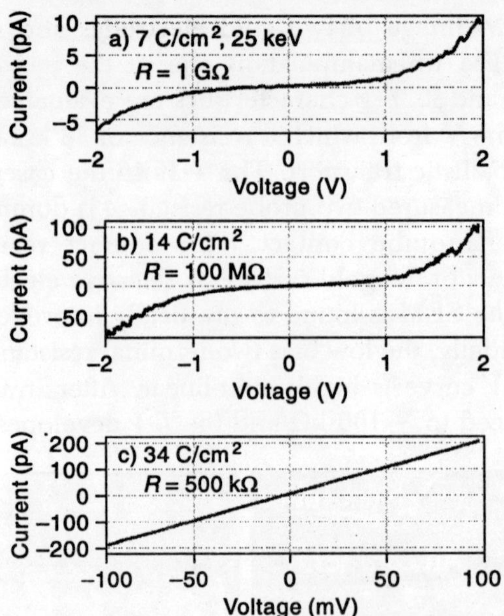

Fig. 11.10 *The effect of irradiation by 25 keV electrons on the current–voltage characteristics of the junction formed by a single nanotube bridging two gold electrodes. The initial resistance was ~1 TΩ. The total dose after each irradiation is indicated on the graph.*

A very interesting development involves the fabrication of novel electronic devices that take advantage of the unique electrical properties of carbon nanotubes. Both room-temperature operation and a conventional switching mechanism such as in field-effect transistors (FETs) are highly desirable.

So, the possibility of using semiconducting tubes as channels in an FET configuration is explored.

As is shown schematically in Fig. 11.11 (top), a nanotube is positioned so as to bridge the gap between two gold electrodes defined lithographically on top of a ~140 nm thick SiO_2 film grown on a silicon wafer. The two gold electrodes can be viewed as the source and drain of the device, while the doped silicon wafer itself can be used as a back gate. Room-temperature results for a ~ 16 Å diameter single-wall nanotube are given in Fig. 11.11 (bottom) which shows the transfer characteristics of the device, i.e., the variation of the source-drain current I as a function of the gate voltage V_G for several source-drain biases V_{SD}. These results clearly demonstrate that the gate can strongly modify the current flow through the nanotube. The enhancement of the current at negative gate bias indicates that positive holes are the main carriers. Magnetoresistance measurements on nanotubes have reached the same conclusion. The nanotube FET is a normally 'on' device which can be switched to the 'off' state by a positive gate bias. As the inset in Fig. 11.11 shows, the gate can modulate the low-bias conductance $G = I/V_{SD}$ of the tube by 5 orders of magnitude.

Large multi-wall nanotubes are, not expected to be useful for field-effect devices since the band gap of semiconducting tubes should decrease as $1/d$. Thus, at room temperature, such nanotubes should effectively behave as if they were metallic.

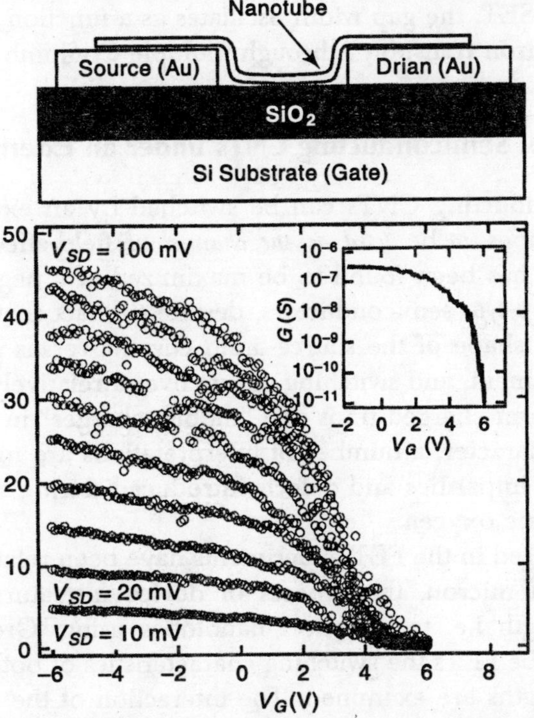

Fig. 11.11 *A field-effect transistor based on a single nanotube. Top: A schematic of the device sructure. Bottom: Source-drain current I as a function of gate voltage V_G for several source-drain biases V_{SD}. The inset shows the variation of the low-bias conductance $G = I/V_{SD}$ of the nanotube as a function of the gate voltage.*

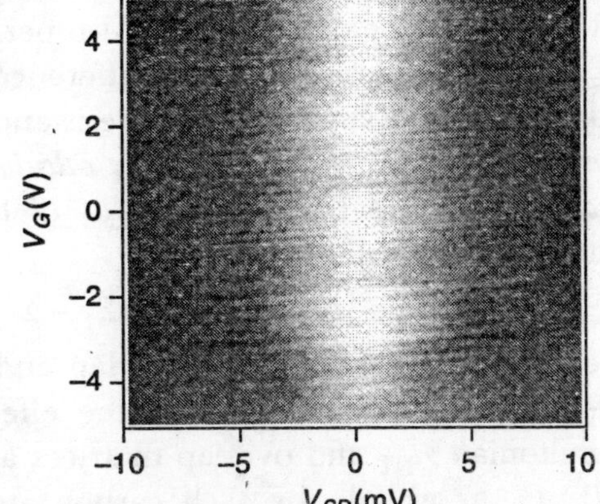

Fig. 11.12 *Single-electron transistor operation using a bundle of single-wall nanotubes as Coulomb island. The differential conductance d I/dV_{SD} of the nanotube bundle is plotted as a function of source-drain voltage and gate voltage (white: low conductance; black: high conductance).*

Using metallic nanotubes as Coulomb islands, single-electron transistor (SET) action has been demonstrated at low temperature. Spreading, bundles of single-wall nanotubes on electrode configurations such as the one sketched in Fig. 11.11 (top), we observed Coulomb blockade at liquid helium temperature. Fig. 11.12 shows a gray-scale plot of the differential conductance dI/dV_{SD} as a function of source-drain voltage and gate voltage. As a consequence of Coulomb blockade, rhombically shaped structures of vanishing, current (white) occur periodically as the gate voltage is varied. The nanotube bundle is charged one by one with dozens of individual electrons as the device operates as a single-electron transistor. In the case of a single metallic island, the width of the Coulomb gap is independent of the gate voltage. In thet nanotube SET, the gap width oscillates as a function of the gate voltage, which is suggestive of single-electron transport , through multiple Coulomb islands formed within the nanotube bundle.

(C) Switching behavior of Semiconducting CNTs under an External Electric Field

The current in semiconducting CNTs can be switched by an external electric field which allows semi-conducting tubes *to* be *used as the channel* of field-effect transistors (FETs). The current through the CNT has been found to be maximized at a negative gate bias, indicating that the CNTs behave as *p-type* semiconductors, despite the fact that the tubes have not been intentionally doped. The shape of the source-drain current versus gate voltage curves varies from experiment to experiment, and switching occurs over a relatively broad voltage range due to complications arising from charged traps and "mobile charges" in the gate dielectric. As for the origin of the *p*-type character, a number of interpretations are made, including interaction with the metal electrodes, imparities and defects introduced during synthesis or processing, or interaction with atmospheric oxygen.

∵, the CNT segments used in the FET experiments have been relatively long, in the order of several hundred nm to a micron, the process of device miniaturization is carried out for reducing the channel length i.e. to get short nanotubes using. Green' function to *oigues* to theoretically mode nanotube FETs the switching characteristics of both intrinsic *p*- and *n*-doped nanotubes of different lengths are examined. The interaction of the nanotubes with the metal (gold) electrodes the generates metal-induced gap states (MIGS)), for switching process, of the short tubes is made to know how short NT FETs can be made calculations are performed on segments of semiconducting (10,0) nanotubes. The tubes are bonded with their dangling-bond bearing ends to two gold electrodes. Each electrode is modeled by a layer of 22 gold atoms in a (111) crystalline arrangement. Also performed are calculations with the tubes lying parallel to the gold electrodes. The main difference between devices with side bonding from those with end bonding is a higher contact resistance (several MΩ as compared *to* a few kΩ). *The electrical transport properties of the devices are calculated with the Landauer–Büttiker formalism using a Green's function.* The Green's function $\mathcal{G}_{NT}$ of the combined electrode-nanotube-electrode system is written in the form

$$\mathcal{G}NT = [E\mathcal{S}_{NT} - \mathcal{H}_{NT} - \Sigma_1 - \Sigma_2]^{-1},$$

where $\mathcal{S}_{NT}$ and $\mathcal{H}_{NT}$ are the overlap and the Hamiltonian matrices, respectively, and $\mathcal{S}_{1,2}$ are self-energy terms that describe the effect of the electronic structure of the two leads. The Hamiltonian $\mathcal{H}_{NT}$ and overlap matrices are determined using the extended Huckel model with s, P_x, P_y, p_z orbitals for each carbon atom and one s orbital for each gold atom [s orbitals dominate the density of states (DOS) of gold near E_F]. The field of the gate is approximated as a capacitor field normal *to* the NT axis. In this case, the Hamiltonian $\mathcal{H}_{NT}$ can be written as

$$\mathcal{H}_{NT} = \mathcal{H}^0_{NT} + \mathcal{H}^1_{NT}$$

where $\mathcal{H}^0_{NT}$ is the Hamiltonian of the unperturbed NT and $\mathcal{H}^1_{NT} = e\mathbf{R}\varepsilon$. Here, e is the electron charge, $\mathbf{R}$ is the position of the atoms, and $\varepsilon(\mathbf{r})$ is the external field. Only diagonal elements of the Hamiltonian $\mathcal{H}^1_{NT}$ on the extended Huckel basis set are taken into account. This approximation assumes that all the charge of the carbon atoms is located at the center of the atoms and this gives accurate results for semiconductors, Finally, the transmission function $T(E)$ is given by

$$T(E) = T_{21} = Tr[\Gamma_2 G_{NT} \Gamma_1 G_{NT}],$$

where $\Gamma_{1,2} = i(S_{1,2} - G_{1,2})$.

Since, in the experiments, the potential difference between the source and drain electrodes is usually very small (typically, less than 100 meV), the conductance at the Fermi energy is evaluated, $G(E_F) = [2e^2 T(E_F)]/h$.

The main panel in Fig. 11.13 shows the variation of the conductance of a 100-Å-Iong (10, 0) NT as a function of the applied gate field. Figures 11.14(a) and (b) show the corresponding total density of states and the DOS projected *on* the P_x and P_y components, for positive and negative gate fields, respectively. The gap of the NT at zero gate field is about 0.8 eV as defined by the two peaks at about –0.4 eV (valence-band edge) and +0.4 eV (conduction-band edge). The *n* (P) doping in Fig. 11.13 corresponds to the conductance calculated with the Fermi level located a few meV above (below) the conduction- (valence-) band edge. In the undoped case the Fermi level is placed in the middle *of* the gap. The balistic limit for a semiconducting tube $[T(E_F) = 1]$ is also indicated by the horizontal dot line. By increasing the gate field in the mV/Å range, the $C(P_x)$ and $C(P_y)$ derived orbitals are shifted by the field to higher (lower) energy for negative

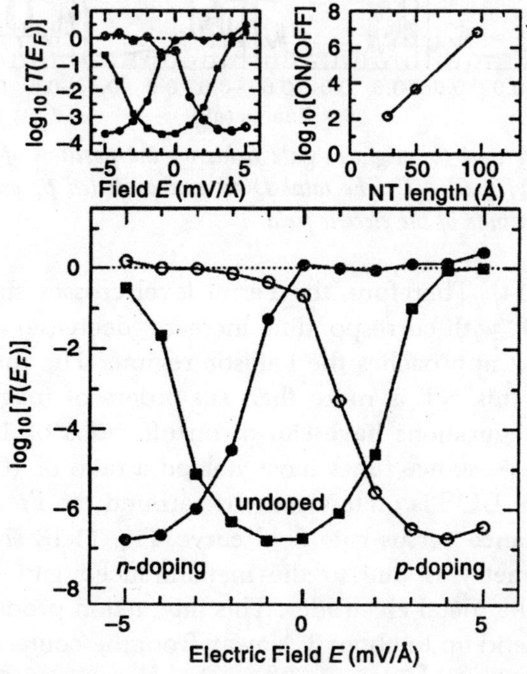

Fig. 11.13 *Variation of the transmission of a 100-Å-long (main panel) and a 50-Å-long (upper-left panel). (10, 0) semiconducting nanotube (NT) as a function of the applied field. Squares: undoped NT, open circles: p-doped, closed circles: n-doped NT. The on/off switching ratio is plotted as a func-tion of the NT length in the upper-righi panel.*

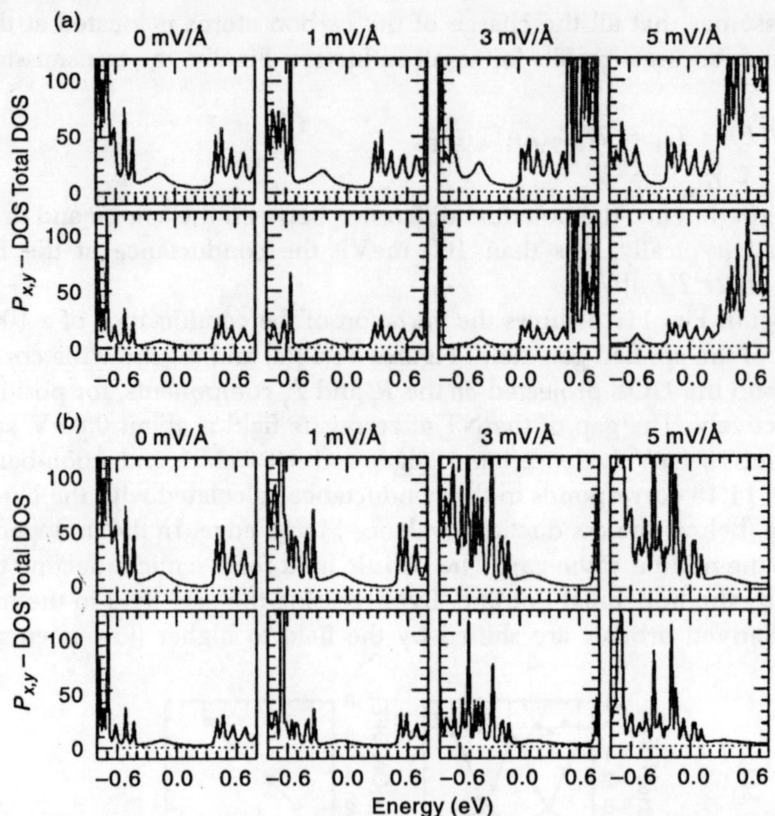

Fig. 11.14 *Effect of (a) positive and (b) negative gate fields on the position of the valence and conduction bands of a 100 Å (10, 0) NT. The total DOS and projected p_x and p_y DOS are plotted as a function of the magnitude of the electric field.*

(positive) fields (see Fig. 11.14). Therefore, the Fermi level crosses states with higher (lower) DOS for the *n*-(*p*-) doped NT, with corresponding increase (decrease) in the conductance. The conductance saturates when it approaches the ballistic regime. The modulation of the current (on/off ratio) achieved with this NT is more than six orders of magnitude. Experimentally, FETs with side-bonding configurations have shown on/off ratios of 10^5. Recent experiments with endbonded low-contact-resistance tubes have yielded a ratio of 10^6.

In undoped nanotubes, the DOS is quite symmetric around the Fermi level and this should lead to a symmetric conductance versus gate field curve. Fig. 11.13 shows a small asymmetry for small fields. This asymmetry is due to the metal-induced gap states produced by the interaction of the tube with the metal electrodes. This interaction produces weak DOS features at about −0.15 e V which extend up to about 8 Å away from the contacts. The computed MIGS features correlate very well with gap features observed in the scanning tunneling spectroscopy (STS) of semiconducting NTs on a gold substrate, whose origin was not previously clarified.

The gap states in the current modulation process is achieved from an undoped NT, the Fermi level (E_F) which is pinned close to the energy position of these states. Further it has been shown experimentally that the band gap of semiconducting tubes does not change significantly down to at least 50 A segments. Figure 11.15 shows the electronic DOS near E F of 30-, 50-, and 100-Å-long NTs connected to gold electrodes. Carbon atoms bonded to gold (C-Au) and gold atoms (Au) pro-vide the main contributions to the gap states. Assuming that small variations can occur in the filing of the MIGS, we have calculated the conductance by varying the Fermi level across the MIGS energy position. - found that neither the switching behavior of the NT, nor the conductance at zero gate field are strongly affected by the movement of *E F* (see the inset of Fig. 11.15. However, a small asymmetry is induced in the switching curve where a negatively biased gate be-comes less effective in switching. This is because, in this case, *E F* is scanned through the MIGS which are localized in the vicinity of the contacts, giving a lower transmission probability. The "metallization" of a NT segment decreases with its increasing length: the ratio of the DOS of MIGS to the DOS of the band edges (van Hove singularities) of the tube is reduced by a factor of 2/3 upon doubling the tube length (see Fig. 11.15. This trend should be valid for other quantum wires, and is in agreement with recent first-principles calculations on Si nanowires. 17 We thus conclude that MIGS do not playa significant role in the operation of NT field effect transistors.

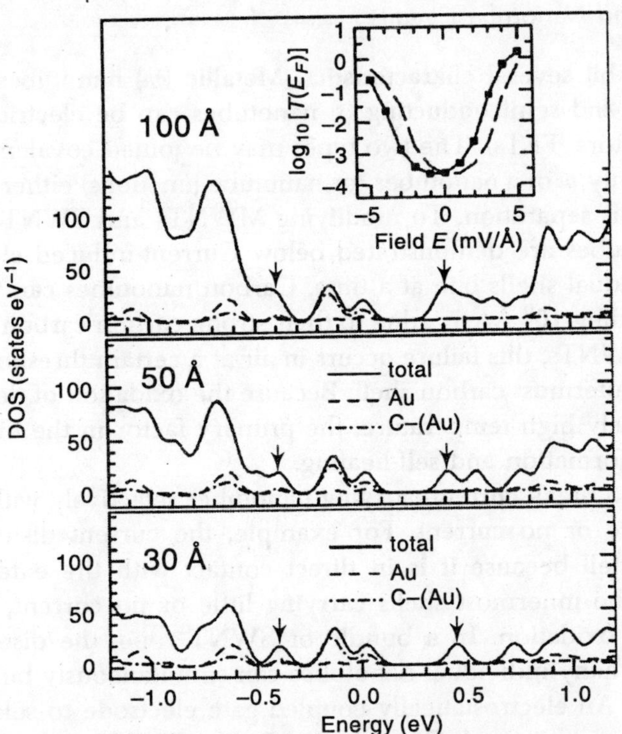

Fig. 11.15 *Energy, composition, and relative intensity of the metal-induced gap states (MIGS) in the vicinity of Fermi energy for a 30-Å-Iong (lower panel), 50-Å-long (middle panel), and a 100-Å-Iong (upper panel) (10, 0) NT (the valence- and conduction-band edges are indicated by the down arrows). The inset shows the effect of displacing the Fermi level (full line) to the energy of the MIGS.(dashed line) on the conductance of the 50 Å NT.*

Next, consider the miniaturization limit of nanotube FETs, calculations and STS measurements indicate that there are no observable finite-size (length) effects on the band gap of semiconducting NTs down to at least 50 Å. The shift of the energy states of the NTs due to the gate field is independent of the length of the tube because the gate field is perpendicular to the NT axis. As a result, the conductance of the NTs at high gate fields should saturate at approximately the same value, irrespective of the tube's length. Thus, the tube length dependence of the on/off current ratio is determined by the off current. Using this ratio as a measure of how well the FET functions, taking MIGS into account and assuming that a ratio of 10^4 is the lowest accept-able for practical applications, then we can see from Fig. 11.13(c) that tubes as short as 50 A can be used as channels of a FET. On this scale transport inside the tube is ballistic, there is no energy dissipation except at the contacts, and THz operation may be possible. These conclusions are not affected by temperature, since, for small source-drain voltages (≤ 0.1 V) the off current is proportional to the conductance via a factor eV/4 kT, which, at room temperature, is of order 1.

(D) Electrical Break Down

(I) Carbon Nanotubes and Nanotube Circuits

Carbon nanotubes exhibit several characteristics. Metallic (*m*) nanotubes can carry extremely large current densities; and semiconducting (*s*) nanotubes can be electrically switched on and off as field-effect transistors (FETs). The two types may be joined covalently. No methods exist for reliably preparing only *m* or *s* nanotubes (or nanotube junctions) either by selective synthesis or through post-synthesis separation. To modifying MWNTs and SWNT ropes, the properties of these bundled nanotubes are demonstrated below Current-induced electrical breakdown is used to eliminate individual shells one at a time. Carbon nanotubes can withstand remarkable current densities, exceeding 10^9 A/cm^2, due to their strong carbon-carbon bonding, but, at high currents they fail. In MWNTs, this failure occurs in air at a certain threshold power through the rapid oxidation of the outermost carbon shell. Because the oxidation of defect-free graphite can proceed only at extremely high temperature, the primary factor in the breakdown initiation is current-induced defect formation and self-heating.

This process is used to break current-carrying nanotubes selectively without affecting parallel nanotubes carrying little or no current. For example, the current distribution in a MWNT favors the outermost shell because it is in direct contact with the external electrodes. This distribution can leave the innermost shells carrying little or no current, which protects them during current-induced oxidation. In a bundle or SWNT rope, the distribution of current is more uniform because many individual nanotubes can simultaneously be in good contact with the external electrodes. An electrostatically coupled gate electrode to select the *s*-SWNTs and deplete them of their carriers is used. Once depleted, the *s*-SWNTs are protected from damage in the same way as the inner shells of a MWNT, and high current densities can be used to initiate oxidation of the metallic SWNTs.

The removal of carbon shells from these composite conductors are observed both electrically, and with microscope. Electrically, the breakdown of a single carbon shell results in a partial conductance drop that typically occurs within a few milliseconds. When stressed at sufficiently high bias, multiple independent drops occur as one carbon shell after another is broken. The resulting staircase of current as a function of time is shown in Fig. 11.16(a), as eight shells of a MWNT are sequentially removed (also see (II) In Multiwalled Carbon Nanotubes). Either atomic force or scanning electron microscopy can resolve the gradual thinning of both MWNTs and SWNT ropes (Fig. 11.16(b) and (c). Using the former method, MWNT diameters can be determined before and after thining; the change in apparent diameter has a linear correspondence with the number of breakdown steps observed. This controlled destruction allows to effectively separate semiconduction SWNTs from mixtures of SWNTs and to easily generate nanotube based FETs. The process allows insight into the complex electronic structure and transport properties of MWNTs and SWNT ropes. By removing the stress at the occurrence of individual conductance steps, these composite nanowires are recharacterized upon the loss of each constituent conductor. Multiple transport measurements can be probed deeper and deeper into the inner shells of a MWNT for comparision.

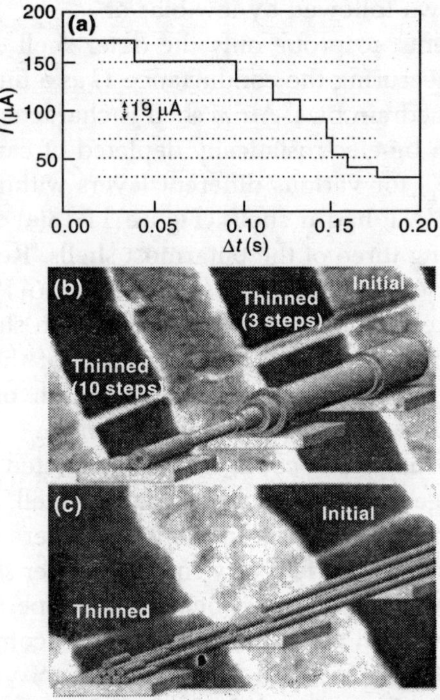

Fig. 11.16 *(a) The partial electrical breakdown of a MWNT at constant voltage stress proceeds in a series of discrete steps corresponds to the loss of individual carbon shells from the MWNT. Equally spaced dotted lines emphasize the surprisingly regular spacing. (b) images of partially broken MWNTs show clear thinning, with a decrease in radius equal to the intershell spacing (0.34 nm) times the number of completed breakdown steps. The two segments of this sample were independently thinned by 3 and 10 shells. (c) Similar thinning, as observed in SWNT ropes.*

For the controlled breakdown technique small arrays of independently addressable SWNTFETs are fabricated. Using standard lithography, an array of electrodes (source, drain, and gate) is defined on top of a dilute film of SWNT ropes (Fig. 11.17, D and E). The SWNT density is adjusted to ensure that at least one rope shorted every set of electrodes while minimizing the unwanted connections between devices. Ropes between source and drain electrodes are converted into FETs by selective breakdown, and stray connections are removed entirely by complete breakdown.

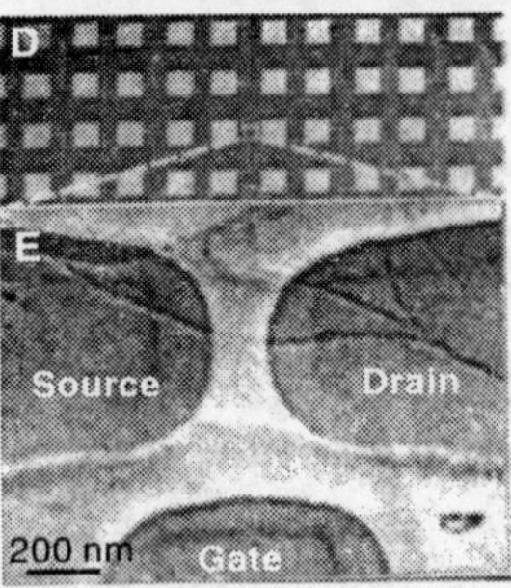

Fig. 11.17 *By combining this technique with standard lithography, arrays of three-terminal, nanotube-based FETs were created (D and E) out of disordered bundles containing both m- and s-SWNTs*

Results

The different shells of a MWNT are presumed to alternate in, a random fashion between *m* and *s* states, by using controlled breakdown followed by low-bias or low-temperature measurements, to probe only the outer shell of a MWNT. Now, differentiate between *s*- and *m*-shell by measuring the conductance *G* as a function of gate voltage V_g using a relatively small 10-mV sourcedrain bias. An *m* shell is characterized by a *G* that is independent of V_g, whereas an *s* shell can be electrostatically depleted of carriers by the gate.

(A) Figure 11.18 shows $G(V_g)$ for various different layers within a particular MWNT. Initially, this 9.5-nm diameter m-MWNT had n shells. Figure 11.18(a) shows the strong modulation in $G(V_g)$ observed after removing three of the outermost shells. Removing a fourth layer results in an *m* $G(V_g)$, and removing yet another produces another $G(V_g)$. This variation is interpreted thus as a signature of the alternating character of the carbon shells being removed.

G does not fall to zero for a particular *s*-shell because of the contribution of inner *m*-shells that remain conducting. Note that the depletion minimum of the $G(V_g)$ curve for shell n-3 coincides with the conductance of the underlying shell n-4.

In this case, the outer, s-shell (n-3) could be fully depleted by the gate, but the measured conductance included leakage through the underlying m-shell. Additional measurements show that this leakage is energetically frozen out in the low-temperature, low-bias limit.

Further breakdown of this MWNT, few very small diameter shells is reached. Upon removing the tenth carbon shell, the MWNT began to behave like a perfect, intrinsic FET with a region of zero conductance even at room temperature due to the complete depletion of carriers (Fig. 11.18(b)). Similar characteristics are found for individual *s*-SWNTs although the SWNTs used are strongly *p*-type (i.e.., hole carriers) and does not have symmetric $G(V_g)$ characteristics. The complete depletion of the MWNT indicates that no *m*-shells remain, and this behavior persists until the 14th, carbon shell is removed, at which time the MWNT circuit finally opens. Thus on the basis of the known intershell spacing of nearly 0.34 nm a MWNT of this diameter can have at most 14 shells.

In Fig. 11.18(b) there is the gradual increase of the zero conductance region as the final *s*-shells are removed. The width of this region is proportional to the band gap of the

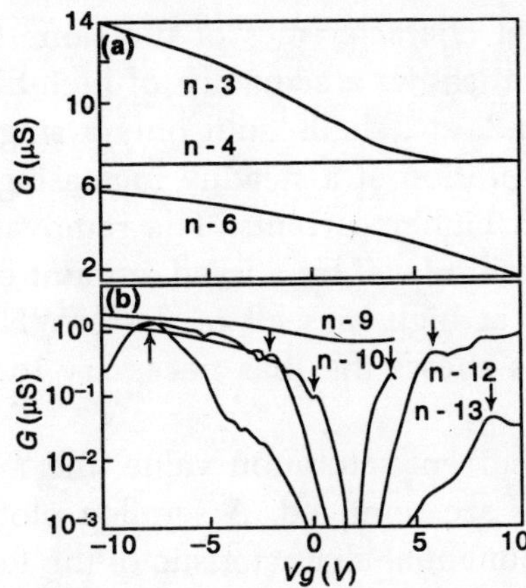

Fig. 11.18 *By halting the stress on each breakdown event, a MWNT may be characterized after the loss of each constituent shell, (a) The low-bias conductance $G(V_g)$ changes between s and m behaviors due to the changing nature of the outermost shell at each stage of breakdown, (b) When the last metallic shell (labeled n-9) has been removed, the remaining semiconducting shells can be completely depleted to give regions of zero conductance. Taking the indirected peaks (arrows) in $G(V_g)$ to correspond to the conduction and valence band edges, we can determine the electronic band gaps of different shells within a constant of proportionally.*

semiconductors, with conduction above and below the gap due to electron and hole carriers, respectively oxidizing environment. A remarkable characteristics of a *s*-carbon nanotubes is that their band gap energy is inversely proportional to their diameter thus smaller carbon shells exhibit large band gaps. So by considering the initial MWNT diameter and the intershell spacing, the diameter of each carbon shell is estimated, so the ratios between band gaps of the innermost shells is calculated. As shown in Table 11.1, the calculated ratios agree well with the experimental ratios, as defined by the conductance peaks on either side of the conductance gap.

TABLE 11.1 Comparison of relative bond gaps from experiments (Fig. 11.18(b)) with calculations based on the expected diameter dependence. The only parameters are the initial diameter of the tube and the 0.34 nm spacing between adjacent shells.

Shell	*Diameter (nm)*	*Relative band gap (eV)*	
		Calculated	*Measured*
n	9.5		
—	—		
n-10	2.7	0.24	0.22
n-11	2.0	0.33	
n-12	1.3	0.49	
n-13	0.7	1000	1.00

(**B**) Next, consider the electrical characteristics of the same 9.5-nm MWNT in the opposite, high-bias regime. Figure 11.19(a) shows a sequence of high-bias current-voltage characteristics (*I-Vs*) of MWNT with n, n-1, n-2, etc. shells until only a single shell remains. Each *I-V* curve shows unmistakable current saturation at a steadily increasing bias similar to that observed in individual SWNTs but at much higher currents. The removal of each shell from the MWNT appeared to decrease this saturation level by a fixed amount of ~20 μA, Fig. 6.47(a). The step-by-step decrease indicates that at high bias all of the MWNT shells contribute to transport. Further the series of *I-V* curves shows the bias necessary to bring each subset of shells into current saturation.

Besides the decrease of the current saturation value, the *I-V* curves in Fig. 11.19 also show increasing nonlinearity as shells are removed. A semilog plot of selected *I-Vs* show the trend from a linear *I-V* toward an exponential characteristic of the form $I = A \exp(V/V_o)$, where $V_o =$ 0.50 V (Fig. 11.19(b)). Apparently, a tunneling barrier dominates the *I-V* of the innermost shells, most likely because these shells can only couple to the external electrodes through a barrier composed of many graphic layers. For intermediate shells the unusual shape of the measured *I-V* curves can be understood qualitatively as the result of a depth-dependent barrier in series with a nanotube's intrinsic longitudinal *I-V* characteristic. This explains the gradual increase of the bias bias required to reach current saturation (in Fig. 11.19). Further the transition from linear to nonlinear *I-V* curves observed and the abundance of similar nonlinear *I-Vs*

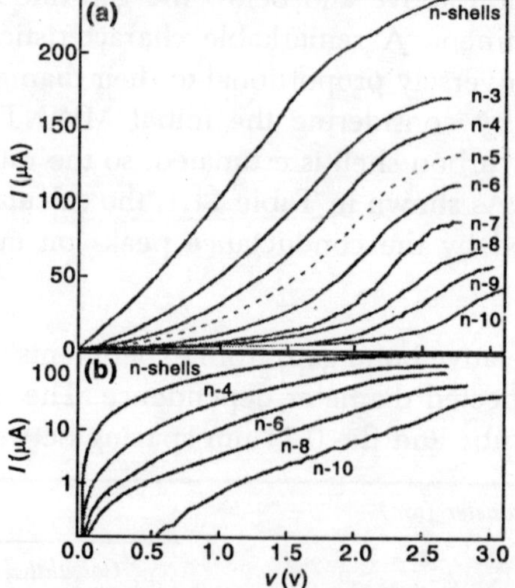

Fig. 11.19 *(a) By recharacterizing the MWNT after the loss of individual shells, the contribution of each shell to the I-V can be determined. On the basis of the uniform spacing of the sequence of I-V curves, it appears that each shell saturates at the same current and that all of the shells contribute to conduction at moderate and high bias. The dashed line marks the estimated position of an I-V curve that was not acquired. (b) A semilog plot of selected I-Vs indicates a tendency toward exponential I-Vs for the innermost shells due to the effective barrier between them and the external electrodes. Similar barriers likely playa role in all shells except the outermost one and help explain the nonlinear I-Vs reported for some MWNTs but not others.*

reported suggest that transport experiments often do not directly contact the current-carrying carbon shell of a MWNT but rather contact partial or incomplete shells as observed by transmission electron microscopy.

The measurements confirm the varying nature of MWNT .shells; quantitatively address the electronic barrier between these shells, and attempt to isolate the contribution loss of a single shell to the overall conductance.

(C) Similar measurements are performed on ropes of SWNTs. A SWNT roper differs from a MWNT in many ways. (1) each SWNT in a small rope can connect independently to the external electrodes. Thus, ropes, much more than MWNTs, may be modeled as independent, parallel conductors with total conductance $G(V_g) = G_m + G_s(V_g)$, where G_m is the contribution of the *m* nanotubes and G_s is the gate-dependent conductance of the s nanotubes (2) multiplies SWNTs within a rope are in contact with air, potentially oxidizing environment, so many carbon shells can fail at once, rather than the uniform, shell-by-shell failure observed in MWNTs. (3) the SWNT within a small rope do not electrostatically shield each other as effectively as the concentric shells of a MWNT. As a result, the gate electrode is quite effective at depleting carriers from the *s*-SWNTs, In this state, diameter of each carbon shell state, current-induced oxidation can be directed solely at the *m*-SWNTs within the rope.

In Fig. 11.20(a) and (b) depict $G(V_g)$ for two small SWNT ropes before and after controlled breakdown. The unperturbed samples had a conductance that could be partially modulated by the gate electrode, much like the case for MWNTs. With V_g held at + 10 V to deplete the predominantly *p*-type *s*-SWNTs, a stress of several volts was applied to the rope. As the

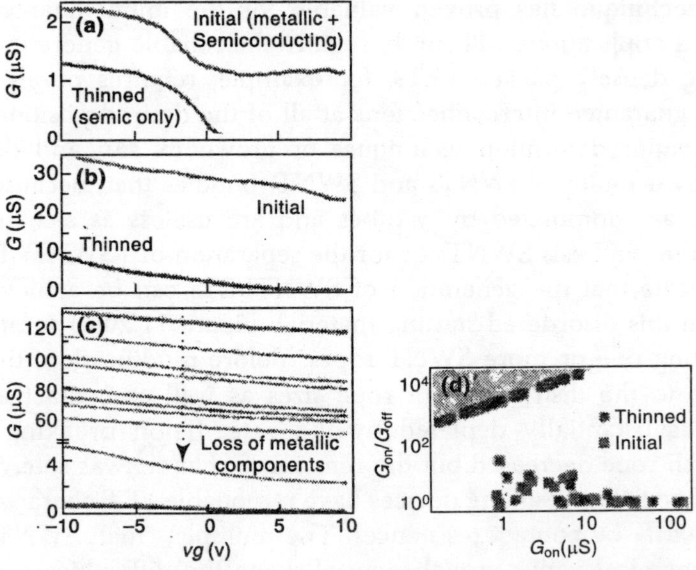

Fig. 11.20 *((a) and (b)) Stressing a mixture of s- and m-SWNTs while simultaneously gating the bundle to deplete the semiconductors of carriers resulted in the selective breakdown of the m-SWNTs. The $G(V_g)$ curve rigidly shifted downward as the m-SWNTs were destroyed. The remaining current modulation is wholly due to the remaining s-SWNTs. (c). In very thick ropes, some s-SWNTs must also be sacrificed to remove the innermost m-SWNTs.*

m-SWNTs in the rope were destroyed, G_m decreased to zero. In contrast, the modulating component $G_s(V_g)$ did not change. The constancy of $G_s(V_g)$ indicates that the depleted *s*-SWNTs are very effectively protected from damage. Further although these bundles initially show little or no switching because of their metallic constituents, final devices with good FET characteristics were reliably achieved (F).

It suggests little electronic interaction between the different types of SWNTs. Measurements of the temperature dependence of $G_s(V_g)$ with and without metallic neighbors could address this issue of interactions and determine at what energy range, if any, such interactions become important.

(D) Now, very thick measured thick ropes containing hundreds of SWNTs; are measured. Even these samples can be converted effectively into FETs (Fig. 11.20(c)). However, in this case the gradual decrease of G_m halted before it reached zero, presumably because *m*-SWNTs at the core of the rope were encased by *s*-SWNTs. The ultimate destruction of these weakly coupled m-SWNTs required higher voltages and a sacrifice of some of the surrounding *s*-SWNTs. As a result, a rope with many *s*-SWNTs and a large, initial modulation of $G_s > 10\ \mu S$ may only result in a FET with $G_s \sim 1\ \mu S$.

The controlled breakdown technique is extremely valuable not only for the study of MWNTs and SWNT interactions but also for the fabrication of nanotube-based electronic devices. Until now, SWNT FETs have been fabricated individually. Typically, SWNTs in solution are spread on top of prefabricated electrode arrays. A very low surface coverage ensures that one SWNT at most connects a source and drain electrode. At this density, the great majority of circuits remain disconnected, but some incorporate an m-SWNT and others have an *s*-SWNT.

Although this technique has proven valuable for the initial characterization of SWNT properties, practical applications will surely require the reliable generation of many devices in parallel. Achieving densely packed FETs, for example, requires purely *s*-SWNTs at a high enough density to guarantee interconnections at all of the desired positions (whether grown in situ by chemical vapor deposition techniques or grown ex situ and deposited). Such high surface densities favor multiple SWNTs and SWNT, bundles that, because of the variability of SWNT properties, are dominated by *m*-tubes and are useless as s-channels. At present, no method exists for the synthesis SWNTs or for the separation of *s*-SWNTs from SWNT mixture.

The results indicate that the generation of SWNTFETs can be achieved with greater than 90% certainty from this disordered starting material. Figure 11.20(d) summarizes results for 32 devices incorporating one or more SWNT ropes. Before modification, the rope conductances varied widely due to the distribution of rope sizes as well as contact effects, and very few devices could be substantially depleted by the gate. Upon breaking the *m*-SWNTs, the conductance of each rope decreased but the remaining channel was solely semiconducting and could be fully depleted. The resulting devices have reasonable FET characteristics with switching ratios limited primarily by contact resistances. The multiple, ⌐mall SWNT bundles, which can be produced by chemical vapor growth may alleviate the difficulties encountered with large bundles and result in FETs having excellent conductivities and switching ratios.

(II) Electrical Breakdown in Multiwalled Carbon Nanotubes

Carbon nanotubes, with their strong carbon-carbon bonds, are much less sensitive to electromigration and have current carrying capacities exceeding 10 $\mu A/nm^2$. Electromigration, is a fundamental problem limiting the further miniaturization of metallic wires. Electromigration is a non-thermal, current-assisted diffusion process initiated at points of current flux divergence such as grain boundaries, and it limits the current carrying capacity of a wire. This current limit is approximately 10 nA/nm^2 in noble metals, severe restriction for nanometer-sized wires. Thus metallic nanotubes, particularly multiwalled carbon nanotubes (MWNTs), are promising as reliable, nanometer-sized metallic wires and microcircuit interconnects.

Consider high bias transport in MWNTs specially the mechanism and the factors that control electrical breakdown in these materials. In contrast to metal wires, MWNTs do not fail via electromigration. Instead, they fail in a series of sharp steps associated with the destruction of individual nanotube shells (See (I) Carbon Nanotubes and Nanotube Circuits). These steps are regularly spaced in current regardless of the applied voltages. This spacing is due to high bias current saturation in MWNTs, similar to that reported for single-walled nanotubes (SWNTs) (A) Figure 11.21 shows the time-resolved electrical breakdown of an 8 nm diameter MWNT in air. Upon raising the (four-probe) stress to 1.7 V, the current becomes a function of time $I(t)$, a portion of which is shown in Fig. 11.21(a). After first decreasing slowly, $I(t)$ drops in a series of seven abrupt steps before stabilizing. The plot shows a remarkably regular staircase with step sizes of about 12 μA.

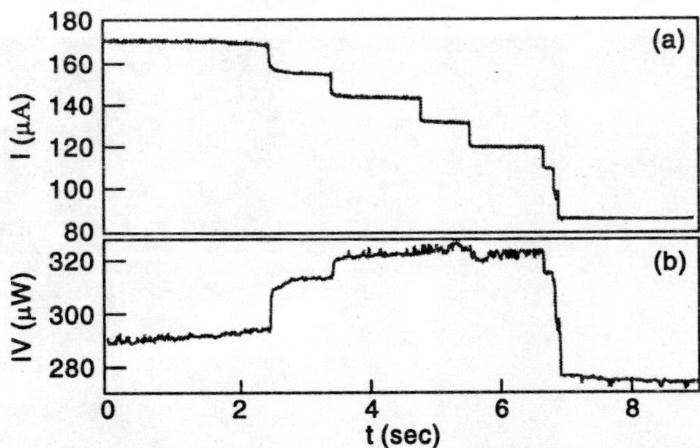

Fig. 11.21 *Time trace of the current I(t) dissipated power P(t) during a MWNT breakdown. (a) The current exhibits a staircase of stable currents separated by 12 μA steps. (b) P(t) indicates that stable values of current (or MWNT resistance) occur only when the power is below 300 μW (note that the four-probe voltage across the segment is not constant as the segment fails).*

Similar plots of the conductance ($G = I/V$) and dissipated power ($P = IV$) for the same segment are not evenly spaced, since in a four-probe configuration the voltage across the failing segment increases from step to step. Nevertheless, the time dependence of P (Fig. 11.21(b)) does contain important additional information. The initial, slowly increasing resistance leads to a gradual increase in P. The first abrupt failure occurs when P reaches a value of about 300 μW, and only after the seventh step does P again fall below 300 μW. At this time, the MWNT resistance also stabilizes, and remains stable unless a larger bias is applied. This suggests that a power threshold P_{thresh} must exceed for the initiation of the stepwise breakdown.

A comparison among many different MWNTs indicates P_{thresh} to be roughly constant for the initiation of breakdown in air, with values falling in a range of 280 to 350 μW independent of MWNT diameter.

(B) As a demonstration, Fig. 11.22 compares the breakdown behavior of two adjacent segments of a 14 nm diameter MWNT, one broken in vacuum and the other in air. First, the MWNT is stressed in high vacuum until failure of one segment (A). After checking the current-voltage

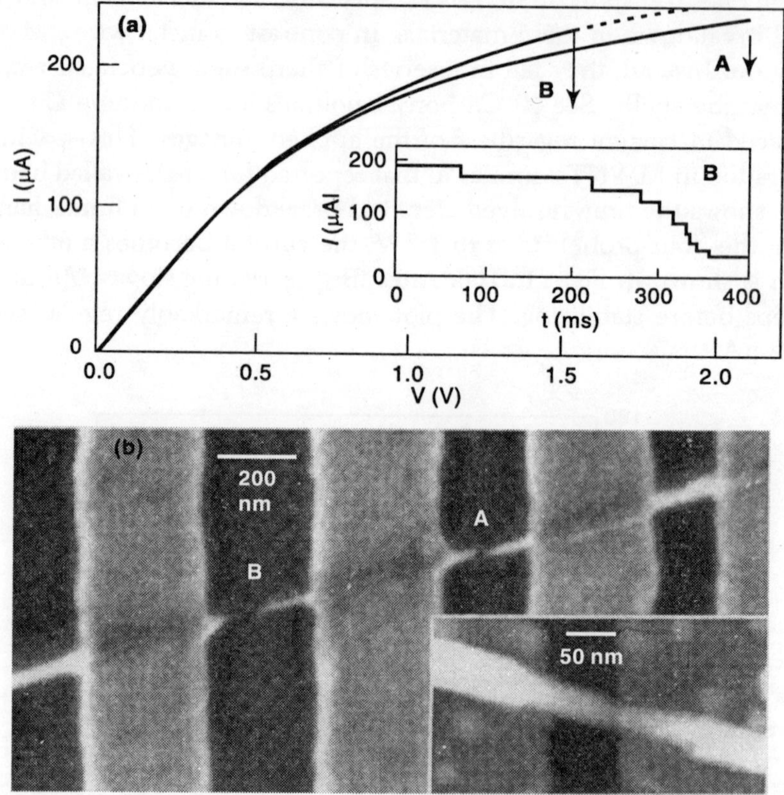

Fig. 11.22 *(a) Four probe I-Vs and I(t) for two adjacent segments of a MWNT. Segment A is electrically broken in high vacuum, while segment B is broken in air. Before failure, both segments have nearly identical I-V characteristics in vacuum, and both segments can withstand more than 400 μW of power (segment B in vacuum shown as dashed line). Vacuum breakdown occurs in segment A at 2.2 V and 235 μA, and at this high power proceeds to completion in less than 1 ms. In air, segment B exhibits a much slower breakdown initiated at a power of 320 μW (inset, the first step at t = 0 is not shown). (b) SEM imaging of the damaged segments shown significant thinning of the remaining MWNT material in both segments, with a wider gap and greater thinning for the segment broken in air. Pausing the stress midway through the breakdown produces a thinned but unbroken MWNT (inset).*

characteristic (*I-V*) of segment B to show that it has been unaffected by the first stress (dashed line in Fig. 11.22(a)), the MWNT is exposed to air and stressed again until segment B fails. The breakdown of segment B (shown in the inset) is substantially the same as for Fig. 6.51, with initiation occurring at a power of 315 μW and a regular staircase in *I*(*t*) with step sizes of both 12 and 24 μA.

Thus by placing the sample in vacuum allows much higher powers to be reached before breakdown. In vacuum, segment B clearly withstands powers exceeding 400 μW with no ill effects, and the vacuum failure of segment A does not occur until 520 μW, at which point the *I-V* is nearly horizontal, i.e., the current is nearly saturated. Once failure is initiated at this high power, it occurs approximately 1000 times more rapidly than in air This can be understood in terms of the theory of unimolecular decomposition reactions, according to which the decomposition rate increases strongly with excess energy above the threshold energy. Of course, the presence of oxidants also slows the breakdown by binding up defects (dangling bonds) in the nanotube.

Because MWNTs have higher power thresholds in vaccum, the *I-V* measurements can be extended to the high current rent, extremely nonlinear regime shown in Fig. 11.22. Above the low bias, linear (or slightly superlinear regime, MWNT *I-Vs* incorporate a point of inflection above which dI/dV decreases towards a sample-dependent saturation current. Similar current saturation at high bias attributed to both electron-phonon scattering processes. Although the exact saturation current of MWNTs is sample dependent, breakdown below the point of inflection in the *I-V*, either in air or in vacuum is never observed. This suggests that the onset of saturation and the eventual breakdown process are linked to a common dissipative process, thus involving the excitation of high energy optical or zone boundary phonons.

In addition image of MWNTs at various stages of breakdown using atomic force microscopy (AFM) and scanning electron microscopy (SEM) is done. Images of failed MWNTs show surprisingly large quantities of lost carbon, in contrast to the small gaps observed when metallic wires fail via electromigration. As seen in Fig. 11.22(b), the remaining MWNT material in the broken segments *A* and *B* both exhibit decreased diameters compared to the adjacent, unbroken segments. Because the breakdown occurs in discrete steps, stabilization of MWNTs at various intermediate steps in done. Imaging these intermediate states shows similar diameter thinning, but without resolvable gaps, as shown in the inset. Contact resistances these intermediate states are usually found unchanged, which agree with the postbreakdown SEM imaging.

(C) The MWNT thinning indicates a loss of individual carbon shells, even far from the point of failure, and suggests that the steps in *I*(1) are due to shell-by-shell failure of the MWNT. If so, then the number of steps should depend the MWNT diameter, which is shown in Fig. 11.23(a). The transmission electron microscopy experiments have resolved similar thinning and the shell-by-shell loss of material due to both oxidation: and repeated electrostatic shocks.

The larger amount of missing material and the excessive thinning in segment *B* are typical of MWNTs broken in air. Combined with the lower P_{thresh} observed in air, this observation strongly suggests that oxidation can assist the breakdown. It is found that MWNT breakdown in inert gases proceeds as in vacuum, with high power thresholds and less loss of material. However, electrical measurements primarily record the initiation of failure; much of the observed

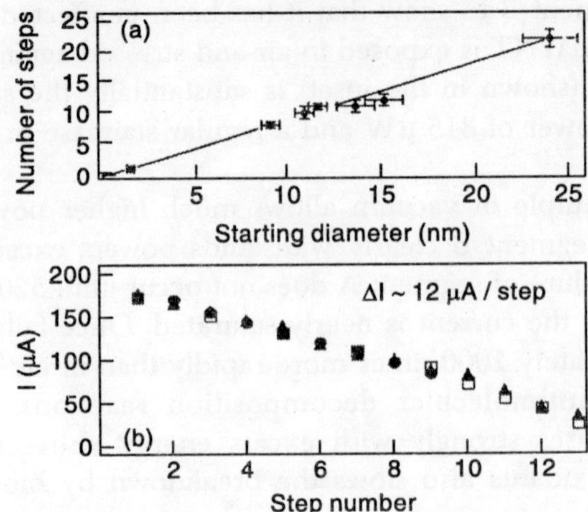

Fig. 11.23 *(a) A comparison of multiple samples shows a proportionality between the number of steps in the breakdown I(t) and the MWNT diameter, consistent with a shell-by-shell mechanism. (b) Further comparison of the current observed at each step of the breakdown reveals remarkable similarities. Here we plot the mean current at each step in I(t) for three different MWNTs with varied segment lengths. The steps are sequentially numbered beginning with the plateau at I = 170 μA, which is the highest current reached by all three samples. Not only do different samples lose current at the same average rate of 12 μA per step, but the current values themselves are nearly identical, independent of the MWNT segment length.*

carbon loss occurs after this initiation, since defects and dangling bonds then provide various chemical pathways for oxidation to proceed.

Images such as Fig. 11.22 also indicate that MWNT breakdown usually occurs midway between two electrodes, which is precisely where dissipative self-heating produces a peak temperature T_p. Taking a power P_{thresh} = 300 μW uniformly distributed along the MWNT, estimate of T_p based on a simple heat transfer analysis is done, the experimental values of MWNT thermal conductivity K range widely, from K = 25 W /mK to K = 200 W/mK. A conservative estimate using the higher K gives T_p = 773–973 K a range which agrees with the known temperatures for thermal oxidation of MWNTs and for the oxidation of damaged graphite.

Therefore, the breakdown of MWNTs may be intimately connected with self-heating and the reaction of nanotube shells with oxygen. However, spontaneous oxidation is probably not the primary cause of electrical breakdown, since oxidation at such low temperatures cannot proceed without the prior formation of defects and thermal vaporization in vacuum would require a T_p exceeding 3073 K. Instead, the initiation of both breakdown and oxidation probably depends upon local, nonequilibrium effects including defect creation due to the extreme current densities exceeding 1 μA/nm^2 (at 12 μA per shell).

The above breakdown behaviors have been observed in many MWNT at 300 and 77 K, both in air and in vacuum. Figure 6.53, the data is obtained in air at 300 K. Figure 11.23(a) shows the total number of discrete steps N observed in the breakdown $I(t)$ as a function of the initial diameter of the MWNT determined by AFM. This figure identities a dependence between

N and the MWNT diameter. Figure 11.23(a) also includes a measurement of an individual SWNT. Here, MWNTs, SWNTs fail in a single, featureless step. This fact, along with the apparent proportionality between *N* and diameter, further supports assigning the observed steps in *I*(*t*) to the failure of individual MWNT shells.

Now, *I*(*t*), rather than *G*(*t*) forms the most regularly spaced staircase, and the spacing is found to be sample independent. Figure 8.23(b) shows the mean current value I_i at each step in the $I(t)$ staircase for three MWNTs of different diameters and four-probe resistances, Here, the three data sets have the same average slope of 12 μA per step, and the I_i values themselves are also in agreement.

The constant step size $\Delta I_i = 12$ μA can be considered a direct effect of the loss of individual carbon shells if taking into account the current saturation observed in Fig. 11.22. In the high bias, current saturation limit, each shell to the single shell saturation of SWNTs. During breakdown, the sequential removal of shells will produce evenly spaced steps ΔI_i as the current carrying capacity of each shell is lost. The constant slope in Fig. 11.23(b) indicates a current carrying capacity of 12 μA per shell, independent of the decreasing shell diameter as expected for an electron-phonon scattering mechanism.

In fact, every shell in a MWNT need not be saturated in order to observe this constant step size. Consider the different shells of a MWNT as independent, parallel conductors, each with a different coupling resistance to the outermost shell (which is in direct contact with the metal electrodes). In this geometry, outer shells contribute more to the total conductance than do inner shells because of their lower coupling resistance to the external electrodes. Similarly, outer shells will carry the greatest proportion of the total current and can saturate before the MWNT as a whole is saturated. At high voltages these shells may be unable to carry additional current, but current injection into the inner shells results in a nonzero *dI*/*dV*. Under these conditions, with an unknown distribution of saturated and unsaturated shells, the breakdown of the outermost shells will still result in fixed drops ΔI_i as we have observed.

The observations indicate that many shells can contribute to the high bias conductance of a MWNT, in contrast to the single-shell model believed to hold in the low bias, low temperature regime. This difference is probably due to the energy dependence of inter shell coupling. At high biases, the effective coupling may improve due to both electron injection at the contacts and also enhanced electron-electron scattering, which will tend to relax momentum conservation requirements among shells. Improved models of MWNT conduction need to be developed in order to further understand our observations of current saturation and the uniform current steps during breakdown.

Thus measured the high bias conduction limits of individual MWNTs. In air, conduction is limited by a power threshold, above which breakdown and oxidation of the MWNT is initiated, In vacuum, the threshold for failure is raised, but conduction remains limited by a sample-dependent current carrying capacity. When these thresholds are exceeded, a MWNT fails via a unique staircase breakdown as its carbon shells are individually destroyed.

BAND STRUCTURE OF CARBON NANOTUBES IN A PERPENDICULAR ELECTRIC FIELD

Description

This interactive simulator performs band structure calculations for zigzag and armchair type nanotubes (CNT) using the nearest-neighbor tight-binding method with a hopping potential 2.5eV. A uniform electric field perpendicular axis is applied, which causes degeneration of eigenvalues to split (Stark effect).

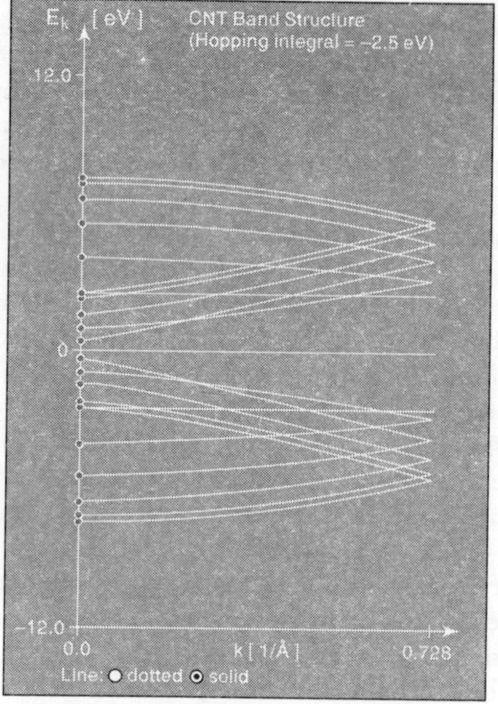

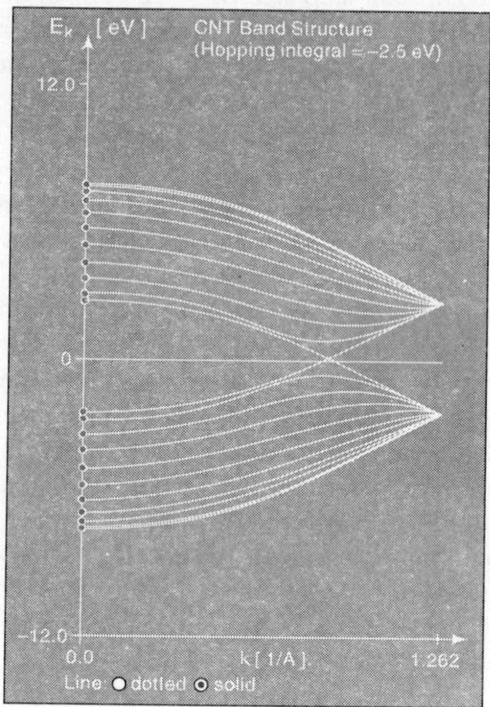

Band Structure of Carbon Nanotube (CNT) in Perpendicular Electric Field

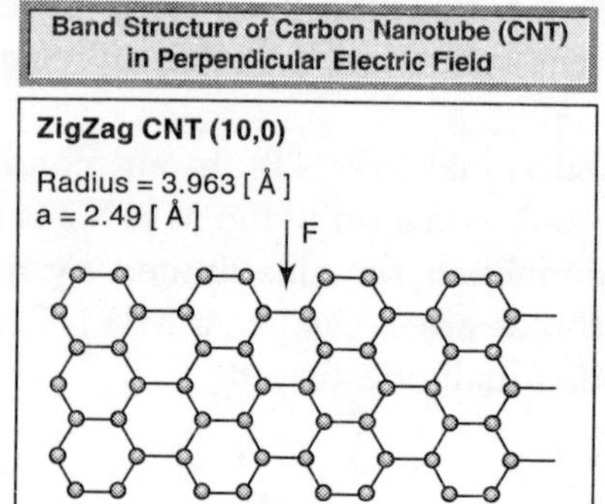

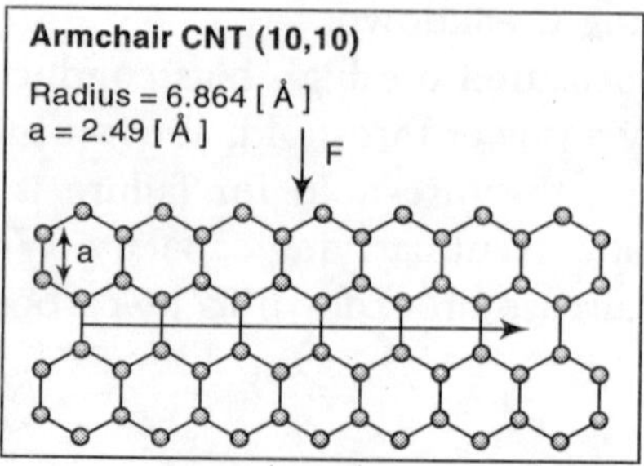

Fig. 11.24

For each structure, the simulator performs eigenvalue solutions, for 51 values of the quasi 'K' evenly spaced in the allowed range. The maximum CNT index is limited by memory that allows the applet to $n = 30$, corresponding to a (30, 0) zigzag CNT and to a (30, 30) armchair CNT.

A solution for very small values of the index n is obtained, even $n = 1$. Solutions are for a strip of graphene sheet of width consistent with n and side periodic boundary conditions. Such hypothetical structures ($n < 5$) are not practical but have educational value. Fig. 6.27 show how the band structure of the nanotube develops as n is increased (similar to building one atom at a time).

EFFECT OF ADSORBATES ON FIELD EMISSION FROM CARBON NANOTUBES

By using DMol^3, we can investigate the interaction of water with a nanotube tip which is critical in designing commercial quality flat panel displays.

It was found that the interaction is weak in the absence of any voltage. However, under field emission conditions, large electric fields present at the tube tip are found to: (1) increase the binding energy significantly, thereby stabilizing the adsorbate; and (2) lower the IP, thereby making it easier to extract electrons. Net binding and IP lowering are both enhanced by an increase in the number of water molecules adsorbed on the tip. In contrast, molecules with small or zero dipole moments are found to interact weakly with tube-tip even in large electric fields. Polar molecules like HCl, HCN and LiH also give DMol^3 calculations as in 1 prediction of carbon nanotubes.

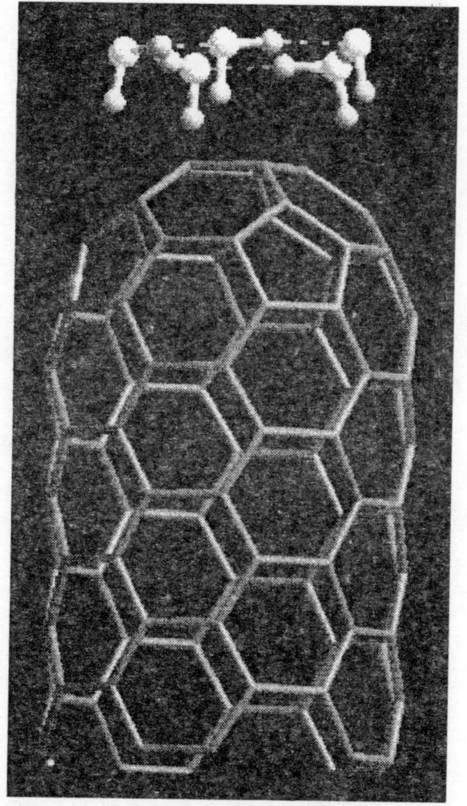

Fig. 11.25 *H-bonded water cluster on a close-capped (5, 5) nanotube stabilized under field emission conditions. The cluster is found to lower the Ionization potential of the tube by almost 0.5 eV.*

Construction of SET on SWNT

A SET consists out of a conducting island that is connected by tunnelling barriers to two metallic leads. For temperatures and bias voltages that are low relative to a characteristic energy required to add an electron to the island, electrical transport through the device is blocked. Conduction is possible by tuning a voltage on a close-by gate. This makes the device into a three terminal transistor.

As with normal CNT transistors the CNTs are doped in an uncovered region with K to make it n-type, the neighbouring regions are covered with PMMA to keep them p-doped. This p-n-p nanotube system behaves as a single QD, investigations showed that the QD are confined in the n region with the two p-n junctions as barriers for the dots. Chemical profiling a single SWNT can lead to on-tube QDs much smaller than the lengths of the nanotubes between metal electrodes (3-5 μm) Fig. 11.26. In order to function_properly the system has to be cooled to <100K.

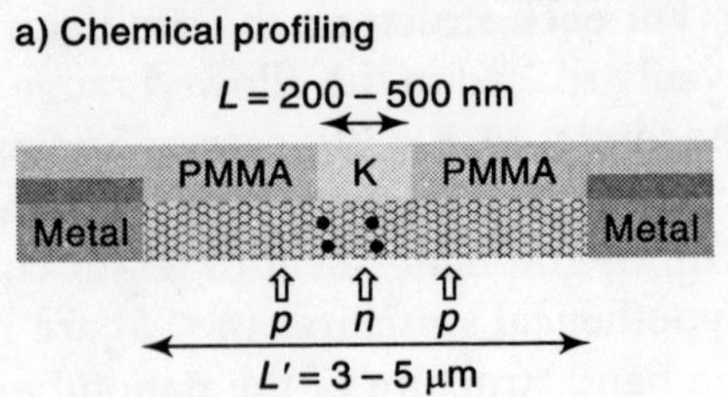

Fig. 11.26 *Schematic p-n-p nanotube device*

CNT—Nanofibers and Nano Textiles

(A) CARBON NANOFIBERS

The catalytic decomposition of hydrocarbons at 773–973 K produces carbon nanofibers (sub-micron diameter carbon filaments) example the catalytic decomposition of methane results in the formation of carbon nanofibers, with Ni catalysts supported on SiO_2, TiO_2 or Al_2O_3. The reaction proceeds fastest at about 773 K, while an increase in temperature (>773 K) dramatically decreases the rate of decomposition.

Carbon nanofibers (CNF) must be distinguished from conventional carbon fibers, which are produced by pyrolysis of pitch or polymers, and from vapor grown carbon fibers (VGCF). The latter are produced by the deposition of carbon atoms on catalystically grown carbon nanofibers during thermal decomposition of carbonaceous gases at 120–1400 K. Conventional carbon fibers typically have diameters around 10 μm and VGCF have diameters up to 10 μm. Carbon nanofibers are used in the same areas of application as those for the carbon nanotubes described above. In addition, the applications of carbon nanofibers include the areas of electromagnetic reflection, electromagnetic interference shielding, surface electrical conduction, DC electrical conduction, electrochemical applications, thermal conduction, porous carbon matrices, and catalyst supports.

The Ni catalysts supported on graphitized carbon fibers (GCF), vapor grown carbon fibers (VGCF), and laboratory made carbon nanofibers (CNF) (denoted as Ni/GCF, Ni/YGCF and Ni/CNF, respectively) are prepared by impregnating the carbon fibers with $Ni(NO_3)_2$ in acetone at room temperature. The metal oxide-supported Ni catalysts (Ni/SiO_2, Ni/TiO_2, and Ni/Al_2O_3) are prepared by impregnating the metal oxides with $Ni(NO_3)_2$ in aqueous solution. The Ni-impregnated samples are reduced with hydrogen at 573 K. The CNF are prepared by the decomposition of methane on Ni (5 wt%)/SiO_2 at 803 K. The weight percent of Ni supported on the carbon materials is adjusted to 5, 10, 20, or 40%.

Extrapure hydrocarbons are obtained (eg; methane, ethane, propane, *n*-hexane, *n*-octane, cyclopentadiene, and benzene). Cyclopentadiene is obtained as a dimer and cracked into its monomer through a vaporizer and the upper part of the reactor furnace. The catalytic decompositions of these hydrocarbons are performed by contacting their vapors with the supported-Ni catalysts at around 773 K. The partial pressures of hydrocarbons are 101 kPa for methane, ethane, and propane and 20 kPa for *n*-hexane, *n*-octane, cyclopentadine, and-benzene using argon as the carrier gas through the catalyst bed. The experiments are performed with a gas flow apparatus, under atmospheric pressure (the total pressure of hydrocarbon and argon is 101 kPa). The products in the exit gas stream from the reactor are analyzed with a gas chromatograph equipped with a thermal conductivity detector. The amount of carbon deposited is estimated either from the weight increase of the catalyst measured after the reaction or from the total amount of hydrogen produced.

The TEM measurements of carbon fibers are performed on a transmission electron microscope (resolution ≈ 1.9 Å). The samples are first dispersed in ethyl alcohol under ultrasonic wave input at room temperature before the TEM measurements.

The X-ray absorption (XANES and EXAFS) measurements are performed. Ni K-edge XANES and EXAFS of supported-Ni catalysts are measured in the fluorescence mode with a Si(111) two-crystal monochromator at room temperature.

Typical SEM images of carbon fibers produced from methane on Ni/SiO_2, Ni/TiO_2, Ni/Al_2O_3, and Ni/CNF catalysts are shown in Figure 12.1. Figure 6.55 shows typical TEM images

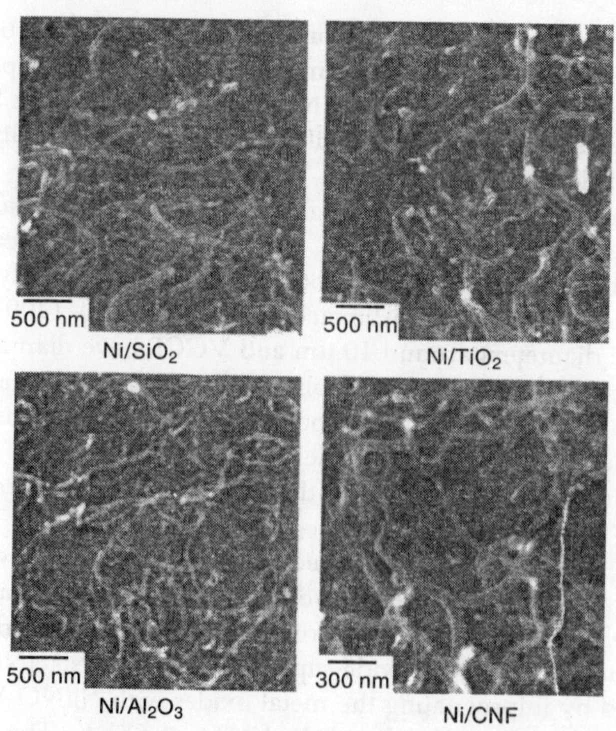

Fig. 12.1 *SEM images of carbon nanofibers produced from methane dccomposition on various supported-Ni catalysts. C/Ni = 200 ~ 210 for Ni (5 wt%)/SiO_2, Ni (5 wt%)/TiO_2, and Ni (10 wt%)/Al_2O_3. C/Ni = 760 for Ni (20 wt%)/CNF*

of carbon nanofibers forced on Ni/CNF. The SEM and TEM images in these figures show that the carbon atoms from methane grow carbon nanofibers. It is well known that the catalytic decomposition of methane on supported-Ni catalysts supported at <873 K produces, in general, fishbone-type nanofibers. This is confirmed by the TEM images in Figure 12.2. Ni metal particles are present at the tip of each carbon fiber. At the walls of the fibers, the edges of the stacking carbon layers are exposed. The structure of the layers is turbostratic graphite in which the layers of atoms, though parallel, are randomly displaced with respect to one another. It is clear that the diameter of the carbon fibers generated on supported-Ni catalysts is roughly the same as that of Ni particles at the tips of carbon fibers. The size of Ni metal particles and the diameters of carbon fibers ranged from about 10 to about 180 nm as can be seen in Figures 12.1 and 12.2. The most effective size of Ni particles for methane decomposition is about 60 to 100 mm.

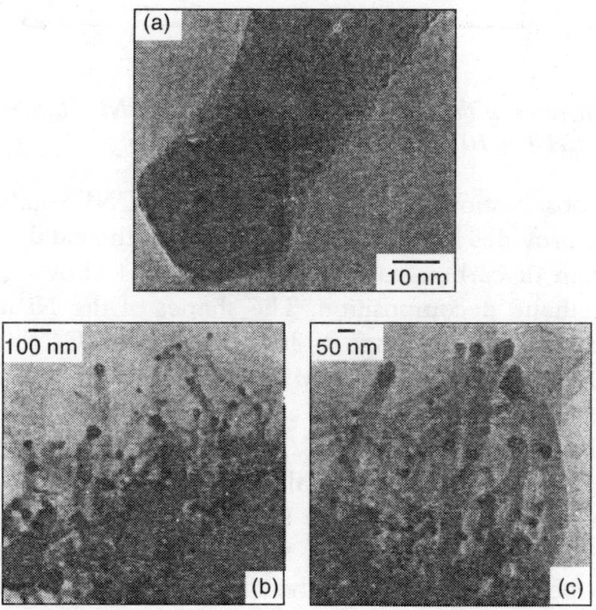

Fig. 12.2 *TEM images of carbon nanofibers produced on Ni (40 wt%)/CNF. C/Ni = 30*

The specific surface area of the carbon fibers made from methane is around 80 m^2 g^{-1}.

In general, the kinetic curves of the decomposition of methane on supported-Ni catalysts at a temperature range 750 ~ 830 K is divided into three stages. Figure 12.3 shows a typical kinetic curve of the one-pass conversion of methane into hydrogen and carbon atoms on Ni/SiO_2 at 773 K. The rate of methane decomposition is fast and steady at the initial stage, but it gradually decreases at the middle stage. At the final stage, the rate of decomposition decreases sharply with the time.

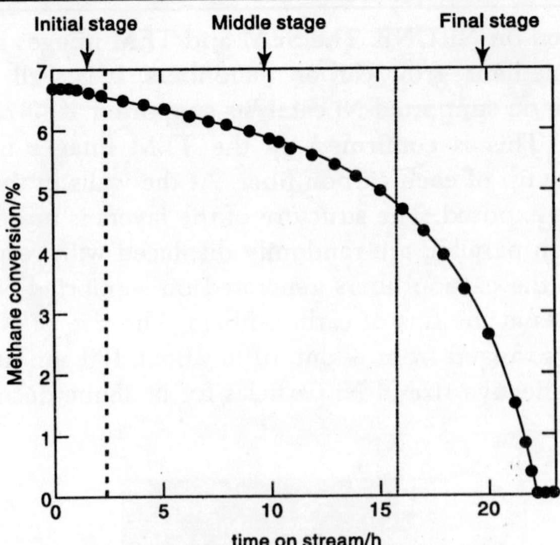

Fig. 12.3 *A typical kinetic curve of the decomposition of methane on Ni (20 wt%)/SiO_2. Reaction conditions: $T = 773$ K, $P(CH_4) = 101$ kPa. $F = 60$ ml. Min^{-1}*

The SEM and TEM observations for the Ni/GCF or Ni/CNF catalysts at the initial stage of methane decomposition provides interesting information on the catalytic functions of Ni as well as the growth mechanism of carbon nanofibers. Figure 12.4 shows TEM images of Ni/GCF before and after the methane decomposition. The shapes of the Ni metal particles in a fresh catalyst are varied over the diameter range of 30 to 100 nm on GCF (diameter 100 ~ 250 nm) as shown in Figures 12.4(a) and (b). The TEM images of Ni/GCF after contact with methane (Figures 12.4(c) and (d)) show the growth of carbon nanofibers. The moles of carbon atoms deposited per mole of Ni atoms (denoted as C/Ni, hereafter) for this sample is about 20. Child carbon nanofibers are firmly fixed to the catalytic support (GCF) and start to grow fishbone type carbon fibers. The TEM images in Figure 12.2 are also of the nanofibers at the initial stage (C/Ni ~ 30). Figures 12.2, 12.4(c), and 12.4(d) suggest that the Ni metal particles at the tips of carbon nanofibers must be faceted, exposing the specific planes which graphitize carbon atoms deposited from methane molecules. Thus, the irregular shapes of Ni metal particles on the support Figures 12.4(a) and (b) before the contact with methane were transformed into pear-shaped ones where shown at the tips of carbon nanofibers after the contact with methane. Now, the average crystallite size of the Ni metal particles increase during the very early stage of the methane decomposition. The carbon atoms formed from methane on a Ni metal particle aggregate and transform the small Ni polycrystals in the Ni particle into a faceted single crystal that catalyzes the decomposition of methane as well as the growth of a carbon nanofiber. The specific facets (probably Ni(111)) of the single Ni crystal promotes growth of the fish-bone type carbon nanofiber.

At the *middle* stage of the decomposition of methane (C/Ni = 40 to 500), suitable size Ni particles decomposes methane and grow carbon nanofibers actively as seen in the SEM images in Figure 12.5. The SEM images in Figure 12.1 were those of the carbon nanofibers (C/Ni = 200–210) at the middle stage.

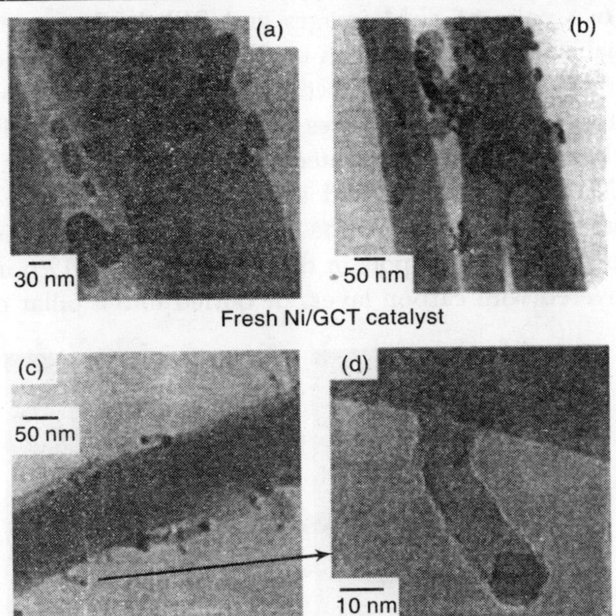

Fig. 12.4 *(a) and (b), TEM images of Ni (10 wt%)/GCT before and after methane decomposition at 773 K. (c) and (d), fresh catalysts after methane decomposition (C/Ni = 20)*

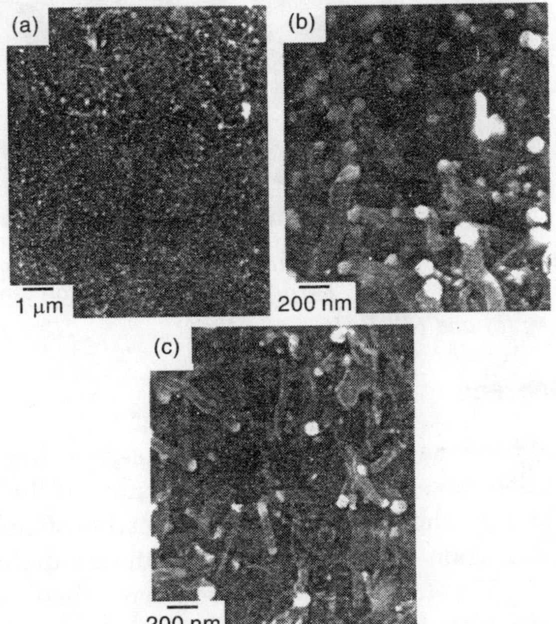

Fig. 12.5 *SEM images of carbon fibers and Ni catalysts at the middle stage. Carbon fibers were formed on Ni (40 wt%)/CNF. For the carbon fibers produced. C/Ni = 40*

Figure 12.6 shows two pairs of SEM images and BEl (backscattering electron images) of carbon fibers at the final stage of their growth on Ni/SiO_2 catalysts. These pictures are for the-samples after complete deactivation of Ni (5 wt%)/SiO_2 and Ni (90 wt%)/SiO_2. The BEIs were measured at the same time as the SEM pictures shown on the left side. Each BEI in Figure 12.6 shows a couple of bright spots. As demonstrated in Figure 12.2, the decomposition of methane on all the supported-Ni catalysts terminated sooner or later because the surface of the Ni particle is covered with carbon layers or the Ni particle would collide into the pillar of a neighboring carbon fiber. The bright spots in the BEIs in Figure 12.6 indicate the presence of Ni particles that are covered with carbon layers or buried into a pillar of a carbon fiber.

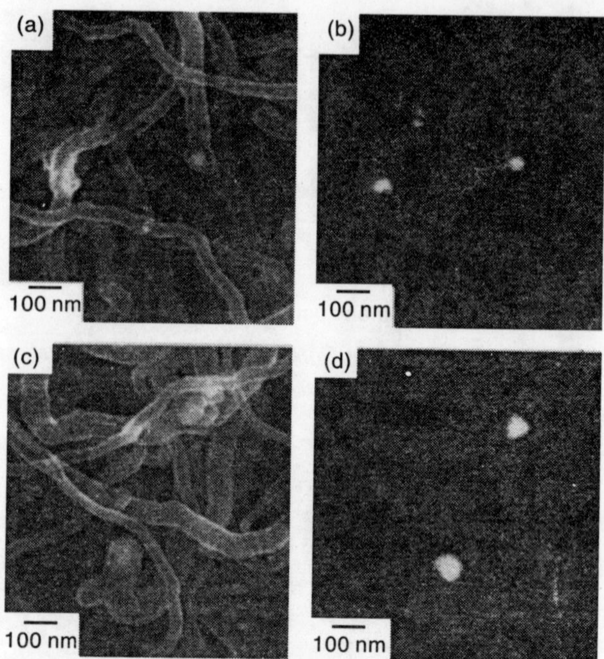

Fig. 12.6 *Carbon nanofibers and Ni catalysts at the final stage. Carbon nanofibers after complete deactivation of the catalysts; Ni (5 wt%)/SiO_2 for (a) and (b), and Ni (90 wt%)/SiO_2 for (c) and (d), (a) and (c), SEM images; (b) and (d), BEIs*

Carbon Fibers from Benzene

The decomposition of benzene also generated thick carbon fibers (diameter 50 ~ 110 nm) together with thin ones that have hollows along the center of the zigzag structure; described earlier. Figures 12.7(a) and (b) show both types of the carbon fibers. As described earlier, the formation of bidirectional carbon fibers was observed during the decompositions of *n*-hexane and *n*-octane (Fig. 12.11). The formation of carbon nanofibers with such bi-pointed nickel particles at the middle of a fiber has been reported by many researchers. Figure 12.7(c) is the TEM image of thin and thick carbon fibers from *n*-hexane decomposition, including a part of the same image in Figure 12.7(a). Both ends of the bidirectional carbon fibers are shown by the

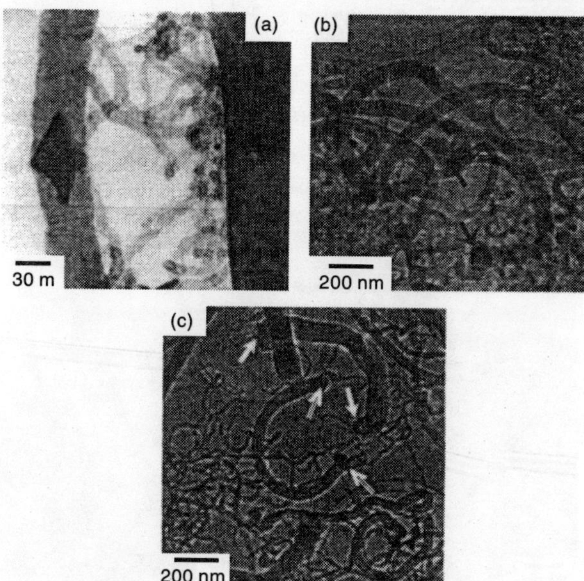

Fig. 12.7 *(a) and (b) TEM images of thick and thin fibers from benzene decomposition over Ni (10 wt%)/VGCF at 773 K. (c) TEM image of thick bidirectional fibers from n-hexane decomposition over Ni(10 wt%)/VGCF at 773 K*

pair of arrows in Figure 12.7(c). A whole shape of a bidirectional fiber formed during benzene decomposition is also seen from one end to the other, pointed out by the arrows in Figure 12.7(b). It is interesting to note that the lengths of the fibers in both directions from the central Ni particle are approximately the same, suggesting that the growth rates in both directions were quite comparable. This further implies that the Ni crystal faces responsible for carbon layer formation, usually Ni(111) faces , should also be present with similar structure and surface areas for both directions of the bi-pyramidal nickel particle.

Figure 12.8(b) is the high-resolution TEM image of the central Ni particle of the bidirectional carbon fiber shown in Figure 12.8(a). The Ni particle at the middle of the carbon fiber shows two hexagonal pyramids with their basal planes in common. The TEM image in Figure 12.8(c) shows the same sample tilted 25 degrees forward compared to the sample in Figure 12.8(b). The image of the Ni cluster shows tapering in both directions. In contrast, tilting the sample by 25 degrees backwards caused the image of the bidirectional pyramids to become quite obscured, as demonstrated in Figure 12.8(d).

It should be noted that the diameters of bidirectional carbon fibers observed by TEM measurements were always greater than 50 nm. In contrast, the diameters of the carbon nanotubes from benzene and cyclopentadiene were less than 20 nm. The bidirectional carbon fibers did not show hollow structures along the central axis of the growing fibers. The Ni metal particles at the middle of the thick fibers were faceted into bidirectional hexagonal pyramids. In contrast, the Ni particles at the tips of thin carbon nanotubes were liquid like and had not been faceted. These observations strongly suggest that the growth mechanisms of both types of carbon fibers are quite different.

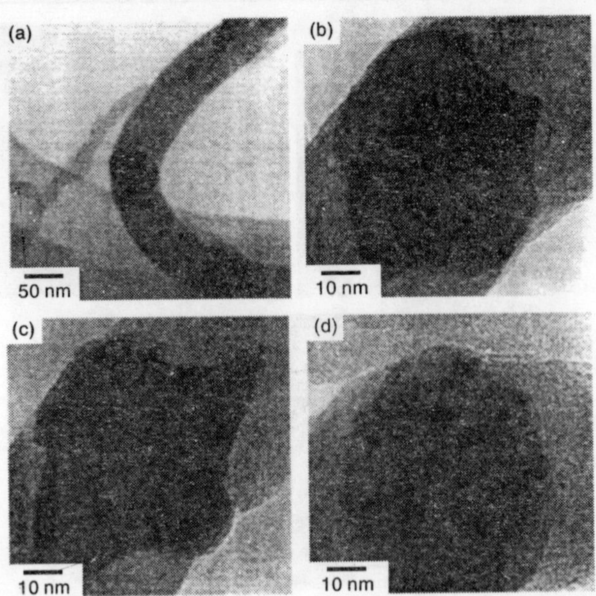

Fig. 12.8 *TEM image of bidirectional Ni particle viewed from differenі directions. (a), a bidirectional nickel particle from n-hexane decomposition at 773 K over Ni (10 wt%)/VGCF; (b), close-up without tilting; (c) tilted forwards 25 degrees fmo (b); (d), tilted backwards 25 degrees from (b)*

We hypothesize that the difference might be caused by the degree of softness of the Ni particles. The surface to volume ratio (S/V) for a Ni particle decreases as the size of the particle increases. Therefore, the dissolution and diffusion of carbon atoms at the near surface layers must become dominant for large metal particles. Under these circumstances, the central body of a large Ni metal stays rigid due to a low bulk carbon concentration. Carbon graphitization occurs through diffusion in the near surface layers for large Ni particles. These surface layers of the metal particles should be softened, thus permitting them to be faceted to Ni(111) in order to maximize the rate of graphitization of carbon atoms or the rate of growth of bidirectional carbon fibers. For small particles, due to the high bulk carbon concentration and consequent large melting point depression, the entire particle takes on a droplet-like shape instead of providing a rigid base for well-defined facets. Therefore, the cross section of the nanotubes shows circular carbon layers, but that of bidirectional carbon fibers must show hexagonal carbon layers.

Thus, catalytic decomposition of alkanes such as methane, ethane, propane, *n*-hexane, and *n*-octane on supported-Ni catalysts at 773 K produced fish-bone type carbon nanofibers with a Ni particle on the tip of each nanofiber. In contrast, catalytic decomposition of benzene and cyclopentadiene formed mainly winding-type, multi-wall carbon nanotubes with uniform diameters (10 ~ 20 nm). Bidirectional, fish bone-type fibers grow outward from a central Ni-catalyst particle are observed during the decompositions of cyclopentadiene, benzene, *n*-hexane, and *n*-octane. Large well-faceted Ni particles (diameter 50 ~ 110 nm) grow the bidirectional fibers. At the early stages of the reaction, the dissolution of carbon atoms into Ni particles

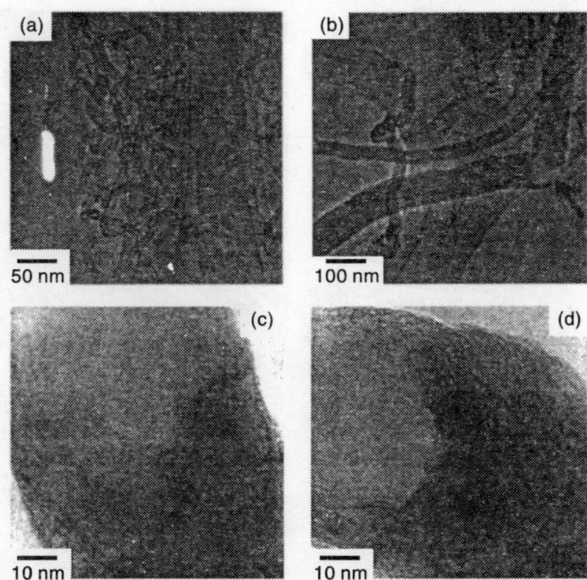

Fig. 12.9 *TEM image of carbon nanotube and bidirectional carbon fibers, after removal of Ni by dissolution in HNO_3 (aq). Carbon nanotubes and bidirecitonal carbon fibers formed by the decomposition of benzene at 773 K. (a) carbon nanotubes (b), carbon nanotubes and bidirectional carbon fibers (c) and (d), magnified views of the larger fiber from which a bi-hexagonal pyramid of Ni particle was removed*

would soften the small size catalysts (10 ~ 20 nm) at 773 K due to melting point depression. The multi-wall carbon nanotubes formed from benzene and cyclopentadiene were caused by these liquid-like Ni particles. These tubular structures could result from the contraction and expansion of the catalytic Ni particles during the decomposition of benzene and cyclopentadiene.

Carbon Nanotube-Fibers

Continuous aligned nanotube fibers and yarns have been developed by directly spinning an aqueous polymer suspension or by directly spinning during chemical vapor deposition synthesis by introducing twist during spinning and by conventional wet spinning in super-acid media without any supporting surfactant or polymeric material.

The first study of macroscopic CNT fibers and ribbons is conducted with an aqueous SWNT/SDS dispersion, followed by extrusion in PYA solution (Fig. 12.10). The resulting fibers with 15 μm radius show higher flexibility and resistance to torsion as compared to commercial carbon fibers. The tensile strength and modulus are 0.15 and 15 GPa, respectively, and with subsequent drawing, the fiber modulus improved by a factor of 4. Macroscopic assemblies of CNTs as a form of fiber and ribbon have been drawn directly from CYD synthesis chamber (Fig. 12.11(a)) without post-processing, and the resulting fibers show tensile strengths as high as 1 GPa.

Continuous well-aligned macroscopic SWNT fibers are produced by conventional wet spinning in 102% sulfuric acid. Fuming sulfuric acid induces protonation of SWNTs and promotes their

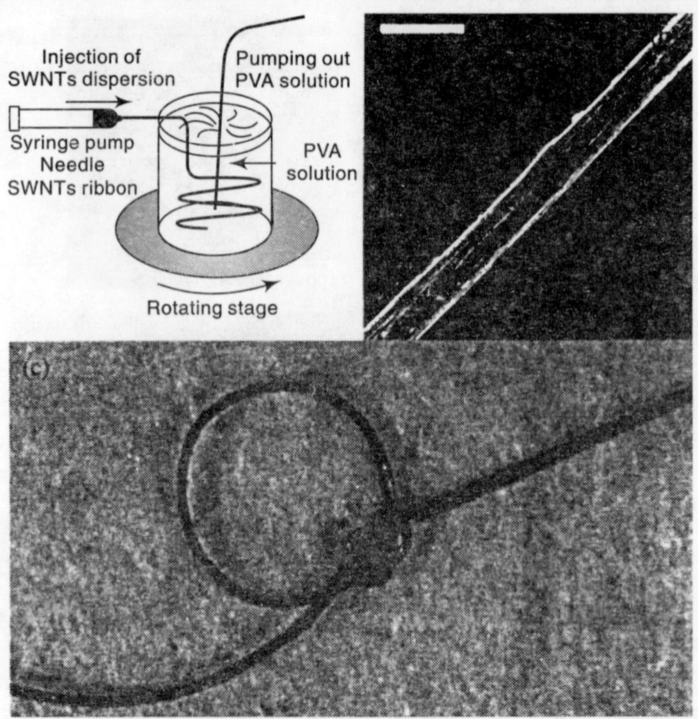

Fig. 12.10 *(a) Schematic drawing of the experimental set-up used to make nanotubes fibers and ribbons; (b) SEM of a nanotube fiber (scale bar = 25 μm); (c) tying knots reveals the high flexibility and the resistance to torsion of 15 μm radius CNT fiber.*

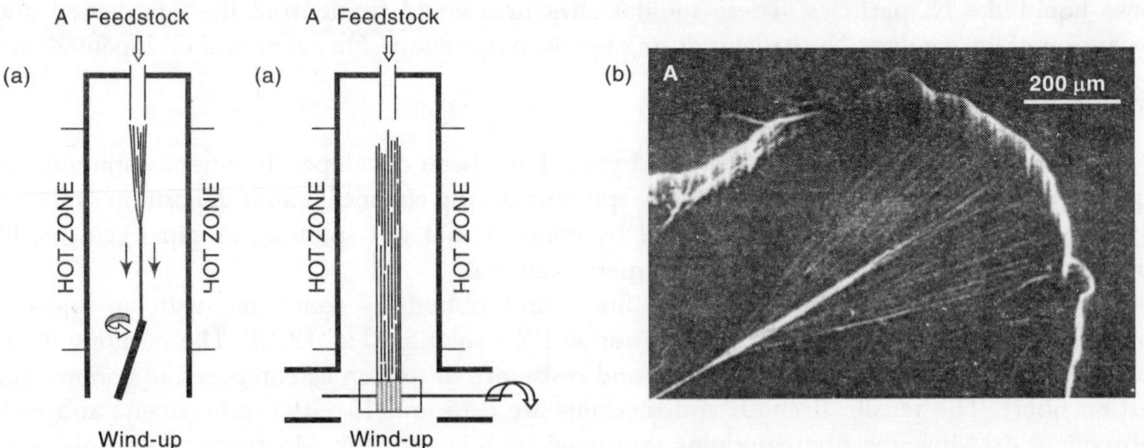

Fig. 12.11 *(a) Schematic description of direct production of CNT yarn from the CVD reactor; and (b) SEM image of direct yarn formation MWNT forest.*

ordering into an oriented phase of individual SWNTs surrounded by acid anions. These SWNT fibers show tensile strength and modulus of 116 MPa and 120 GPa, respectively.

Multifunctional CNT yarns of 2 to 10 μm diameter have been produced by simultaneous drawing and twisting tubes from MWNT forest (Fig. 12.11(b)). These yarns exhibit tensile strength of 460 MPa, electrical conductivity of 3×10^4 S/m, and Poisson's ratio in the range of 2 to 4.2.

CNT assemblies such as fibers and ribbons are of great interest as multifunctional materials in electronic applications, including actuators, supercapacitors, and batteries, by providing specific abilities of radio or microwave absorption, textile heating, or wiring for electronic devices.

Single-wall CNTs Intertube Coupling in Ropes

While the electronic structure of individual tubes has been characterized using scanning tunneling spectroscopy and found to be in agreement with the theoretical predictions the interaction between tubes in ropes has received much less attention. Though coupling between SWNTs is weak but this interaction is measured directly by bridging single tube metal contacts. However, the extensive use of nanotubes in nanoelectronics also require a knowledge of the tube electronic coupling. Below is an approach that allows to determine the electrical coupling between tubes in a rope using four terminal transport measurements. The ropes are self-assembled bundles of carbon nanotubes, in which the tubes line up parallel to each other. The tubes in ropes have diameters close to 1.4 nm and form a regular triangular lattice with a lattice constant of $d_0 = 1.7$ nm. Both semiconducting and metallic tubes are present in a rope in a random distribution. The ropes are dispersed in dichloroethane using mild sonication, followed by several stages of centrifugation. The purified nanotube ropes are dispersed on an oxidized Si substrate and gold electrodes are subsequently fabricated on top of the ropes (see Fig. 12.12(a)). By a sputtering of the rope before deposition of the electrodes by an Ar^+ ion beam at an energy of 500 eV., is to introduce defects into the top nanometers of the rope to enable us to vary the path taken by the electric current in a well-defined manner.

To estimate the extent of the sputter damage, a Monte Carlo simulation is performed on a carbon target with the mass density of nanotube ropes. The ions are found to produce about eight carbon vacancies per ion in a range 6 ± 1 nm deep into the rope (Fig. 12.12(b)). From sputter conditions (Ar^+ flux $\approx 1.5 \times 10^{15}$ $m^{-2}s^{-1}$ for 20 s) a defect density of about one defect per 1000 atoms, is obtained which gives a distance of 5–10 nm between defects along the tubes. This defect density is high enough to have a significant influence on the electrical properties of the tubes, while at the same time, it is low enough for the damaged tubes to preserve their structural integrity. The damage in the upper part of the rope is confined to the area directly underneath the gold contacts (note the shading in the schematic of the rope in Fig. 12.12(a)), while the main part of the rope between the electrodes is not exposed to the ion beam and is thus undamaged. Electronic transport in the damaged metallic tubes is strongly affected by the defects, while contributions from semiconducting tubes are negligible at the low temperatures used in the experiments (the typical band gap for semiconducting tubes of ~ 1.4 nm diameter is ~500 meV).

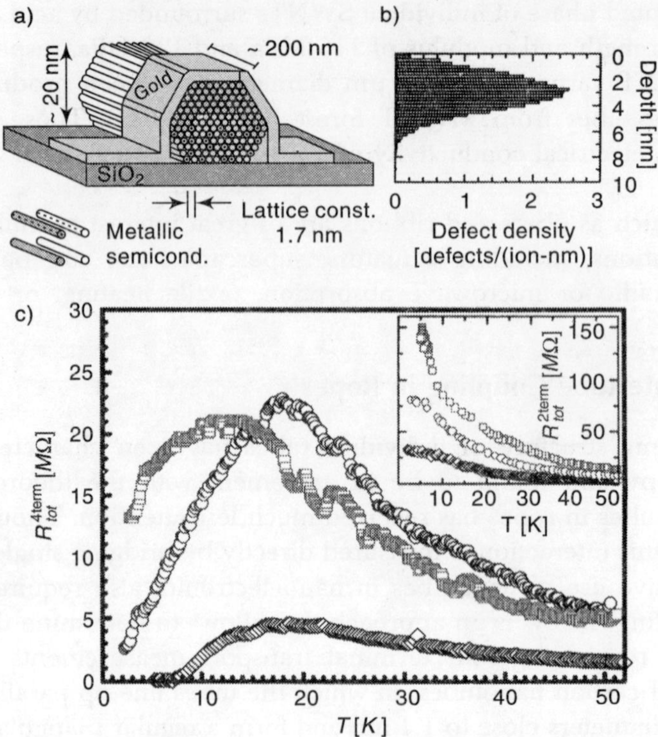

Fig. 12.12 *Schematic of a section of rope (a) together with a simulation of the distribution of the defects introduced by the sputter treatment (b). The shading marks the damaged tubes. (c) four (4t) and two (2t) terminal resistances versus temperature for three different samples*

The resistances are measured in the Ohmic regime using standard lock-in techniques while the sample is cooled in a ^{4}He continuous flow cryostat with a base temperature of 1.5 K.

Typical results of R vs T curves are presented in Fig. 12.12(c). The two terminal (2t) resistances are found to increase with decreasing temperature. Also, the four terminal (4t) measurements show a pronounced resistance maximum at temperatures around 20 K. A 4t measurement of an undamaged rope is shown in the lower inset of Fig. 12.13. Here, a decrease in resistance with decreasing temperature over the whole temperature is observed. Thus, the undamaged ropes show a metallic behavior as is expected for ropes consisting, at least in part, of metallic tubes. It is important to rote the very different values of the resistance in the damaged and undamaged ropes, the latter being only of the order of 1 kΩ, while the former is in the range of several MΩ. Thus, the damage greatly decreases the resistance, but it does not block the electrical transport. As the damaged metallic tubes in direct contact to the gold electrodes carry most of the current, since only a few kΩ would be expected for undamaged nanotubes in the rope, and the semiconducting ones, damaged or undamaged, are insulating. The metallic tubes are one-dimensional systems with two degenerate modes at the Fermi energy hence the resistance of a segment of length L containing defects with an average distance L_0 is given by $R = (h/4e^2)\ (L/L_0)$. The extent of the damaged areas along the direction of current transport, i.e., the width of

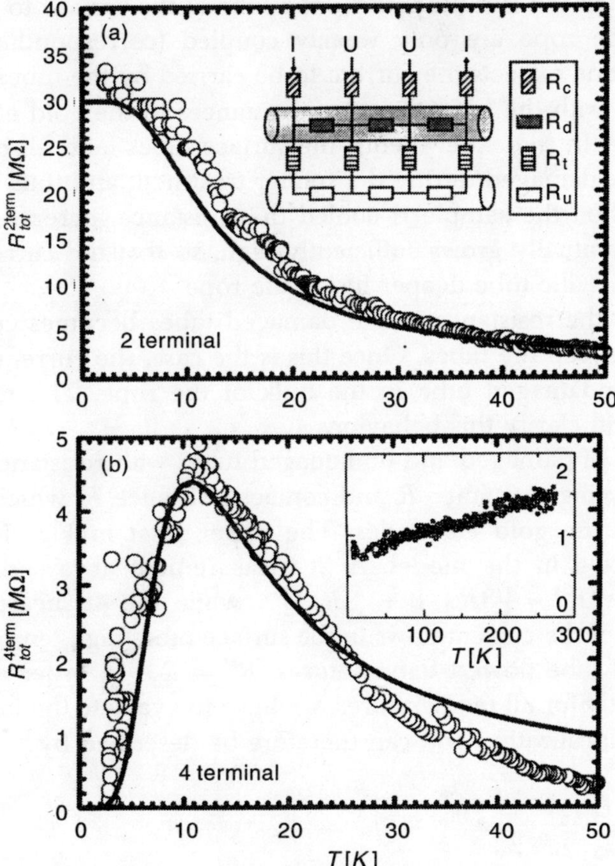

Fig. 12.13 *Comparison between measurement (symbols) and fit (solid line) (a) The insets show the resistance network and (b) the temperature dependence of the 4t resistance of an undamaged rope. (Note the very different resistance scales of the 4t plots)*

the gold electrodes, is typically 200 nm. (The rope segments between the contacts are undamaged and thus their contribution to the resistance is negligible.) At room temperature, the resistances of the ropes is around 200 kΩ, hence a mean free path of $L_0 \simeq 6$ nm, (this agrees with the average distance between defects obtained by the Monte Carlo simulation.)

In Fig. 12.12(c), both 2t and 4t measurements show an increase in resistance when cooling the sample. This increase is caused by electronic localization in the damaged tube, an interference effect which increases the resistance by coherent backscattering of electrons at the defects. When the phase coherence length exceeds the localization length ML_0 (M number of modes) the resistance increases exponentially with decreasing temperature.

Below a sample-dependent temperature around 20 K (11 K in the example of Fig. 12.13, the resistance obtained by the 4t measurement starts to decrease. This effect is not used by gold-tube contact resistance, since these do not contribute in a 4t measurement.

As disorder can switch the current path from a tube at the surface to another tube inside the rope. The tubes in the rope are only weakly coupled (corresponding to a large coupling resistance), so usually one expects the current to be carried by the tube with the lowest contact resistances to the tube with the lowest contact resistances to the gold electrodes, i.e., a tube at the surface at the rope. In the experiments, the surface tubes (and all other tubes about 6 nm deep into the rope) are damaged during the sputter treatment and thus show high resistance at room temperature. When the sample is cooled the resistance increases due to localization in damaged tube, and eventually grows sufficiently high, so that the current switches its path to another undamaged metallic tube deeper inside the rope.

This happens when the resistance in the damaged tubes becomes comparable to the large coupling resistance between the tubes. Once this is the case, the current favors the "new" path and switches to the undamaged tube in the bulk of the rope. The model presented in the following paragraph will clarify this behavior.

Consider a network of damaged and undamaged tubes with resistances R_d and R_u $(R_d \geq R_u)$, with the intertube coupling resistance R_t and contact resistance Rc which connect the damaged tube at the surface to the gold electrodes. The upper inset in Fig. 12.13 shows how these resistances are connected in the model. A 2t measurement always detects the two parallel current paths and thus $R^{2t} \approx [(3R_d)-1 + (2R_t)^{-1}]^{-1}$, while the 4t measurement is sensitive to whether the main part of the current flows in the surface tube (high temperatures, $R^{4t} \approx R_d$) or in the undamaged "bulk" tube (lowest temperatures, $R^{4t} = R_u$). In order to calculate the total $2t$ and $4t$ resistances exactly for all temperatures, we have to evaluate the individual resistances. R_d is governed by strong localization and can therefore be described by

$$R_d = \frac{L}{L_\Phi}\frac{h}{2e^2}\frac{1}{2}\left(e^{\frac{2L_\Phi}{ML_0}} - 1\right)$$

L_Φ follows a power law dependence on temperature, $L_\Phi \sim T^{-\alpha}$, and we will use this to describe R_d. The intertube coupling resistance R_t will be taken to be independent of temperature. The coupling resistances are placed underneath the electrodes, i.e., connected to the damaged areas of the surface tube, since a change in current path occurs only where transport inside the surface tube gets "blocked" by localization. The resistance of the undamaged tube R_u is much smaller than R_t and R_d (cf. the lower inset in Fig. 12.13 and since it is always connected to R_t, it does not play any significant role. The last resistance to be discussed is the contact resistance R_c between the gold electrode and the surface tube. This resistance will only show up in $2t$ measurements but will not contribute to the 4t resistance.

Figure 12.13 shows experimental data, together with the fits based on the sample model. The results of both 4t and 2t measurements are reproduced. First, note that no additional contact resistances R_c are necessary to describe the 2t measurements. Second, from the fits in Fig. 12.13 L_Φ (T) as a function of temperature is found. The coherence length turns out to be about 200 nm at the lowest temperature.

The temperature dependence of L_Φ can be described by $L_\Phi \sim T^{-\alpha}$ with $\alpha = 0.33 - 0.5\alpha$ but $\alpha = 1/2$ suggests electron-phonon scattering.

Next the coupling resistance R_t between the tubes is extracted from the data. It turns out to be the most reliable and stable parameter. It is determined by the *value* of the resistance maximum in the 4t measurement. The temperature dependence of R_d determines the *shape* and *position* of the maximum in temperature $[R_d(T_{max}) \simeq R_t]$. Since slight variations in sputter damage (L_0) significantly affect R_d there is no strict correlation between R_t and T max, but in spite of this R_t can be obtained from the value of the resistance maximum. Analyzing R_t values ranging from 2 to 140 MΩ are obtained. Any proposed coupling mechanism must be able to account for this large range of transfer resistance values.

This mechanism involves transport by hopping through intermediate states (e.g., via other tubes). A single transfer corresponds to a resistance of about 2 MΩ, and thus the highest value of 140 MΩ would needs 70 transfers, with only 100 tubes in the rope. The hopping processes are thermally activated and thus the coupling between tubes eventually freeze out, leaving the rope insulating at the lowest temperatures.

A small range of distances between the tubes leads to a large range of resistances due to the exponential dependence typical for tunneling between tubes as this process does not freeze out even at the lowest temperatures.

Now, the cross section of a rope consists of typically 100 single tubes. A fraction of about 2/3 of the tubes is semiconducting, while the remaining 1/3 is said to be metallic. The switching of the current path from damaged to undamaged tubes involves tunneling over some distance d within the triangular lattice of the rope. The depth of the damage of the sputter treatment (about 6 nm) sets a lower limit for the distance d within which an undamaged metallic tube can be found. To describe the coupling resistance R_t that is caused by the tunnel process, consider the coupling of two one-dimensional wave-guides separated by a tunnel barrier $R_t = (h/4e^2)(v_F/v_\perp)(1/T) \simeq (h4e^2)e^{2\kappa d}$. The velocity perpendicular to the tube axis $v_\perp$ can be approximated by the Fermi velocity v_F when transport along the damaged tube is blocked by localization. The transmission T in the tunnel process is determined by the overlap of the wave functions of the tubes, with κ being related to the barrier height. Given the linear dispersion relation for the metallic nanotubes $\in(k) = hv_F(k - k_F)$ around the Fermi energy and the barrier height Φ, κ is calculated as $\kappa = \Phi/(hv_F)$. Since the electrons tunnel through the other tubes, which are mostly semiconducting, the barrier height is given by the conduction band edge of these semiconducting tubes. All nanotubes share the same graphene structure, hence their work function is expected to be nearly the same, and the Fermi level of the metallic tubes is expected to align midgap the semiconducting energy gap. With an average band gap of 500 meV; $\Phi = E_{gap}/2 = 250$ meV. Using $v_F = 10^6$ m/s, a penetration depth of 1/2K = 1.25 nm is obtained.

By comparing theoretical and experimental data (Fig. 12.14), where the theoretical predictions result from evaluating the above formula for the discrete distances d realized in the triangular lattice of the rope (the tunnel distance d is the distance between the centers of the involved tubes). Using $\Phi = 225$ meV. Since $k_BT << \Phi$, the tunnel resistance is indeed temperature independent. Thus, all data points coincide with values allowed by the model. Also, tunneling into tubes closer than about 8 nm to the damaged surface tube is not there. This is because tunneling into another damaged metallic tube is not favorable.

Thus, the theoretical assumption of direct tunneling yields a consistent picture for the electronic coupling in nanotube ropes.

In conclusion, by Ar^+ sputtering of single-wall carbon nanotube ropes before making electrical contact to them, defects are introduced that led to strong scattering in the current carrying tubes at the surface of the rope. Below a specific threshold temperature, the current tunnels into an undamaged metallic tube in the bulk of the rope, leading to a dramatic reduction of the four terminal resistance. The value of the resistance maximum is related to the intertube coupling resistance between the involved tubes.

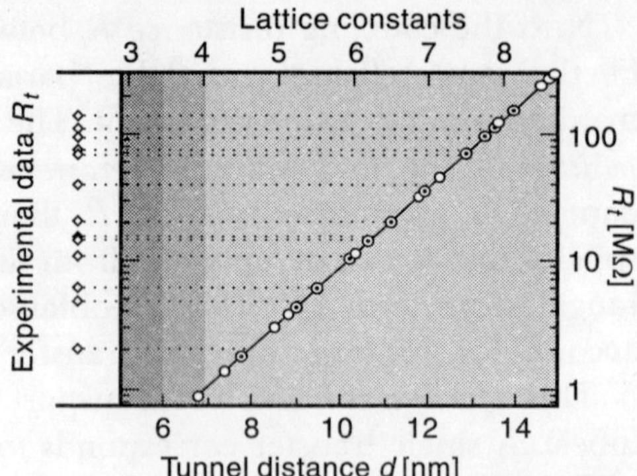

Fig. 12.14 *Comparison between experimentally found coupling resistance R_t (symbols on the left-hand side) and values allowed by the theory (circles) for tunneling between tubes. Filled circles mark coincidences. The shaded area marks the depth of the sputter damage, which sets a lower limit for the observable tunnel distances*

CARBON NANOFIBRES FROM CYCLOPENTADIENCE AND BENZENE

The decompositions of cyclopentadiene and benzene do not continue for more than 20 min at 773 K, i.e., supported-Ni catalysts are deactivated quickly by the deposition of carbon layers on the surface of Ni particles, although the initial rate of decomposition of these hydrocarbons is of the order of magnitude greater than that of methane decomposition. However, these unsaturated hydrocarbons produced multiwall carbon nanotubes as the main carbon products.

The carbon fibers formed from the decomposition of cyclopentadiene are shown in Figure 12.15(a). It should be noted that these fibers have relatively uniform diameters (10 ~ 20 nm and are comprised of hollow tubes. A high resolution TEM image of the nanotubes is shown in Figure 12.15(b). The graphitic layers run with small angles against the longitudinal axis of the nanotubes. The inner cavities seem to be blocked in places.

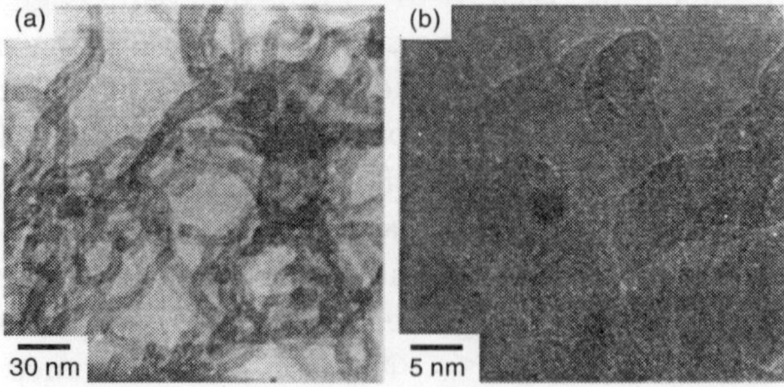

Fig. 12.15 *TEM images of the carbon nanotubes formed from cyclopentadiene at 772 K. (C/Ni = 30). Catalyst = Ni (10 wt%)/GCF*

Typical carbon nanofibers obtained from benzene decomposition at 723 and 923 K are indicated in Figures 13(a) and (b), respectively. The sizes of the carbon fibers are between 10 to 20 nm. The zigzag carbon fibers clearly show hollow structures along the center. The shape and the structure of these carbon nanotubes did not change appreciably when the decomposition temperature was raised from 723 K (Fig. 13(a)) to 923 K (Fig. 13(b)). Several TEM images of the Ni particles at the tips of carbon fibers from benzene are demonstrated in Figures 13(c) to (f). Figure 13(c) is a further magnification of the fiber tip highlighted by the rectangle in Figure 13(a). All the Ni particles in Figures 13(c) to (f) seem to have tapered their bodies as they grew their carbon tubes. The particles of Ni metal appear to have been liquidized during the decomposition of benzene due to melting point depression of Ni metal, which must be caused by the dissolution of carbon into the Ni particles.

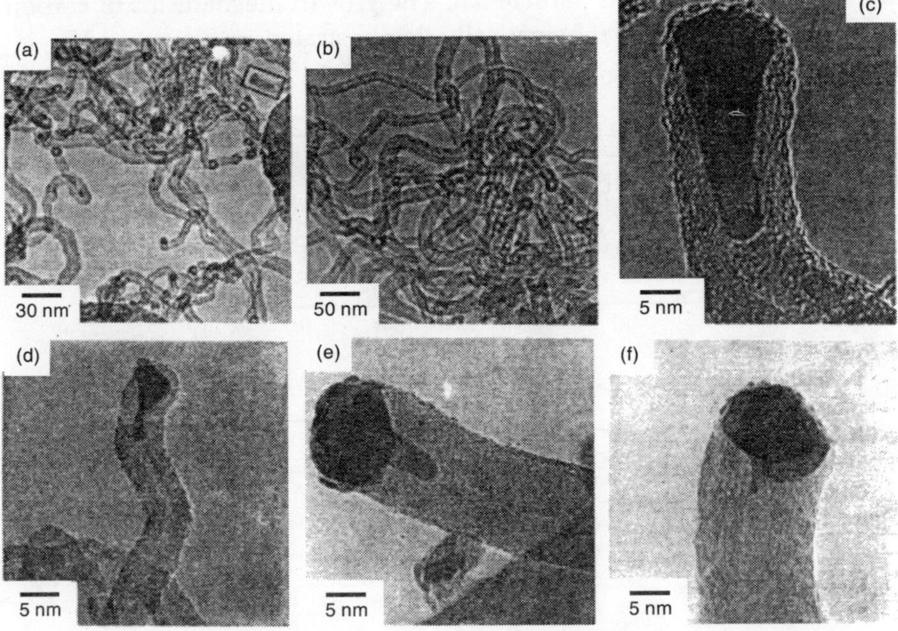

Fig. 12.16 *TEM images of the multiwall carbon nanotubes formed from benzene on Ni (10 wt%(/VGCF at 773 K. The range of C/Ni = 30 ~ 160*

Uniform carbon nanotubes can be synthesized by using cyclopentadiene and benzene at relatively low temperature (773 K). The TEM images shown in Figures 12.16(c) to (g) suggest that flexible Ni particles may expand (Fig. 12.16(c)-(e)) and contract (Fig. 12.16(f)) their tails during the growth of carbon nanotubes. This tail shape and its behavior may be ascribed to the generation of these nanotube structures. The direction of growth of the nanotube suddenly changes as can be seen in Figures 12.16(c) and (d). This explains the formation of zigzag carbon nanotubes indicated in Figures 12.16(a) and (b) as well as in Figures 12.16(a) and (b). Decomposition of methane and other alkanes do not give carbon nanotubes at 773 K.

The zigzag shape and tubular structure observed for the carbon fibers from cyclopentadiene and benzene is due to softening of Ni particles from the dissolution of carbon atoms into the metal. The initial rates of carbon deposition on Ni (10 wt%)/VGCF from benzene and cyclopentadiene are one order of magnitude greater than that of carbon deposition from methane. This implies that the concentrations of carbon atoms deposited from benzene and cyclopentadiene on the on the surface as well as in the bulk of active Ni metal are greater than those for methane decomposition. Therefore, the melting point of Ni metal is more depressed during decomposition of benzene or cyclopentadiene than that for alkane decompositions. Liquid-like Ni metal work as a catalyst during the decomposition of benzene and cyclopentadiene. Long and thin nickel metal are squeezed out at the tip of the carbon nanotube, resulting in the opening of fresh Ni surface to benzene molecules and cyclopentadiene molecules. The growth direction of the nanotube changes at this moment, and a tubular structure results from the contracting and expanding of the Ni particle tail. The growth mechanisms of carbon nanotubes during benzene decomposition are schematically demonstrated in Figure 12.17.

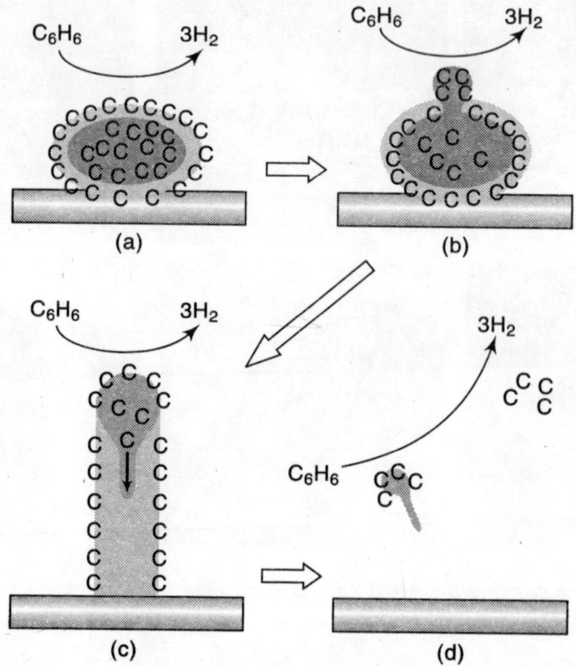

Fig. 12.17 *(a) Rapid decomposition of benzene saturates the bulk of small Ni particle (~10 nm) with carbon atoms, softening the particle by dissolved carbons. (b) The metal particle is squeezed out because of a strong pressure buildup due to the formation of graphite layers at the internal surface of the graphite envelope. (c) The fresh surface of the squeezed-out Ni particle would then further decompose benzene and continue to grow into a carbon nanotube. (d) Tapering and contrasting of the particles causes the formation of zigzag nanotubes. The dissolved carbon atoms graphitize at the tail wall of Ni leaving hollow structures behind*

NANO TEXTILES

Textile manufacturing is an old technology, but it can provide the bridge to making nanotechnology textiles, by the two technologies (textile technology and nanotechnology) with new development in science. *Nanofibers* that are capable of filtering out viruses, bacteria and hazardous particles, which are too small to see with naked eye naked eye are produced. These nanofibers are produced by a process called electrospinning. Biofilters are made from these nanofibers.

Electrospinning

Electrospinning is a method for producing nanometer diameter polymer fibers. The instrumentation consists of a nozzle of capillary dimensions, a metal ground, a collecting mandrel and a high voltage source, In the most common scenario the collecting mandrel coincides with the metal ground. The electric field created between the capillary and the metal ground pulls the polymer solution from the nozzle.

The technique is able to produce nanometer diameter fibers. Most prototype devices are very slow (produce only grams per hour), while several companies succeeded to mass produce electrospun fibers.

The main application of the technique itself and of electrospun fibers are in the fields of biomedicine (especially tissue engineering), filtration media (e.g. protective clothing), and nano-electronics (e.g. conductive polymers for micro and nano sensors).

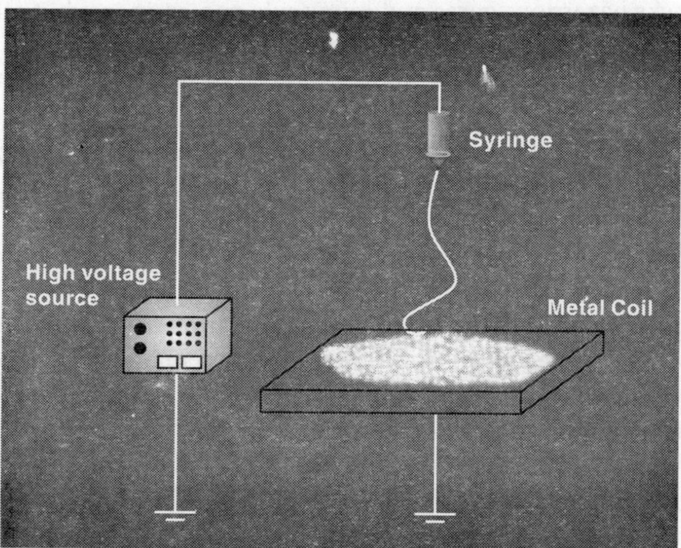

Fig. 12.18 *The electrical field applied between the needle and the collector determines the emergence of a thin jet of polymer solution. The jet has an unstable trajectory through air determining a random alignment of the fibers on the metal collector*

Electrospinning involves dissolving a polymer– either a natural polymer derived from plant-based cellulose or synthetic polymer such as nylon or polyester–in a solvent, squeezing the applying high voltage to the pinhole. The electrical field pulls the polymer solution through the air, stretching it into a tiny fiber. An electron microscope is needed to see these fiber, which are less that 100 nanometers in diameter, or 1,000 times smaller that conventional fibers.

Once you get to a really small-size scale like this, the forces that dominate on a large scale suddenly become much less important, and properties like friction and flow past objects that are less significant on a large scale become the dominant terms. Nanoscale fibers are roughly equivalent in size to air molecules. So as filtration devices, nanofibers make very effective filters because they don't get in the way of the air flowing past them. And less power is need to push the air through the filter because the volume of the fibers isn't blocking the air flow."

Ingenious modification are made to the electrospinning process to create supersensitive biofilters. By manipulating magnetized nanoparticles in a magnetic field during electrospinning, one can direct the flow of the polymer and target chemical molecules to deposit layer after layer onto the surface of the electrospun fibers. The advantage is that both the size and the position of the nanofibers can be precisely controlled.

Because of the combination of the small size of the fibers and the electrical fields created between the fibers, the particles in the 100- to 300-nanometer range (the size of viruses and bacteria) are captured.

These novel filters provide protection against infectious agents that cause severe acute respiratory syndrome (SARS) and avain flu. However, the process is customized to serve a variety of purposes. By applying nanolayers to natural fibers, for instance, protective clothing is produced, which blocks hazardous gases and minuscule contaminants but allows air and moisture to flow through so that the wearer can sweat and feel comfortable. Such clothes are used by firefighters, emergency responders controlling a biohazard threat, or military personnel on duty in hot climates.

Another application is controlling the movement of medicine through such fibers. This technology is used to administer time-released antibacterial and antiallergenic compounds, (arthritis medicine by wearing gloves can be given). Or have antibacterial sheets in hospitals or military barracks that could help prevent the spread of illnesses where lots of people are confined to small, common spaces.

Even the law is going nano. Also embed magnetic nanoparticles inside a garment or paper document that create a unique but invisible signature are made, which could be scanned, for example, to detect counterfeit currency, fake passports, or even knockoff designer clothing.

Textiles as Sensors Tunique characteristics of nanofibers–such as greater surface area and porosity has resulted in fabricating a textile that could act as a sensor. A material that is developed, perhaps, something that looks like a sponge, that anyone could easily use to swab a piece of fruit or meat, or wipe a food or surgical preparation surface, to detect the presence of hazardous bacteria such as *E. coil* or anthrax. If the contaminant is detected, the fibers would capture it and alert the user by changing color or becoming fluorescent.

Single-walled carbon nanotubes (SWNTs) show promise for applications ranging from ultra-small electronic and sensing devices to multifunctional materials. To date, a number of SWNT-

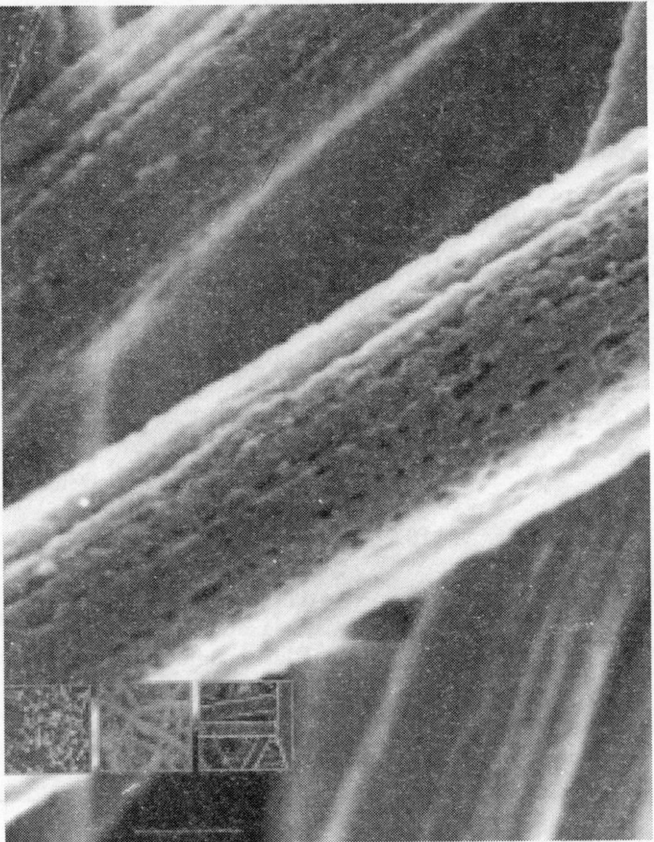

Fig. 12.19

based proof-of-concept devices such as field effect transistors field emission displays and chemical sensors have been fabricated, and the integration of the nanotube materials in all cases relies on one's ability to control the placement, orientation, and shape of the nanotube components within the context of the device on the micrometer-to nanometer-length scale. Such positional control over large areas is extremely challenging and currently quite limited. Depending on the intended application, one wants to be able to pattern SWNTs as individual tubes, small bundles or thin films.

QUICK FACTS ABOUT CARBON NANOTUBES

Equilibrium Structure	Å
Average Diameter of SWNT's nm	1.2-1.4
Distance from opposite Carbon Atoms (Line 1)	2.83
Analogous Carbon Bond Separation (Line 2)	2.456
Parallel Carbon Bond Separation (Line-3)	2.45
Carbon Bond Length (Line 4)	1.42
C-C Tight Bonding Overlap Energy eV ~2.5	
Group Symmetry (10,10) C5V	
Lattice: Bundles of Ropes of Nanotubes:	
Triangular Lattice (2D)	
Lattice constant	17
Lattice Parameter:	
(10,10) Armchair	16.78
(17,0) Zigzag	16.52
(12,6) Chiral	16.52A
Density in gm/cm^3	
(10,10) Armchair	1.33
(17,0) Zigzag	1.34
(12,6) Chiral	1.40
Interlayer Spacing	
(n,n) Armchair	3.38 A
(n,0) Zigzag	3.41
(2n,n) Chiral	3.39

Optical properties	
Fundamental Gap:	
For (n,m):n-m is divisible by 3] [metallic}	0 eV
For (n,m): n-m is not divisible by 3 [Semiconducting]	~0.5eV
Conductance Quantization	n x (12.9k Ω)$^{-1}$
Resistivity	10^{-4} Ω · cm
Maximum Current Density	10^{13} A/m^2
Thermal Transport	
Thermal Conductivity (Room Temperature)	~2000 W/m · K
Phonon Mean Free Path	~100nm
Relaxation Time	~10^{-11}s
Elastic Behavior	
Young's Modulus (SWNT)	~1 Tpa
Young's Modules (MWNT)	1.25 Tpa
Maximum Tensile Strength	~30 Gpa

Young's modulus, tensile strength and density of carbon nanotubes compared with some other materials

Material	*Young's modulus (Gpa)*	*Tensile Strength (Gpa)*	*Density (g/cm^3)*
Single wall nanotube	1054	150	
Multi wall nanotube	1200	150	2.6
Steel	208	0.4	7.8
Epoxy	3.5	0.005	1.25
Wood	16	0.008	0.6

CARBON NANOTUBE-A NEW CARBON ALLOTROPE

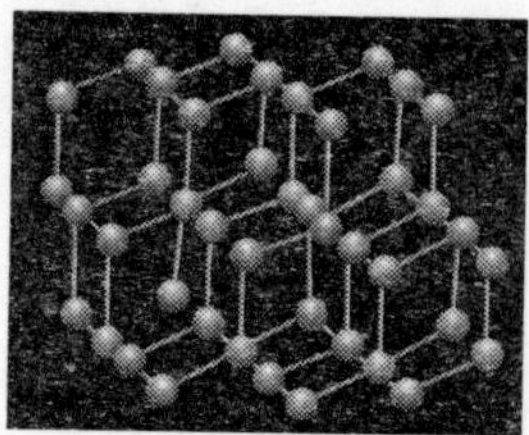

Diamond sp^3 bonding, hard and insulation

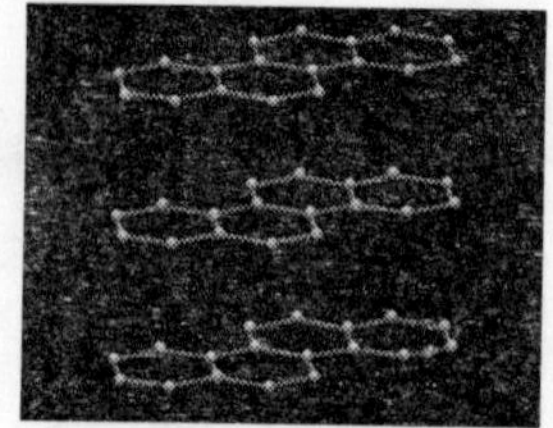

Graphite sp^2 bonding, soft between graphite layers

Nanotubes's Properties

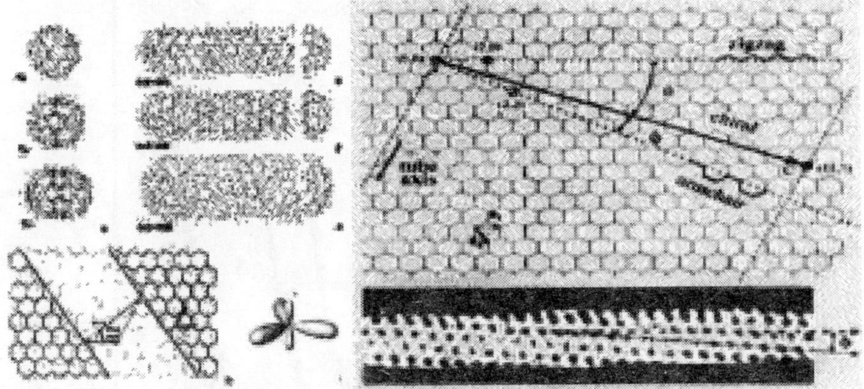

Carbon Nanotube can be either Metallic or semiconducting depending on different chirality Mertalic: 1. (n,n) armchair 2. Nm=3i; Semiconducting: 1. (n,0) Zigzag 2. all of others 1/3 of carbon nanotubes are metallic if chirality distribution is random.

Mechanical Properties

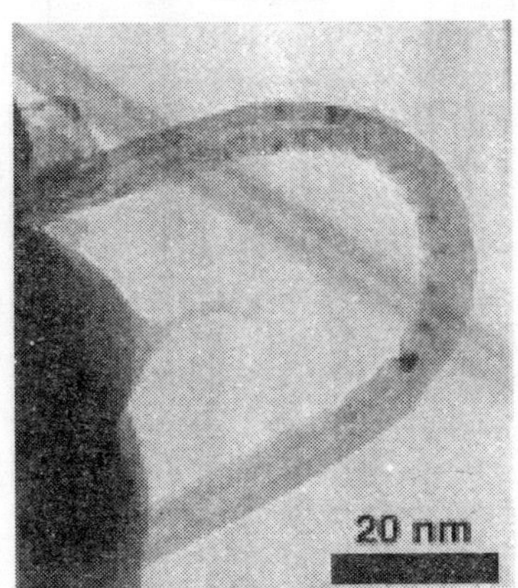

High modulusasdiamond

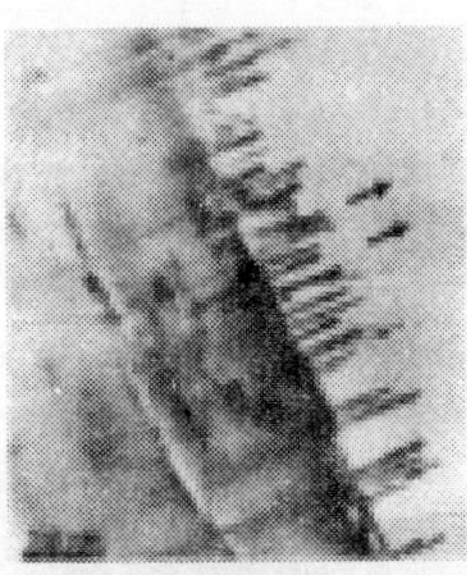

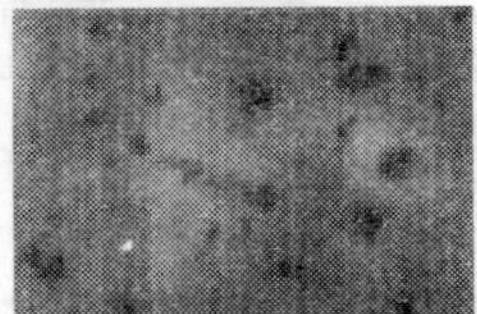

High strength of composite

Typical I-V Characteristics of Carbon Nanotubes

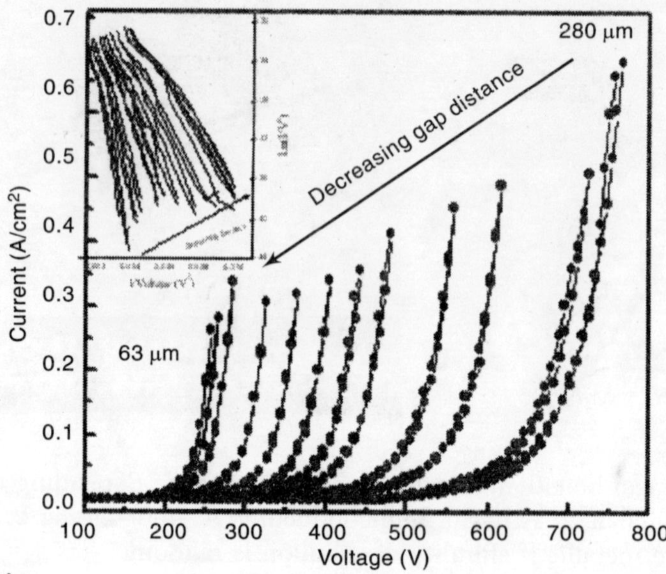

Electronic Properties

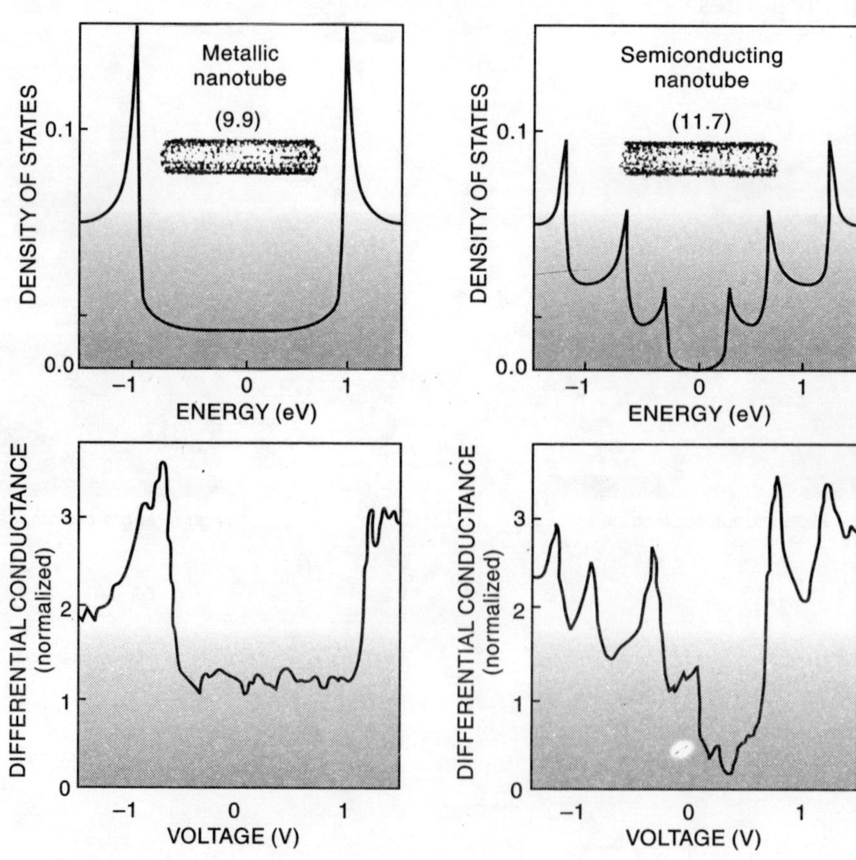

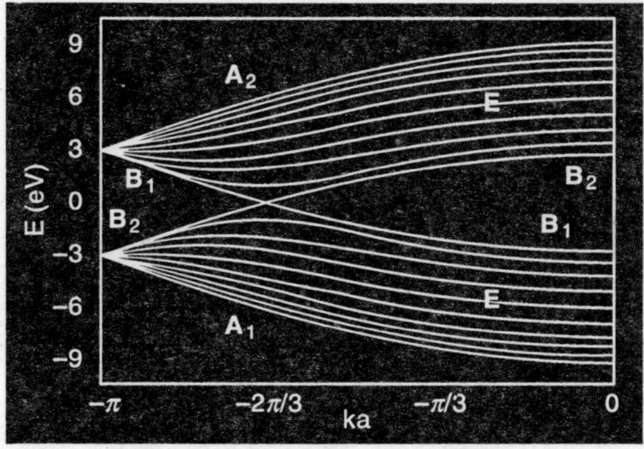

Band Structure of the (10,10) tube

Thermal Properties

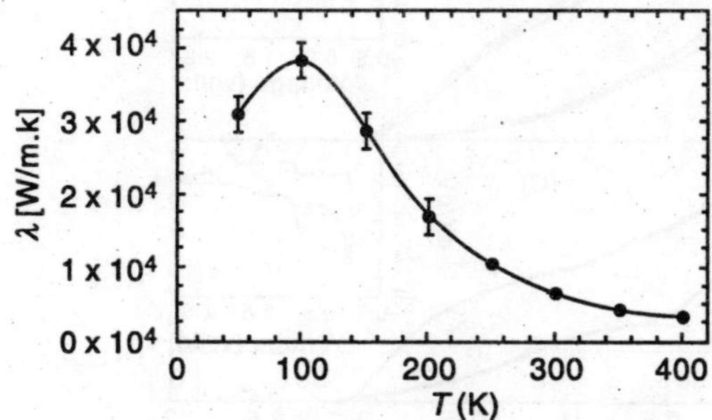

Thermal conductivity vs. T for (10, 10) Tube

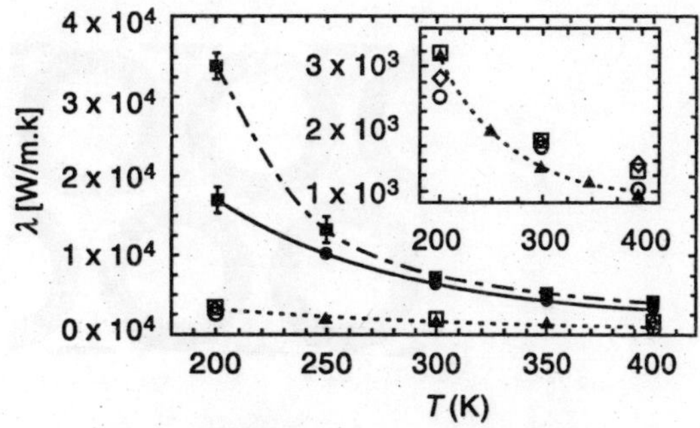

Comparison of (10, 10) tube with graphite and graphite monolayer

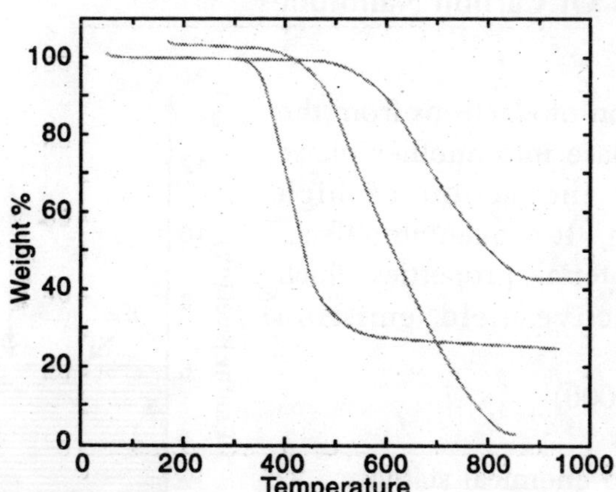

Carbon Nanotubes are stable at high temperature

- Carbon nanotubes are stable in Ar environment.
- Carbon nanotubes are very strong against strong acid and high temperature because of their perfect conjugated system. Acid and heat are often applied to purity carbon nanotubes.

Electrochemical Properties

New negative electrode for rechargeable Li batteries
Anode: Li Cathode: CNT/ graphite film
Discharge: Intercalation of Li into CNT/graphite film
Charge: De-intercalation out of CNT/graphite film

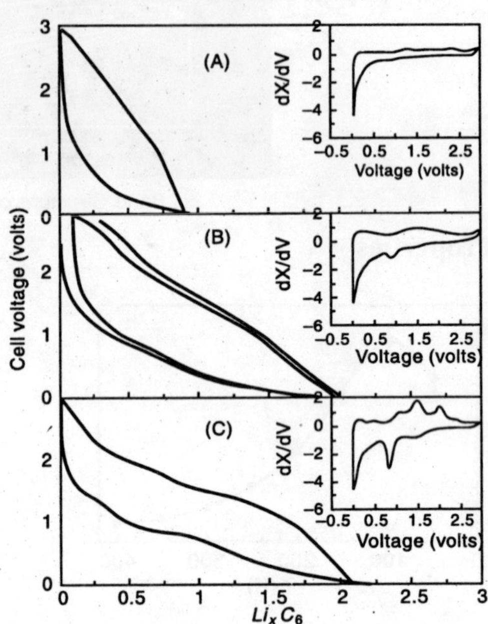

Field Emission Properties Of Carbon Nanotubes

What is Field Emission?

Filed emission is emission of electrons from the surface of a condensed phase into another phase usually vacuum, under the action of high electrostatic field (10^8V/cm). It is quantum effect.

CNTs have excellent materials properties which make them have attractive field emission characteristics.

- Large aspect ratio(>1000)
- Atomically Sharp tips
- High temperature and chemical stability
- High electrical and thermal conductivity

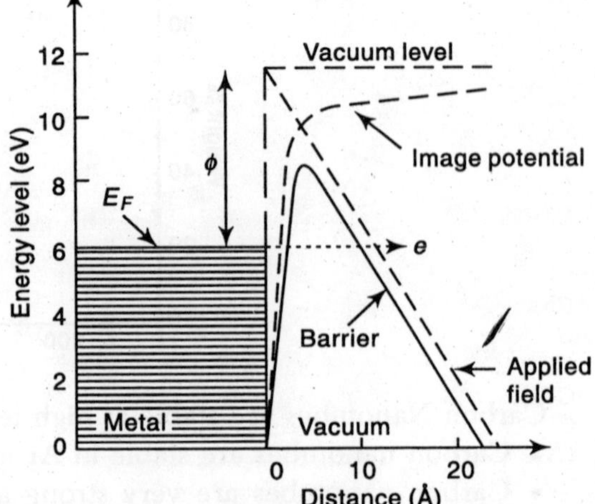

Useful Acronyms

μCP	Microcontact printing
aa	amino acid
Ab	antibody
ADMET	Adsorption, distribution, metabolism, excretion and toxicology
AES	Auger electron spectroscopy
AFM	Atomic Force Microscope (Microscopy)
Ag	antigen
ALD	Atomic Layer Deposition
ALE	Atomic Layer Epitaxy
ASIC	Application-specific integrated circuit
ATMP	Advanced Therapy Medicinal Product
ATP	adenosine triphosphate
BBB	blood brain barrier
BCC	Body-centred Cubic (Crystal structure)
BEEM	Ballistic Electron Emission Microscopy
BET	Brunauer, Emmett and Teller method of measuring surface-area
BioNEMS	biofunctionalised nanoelectromechanical systems
BMR	Ballistic Magnetoresistance
BSF	Back surface field
BZ	Brillouin Zone
CARL	Chemically amplified resist lithography
CCD	Charge-coupled device
Cermet	Ceramic-Metal Composite
CFC	Chloroflourocabrons
CMOS	Complementary Metal-Oxide Semiconductor
CMP	Chemical Mechanical Polishing
CNS	central nervous system
CNT	Carbon Nano Tube
CPC	Condensation Particle Counter
CRD	chronic respiratory disease
CT	computer tomography
CVD	cardiovascular disease
CVD	Chemical Vapour Deposition
CW	Continuous wave
CWA	CEN Workshop Agreement
Cz	Czochralski

DBQW	Double-barrier quantum-well
DFB	Distributed feedback (QCL)
DFT	Density Function Theory
DLC	diamond-like carbon
DLTS	Deep level transient spectrocopy
DMA	Differential Mobility Analyzer
DNA	dexoyribonucleic acid
DOF	Depth of focus
DPN	dip-pen nanolithography
DRAM	Dynamic random access memory
DUV	Deep ultraviolet
EBIC	Electron beam induced current
ECL	Emitter-coupled logic
ECM	extracellular matrix
ECR	Electron cyclotron resonance (CVD, plasma etching)
EDP	Ethylene diamine/pyrocatechol
EDX	Energy Dispersive X-ray analysis
EELS	Electron Energy Loss Spectroscopy
EEPROM	Electrically erasable programmable read-only memory
EFA	Field emitter cathode array
EL	Electroluminescence
ESCA	(also called XPS) Electron Spectroscopy for Chemical Analysis
ESEM	Environmental Scanning Electron Microscope
ESR	Electron spin resonance
ESTOR	Electrostatic data storage
Et	Ethyl
EUV	Extreme ultraviolet
EUV	Extreme Ultra-Violet
EUVL	Extreme ultraviolet lithography
EXAFS	Extended X-ray Absorption Fine Structure
Fab	fragment antigen binding (of an antibody molecule)
FCC	Face Centred Cubic (Crystal structure)
FEG-SEM	Field Emission Gun Scanning Electron Microscope
FET	Field-Effect Transistor
FIB	Focused ion beam
FP	Fabry-Perot
FRET	fluorescence resonance energy transfer
FTIR	Fourier transform infrared
FWHM	Full Width, Half Maximum
GDOES	Glow Discharge Optical Emission Spectroscopy
GMR	Giant Magneto Resistance
HBT	Hetero bipolar transistor
HEL	Hot-embossing lightography
HEMT	High electron mobility transistor
HIT	Heterojunction with intrinsic thin layer
HOMO	Highest occupied molecular orbital
HREELS	High-Resolution Electron Energy Loss Spectroscopy
HREM	High-Resolution Electron Microscopy
hTEP	human tissue-engineered product
HVOF	High Velocity Oxygen Fuel
IC	Integrated circuit
ICP	Inductively coupled plasma
IMPATT	Impact ionization avalanche transit time
IPG	In plane gate
IR	Infrared

ITO	Indium-tin-oxide
ITRS	International technology roadmap for semiconductors
Laser	Light amplification by stimulated emission of radiation
LBIC	Light beam induced current
LCD	Liquid crystal display
LDD	Light-emitting diode
LED	Light-emitting diode
LED	Light Emitting Diodes
LEED	Low energy electron diffraction
LISA	Lithographically induced Self-Assembly
LMIS	Liquid metal ion source
LOC	Lab-on-a-chip
LPE	Liquid phase epitaxy
LSS	Lindhardt, Scharff, SchiØtt (Researchers)
LUMO	Lowest unoccupied molecular orbital
M	Metal
MAL	Mould-assisted lithography
MBE	Molecular Beam Epitaxy
MCT	Mercury cadmium telluride
Me	Methyl
MEMS	Micro Electro Mechanical Systems
MEMS	Microfabricated Electro Mechanical Systems
MFH	Magnetic Fluid Hypothermia
MFM	Magnetic Force Microscopy
MIS	Metal-insulator-semiconductor
MMIC	Monolithic microwave integrated circuit
MNT	(Forum)-Micro and nano technologies
MNT	Molecular Nanotechnology
MOCVD	Metallo-organic chemical vapor deposition
MODFET	Modulation-doped field-effect transistor
MOLCAO	Molecular orbitals as linear combinations of atomic orbitals
MOS	Metal-oxide-semiconductor
MOSFET	Metal-oxide-semiconductor field effect transistor
MPU	Microprocessor unit
MQW	Multi wall nanotubes
MRI	Magnetic Force Microscopy
MST	Micro Systems Technology
MWCNT	Multi-Walled Carbon NanoTube
MW	Molecular weight
MWNT	Multi wall nanotubes
NA	Numerical aperture
NAND	Not and
Nd:YAG	Neodymium yttrium aluminum garnet (laser)
NDR	Negative differential resistance
NEMS	Nano Electro Mechanical Systems
NEMS	Nanofabricated ElectroMechanical Systems
NIL	Nanoimprint lithography
NIL	NanoImprint Lithography
NM	Nuclear Medicine
NMOS	n-Channel metal oxide-semiconductor (transistor)
NMR	Nuclear magnetic resonance
NMR	Nuclear Magnetic Resonance
NOR	Not or
NOx	Nitrogen Oxides
NQR	Nuclear Quadrupole Resonance

NSOM	Near-field Scanning Optical Microscopy
OCL	Quantum cascade laser
OLEDs	Organic Light Emitting Diodes
PADOX	Pattern-dependent oxidation
PCR	Polymerase Chain Reaction
PDMS	Polydimethylsiloxane
PE	Plasma etching
PECVD	Plasma-enhanced chemical vapor deposition
PEEM	PhotoEmission Electron Microscopy
PEG	Polyethylene glycol
PET	Polyethyleneterphthalate
PET	Position Emission Tomography
PL	Photoluminescence
PLAD	Plasma doped
PMMA	Polymethylmethacrylate
PREVAIL	Projection reduction exposure with variable axis immersion lenses
PTFE	Polytetrafluorethylene (Teflon®)
PVD	Physical vapor deposition
QDNs	Quantum Dot Nanotcrystals
QSE	Quantum size effect
QWIP	Quantum well infrared photodetector
RAM	Random access memory
RBS	Rutherford backscattering spectrometry
RCA	Radio Corporation of America (Company)
RF	Radio frequency
RGD	arginine, glycine and aspartic acid motif
RHEED	Reflection high-energy electron diffraction
RIE	Reactive ion etching
RITD	Resonant interband tunneling diode
RLS	Resonance Light Scattering
RNA	Ribonucleic acid
RTA	Rapid thermal annealing
RTBT	Resonant tunneling bipolar transistor
RTD	Resonant tunneling diode
SAMMS	Self Assembled Monolayer on Mesoporous Supports
SAM	Self-Assembled Monolayer
SCALPEL	Scattering with angular limitation projection electron beam lithography
SCM	Scanning Capacitance Microscopy
SCZ	Space charge zone
SECM	Scanning Electro Chemical Microscopy
SEM	Scanning Electron Microscope (Microscopy)
SET	Single Electron Transistor
SFIL	Step and flash imprint lithography
SFM	Scanning Force Microscope
SHT	Single hole transistor
SIA	Semiconductor Industry Association
SICM	Scanning Ion Conductance Microscopy
SIMOX	Separation by implantation of oxygen
SIMS	Secondary Ion Mass Spectrometry
SL	Superlattice
SMD	Surface-mounted device
SMPS	Scanning Mobility Particle Sizer
SNP	Single Nucleotide Polymorphisms
SOI	Silicon on insulator
SOS	Silicon on sapphire

SPM	Scanning Probe Microscope (Microscopy)
SPR	Surface Plasmon Resonance
STM (STEM)	Scanning Transmission Electron Microscope
STM	Scanning Tunnelling Microscope (Microscopy)
SWCNT	Single-Walled Carbon NanoTube
SWNT	Single-Wall NanoTube
TA	Thermal analysis
TED	Transferred electron device
TEM	Transmission Electron Microscope (Microscopy)
TEOS	Tetraethylorthosilicate
TFT	Thin film transistor
TMAH	Tetramethylammonium hydroxide
TOFSIMS	Time of Flight SIMS
TSI	Top surface imaging
TTL	Transistor-transistor logic
TUBEFET	Single carbon nanotube field-effect transistor
UHV	Ultrahigh vacuum
ULSI	Ultra large scale integration
UPS	UltraViolet Photoelectron Spectroscopy
UV	Ultraviolet
VHF	Very high frequency (30–300 MHz; 10–1 m)
VLSI	Very large scale integration
VMT	Velocity-modulated transistor
V-PADOX	Vertical pattern-dependent oxidation
VPE	Vapor phase epitaxy
XOR	Exclusive or
XPS	X-Ray Photoelectron Spectroscopy
XRD	X-ray diffraction
ZME	Zeolite modified electrode
ZVI	zerovalaent ion

Web Addresses

http://www.rpi.edu/dept/materials/COURSE/NANO/crawford
IBM Research | Almaden Research Center | Science and Technology |
ROBERT R. McCormick School of Engineering and Applied Science
http://web.mit.edui/isn/research
http://online.nanopolis.net/viewer.phb?subject.
The Royal Society & The Royal Academy of Engineering
http://www.zurich.ibm.com/images/st/atomic_manipulation/manipulation.gif
http://www.nottingham.ac.uk/physics/research/nano/instrumentation.htm
http://en.wikipedia.org/wiki/Carbon_nanotubes
Functional Nanomaterials
Edited by Kurt E. Geckeler and Edward Rosenberg
Wikipedia, the free encyclopedia
http://www.pnas.org/cgi/content/full/101
http://www.nano.gov/html/facts/whatIsNano.html
Advanced Semiconductor and Organic Nano-Techniques, Parts I, II and (Part III) H. Morkoç (Ed.)
K.B. Blodgett, I. Langmuir, Phys. Rev. 51 : 964–92, 1937.
B. Moore, C.M. Knobler J., Phy. Chem. 94 : 4588–4595, 1990.
http://www.scientific americans.com.
Lucent Technologies (2002)_http://wwwbell_labs.com/project/SCALPEL/mask format.html.
http:/www.research.ibm.com/journal/rd/411/seeger.html.
For further reference please refer to website.
R.R.H. Coombs and D.W. Robinson (Eds), *Nanotechnology in Medicine and Biosciences* (Gordon and Breach, New York, 1996)
A.S. Edelstein and R.C. Cammarate (Eds), *Nanomaterials: Synthesis, Properties and Applications* (Institute of Physics, 1998)

Bibliography

Nanotechnology and Nano Science Nanowork.
The Nanotechnology site.
JPK Instruments AG-Scanning Probe technology nanoHUB.
Physics Education.
Timeline of Carbon nanotubes–Wikipedia, the free encyclopedia.
Nanotechnology–Seeing Nano–Wikibooks.
Stable and Robust polymer nanotubes streched from polymer some–Reiner et al.
Instrumentation and tools for nanotechnology.
Nanostructured materials–Yahoo search.
www.nano engineer–1.com–Nano Engineer–1–
Gallery–SRG III speed reducer gear.
blog–nano Nanoscale materials.

Index